AF572757

# MICROWAVE HEATING

Edited by **Usha Chandra**

INTECHOPEN.COM

**Microwave Heating**
Edited by Usha Chandra

**Published by InTech**

Janeza Trdine 9, 51000 Rijeka, Croatia

**Edition 2014**



**Publishing Process Manager** Petra Nenadic

**Technical Editor** Teodora Smiljanic

**Cover Designer** InTech Design Team



Additional hard copies can be obtained from orders@intechopen.com

Microwave Heating, Edited by Usha Chandra
p. cm.
ISBN 978-953-307-573-0

INTECH
open science | open minds

# Contents

**Preface IX**

**Part 1 General 1**

Chapter 1 **Microwave Synthesis: A Physical Concept 3**
V.K. Saxena and Usha Chandra

**Part 2 Medical Sciences 23**

Chapter 2 **Microwave Radiometry as a Non-Invasive Temperature Monitoring Modality During Superficial Hyperthermia 25**
Svein Jacobsen

Chapter 3 **Role of Microwave Heating in Antigen Retrieval in Cryosections of Formalin-Fixed Tissues 41**
Hrvoje Brzica, Davorka Breljak, Ivana Vrhovac and Ivan Sabolić

Chapter 4 **Microwave Assisted Disinfection Method in Dentistry 63**
Carlos Eduardo Vergani, Daniela Garcia Ribeiro, Lívia Nordi Dovigo, Paula Volpato Sanitá and Ana Claudia Pavarina

**Part 3 Health and Environment 89**

Chapter 5 **Microwave Enhanced Advanced Oxidation in the Treatment of Dairy Manure 91**
K.V. Lo and P.H. Liao

Chapter 6 **Microwave Processing of Meat 107**
M.S. Yarmand and A. Homayouni

**Part 4 Technology 135**

Chapter 7 **Reproducibility and Scalability of Microwave-Assisted Reactions 137**
Ángel Díaz-Ortiz, Antonio de la Hoz, Jesús Alcázar, José R. Carrillo, M. Antonia Herrero, Juan de M. Muñoz, Pilar Prieto and Abel de Cózar

Chapter 8 **Use of Microwave Heating in Coal Research and in Materials Synthesis 163**
Mohindar S. Seehra and Vivek Singh

Chapter 9 **Synthesis and Processing of Biodegradable and Bio-Based Polymers by Microwave Irradiation 181**
Guido Giachi, Marco Frediani, Luca Rosi and Piero Frediani

Chapter 10 **Microwave Pyrolysis of Polymeric Materials 207**
Andrea Undri, Luca Rosi, Marco Frediani and Piero Frediani

**Part 5 Synthesis 233**

Chapter 11 **Synthesis of SiC Powders and Whiskers by Microwave Heating 235**
Jianlei Kuang, Fu Wang, Qiang Wang, Matiullah Khan and Wenbin Cao

Chapter 12 **Microwave Synthesis of Core-Shell Structured Biocompatible Magnetic Nanohybrids in Aqueous Medium 265**
Ling Hu, Aurélien Percheron, Denis Chaumont and Claire-Hélène Brachais

Chapter 13 **Synthesis of Carbon-Based Materials by Microwave Hydrothermal Processing 291**
Marcela Guiotoku, Claudia M. B. F. Maia, Carlos R. Rambo and Dachamir Hotza

Chapter 14 **Sucrose Chemistry: Fast and Efficient Microwave-Assisted Protocols for the Generation of Sucrose-Containing Monomer Libraries 309**
M. Teresa Barros, Krasimira T. Petrova and Paula Correia-da-Silva

**Part 6 Applications 333**

Chapter 15 **Microwave Irradiation Effect in Water-Vapor Desorption from Zeolites 335**
Hongyu Huang, Seiya Ito, Fujio Watanabe, Masanobu Hasatani and Noriyuki Kobayashi

Chapter 16 **Microwave Heating in Organic Synthesis and Drug Discovery 351**
Hong Liu and Lei Zhang

# Preface

Electromagnetic (EM) environments in the microwave region have become very complex, because of the wide and rapid spread of a large number of electronic devices in the area of cellular telephony. The associated effect of the radiations -heating- observed after the serendipity experiment of chocolate melting by Dr Percy Spencer exposed a new vista in microwave philosophy opening up research activities spanning into multiple areas including medicine and public health. Thus the obligation to develop a new book was realized which would encourage further research in the unique area of RF/microwave interaction. The main objective of this new book is to serve both as a text book and as a reference monograph. As a result, each chapter is designed to cover both basic understanding and advanced aspects.

The book begins with the basics of the microwave heating touching upon various fields like synthesis in chemical industry, health and environment, medical sciences etc. Time is valuable – this quote is certainly also true for science especially in the field of chemical synthesis. Microwave heating does not only reduce the reaction time but also suppresses side reactions, improving its yield and reproducibility. The technique has found its place not only in controlling quality of meat and meat products but also effectively been employed in pyrolysis processes of plastics, complex polymeric materials such as tires or mixed waste products. When microwave methodology was introduced, domestic ovens were used without appropriate temperature controls. The reproducibility of results between mono-mode and multi-mode microwave instruments were difficult therefore one chapter is dedicated to the problems concerning reproducibility and scalability of microwave assisted processes. Microwave radiometry derives information of internal body temperature patterns. Knowledge of such thermal patterns gives the valuable information in clinical disease detection and diagnosis; also providing quantitative temperature feedback in monitoring thermal therapeutical processes- use of microwave radiometry as a non-invasive temperature monitor is discussed. The effectiveness of microwave-enhanced-advanced-oxidation process for disintegrating solids and releasing nutrients into solutions from dairy manure as a potential effective pre-treatment method for nutrients and energy recovery is described. The transmission of infectious-contagious diseases and control of cross infection has been a subject of concern in the dentistry area. The constant exposure of dental personnel and instruments to saliva and blood in every procedure indicated a hazard; therefore the use of an adequate disinfection procedure was required. Microwave radiation has been proving to be a practical physical sterilization technique in preventing cross-infection and treating denture stomatitis.

Indeed, the field has grown up so rapidly in the past few years that it is difficult as well as necessary to keep up with the current developments. Considering what is easiest and at the same time very difficult way out, I have entrusted the collaboration of those contributors whose opinions seem to be the best in their individual fields. One of the inevitable consequences of such a collaborative effort is the fact that writing style and method of approach to a given topic would vary rather widely and will reflect the individualities involved. As an editor, I have not attempted to fit everyone in the similar mold; consequently the imprints of the individual authors have survived.

I am grateful to the authors for their valuable time and sincere efforts towards preparation of the respective contribution. I am especially indebted to Dr V. K. Saxena, who not only authored the chapter but also supported me throughout the endeavor.

Jaipur, 27 June 2011

**Usha Chandra**
University of Rajasthan,
India

# Part 1

## General

# 1

# Microwave Synthesis: A Physical Concept

V.K. Saxena and Usha Chandra
*University of Rajasthan, Jaipur*
*India*

## 1. Introduction

The term 'microwaves' is used for those wavelengths measured in centimetres roughly from 1 m. to 0.1 cm or the bands of frequencies between 300 MHz to 300 GHz. Microwaves are most popularly used in point to point communication, TV broadcasting via satellites and in RADAR systems. Besides these, they are also being used in industrial, biomedical, chemical and in scientific research applications. The latest wide acceptance of the microwave application is as "microwave heating" in synthesizing various compounds (organic and inorganic) in research laboratories as well as in industries. The serendipity experiment of radar-orientated research by Dr Percy Spencer in 1946 discovered the heating property of microwave (the candy bar melted in the pocket). Since then microwave radiation has opened up a new technology. The application is no longer confined to food industry, but shown impacts in other areas also. Inorganic synthesis, polymer curing, textile drying are growing industrial areas where microwaves are being used. The first published reports on the use of microwave irradiation to carry out organic chemical transformation by the group of Gedye (1986) and Giguere et al.(1986), it has emerged as an promising technology in medicinal chemistry, polymer synthesis, material sciences, nanotechnology and biochemical processes. In the early days of microwave synthesis, experiments were carried out in sealed Teflon or glass vessels in a domestic oven without any pressure or temperature measurements, sometimes resulting into violent explosion due to rapid uncontrolled heating in organic solvents under closed vessel condition. In 1990's synthesis started with solvent-free reactions eliminating the danger of explosion eventually improvising the device designs. Microwave heating not only provides an alternative tool but also improves the yield and reproducibility .Moreover it may lead to novel products by permitting access to the alternative kinetic pathways and stabilizing different energy minima in a reaction vessel. Various aspects of microwave application especially in chemical synthesis have been reported and reviewed since then (Rao et al. 1999,Vanetsev and Tretyakov 2007, Tsuji et al. 2005,Yoshikawa 2010,Bilecka & Niederberger 2010).

There are many distinct frequency bands which have been allotted for industrial, scientific and medical uses. 24.15 MHz, 2.45 and 5.80 GHz are world wide accepted frequencies. Still there are few countries where different frequencies are used. In United Kingdom 896 MHz and 40.6 MHz are popular whereas the Netherlands operate on 3.39 and 6.78 GHz. (Metaxas and Meredith 1983)

## 2. Experimental set up

A lot of work has been done on the synthesis of materials due to enhancement of the speed of chemical reactions, demanding improvement in the experimental technology during microwave heating. In most cases domestic microwave oven operated in multimode configuration is being used to perform the experiment. These microwave ovens have certain limitations. Different modes are developed in the main chamber or cavity of the oven, disturbing the power matching when the sample is placed inside. The use of domestic microwave ovens was impeded by possible heterogeneity of the magnetic field, sometimes inducing insufficient reproducibility, non-uniform heating , mixing and precise determination of the reaction temperature. The size, shape and power are the major constraints which should be improved according to the need to get better and efficient results. The use of mono-mode system enabled microwave beam to be focussed on the sample. Therefore it is advisable that the researcher be familiar with different components of the experimental set up used for microwave heating. The block diagram of microwave processing unit is shown in figure 1a. The main chamber as well as parts of this unit can be modified according to the need of individual researcher. In microwave heating high power source is used therefore one should be careful about the hazardous effect of microwaves. The construction and functioning of different components are described in the following sub-sections.

### 2.1 Microwave source: magnetron

Magnetrons, Klystrons, Gyrotrons and Travelling wave tubes (TWT) are used to generate microwave power. These are able to generate high power of microwaves. Each has its own advantages. Klystron offers precise control in amplitude, frequency and phase, while Gyrotron provides much higher power output and beam focussing. TWTs can provide variable and controlled frequency of microwave energy. Solid state devices are also used wherever low power microwaves are needed. Magnetrons are widely popular in microwave heating due to their easy availability and low cost. The size and configuration of a cylindrical magnetron is much suited for microwave ovens as well as for other applications of microwave heating. Magnetron can generate either continuous or pulsed power up to megawatt and frequency between 1-40 GHz. Its power efficiency is around 85% and life time is ~ 5000 hours. The basic structure of a cylindrical magnetron consists of a number of identical cavity resonators arranged in a cylindrical pattern around a cylindrical cathode (Figure 1b).

Two large pole pieces of permanent magnets are used to produce a strong magnetic field normal to the plane of cavities. The anode is kept at higher potential relative to cathode. The electrons emitted from the central cathode are accelerated towards the anode but the presence of transverse magnetic field exerts a torque which causes the electrons to move in a curved path in the drift space. Due to initial perturbation or interaction of electrons an r.f. field is induced in the cavities which propagate in the azimuthal direction with a certain phase velocity. For a constant value of magnetic field the potential difference between anode and cathode is applied to obtain synchronism between the radial velocity of electrons and phase velocity of $n^{th}$ spatial harmonics of r.f. field in azimuthal direction. The condition is achieved in which the electrons continuously interact with r.f. field in retarding phase and give up their energy to the field. Waveguides are used to couple the microwave power from the magnetron to the main chamber of the oven. The conducting walls of the guides /

resonators confine the electromagnetic fields and couple the power from magnetron to the cavity of the oven. A number of distinct field configuration or modes can exist in wave guides. Inside the wave guide either TE or TM modes can propagate but not TEM. These modes are labelled by two identifying integer subscripts 'n' and 'm'i.e. $TE_{nm}$ or $TM_{nm}$. The integer n and m shows the number of half variation of the respective fields along the two transverse coordinates. Each mode has associated with it a characteristic cut-off frequency $f_{c,nm}$ below which the mode does not propagate or a characteristic cut-off wavelength $\lambda_{c,nm}$ above which the mode does not propagate. The dominant mode in a particular guide is that mode having the lowest cut off frequency or highest cut off wavelength. The field configurations for different modes in a rectangular and a circular waveguide are shown in the figures 1c & 1d.

### 2.2 Circulator

A circulator is a multi-port junction providing one way sequential transmission of power between its ports (Figure 1e) . Port 1 couples to port 2 but not to any other ports similarly port 2 couples with port 3 only and so on. The direction of circulation of the circulator is shown by an arrow marked on the device. In the simplest form the circulator is fabricated by a three port Y junction of rectangular wave guides that is loaded with ferrite cylinder or discs magnetized in the direction normal to the plane of the junctions. The three port junction is completely matched at all its ports. In microwave heating set up, the power is allowed to transmit from port 1 to port 2 and port 3 is terminated by a dummy load which can absorb the reflected power from mismatched load.

### 2.3 Tuner

Tuner is a passive device used to match power from source to unmatched circuit load. Tuners provide variable equivalent inductive or capacitive load by varying the depth of the sliding screws in the wave guide. A typical construction of the sliding screw tuner is made compatible with the experimental set up (figure 1f). The main characteristic of a tuner is its matching performance, expressed as maximum magnitude of load reflection coefficient that can be perfectly matched irrespective of phase. The limit of maximum magnitude depends upon insertion depth of screws and the power level of generator. As input power increases beyond the maximum limit power dissipation in the tuning stub also increases which leads to overheating of the stub, thus melting and damaging the softer surrounding parts. It is therefore important to know safe limits of the input power for the tuner.

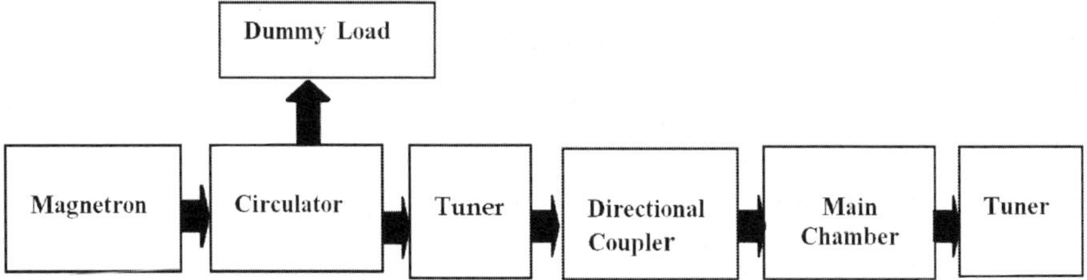


Fig. 1. (a) Block diagram of Microwave Processing Unit

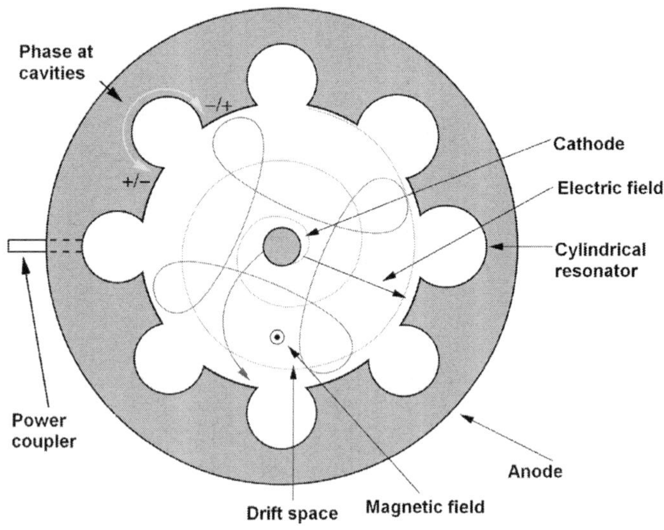


Fig. 1. (b) Construction and working of Magnetron

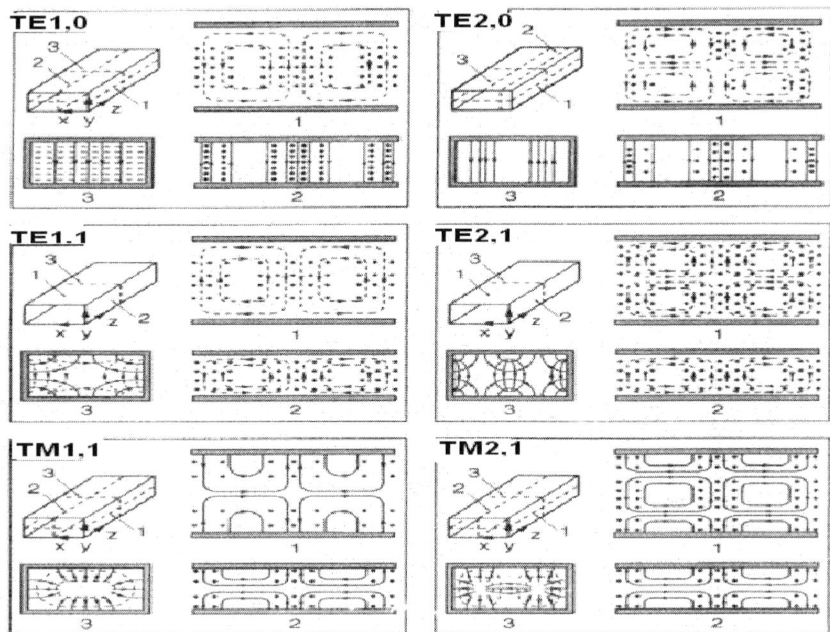


Fig. 1. (c) Field distribution for different modes in a rectangular wave guide

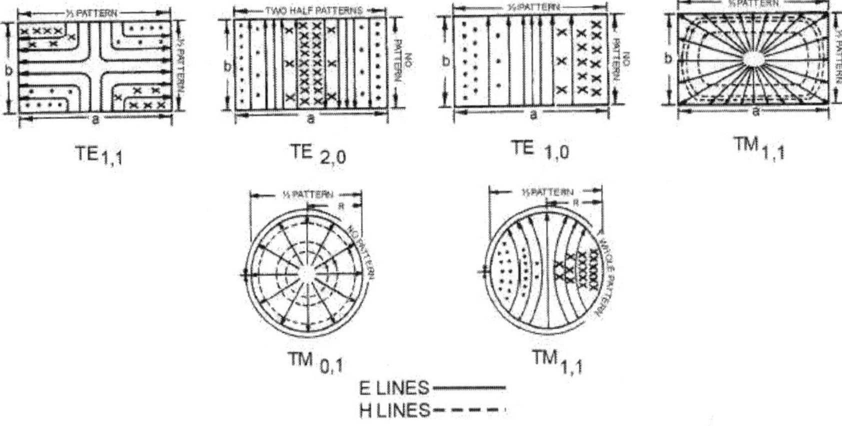


Fig. 1. (d) Field distribution for different modes in a Circular wave guide

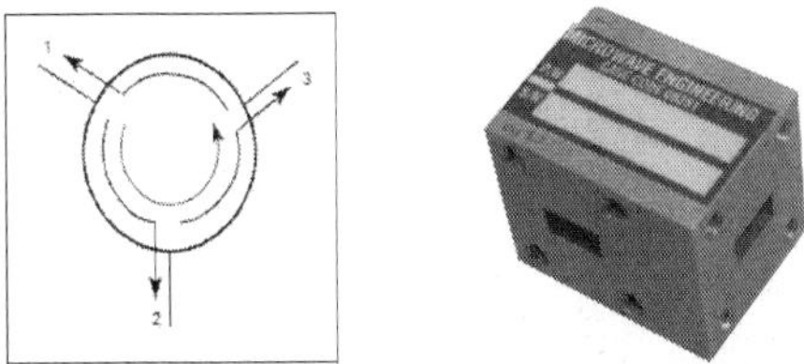

Fig. 1. (e) Schematic diagram of Circulator with its commercial design

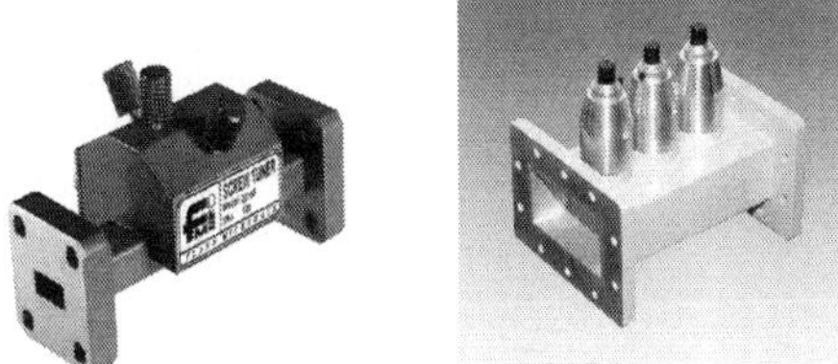

Fig. 1. (f) Types of commercial Tuner

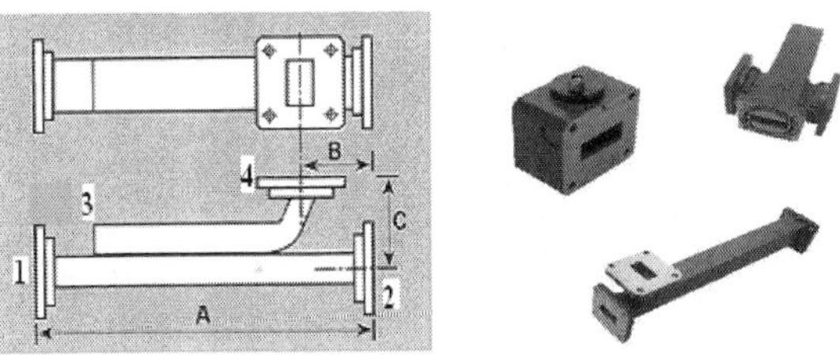


Fig. 1. (g) Schematic diagram of Directional coupler with commercial models

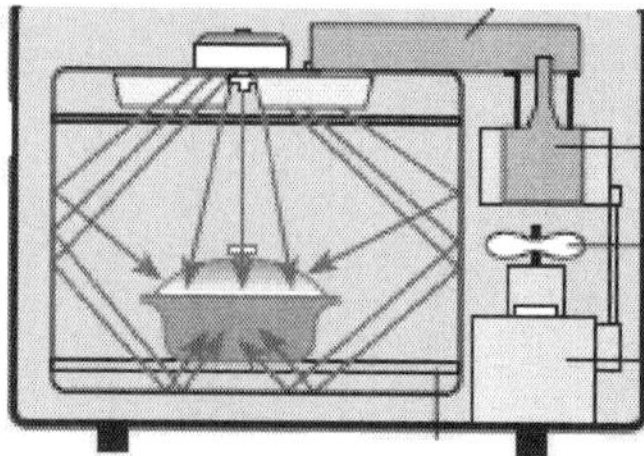

Fig. 1. (h) Main Microwave chamber

### 2.4 Directional coupler

A directional coupler is a four port wave guide junction with the property that power incident in port 1 couples to port 2 with a fraction of incident power coupling to port 4 and no power coupling to port 3. Further power in the back direction entering port 2 couples to port 1 with a fraction of its coupling to port 3 and none of it coupling to port 4 ( figure 1g). The line 1-2 is called primary line and the line 3-4 is called coupled line. All the ports of the directional coupler are well matched. The directional coupler is used to measure incident and reflected power from mismatched load.

### 2.5 The Main chamber

The Main chamber where material to be microwave processed is placed, is like a cavity resonator and microwave power is coupled to it through wave guides (figure 1h). Once the microwave power enters the chamber, standing waves are formed due to multiple reflections from metallic walls. The size of the chamber is of the order of wavelength of microwave used so that multiple modes are excited. In the common domestic ovens the power distribution is uneven inside the chamber resulting into uneven heating of the material. The rotating turn table, therefore, is used inside the oven to make the power distribution equally on the sample. According to the need of microwave-materials interaction, the construction of the main chamber can be modified. In some applications multi chamber ovens are used and energy is coupled through a slot, an array of resonant slots, a radiating horn etc. In some cases modified wave guides are used to excite power in the main chamber. These are called applicators or exciters. Main chamber is loaded with different materials and in different amounts which may change the impedance of the chamber. These applicators are designed always to match the impedance of microwave source and main chamber. Mode stirrer technique is also used to spread the power uniformly in all the directions. The proper cooling arrangements are required to exhaust the unwanted heat from the main chamber. Vollmer (2004) in his special feature demonstrated the intensity distribution within an oven of 29x29x19cm$^3$ at a height of 8 cm. A horizontal glass plate covered with a thin film of water was placed inside the chamber (without its rotating turn table) on full power for a few seconds. The colour image thus obtained with a thermal infrared camera is shown in figure 2a. With only a thin film of water present, the image beautifully illustrates the microwave intensity distribution in a nearly empty chamber. A pronounced horizontal mode structure also shows heterogeneous heating of material. Cheng et al. (2001) studied the distribution of fields in the microwave cavity. The electric (E) field was found maximum in the centre, where the magnetic field is minimum. On the other hand maximum magnetic field is found near the wall with minimum electric field (Figure 2b).

## 3. Microwave heating

### 3.1 Microwave versus conventional heating

Conventional heating usually involves the use of a furnace or oil bath which heats the walls of the reactors by convection or conduction. The core of the sample takes much longer to achieve the target temperature. On the other hand Microwave penetrates inside the material and heat is generated through direct microwave-material interaction. Moreover volumetric heating, reaction rate acceleration, higher chemical yield, lower energy usage and different reaction selectivity the advantages microwave heating has over conventional methods. Combustion synthesis has been one of the methods used to obtain powder with compositional uniformity. Figure 3 illustrates merit of microwave heating over conventional method during the synthesis of $LaCrO_3$ powder as homogeneous fine particles (Park et al.1998). The mixer solution of the parent constituents was equally divided and kept for combustion separately on a hot plate and in a microwave oven. Combustion products thus obtained were treated in a similar way and morphology was examined by scanning electron microscopy (SEM). The product obtained by hot plate combustion is found to be consisting of hard agglomerates formed by the two dimensional interconnection of spherical particles. On the other hand product of microwave induced combustion appears to be smaller in size and with reduced agglomeration. Kappe and Dallinger (2006) measured temperature of the 5 ml volume of ethanol and observed that throughout the volume temperature rises simultaneously when heated using microwave radiation while sample in oil-heated tube in contact with the vessel wall heats up first ( figure 4). Gedye et al. (1988) carried out different types of organic reactions both in sealed teflon vessels in a microwave oven and under traditional reflux conditions and approximate rate enhancement of between 5 and 1240 times were recorded. Derivatives which normally required 2-3 hours of preparation, effectively synthesized in 2-3 minutes in microwave oven.

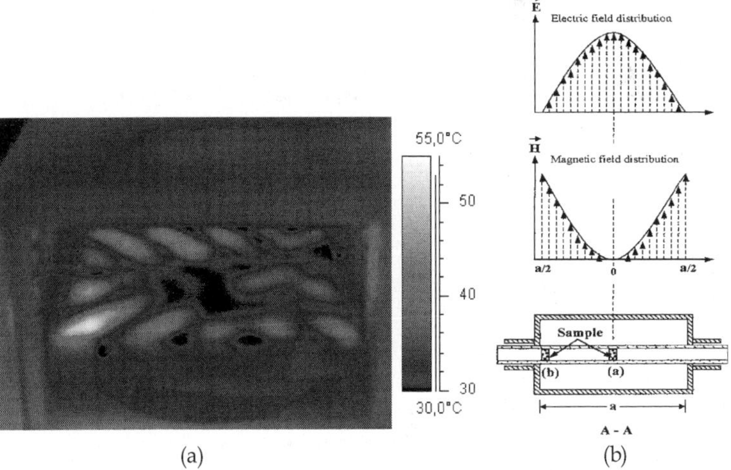


Fig. 2. (a) Infra red thermal imaging inside microwave oven (Vollmer 2004)
(b) The schematic of the microwave field distribution (E & M)in the microwave cavity (Cheng et al.2001)

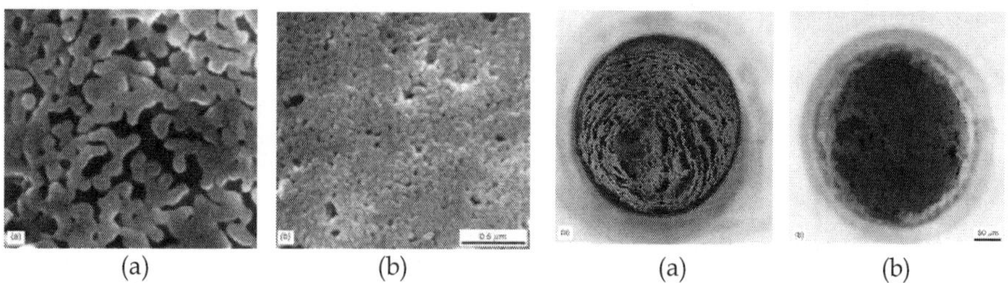

(a) (b) (a) (b)

Fig. 3. Morphology with SEM micrograph of the combustion process (a) hot plate (b) microwave radiation (Park et al. 1998)

A few reactions which were carried out using microwave heating and compared with conventional heating indicating time and energy efficiency of the technique is compiled in Table 1.

| Compound synthesized | Reaction time - microwave | Reaction time - conventional | references |
|---|---|---|---|
| Esterification (benzoic acid with methanol) | 5 min. | 8h | Gedye et al (1988) |
| 4-nitrobenzyl ester | 2 min | 1.5 h | Gedye et al (1988) |
| $CuBi_2O_4$ | 5 min | 18h | Jones & Akridge (1995) |
| $Bi_2Pd$( Intermetallic) | 4 min. | 12 h | Lekse et al. (2007) |
| $Ag_3In$( intermetallic) | 2 min. | 48 h | Lee & So. (2000), |
| Layered Al and Zn double hydroxide with Na-dodecyl sulfate | 1-2 h | 2-3 days | Hussein et al. (2000) |
| Bronzes ($Na_xWO_3$) | 13-15 min. | - | Guo et al. (2005) |
| Ti N | 30 min | - | Vaidhyanathan & Rao (1997) |
| Cubanite $CuFe_2S_3$ | 3min | 3 days | Chandra et al. 2010 |
| $La_{2-x}Sr_x Mn_2O_4$ | 30 s | - | Mingos & Baghurst. 1991 |
| High Tc superconductors YBCO | 12 h | 72 h | Binner & Al-Dawery (1998) |
| Zeolite synthesis | 170 $^0$C in 30s | 170$^0$C in 60min | Jansen 2004 |
| $MgB_2$ | 11min | - | Dong et al. 2007 |
| $NaAlH_4$ | 2h | 8h | Krishnan et al. 2009 |
| $La_{0.2}Sr_{0.8}Mn_{0.8}Fe_{0.2}O_{3+\Delta}$ | 3h | 3 days | Chandra (unpublished) |

Table 1. Comparison of the reaction times using microwave versus conventional heating.

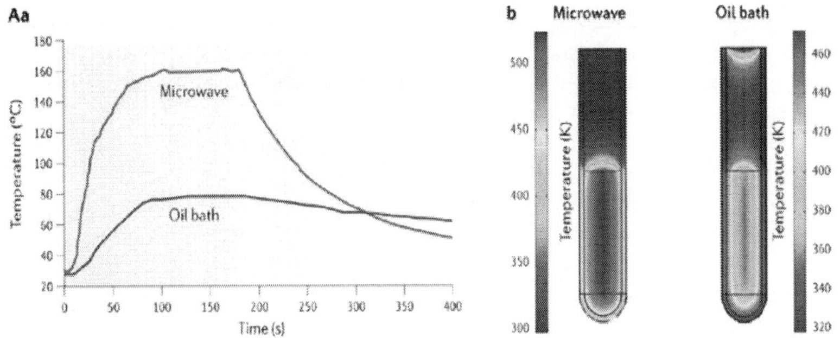


Fig. 4. Temperature profiles for ethanol under microwave radiation and open vessel oil bath condition and temperature gradient 1 min. after heating (Kappe and Dallinger 2006).

### 3.2 Interaction of microwaves with materials

Two factors are important to select the frequency of microwave radiation to heat the materials (i) power absorption in the matter, and (ii) depth of penetration. In electromagnetism, materials are divided into two categories: (i) conductors and (ii) insulators or dielectrics. The distinction between them is not very sharp. The same material may behave as a conductor in one part of electromagnetic frequency and as a dielectric in another. According to Maxwell's theory, the ratio ($\sigma / \omega\varepsilon$) is considered to be a dividing factor where $\sigma$ is the electrical conductivity; $\omega$ is the angular frequency and $\varepsilon$ is the permittivity/ dielectric constant. For good conductor, this ratio is much greater than unity while in dielectrics, it is much smaller than unity. In dielectrics the electrostatic fields can persist for a long time offering very high resistance when static voltage is applied whereas in good conductors (metals) the static current feels negligible resistance. At higher frequencies, the behaviour depends upon frequency of the electromagnetic waves and corresponding conductivity, permittivity, permeability of the material. The electromagnetic energy propagation through a material medium is associated with the numerical values of permittivity or dielectric constant at that frequency.

The absorption of electromagnetic energy depends upon complex permittivity $\varepsilon$ of that material and can be expressed as

$$\varepsilon = \varepsilon' + j\,\varepsilon'' \tag{1}$$

where $\varepsilon'$ = the real component of permittivity

$\varepsilon''$ = the imaginary component of permittivity

The real part or relative permittivity represents the degree to which an electric field may built up inside a material when exposed to the electric field while the imaginary part or dielectric loss is a measure of amount of the field transformed into heat.

The Loss angle $\delta$, the phase difference between the electric field and the polarization of the material is related to the complex dielectric constant as

$$\tan\delta = \varepsilon'' / \varepsilon' \tag{2}$$

Thus the tan $\delta$, the dissipation factor determines the ability of material to transform absorbed energy into heat.

In terms of microwave interaction, the materials can be classified into three categories:

i. Microwave reflectors e.g. metals
ii. Microwave transmitters - transparent to microwave radiations e.g. fused quartz, ceramics, zircon etc.; tan $\delta < 0.1$
iii. Microwave absorbers taking up energy from the microwave field and heating the materials rapidly; tan $\delta > 0.1$

The electromagnetic energy absorption in dielectric materials primarily is due to the existence of permanent dipole moment of the molecules which tend to orient and reorient under the influence of electric field of microwave. The reorientation loss mechanism originates from the inability of the polarization to follow extremely rapid reversals of the electric field. In the low frequency (up to 100 MHz) electric field, the dipoles easily follow the changes in the field and their orientation changes in phase with the field. At higher frequencies the inertia of molecules and their interactions with neighbours make changing orientation more difficult and the dipoles lag behind the field. As a result, the conduction current density has a component in phase with the field and therefore power is dissipated in the dielectric material. At very high frequencies (1- 10 THz), the molecules can no longer respond to the electric field. At GHz frequency (ideal working range) the phase lag of the dipoles behind the electric field absorbs power from the field and therefore pronounced as dielectric loss due to dipole relaxation.

Another important parameter for microwave heating ,penetration depth $D_{ph}$ is defined as the depth into the material where the power is reduced to ~ 1/3 of the original intensity ( Jansen 2004) The absorption coefficient $\alpha$ of dielectric material is related to imaginary parts of dielectric constant $\varepsilon''$ or the refractive index. The depth of penetration $D_{ph}$ of electromagnetic waves in matter is related to $\alpha$ as $D_{ph} = 1/\alpha$. Thus a material with higher dissipation factor will have lower penetration depth. The wavelength of the radiation also has influence on penetration depth. For microwave prone materials the absorption coefficient at 2.45 GHz is moderate and depth of penetration is of the order of 10 cm to 1m which results in absorption of microwave everywhere in material. The intensity of heat evolution in a sample depends on electrophysical properties of the materials, frequency, intensity of the applied field, penetration depth of the electromagnetic waves into the substance under treatment and geometric size of the sample. The dielectric parameter of the matter and the penetration depth depend strongly on temperature and therefore varies during heating.

Water is an excellent example of polar molecule. The principle of microwave heating of polar molecules (for the case of water) is illustrated very nicely in figure 5 (Tsuji et al. 2005). These molecules absorb microwave energy rapidly and the rate of absorption varies with the dielectric constant of the material. The electric dipoles present in the dielectric materials respond to the applied electric field of microwave. Resistive heating occurs when the dipolar orientation is unable to respond to the applied microwave field resulting into a phase lag.

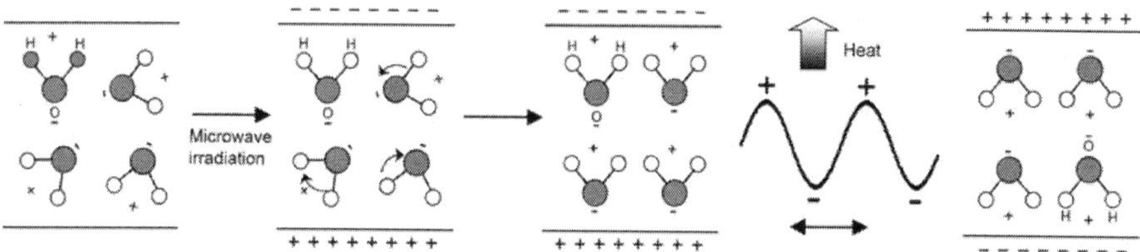


Fig. 5. Heating mechanism of water due to microwave field. (Tsuji et al. 2005)

In general, the dipolar species in any medium possess a characteristic relaxation time $\tau$ and frequency dependent dielectric constant. If the dipolar relaxation is characterized by a single relaxation time, then dependency with frequency is given by Debye equations (Rao et al. 1999):

$$\varepsilon' = \varepsilon_{in} + (\varepsilon_s - \varepsilon_{in}) / (1+\omega^2\tau^2) \quad (3)$$

$$\varepsilon'' = (\varepsilon_s - \varepsilon_{in})\, \omega\tau / (1+\omega^2\tau^2) \quad (4)$$

Where $\varepsilon_s$ and $\varepsilon_{in}$ are the zero frequency and the infinite frequency dielectric constants respectively. $\varepsilon''$ varies with frequency giving rise to a characteristic frequency at $\omega=1/\tau$. For water , $\varepsilon''$ being significant at 2.45 GHz , a rapid dissipation occurs as heating of water.

Microwave interaction with materials can be predicted if dielectric properties ( relative dielectric constant $\varepsilon'$ and relative dielectric loss factor $\varepsilon''$) are known. The loss tangent $\tan\delta = \varepsilon''/\varepsilon'$ is an indicator of the ability of the material to convert absorbed energy into heat. A good absorber has $\tan\delta \geq 0.1$ while those with $\tan\delta \leq 0.01$ are transparent to microwave. Materials with high loss factor at the frequency of the incident radiation will heat at a faster rate from core to surface. Fused quartz, zircon, several glasses, ceramics and Teflon are good transmitter while in $SiO_2$ ,dielectric constant and losses do not show much dispersion hence no heating occurs. Tap water is microwave active as compared to distilled water. Alumina is microwave transparent at room temperature at 2.45 GHz but heats up efficiently at $1000^0$C becoming highly susceptible at $1500^0$C. SiC, on the other hand usually heat well at moderate frequency (Kubel 2005). Microwave susceptors are fabricated by depositing very thin layer of aluminium on polyster (PET) sheets. The thin layer of aluminium absorbs part of microwave energy creating currents in the metal. The thickness of the layer limits the currents hence preventing arcing however the currents are high enough to heat the susceptor to a high temperature. Such arrangements are popular in food industry. A list of materials useful during microwave heating process along with $\varepsilon'$ , $\varepsilon''$ , $\tan\delta$ and $D_{ph}$ is tabulated in Table 2.

### 3.3 Microwave interaction in metals

Skin depth and penetration depth are important parameters in microwave heating of metals. When microwaves are incident perpendicularly on the surface of the materials, its intensity decreases progressively due to dissipation inside the volume of the materials. Therefore the term 'penetration depth $D_{ph}$ is defined as the distance in the direction of penetration at which the incident power is reduced to half of its initial value (Rao et al. 1999):

$$D_{ph} = 3\lambda_0 / [\, 8.686\pi \tan\delta\, (\varepsilon')^{1/2} \,] \quad (5)$$

Where $\lambda_0$ is the wavelength of the microwave radiation.

In metals microwave propagation is usually described in terms of 'skin depth'. The skin depth is defined as the depth at which the magnitude of electric field drops to 1/e of the value at the surface and is given as (Newham et al. 1991):

$$\delta = 1/ (\nu\pi\mu\sigma)^{1/2} \quad (6)$$

where $\nu$ is the microwave frequency ($\omega = 2\pi\nu$), $\mu$ is permeability of the free space and $\sigma$ is the electrical conductivity.

The behaviors of bulk metal pieces and metal powder under microwave radiation are different. Skin depth in metal is very low and varies as $(\sigma)^{-1/2}$ showing less penetration of microwaves. In large metals and metal films, electric field gradients occur in the microwave cavity giving rise to electric discharge. In metal powder, due to eddy currents and plasma effects, very rapid heating takes place without any discharge.

| Material | ε′ | ε″ | **tan δ** | **$D_{ph}$(cm)** | **Reference** |
|---|---|---|---|---|---|
| Silicon Carbide | 10.5 | 11.0 | 1.048 | 0.28 | Clark & Sutton (1996) |
| Ethylene Glycol | 41 | 41 | 1.350 | - | Kappe & Dallinger ( 2006) |
| Active C | 7 | 2 | 0.286 | - | Vos et al. 2003 |
| Ethanol | 24.3 | 6.08 | 0.250 | - | Kappe & Dallinger ( 2006) |
| Methanol | 32.7 | 20.9 | 0.240 | - | " |
| Wood<br>Wood ( 40% water) | 1.591<br>5.1 | 0.033<br>1.12 | 0.0207<br>0.2196 | 3.57<br>4.0 | Rattanadecho(2006)<br>Starck et al (2005) |
| Soda Lime Glass | 6.0 | 1.20 | 0.200 | 1.5 | |
| Water ( Distilled 25$^0$C)<br>" (95$^0$C)<br>Water ( Deionized)<br>Water (0.5% salt)<br>Ice<br>Water ( Tap Water) | 78.3<br>-<br>78.2<br>75.8<br>3.2<br>67.5 | 12.3<br>-<br>10.3<br>15.6<br>0.0029<br>6.0075 | 0.157<br>-<br><br><br>0.0009<br>0.089 | 1.5<br>5.7<br>1.68<br>1.09<br>1.80 | Shulman (2002)<br>'<br>Schiffmann (1986)<br>"<br>Starck et al.(2005)<br>Hasna (2009) |
| SiC | 10.4 | 0.9 | 0.0865 | 7 | Starck et al (2005) |
| $SiO_2$ | 3.066 | 0.215 | 0.0701 | - | Vos et al. 2003 |
| Soot | ~10 | ~3 | 0.03 | - | Vos et al.2003 |
| Lithium Disilicate Glass | 7.5 | 0.17 | 0.023 | 1.30 | Clark & Folz |
| Silicon Nitride | 0.68 | 0.015 | 0.022 | 8.13 | Starck et al (2005 |
| Zirconia | - | - | 0.015 | 100 | Hasna (2009) |
| PVC | 2.9 | 0.016 | 0.0055 | 200 | Starck et al (2005 |
| Natural Rubber-<br>With S (3.0pph)<br>With S (3pph) & C black | <br>2.27<br>3.366 | <br>0.004<br>0.0142 | <br>0.002<br>4.219 | <br>173400<br>274.0 | <br>Makul et al.(2010)<br>" |
| Boron Nitride | 4.35 | 0.0131 | 0.003 | 0.5 | Starck et al (2005 |
| pyrex | 4.0 | 0.005 | 0.0013 | 1.8 | " |
| Aluminum Nitride | 9.0 | 0.008 | 0.00089 | 1.18 | " |
| Polyethylene | 2.3 | 0.001 | 0.00040 | 2.57 | " |
| PTFE | 2.1 | 0.0006 | 0.0002857 | 4.7 | " |
| Teflon | 2.1 | 0.0003 | 0.00014 | 2.73<br>9200 | "<br>Shulman (2002) |
| Alumina<br>γ - $Al_2O_3$ | 8.9<br>3.006 | 0.009<br>0.1720 | 0.00010 | 1.46 | Starck et al (2005)<br>Vos et al. 2003 |
| Fused Quartz | 3.8 | 0.0001 | 0.00003 | 1.9 | - |
| $Cr_2O_3$ | 10.3 | ~ .0025 | 0.0002427 | | Dube et al. 2007 |
| Graphite | - | - | - | 0.0038 | Vatensev & Tretyakov 2007 |
| Copper | | | | 0.00026 | Vatensev & Tretyakov 2007 |
| Epoxy resin | | | | 7300<br>4100 | Vatensev & Tretyakov 2007<br>Shulman (2002) |
| Porcelain | | | | 56 | Shulman (2002) |
| Quartz | | | | 1600 | " |

Table 2. Various factors of the materials useful for microwave heating

Metals are prohibited inside microwave oven due to discharges it produces inside damaging the magnetron supply. However microwaves are being used efficiently in melting of metals. Domestic microwave appliances are based on magnetron, cooled by a stream of air from a fan. The microwaves are reflected by the metal walls till all the energy is absorbed and converted into heat. To be used for metal casting, the domestic oven needs slight modifications (David Reid). The rotating glass plate must be removed, the vents admitting air into the chamber must be taped over and the air from the magnetron cooling redirected to the exterior. The insulation of the sample chamber, a critical factor, has two roles: one it contains the microwave energy within the shell so that the temperature rises to the melting point of the metal; second it protects the walls of the oven. Carbon, magnetite, ferrites are good susceptors of microwave radiation. Early experiments with Carbon ( considered a good susceptor of microwaves) for melting of silver were very discouraging. The un-insulated crucible barely attained red hot. Though insulation helped but it was obvious to find a more efficient absorber. 8 mm of SiC paste mixed with clay, applied to inside of ceramic crucible also did not show any results even after 10 minutes of heating. Another crucible with clay-ferrite paste capped with carbon lined shell showed some improvement but silver did not melt. On lining the cap with magnetite-clay mixture and after 15 minutes of firing, silver started melting. A double susceptor- a carbon (graphite) loaded primary coat with magnetite was used successfully to melt small amount of cast iron. The caveat of metal melting is oxidation; hence all melting must be performed in an inert atmosphere or under vacuum.

Agrawal et al. (2006) attempted joining of metal/metal systems by inserting metal powder between metal/ ceramic pieces and irradiating with microwaves. The powder was heated preferentially and thus the steel joints were formed. Attempt of Microwave sintering of metal glasses without crystallization and with addition of Sn for sintering have been reported ( Xie et al. ,2009).

### 3.4 Role of magnetic field in microwave heating

So far the microwave interaction with nonmagnetic material is discussed in which electromagnetic energy loss is associated with complex permittivity of the material. The total microwave power dissipation in a material is given by

$$P= 2\pi \left[ \varepsilon_0 \varepsilon' \tan\delta E^2 + \mu_0 \mu' \tan\varphi H^2 \right] \qquad (7)$$

Where $\varepsilon_0$ and $\mu_0$ are dielectric permittivity and magnetic permeability of free space, $\varepsilon'$ and $\mu'$ are the real part of the dielectric permittivity and magnetic permeability of the sample, tan $\delta$ and tan $\varphi$ are the dielectric and magnetic loss factor values and E and H are the electric and magnetic field values inside the sample. Although E and H appear in the equation, mostly it is assumed that entire reaction is due to the losses by the electric field vector. Cherradi et al.( 1994) showed substantial contribution of magnetic field towards the heating of alumina ( at high temperature ) , semiconductors and metals. Different materials seem to have different heating behaviours in the E and H microwave fields (Cheng et al. 2001). The high conductivity samples like metallic powders could be efficiently heated up in the magnetic field while the low conductivity samples (ceramics, alumina) got much higher heating in the pure electric field. For the same material (Cu), the compact sample absorbed lot of microwave energy in the magnetic field while the solid sample did not respond in the same condition. Roy et al.(2002) demonstrated that the common crystalline phase can be made non crystalline and hard magnets can be softened in the solid state by heating them in magnetic component of microwave radiation. The decrement in the resistance for polycrystalline LCMO films observed under influence of microwave radiation was explained by tunnelling of carriers through inter-grain boundaries (Lučun et al. 2007). The induced resistance depended on the sample position in the wave guide thereby near the narrow wall of the wave guide, magnetic component of the radiation had significant influence on the magneto-resistive properties of the film. While studying the dependence of electron paramagnetic resonance (EPR) of some manganites on microwave power, it was observed that the sample became unusually hot at high microwave power (Singh 2007) due to the interaction of the spin of $e_g$ electron of $Mn^{3+}$ with the magnetic component of microwave field.

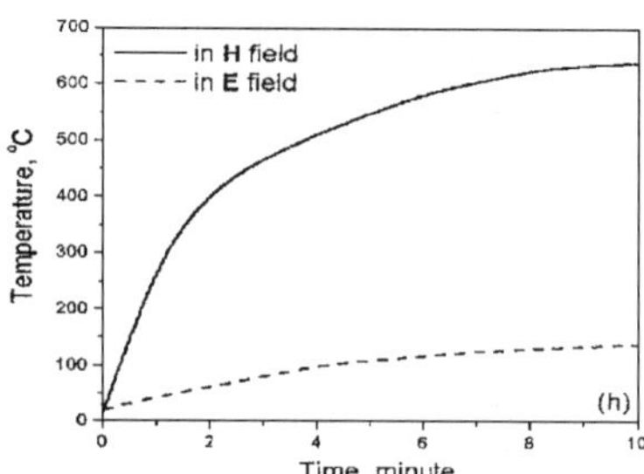


Fig. 6. Comparison of the heating rate of powdered tungsten carbide- cobalt (WC-Co) sample (microwave power 200W) (Cheng et al. 2001)

## 4. Examples of microwave synthesis

The nature of the minerals as they exist at atmospheric pressure represents only a part of their real nature. The pressure and temperature conditions impart strong influence on the structure and characteristics of the mineral. As one travels from Earth's mantle to the core, both temperature and pressure conditions changes drastically. The range of pressure and temperature prevailing at the surface of the Earth can be compared to those on other planets of the solar system. For example Sulfide minerals occur in the Earth's as well as in the Mars's mantle. Determining the thermodynamical properties of sulfides by synthesizing minerals under various laboratory conditions and comparing their behavior with natural one would be very important to understand the paragenesis of their formation in the planetary system. One of the examples described here is orthorhombic cubanite $CuFe_2 S_3$ which occurs in mineral deposits along with chalcopyrite ($CuFeS_2$) and pyrrhotite ($Fe_{1-x}S$) under certain conditions. Orthorhombic cubanite is very sensitive to temperature and pressure. At ~ 200$^0$C or ~ 4 GPa, the orthorhombic structure transform to cubic isocubanite. The discovery of cubanite in CI chondrites and stardust in comet wild 2 clarified the mechanism of large scale mixing of low-temperature assemblages in early solar system. The phase relationship in Cu-Fe-S determines the co-existing phases as a function of temperature and point to the unique combination of temperature and composition (Yund and Kullerud 1966). Owing to the unique physicochemical conditions needed, orthorhombic cubanite is difficult to synthesize under normal laboratory environment. In this direction, synthesis of this rarely found mineral, cubanite ($CuFe_2S_3$) was successfully attempted using conventional as well as microwave heating. The rapid microwave heating seems to be advantageous over conventional technique showing a characteristic peak of cubanite at d= 3.22 Å (Pareek et al. 2008, Chandra et al. 2010,2011). It is reported that synthesis of orthorhombic cubanite depends not only on the preparatory conditions (controlled heating and cooling rate) but also on compositional variation. Figure 7 shows the dependency of synthesis process as well as proportional of sulfur. X-ray diffraction pattern of resistive heated sample slowly heated and cooled to room temperature (a) with stoichiometry Cu:Fe:S::1:2:3 has been analyzed as isocubanite ( cubic cubanite) , pyrrhotite and chalcopyrite while rapidly heated microwave synthesized sample (b) with Cu:Fe:S::1:2:1.5 showed orthorhombic cubanite indicated by the arrows in the figure 7b. It is to be noted in figure 7b that the proportion of sulfur is an important component for the formation of orthorhombic cubanite. The sample was prepared with the same conditions but with increased proportion of Sulfur (Cu:Fe:S::1:2:3). The conspicuously absence of prominent pyrrhotite peak at ~ $55^0$ and formation of cubic isocubanite is observed.

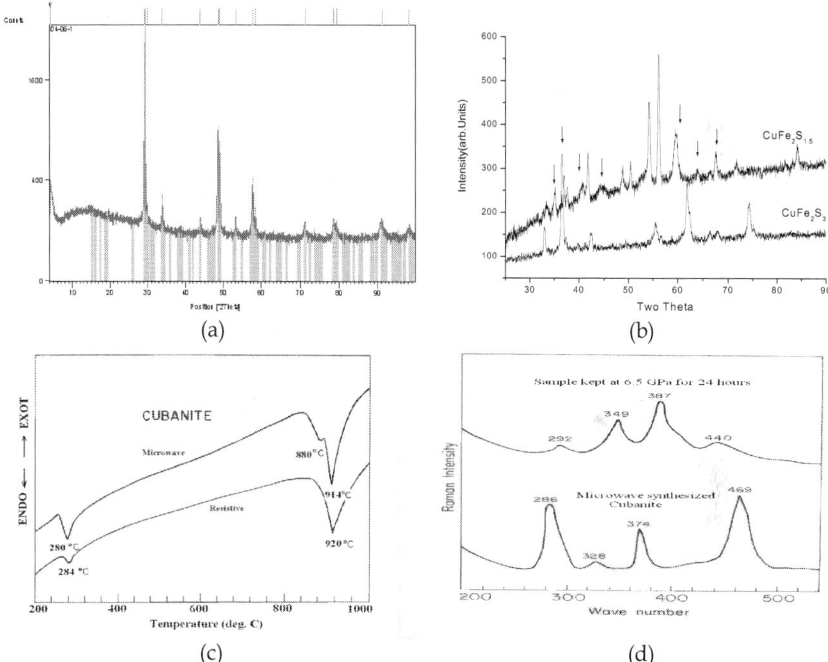


Fig. 7. X-ray diffraction patterns of (a) resistive heated (b) microwave synthesized cubanite. The arrows mark the orthorhombic cubanite. (c) DTA and (d) Micro-Raman spectroscopic studies of - resistive heated and microwave synthesized cubanite

The intensity at endothermic peak at $280^0$C in the differential thermal analysis (DTA) patterns of synthetic cubanite samples prepared by resistive and microwave heating is a quantitative indicator of the presence of orthorhombic cubanite (Figure 7c). The peak at $880^0$C represents pyrrhotite phase. The absence of an endothermic peak in the region $550^0$C-$650^0$C confirms the absence of chalcopyrite in the sample. The Micro-Raman spectroscopic studies on the as prepared synthetic cubanite show strong Raman modes at 286,374 and 469cm$^{-1}$ and a weak peak at 328cm$^{-1}$ corresponding to the desired cubanite (Fig 7d). The measurements repeated again after pressurizing the sample at 6.5 GPa for 24 h indicates isocubanite. Though x-ray diffraction pattern on conventionally synthesized sample does not indicate orthorhombic cubanite, Mössbauer spectroscopic studies confirm the presence of the phase through its sensitive parameters- magnetic hyperfine interaction and isomer shift. This is first ever study of successfully synthesized cubanite (Chandra et al. 2010).

The microwave dielectric heating effect originates from the natural ability of certain substances to efficiently absorb and to transform the electromagnetic energy into heat. If sufficient heat can be generated at a local level, then chemical reaction may be initiated at a rapid pace. To initialize the process, sometimes, it is necessary to introduce good susceptors of microwave radiation along with the constituents. The reaction can accelerate even if one of the constituents is a good susceptor. Baghurst et al. 1988 carried out pioneering work of the use of microwave dielectric heating for the synthesis of advanced ceramic materials- high temperature superconducting cuprates and $La_{2-x}Sr_xCuO_4$. The strong absorbing property of CuO ( a good susceptor)leads to a rapid and effective pathway for the reaction achieving $700^0$C in 30 sec using domestic oven. However synthesis of ternary CMR perovskites [(La,Sr)$MnO_3$] are difficult to synthesis because $La_2O_3$ , $MnO_2$ etc attain only $107^0$ and $321^0$ C respectively after 1800 s ( Gibbons et al. 2000) while $SrCO_3$ does not absorb microwaves to any significant degree (Mingos & Baghurst 1991).The addition of efficient absorber ( fine grained graphite) in the sample promotes the desired effect by significantly increasing the temperature of the reaction mixture. $Ca(NO_3)_2$ and $Mn(NO_3)_2$ are better receptors of microwave as compared to $MnO_2$.In the synthesis of mixed valent perovskites, these nitrates decompose to yield highly oxidising $NO_2$ facilitating the formation of $Mn^{3+}$/ $Mn^{4+}$ ( Gibbons et al. 2000). In the synthesis of $MnCo_2O_4$, the aqueous solution of nitrates of Mn and Co (non susceptible to microwave radiation) are mixed with carbon powder (good absorber of microwave)which not only acts as a surface enhancing agent but also decreases the decomposition temperature of the nitrates by~100 $^0$C ( Nissinen et al 2003)

Bulk $MgB_2$ with highest $T_c$ value is not easy to prepare because Mg is extremely volatile and susceptible to oxidation at elevated temperature. Usually Mg $B_2$ bulk sample are prepared in a sealed quartz / tantalum tubes or by high pressure sintering. However using Microwave heating, the evaporation and oxidation of Mg can be significantly reduced. Pellets of the constituent mixture in an alumina boat filled with SiC powder (used as MW susceptor ) placed in a modified domestic oven operated at 800 W for 11 min resulted into good crystals of $MgB_2$ (Dong et al 2004)

Bosi et al (1992) predicted the possibility of improved microstructure on superconductors while processing the thin film via microwave heating. If the film thickness is less than the skin depth, "volumetric" heating similar to dielectric might be possible through microwave heating. Also in some thin film applications, where film needs to be deposited on the substrate (such as Si ,a potential contaminant of superconductors) microwaves might be useful in the final post-processing stages because selective heating would heat the film keeping the substrate cool, suppressing the diffusion of silicon through the film. Conventional processing of bulk ceramic high $T_c$ YBCO superconductors results in low and non-uniform oxygen contents throughout the sintered and annealed bodies, limiting the value of $T_c$. In contrast microwave heating offers the ability to achieve perfectly uniform and full oxygen content (x=7) and almost full densities (98%) throughout the bulk ceramic samples. The critical transition temperature was also improved, grain size was uniform and finer than in the conventionally processed sample. In addition the total processing time was approximately a factor of six faster. (Binner & Al-Dawery 1998)

Mercury sulphide is a useful material and is used widely in many fields such as ultrasonic transducers, image sensors, photoelectric devices, conventional preparation of the item is

not popular because of the toxicity of mercury. Microwave heating found to be a fast (5-10 min) , convenient , energy efficient and environment friendly route to the synthesis of nano-crystalline Hg S particles with controllable sizes. The size of the particles is found to be dependent on various solvent used e.g. Absolute ethanol, tetra-hydro-furan, distilled water, dimethyl formamide (DMF), and 20% DMF solutions respectively (Wang et al 2001)
The unprecedented flexibility of ink-jet printing makes it very suited for rapid prototyping applications. The ink-jet printing requires conducting tracks by using inks based on organic silver or copper precursors. The precursor is reduced to corresponding metal via a post-printing annealing step. Conventional radiation-convection-conduction heating is the most commonly used method wherein sintering temperatures are ~ $200^{0}$C. Therefore many potentially interesting substrate materials such as polymers cannot be used. The required long sintering time (~60 min.) also implied the non feasibility of the method for fast industrial application. Replacement of microwave heating fulfils the requirement for a fast, simple, cost effective and selective heating of only the printed components (Perelaer 2006)
In the manufacturing of diamonds, the processing through carbon is either by high temperature / high pressure (HTHP) or by high temperature/ low pressure deposition process such as Chemical Vapour Deposition (CVD). Crystalline diamonds predominantly composed of {100} and {111} faces was grown on a non-diamond substrate from a gaseous mixture of hydrogen and methane under microwave glow discharge condition (Kamo et al. 1983) Diamond wafers with a thickness ranging to several microns is a promising material for Infra-red optics as well as for x-ray and particle detector. To be economically viable, the chemical vapour deposition of diamond should be performed at a large area. Compared with hot filament CVD and other techniques, microwave plasma can provide a large and good quality diamonds using hydrocarbon radical and atomic hydrogen (Weidong 2006).
Use of microwave heating in drying of wood is becoming popular. Rattanadecho(2006) studied the influence of irradiation times, working frequencies and sample size dependency of microwave heating patterns within the wood under various conditions. At 2.45 GHz, the power distribution as well as temperature distribution within the sample displays a wavy nature due to the thickness of the sample being close to the penetration depth. Most of the heating seems to occur at the centre of the test sample due to maximum electric field distribution.

## 5. Hazards of microwaves

It is well established fact that microwave radiations cause serious health problems like loss of appetite, irritation, discomfort, fatigue and headaches known as microwave syndrome. Dr. Neil Cherry, a pioneer worker in the field of r.f. radiation hazards in a detailed study had proved beyond doubt that r.f. radiation cause sleep disturbance, melatonin reduction and cancer in many parts of the body. Effect of microwave radiation on the human biochemistry and physiology depend upon frequency, intensity and duration of exposure of radiation. The safe limit for microwave radiations as per WHO and ICNIRP guidelines is ~2.5 $mW/cm^2$ or less. A simple device called microwave leakage meter should be used to check the level of microwave radiations around the microwave heating system and if it is more than the prescribed limit arrangements must be made to shield the system or proper absorbing materials should be put to reduce the level below tolerable limit. Nothing in the life is absolutely safe. Even ultra violet (UV) part of solar radiation is known to pose serious hazards. It is good practice to always wear microwave safe gloves, apron, and goggles now available in the market while working near microwave heating systems.

## 6. Conclusion

In modern days where scientific finding and technology are going hand in hand, any new synthesis technique which would save time in synthesis of new materials or improving the thermo-dynamical properties of the materials by sintering would be extremely beneficial. In this respect microwave heating is faster, eco-friendly and has potential to contribute towards the synthesis of nano particles. This Green Chemistry approach not only reduces the synthesis time but also suppresses the side reactions thus improving the yield and reproducibility. Familiarity with basic physical concepts and practical aspects of microwave heating process in chemical industry may definitely increase the speed and efficiency of yield.

## 7. Acknowledgment

We acknowledge DST, CSIR- New Delhi and PLANEX (ISRO), Ahmedabad for providing the financial support. We thank Ms Pooja Sharma for the help rendered by her during manuscript typing.

## 8. References

Agrawal, D., Cheng, J., Peelamedu, R., Fang, Y., & Roy, R. (2006). Materia Japan. *Bulletin of Japan Institute of Metals. 45, 574-576.*

Baghurst, D. R., Chippindale, A. M., & Mingos , D. M. P., ( 1988). Microwave synthesis for superconducting ceramics, *Nature* 332, 311.

Binner, J. G. P. & Al-Dawery, I. A. H. (1998). Bulk YBCO high $-T_c$ superconductors with uniform and full oxygen content via microwave process, *Superconductor Science and Technology,*11, 449-457.

Bilecka, I., & Niederberger, M., (2010). Microwave chemistry for inorganic nanomaterial synthesis, *Nanoscale* 2, 1358-1374.

Bosi, S., Beard, G., Moon, A. & Belcher, W. (1992). Microwave preparation of the $YBa_2Cu_3O_{7-\delta}$ superconductor. *Journal of Microwave power and Electromagnetic Energy,* 27, 2, 75-80.

Chandra, U., Parthasarathy, G. & Sharma, P. (2010). Synthetic cubanite $CuFe_2S_3$:Pressure induced transformation to isocubanite. *Canadian Mineralogist* 48 : 1137-1147

Chandra, U., Pooja sharma & Parthasarathy, G., (2011). High-pressure electrical resistivity, Mössbauer, thermal analysis and micro-Raman spectrscopic investigations on microwave synthesized orthorhombic cubanite ($CuFe_2S_3$), *Chemical Geology* (in Press).

Cheng, J., Rustum Roy & Dinesh Agrawal, (2001). Experimental proof of major role of magnetic field losses in microwave heating of metal and metallic composites, *Journal of Materials Science Letters,* 20, 1561-1563.

Cherradi, G., Desgardin, J., Provost & Raveau, B. (1994). In: Electroceramics IV, Vol.II (edited by Wasner R.,Hoffman S.,Bonnenberg and Hoffman C., RWTN), Aschen , p1219.

Clark, D. E. & Sutton, W. H., (1996). Microwave processing of materials, *Annual Review of Material Science,* 26, 299-331.

Dong, C., Guo, J., Fu, G. C., Yang, L.H. & Chen, H. (2007). Rapid preparation of $MgB_2$ superconductor using hybrid mirowave synthesis, *Superconductor Science and Technology,*17, L55-L57.

Dube, D. C., Agrawal, D., Agrawal , S. & Roy, R, (2007). High temperature dielctric study of $Cr_2O_3$ in microwave region, *Applied Physics Letters,* 90, 124105-1to 3

Gedye, R., Smith, F., Westaway, K., Ali, H., Baldisera, L., Laberge, L. & Rousell, J, (1986). The use of microwave ovens for rapid organic synthesis, *Tetrahedron Letters,* 27, 279- 282.

Gedye, R. N., Smith, F.E. & Westaway, K. C., (1988). The rapid synthesis of organic compounds in microwave ovens, *Canadian Journal of Chemistry* 66:17-26.

Giguere, R. J., Bray, T.I., Duncan, S.M. & Majetich, G. (1986). Application of commercial microwave ovens to organic synthesis. *Tetrahedron Letters,* 27:4945-4948

Gibbons, K. E., Jones M.O., Blundell S.J., Mihut A.LI, Gameson I., Edwards P.P., Miyazaki Y.,Hyatt, N.C. & Porch A. (2000). Rapid synthesis of colossal magnetoresistance manganites by microwave dielectric heating. *Chemical Communications* 159-160

Guo, J., Dong,C.,Yang,.& Fu,G.(2005) A green route for microwave synthesis of sodium tungsten bronzes $Na_xWO_3$ ( $0<x<1$). *Journal of Solid State Chemistry* 178: 58-63

Hasna, A.M. (2009), Composite Dielectric heating and drying :The computational process, *Proceedings of the World Congress on Engineering (WCE 2009). Vol I,* pp679- 686, ISBN 978-988-17012-5-1.

Hyatt, N. C. & Porch, A., (2000). Rapid synthesis of colossal magnetoresistance manganites by microwave dielectric heating, *Chemical Communications,* 159-160.

Hussein, M. Z. B., Zainal, Z. & Ming, C. Y., (2000). Microwave-assisted synthesis of Zn-Al layered double hydroxide-sodium dodecyl sulphate nano composite, *Journal of Materials Science letters,* 19, 10, 879-883.

Jansen , K.. (2004). Verified synthesis of zeolite materials, In: (2nd Revised edition), *Microwave Technology in Zeolite Synthesis. http://www.ionline.org/synthesis/VS_2ndEd/MicrowaveTech.htm*

Jones, S.D., & Akridge, J. R., (1995)., In: *Handbook of solid state batteries and capacitors.* Munshi, M.Z.A., Ed. World Scientific , Singapore , 209.

Kamo,M., Sato,Y., Matsumoto, S., & Setaka, N., (1983), Diamond synthesis from gas phase in microwave plasma, *Journal of Crystal Growth,* 62(3) , 642-644.

Kappe, C. O., & Dallinger, D., (2006). The impact of microwave synthesis on drug discovery, *Nature Reviews Drug Discovery,* 5, 51-63

Krishnan,R., Agrawal, D. & Dobbins, T., (2009). Microwave irradiation effects on reversible hydrogen desorption in sodium aluminium hydrides ($NaAlH_4$), *Journal of Alloys and Compound,* 470, 250-255.

Kubel ,E., (2005). Advancements in Microwave heating technology, *IndustrialHeating.com* (January issue) , 43-53.

Lee, C. C., & So, W.W., (2000). High temperature silver- indium joints manufactured at low Temperature, *Thin solid films,* 366, 196-201.

Lekse, J. W., Stagger, T.J., & Aitken, J.A., (2007). Microwave metallurgy: synthesis of intermetallic compounds via microwave irradiation, *Chemistry of Materials,* 19, 3601- 3603

Lučun, A., Butkute, A., Maneikis, O., Kiprijanovic, O. Anisimovas, F., Gradauskas, J., Suziedelis,A., Vengalis ,B., Kancleris,Z.,& Asmontas ,S., (2007). Magnetoresistance of polycrystalline $La_{0.7}Ca_{0.3}MnO_3$ films in a microwave magnetic field, *Acta Physica Polonica* 111: 147-152

Makul, N. & Rattanadecho, P., (2010). Microwave Pre-curing of natural rubber compounds using a rectangular wave guide, *International Communication in Heat and Mass Transfer,* 37, 914-923.

Metaxas, A.C. & Meredith, R. J., (1983). *Industrial Microwave Heating,* Peter peregrinus Ltd., London, UK (for IEE).

Mingos D. M. P., & Baghurst, D. R., (1991). Application of microwave dielelctric heating effects synthetic problems in chemistry, *Chemical Society Reviews,* 20, 1, 1-47.

Newham, R. E., Jang, S. J., Xu, M. & Jones, F., (1991). Fundamental interaction mechanisms between microwave and matter, *Ceramic transactions,* 21, 51.

Nissinen, T., Kiros, Y., Gasik, M. & Leskela, M., (2003). Microwave synthesis of $MnCo_2O_4$ nanoparticles on carbon, *Chemistry of Materials,* 15, 4974.

Pareek, S., Rais, A., Tripathi, A., & Chandra, U., (2008). Mössbauer study of microwave synthesized (Cu,Fe) sulfide composite and correlation with natural mineral cubanite, *Hyperfine Interaction,* 186, 113-120.

Park, H. K., Han, Y. S., Kim, D. K., & Kim, C. H. (1998). Synthesis of $LaCrO_3$ powders by microwave induced combustion of metal nitrate-urea mixture solution, *Journal of Materials Science Letters,* 17, 785-787.

Perelaer, J., Berend-Jan de Gans & Schubert, U. S., (2006). Ink-Jet printing and microwave sintering of conducting silver tracks, *Advanced Materials,* 18, 2101-2104.

Rattanadecho, P., (2006). The simulation of microwave heating of wood using a rectangular waveguide: influence of frequency and sample size, *Chemical Engineering Science,* 61, 4798-4811.

Rao, K. J., Vaidhyanathan, B., Ganguli, M. & Ramakrishnan, P.A., (1999). Synthesis of inorganic solds using microwaves, *Chemistry of Materials,* 11, 882-895.

Reid David, *Melting metals in a domestic microwave. home.c.2i.net/metaphor/mvpage.html.*

Roy, R., Peelamedu, R., Grimes, C., Cheng, J. & Agrawal D., (2002). Major phase transformations and magnetic property changes caused by electromagnetic fields at microwave frequencies, *Journal of Materials Research,* 17, 12, 3008-3011.

Schiffmann, R. F., (1986). Food Product development for microwave processing, *Food Technology,* 40, 6, 94-98.

Shulman, H. S., (2002). *Microwave heating ceramic.* www.ceralink.com

Singh R. J., (2007). Heating of manganites by magnetic component of microwave., *Indian Journal of Pure and Applied Physics,* 45, 454-458.

Starck, A.V., Muhlbauer, A. & Kramer, C., *Hand book of thermoprocessing Technologies : Fundamental processes component safety,* (2005). ISBN3-8027-2933-1 , Vulken-Verlag GmbH.

Tsuji, M., Hashimoto, M., Nishizawa, Y., Kubokawa, M. & Tsuji, T., (2005)., Microwave-assisted synthesis of metallic nanostructures in solution, *Chemical European Journal,* 11, 440-452.

Vaidhyanathan, B. & Rao, K. J., (1997). Synthesis of Ti,Ga, and V Nitrides: microwave-assisted carbothermal reduction and nitridation, *Chemistry of Materials,* 9, 1196-1200.

Vanetsev, A. S. & Tretyakov, Yu.D., (2007). Microwave-assisted synthesis of individual and multicomponent oxides, *Russian Chemical Reviews,* 76, 5, 397-413.

Vollmar, M. (2004). Physics of microwave oven, *Physics Education,* 39, 1, 74-81.

Vos, B., Mosman, J., Zhang,Y.,Poels,E., & Bliek, A. (2003). Impregnated carbon as a susceptor material for low loss oxides in dielectric heating, *Journal of Materials Science,* 38, 173-182.

Wang, H., Zhang, J. & Zhu, J. (2001). A microwave assisted heating route method for the rapid synthesis of sphalerite- type mercury sulfide nanocrystals with different sizes, *Journal of Crystal Growth,* 233, 829-836.

Weidong, M., Chuanxin, W. & Shenggao, W., (2006). Microwave chemical deposited thick diamond film synthesis using $CH_4$/ $H_2$/ $H_2O$ gas mixture, *Plasma Science and Technology,* 8, 3, 329- 334.

Xie, G., Li, S., Louzguin, D.V., Cao, Z., Yoshikawa, N., Sato, M.& Inoue, A., (2009). Fabrication of Ni-Nb-Sn metallic glassy-alloy powder and its microwave induced sintering behaviour, *Journal of Microwave Power and Electromagnetic Energy,* 43, 17- 22.

Yoshikawa, N. (2010). Fundamentals and applications of microwave heating of metals, *Journal of Microwave Power and Electromagnetic Energy,* 44, 1, 4-11.

Yund, R.A. & Kullerude, G., (1966). Thermal stability of assemblages in Cu-Fe-S system, *Journal of Petrology,* 7, 454 – 488.

# Part 2

# Medical Sciences

2

# Microwave Radiometry as a Non-Invasive Temperature Monitoring Modality During Superficial Hyperthermia

Svein Jacobsen
*University of Tromsø*
*Norway*

## 1. Introduction

The application of microwaves in medicine has recently obtained renewed attention within the scientific community through emerging techniques in breast cancer detection (microwave tomography and ultra-wide-band radar imaging), high power ablation as well as microwave angioplasty and lipoplasty (Rosen et al., 2002). Nevertheless, more established techniques reported in the literature based on RF/microwaves are hyperthermia and medical radiometry. During the last decades, a number of research groups have studied various clinical applications based on radio-thermometry for detection of thermal anomalies in subcutaneous, or invasively more deeper, parts of the human body.

The possibility of using microwave radiometry for non-invasive thermometry was originally suggested back in the early 1970's (Edrich & Hardee, 1974; Enander & Larson, 1974) and the sensing principle was denoted *microwave thermography*. Prospected medical applications with highest potential include detection of breast cancer (often in conjunction with infrared thermometry) (Mouty et al., 2000), non-invasive temperature control of superficial (Ohba et al., 1995) and interstitial (Camart et al., 2000) hyperthermia, and control of brain temperatures of newborn infants during mild hypothermia (Hand et al., 2001; Maruyama et al., 2000). Additional applications, which have been investigated, comprise detection of inflammatory arthritis (MacDonald et al., 1994), extravasation rate of drugs (J. Schaeffer & Carr, 1986), changes of blood flow (Gabrielyan et al., 1992) or amount of lung water (Iskander et al., 1984) as well as post-mortem or cerebral temperature monitoring (Al-Alousi et al., 1994).

Microwave radiometry in clinical medicine aims at deriving information on internal body temperature patterns by measurement of natural thermal black-body radiation from tissue in the lower part of the microwave region (<5 GHz). Knowledge on such thermal patterns can give valuable information in clinical disease detection and diagnosis as well as providing quantitative temperature feedback in monitoring of thermal therapeutic processes.

The application of microwave radiometry raises several issues in determination of subcutaneous temperature heterogeneities. Most important is how to optimize design of the receiver hardware by identification of mutual compatible microwave devices with sufficiently low noise figures. The extremely low power levels (-174 dBm/Hz at $37^{o}$C body temperature) of electromagnetic thermal noise limit the applicability of microwave radiometers as the usable thermal signal competes both with external electromagnetic interference (EMI) and the internal noise produced by the hardware. When operated in, or close to, certain communication bands, proper electromagnetic shielding is imperative to stabilize the signal. As for internal noise interference, certain design schemes can be utilized to minimize the overall system noise.

Advances in controllable heating equipment, combined with improvements in non-invasive as well as invasive tissue thermometry techniques (Rhoon & Wust, 2005), have increased the feasibility of using hyperthermia (41-45 $^{o}$C for 60 min) as a complementary agent in the treatment of cancer. The effectiveness of hyperthermia treatment is related to the induction of elevated temperatures integrated over time and is often quantified by cumulative equivalent minutes at 43 $^{o}$C (CEM43$^{o}$C). The optimal hyperthermia dose for combinations with radiation or chemotherapy is widely accepted to be 60 min treatment at 43 $^{o}$C throughout 90% of measured tumor target temperatures, which represents a CEM43$^{o}$CT90 thermal dose of 60 min (Sapareto & Dewey, 1984). Temperature homogeneity within the target tissue is an advantage to improve the efficacy of hyperthermia, since a narrow temperature window should bring the response-related minimum tumor temperature closer to the maximum tolerable temperature ($\sim$45 $^{o}$C) which avoids undue patient discomfort and complications. Maintaining the targeted temperature distribution within the therapeutic window is a major challenge in hyperthermia in order to attain effective radio- and/or chemosensitization of the treated volume. The therapy goal of obtaining cytotoxic temperatures (>42 $^{o}$C) for periods of 1 h represents the classical dosimetry view and is supported in studies by Dewhirst *et al.* (1984) and Kapp and Cox (1993). In recent years, however, the biologic rationale for hyperthermia has been subject to discussion (Corry & Dewhirst, 2005). Physiological and cellular effects of mild temperature hyperthermia (39-42 $^{o}$C for 1-2 h) are currently under investigation (Calderwood et al., 2005; Dewhirst et al., 2005). The importance of mild temperature hyperthermia related to positive thermo-radiotherapy treatment response is documented in published clinical trials (Jones et al., 2005; Thrall et al., 2005).

In practice, general characterization of tissue temperature distributions with bulk properties like volume average temperature or maximum temperature and location of the maximum can provide very useful realtime feedback information and clinical documentation to an operator. This information is fed back into the heating system for adjustment of antenna power in order to maintain a therapeutic temperature range and avoid side effects from excessive temperatures. Modeling with numerical computations can alternatively or complementary provide additional insight for dose-planning, although unknown patient-specific data (e.g. blood perfusion and power absorption characteristics) limit the reliability of such simulations. Microwave applicators are commonly used for heating of cancerous tissue. Hyperthermia applied to recurrent breast cancer resulted in a complete response rate increase from 31% to 65% in tumors of size smaller than 3 cm (Van der Zee *et al.* 1999). Multi-institutional randomized trials on the same disease all showed similar improvement from adding hyperthermia to radiotherapy (Jones et al., 2005; Vernon et al., 1996). Hyperthermia applicators used in previous efforts have in general been rather bulky and restricted to heating of smaller regions and flat anatomy (Lee, 1995; Stauffer, 2005). However, low cost and expandable printed circuit board (PCB) array construction facilitates development of antennas without the shortcomings of early superficial heating devices. Now lightweight and low profile multi-element array designs have been introduced such as the current sheet applicator (CSA) (Gopal et al., 1992), the contact flexible microstrip applicator (CFMA-12) (Lee et al., 2004), the microstrip spiral applicator (Lee, 1995), the annular aperture and horseshoe applicators (Carlier et al., 2002), and the Dual Concentric Conductor applicator (DCC) (Rossetto & Stauffer, 2001).

In the field of bio-electromagnetics, thermal considerations are crucial to evaluate dosimetry levels for bio-experiments. A multitude of factors determine the degree of temperature homogeneity in tissue during hyperthermia. This includes heterogeneity in tissue composition which in turn implies local variations in blood flow rate together with thermal and electric conductivity. As a result power absorption, effective heating depth, and efficiency of coupling electromagnetic power into the volume under treatment will vary spatially.

While the vast majority of thermal dosimetry for hyperthermia has been performed using invasive temperature probes to sample a small number of points, there are a number of non-invasive approaches under investigation which can quantify more complete 2-D and 3-D temperature distributions. These techniques include infrared thermography (Tennant & Anderson, 1990), computerized axial tomography (Rutt et al., 1986), ultrasound time-of-flight tomography techniques (Seip & Ebbini, 1995), electrical impedance tomography (Moskowitz et al., 1995), microwave tomography (Chang et al., 1998), magnetic resonance imaging (Samulski et al., 1994), and microwave radiometry (Ohba et al., 1995).

Whereas infrared thermography is able to map thermal emissions from the body surface only, radiometry in the lower microwave region has the potential to detect thermal radiation emitted by subcutaneous tissue up to a depth of several centimeters. Multispectral radiometry has been under investigation as a measurement technique to provide information on temperature depth distributions in tissue. As the radiometric signal strength from a volume element at a particular depth of the medium is correlated with frequency, a multi-frequency scan of a broader band can (at least in principle) be used to map depth temperature gradients. An important limitation of the radiometric observation principle is however the extremely weak signal level of the thermal noise emitted by the lossy material. Consequently, requirements of long integration time ($\sim$5 s) and wide integration bandwidth ($\sim$500 MHz) result in a maximum of 5-6 radiometric bands per antenna within the usable frequency scan range. Even with long data acquisition and signal post-processing times combined with increasingly complex system hardware for multifrequency radiometry, only a few data samples of depth related information are available. For a single antenna system, these sparse data sets can nevertheless produce useful depth-temperature information. Combined with *a priori* information on antenna radiation patterns and tissue dielectric and thermal properties, viable estimates of average temperature in selected tissue volumes can be obtained (Bardati et al., 1991; Camart et al., 2000; Jacobsen & Stauffer, 2003). However, for a full scan of multiple frequencies in a multiple antenna system, the temperature scanning time increases rapidly and can lead to unacceptable cooling during the radiometric listening period while microwave heating power is switched off. Thus from clinical, practical, economical, and technical considerations, the number of radiometric bands should be kept to a minimum (preferably single-band) for temperature monitoring of clinical hyperthermia procedures.

The paper is organized as follows. Section 2 describes the basic radiometric theory (detection scheme, signal model and inversion algorithm) and experimental setup. Section 3 reports experimental phantom results including heating and radiometric antenna layouts, measured antenna return loss as well as radiometric temperature scans of a dynamic heating process. Finally, in Section 4, the findings are evaluated and Section 5 draws conclusions from the research.

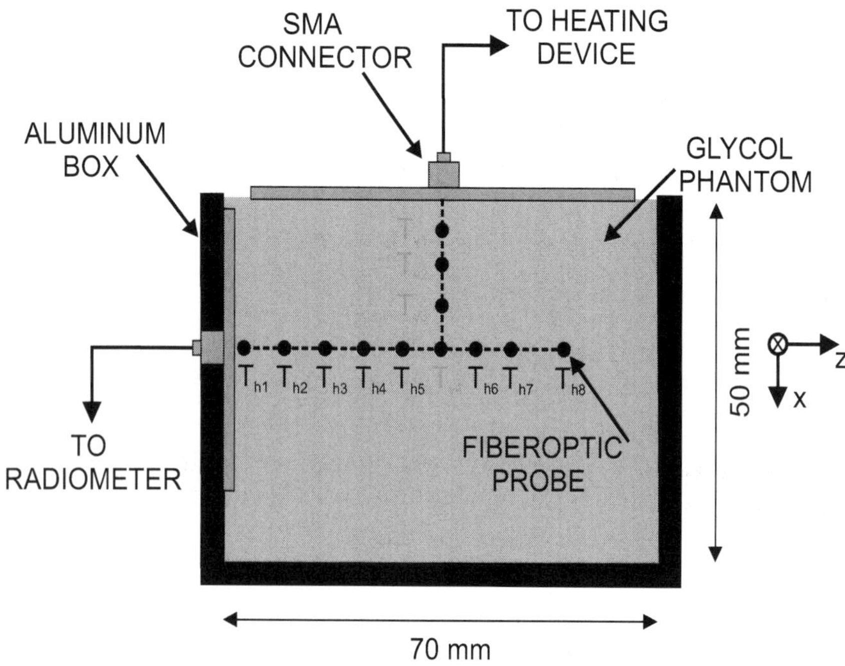


Fig. 1. Pitch-catch mode setup for 915 MHz heating and 3.5 GHz radiometric temperature reading.

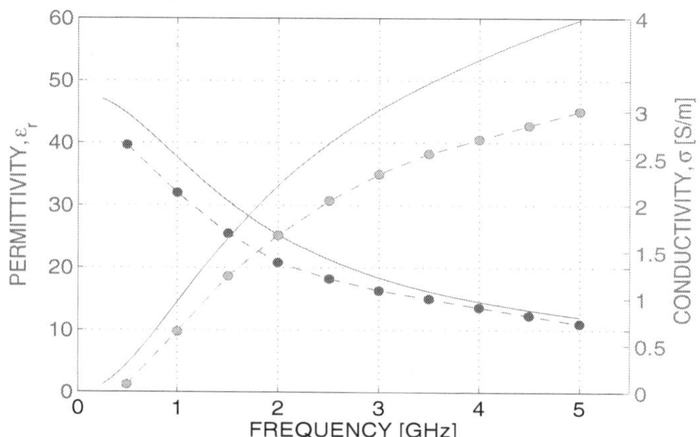


Fig. 2. Dielectric properties of glycol based phantom. Solid line: 90% glycol-10% water mixture. (•): solidified phantom of same mixture.

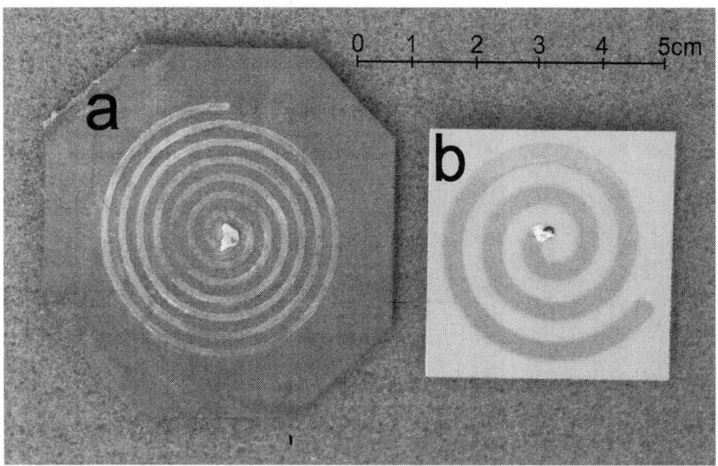


Fig. 3. Antennas used in experiment. a) 915 MHz heating spiral antenna, and b) 3.5 GHz radiometric spiral antenna.

## 2. Methodology

### 2.1 Antenna configuration and setup

Fig. 1 depicts the combined heating and temperature measurement configuration considered in the present study. A glycol-water phantom load was solidified with gelatine in a metal box of size 50×70×50 $mm^3$. We notice from Figure 2, that solidification using gelatine (low permittivity material) lowers both the overall permittivity and conductivity of the applied fluidic phantom. The phantom dielectric properties is similar to that of fat infiltrated mammary tissue.

The five sidewalls of the supporting box, of highly electrically conducting material, provide electromagnetic shielding against competing electromagnetic sources (e.g. cellular phones). However, the aluminum box was not thermally insulated from the surroundings and thus participates in the cooling process as the ambient temperature will influence thermal fluxes within the phantom. Furthermore, the phantom structure was topped with a 50×50 $mm^2$ spiral heating antenna operated at the medical 915 MHz frequency and driven by a 9 W power source. Arranged in a pitch-catched mode, a smaller 35×35 $mm^2$ spiral antenna was mounted inside one adjacent sidewall to monitor temperature gradients within the heated volume. The small spiral antenna was connected to a 3.5 GHz single band radiometer with an integration bandwidth of 500 MHz. A photograph of the respective antennas is shown in Figure 3 and measured return loss is depicted in Figure 4. Notice that the small spiral is well matched at the radiometric frequency of 3.5 GHz ($S_{11}$ better than -10 dB) whereas the heating antenna has a return loss of -14 dB at 915 MHz. For correlation with radiometric readings, two perpendicular 1-D temperature profiles (see Figure 1) were established by means of twelve fiberoptic probes rack mounted in a LumiTherm X5R system [1]. The custom made Dicke radiometer used in the experiments has been described in the literature before (Klemetsen et al., 2011). Using miniature surface mount devices (amplifiers, switches and circulators), the front-end printed

[1] www.ipitek.com

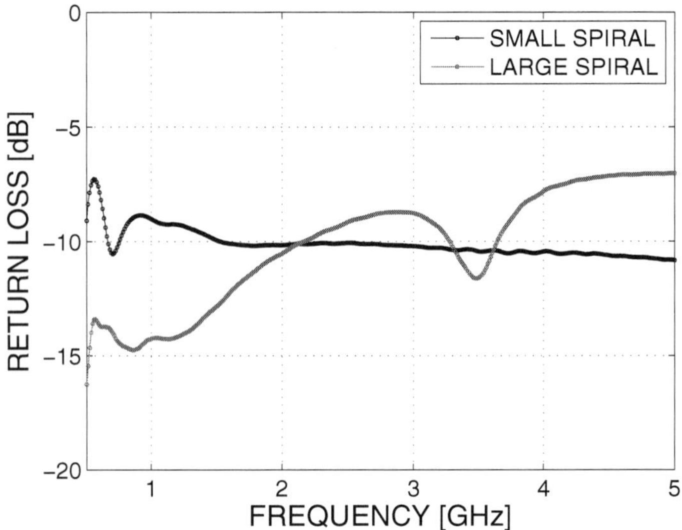


Fig. 4. Measured antenna return loss vs frequency.

circuit board is only $40 \times 50$ mm$^2$ in size. With a detector intergration time of 2 secs and integration bandwidth of 600 MHz, the brightness temperature resolution and theoretical lower accuracy are as low as $\sigma_{T_B}$=0.029$^o$C.

### 2.2 Brightness temperature integral evaluation

Various approaches have been suggested to convert measured radiometric brightness temperatures to an actual temperature distribution within the volume under observation. These methods include parametric (Hand et al., 2001; Ohba et al., 1995) and non-parametric (Bardati et al., 1993; Jacobsen & Stauffer, 2003) modeling. Below we present a 3-D parametric model which leads to a proportionality between the observed and sought parameter.
Consider the general definition of the brightness temperature $T_B$:

$$T_B = \int_{\Xi} T(\underline{r}) W(f_c, \underline{r}) dV \tag{1}$$

where $W$ is the weighting function (WF) dependent on the spatial variable $\underline{r}$ and center frequency $f_c$ within the integration bandwidth. $\Xi$ is the volume under investigation and $dV$ is an infinitesimal volume element.
The weighting function is normalized according to:

$$\int_{\Xi} W(f_c, \underline{r}) dV \equiv 1 \tag{2}$$

Next, we evaluate the term in Equation (1), which describes the contribution to the lossy load brightness temperature by introducing a parametric model. We assume that both the temperature distribution and weighting function are separable in space with respect to a

Cartesian coordinate system; i.e. $T = T_x(x) \cdot T_y(y) \cdot T_z(z)$ and $W = W_x(x) \cdot W_y(y) \cdot W_z(z)$. Hence, the integral in Equation (1) expands to:

$$T_B = \int_{L_x} T_x(x) W_x(f_c, x)\, dx \int_{L_y} T_y(y) W_y(f_c, y)\, dy \int_{L_z} T_z(z) W_z(f_c, z)\, dz \tag{3}$$

where $L_{x,y,z}$ are the spatial integration lengths within the tissue volume $\Xi$ in x-, y-, and z-direction, respectively. Using the following model functions for the respective factors in Equation (3):

$$T_x = \exp\left(-x/\gamma_{T_x}\right) \tag{4}$$
$$T_y = \exp\left(-y^2/\sigma_{T_y}^2\right) \tag{5}$$
$$T_z = \frac{T_{\max}}{1 + \alpha(z - z_m) + \beta(z - z_m)^2} \tag{6}$$
$$W_x = \frac{1}{\sqrt{\pi}\sigma_{E_x}} \exp\left(-x^2/\sigma_{E_x}^2\right) \tag{7}$$
$$W_y = \frac{1}{\sqrt{\pi}\sigma_{E_y}} \exp\left(-y^2/\sigma_{E_y}^2\right) \tag{8}$$
$$W_z = \frac{1}{d} \exp\left(-z/d\right) \tag{9}$$

where $\sigma_{E_{x,y}}$ are the lateral spatial widths of the weighting function in the $x, y$-direction, $d$ is the (z-direction) spiral antenna 1/e-sensing depth, $\gamma_{T_x}$ and $\sigma_{T_y}$ are the lateral widths of the temperature distribution in the $x$ and $y$-direction, $T_{\max}$ is the maximum temperature in tissue, and $\{z_m, \alpha, \beta\}$ is a parameter set modeling curvature and asymmetry of the 1-D depth temperature profile. Notice that the required normalization $\int_\Xi W(\underline{r})dV \equiv 1$ is satisfied through the prefixed normalization factors. The 3-D model given in Equations (4)-(9) is a direct generalization of the 1-D model used by Jacobsen and Stauffer (2003) and similar to the 3-D model in Jacobsen and Stauffer (2007).

By carrying out the integration along all directions, and assuming that $L_x = L_y = (-\infty, \infty)$ and $L_z = (0, \infty)$, it is readily shown:

$$T_B = \frac{1}{d\sqrt{1 + \sigma_{E_y}^2/\sigma_{T_y}^2}} \frac{1}{\sqrt{\pi\sigma_{E_x}^2}} \int_{-\infty}^{\infty} \exp(-x/\gamma_{T_x}) \exp(-x^2/\sigma_{E_x}^2)\, dx \cdot \tag{10}$$
$$\int_0^\infty \frac{T_{\max}}{1 + \alpha(z - z_m) + \beta(z - z_m)^2} \exp\left(-z/d\right) dz \tag{11}$$

which might be semi-analytically expressed:

$$\begin{aligned} T_B = {} & \frac{T_{\max}}{d} \frac{1}{\sqrt{1 + \sigma_{E_y}^2/\sigma_{T_y}^2}} \frac{1}{\sqrt{\alpha^2 - 4\beta}} \cdot \exp\left(-\frac{\sigma_{E_x}^2}{4\gamma_{T_x}^2}\right) \cdot \\ & \left[\exp\left(\frac{1}{2\beta d}(\alpha - 2\beta z_m - \sqrt{\alpha^2 - 4\beta})\right) E_1\left(\frac{1}{2\beta d}(\alpha - 2\beta z_m - \sqrt{\alpha^2 - 4\beta})\right) - \right. \\ & \left. \exp\left(\frac{\alpha}{2\beta d}(\alpha - 2\beta z_m + \sqrt{\alpha^2 - 4\beta})\right) E_1\left(\frac{1}{2\beta d}(\alpha - 2\beta z_m + \sqrt{\alpha^2 - 4\beta})\right)\right] & (12) \\ = {} & \nu(\alpha, \beta, z_m, \gamma_{T_x}, \sigma_{T_y}, \sigma_{E_y}, \sigma_{E_x}, d) T_{\max} & (13) \end{aligned}$$

| Characteristic | Parameter | Nominal value |
|---|---|---|
| WF sensing depth | $d$ | 6.0 mm |
| WF lateral width, x-direction | $\sigma_{E_x}$ | 7.0 mm |
| WF lateral width, y-direction | $\sigma_{E_y}$ | 7.0 mm |

Table 1. Parameter values of radiometric antenna.

where $E_1(z) = \int_1^\infty \exp(-zt)t^{-1}dt$ is the exponential integral of first order (Gradshteyn, 1980) which can be tabulated. Here, $\nu \leq 1$ is a proportionality factor that relates the observed and sought parameter.

$\nu$ can be derived given that the thermal, dielectric, and geometric parameters are *a priori* known for the setup. From the numerically generated brightness temperature $T_B$, and using Equation (13), the maximum temperature within the volume can be calculated.

Applying the commercial software package CST Microwave Studio[2], full wave simulation of the radiometric antennas' radiation properties was used to establish the model parameters in Equations (7-9). Numerical values of these runs for the specific phantom are depicted in Table 1.

## 3. Results

Figure 5 displays the probe-measured temperatures within the phantom as a function of time. We notice in Figure 5(a) the marked dynamic heating bursts displayed by $T_{v1}$ for times less than 2000 secs. The probe closest to the heating antenna reaches a maximum temperature of 40$^o$C at 1900 secs. Deeper down into the phantom, lower temperatures are observed as intuitively expected. Maximum temperatures at depth are also slightly time-delayed compared to more shallow temperatures due to some minor effects of thermal conduction.

As for the horizontal probes, it is seen in Figures 5 (b) and (c) that the maximum temperature along the z-axis occurs at a depth of 20 mm and 30 mm (27.2$^o$C). Since probe $T_{v4}$, located in between $T_{h5}$ and $T_{h6}$, shows a maximum of 27.4$^o$C, the z-profile horizontal maximum is located (as anticipated) at the boresight symmetry line of the heating antenna.

To get an overview of the dynamic heating process taking place within the phantom, perpendicular to and in the observation direction of the radiometric antenna, temperature profiles along the z- and x-axis are plotted in Figure 6 at selected points in time. The shape of the profiles in both directions is modeled accurately by Equations (4) and (6). Notice the smooth shape of the model function in the z-direction with a maximum plateau underneath the center of the heating antenna. In addition, the x-direction temperature gradient is significant (up to 1$^o$C/mm), especially subsequent to the heating sequences taking place for times less than 1800 secs. During the cooling process, the decay curves flatten and eventually become spatially constant.

Table 2 quantifies the parameter values obtained for the snapshots in Fig 6. Observe the relatively small values of $\gamma_{T_x}$ in the heating phase. During the cooling process (time$\geq$2150 secs), large values of $\gamma_{T_x}$ model the x-direction temperature profile, meaning that it is virtually flat (homogeneous temperature) lateral to the radiometric observation axis.

Based on Table 1 and Equations (4)-(9), the brightness temperature $T_B$ can be estimated through Equation (12) for the dynamic run discussed above.

[2] www.cst.com

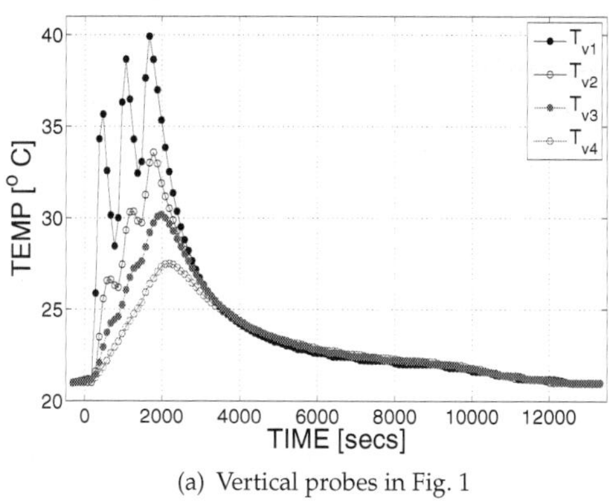


(a) Vertical probes in Fig. 1

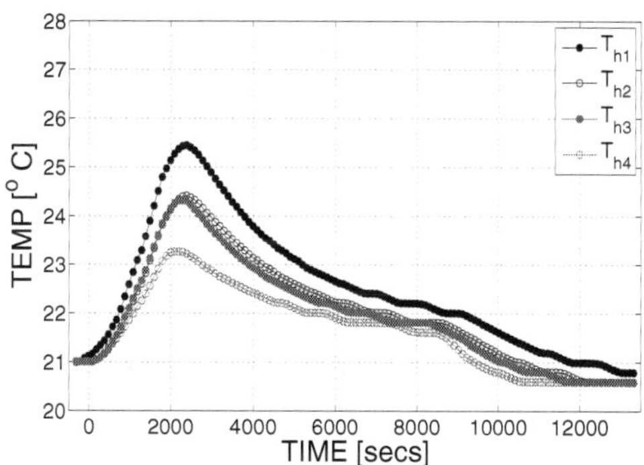


(b) Horizontal probes (first 4) in Fig. 1

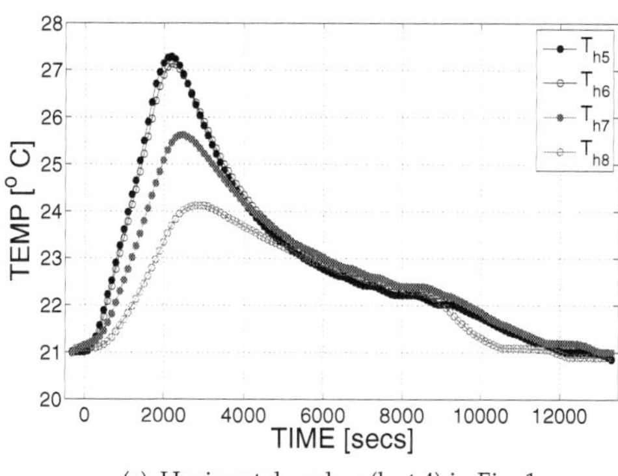


(c) Horizontal probes (last 4) in Fig. 1

Fig. 5. Fiberoptic probe temperature measurements vs time.

To evaluate the validity of the radiometric model, comparisons are made with the measured temperatures derived from the miniature radiometer. Figure 7 (upper panel) shows this comparison for two different models. This includes a simplified 1-D parametric model where variations along the transversal axis (x-direction) are neglected. In this case it is implicitly assumed that $\gamma_{T_x} = \infty$. In the other 2-D model, also temperature variations in the transverse direction are adopted into the model.
From the deviation plot in Figure 7 (lower panel) we notice improved performance (smaller deviation between measurement and model) in the heating phase (times less than 2000 secs) when the more advanced 2-D model is applied. A quantitative analysis of the respective models shows an improvement in the standard deviation of the instantaneous error from $\sigma_{T_{1-D}}$=0.28$^o$C in the 1-D case to $\sigma_{T_{2-D}}$=0.21$^o$C in the 2-D case.
During the cooling phase, $\gamma_{T_x} \gg 1$ and the difference in performance between the 1-D and 2-D model should be negligible. This is verified in Figure 7 (lower panel) as only marginal deviations in performance are observed for times greater than 3000 secs.
One pertinent question is to what extent the maximum temperature within the heated volume can be predicted given a measured brightness temperature; information that can be valuable to a system operator. The numerical span of the model parameters in Table 2 indicates that there exists no unique parameter combination that provides an overall proportionality factor $\nu$ in Equation (13). However, as seen in Figure 8, a constant $\nu$ still provides correlated estimates between radiometrically estimated and probe measured maximum temperatures within the heated volume.

## 4. Discussion

Hyperthermia has been demonstrated to be an efficacious adjuvant to other conventional approaches for cancer therapy. Breast cancer is one of the most prevalent forms of cancer in women and local control of this disease remains an issue of continually increasing magnitude. Due to uncertainties associated with the actual electromagnetic power deposition and thus the resulting temperature distribution in an arbitrary, heterogeneous, and temporally varying tissue configuration, some type of temperature feedback control is essential for safe and reliable heating performance of any hyperthermic system. With its non-invasive character and being a totally passive sensing principle, microwave radiometry has the potential to allow thermometry of subcutaneous tissues to a depth of some centimeters. As opposed to IR radiometry, microwave radiometry does not provide high-resolution thermographic mapping of the tissue, since lateral spatial resolution is limited by antenna size and spacing, which are

| Time [s] | $\alpha[mm^{-1}]$ | $\beta[mm^{-2}]$ | $z_m[mm]$ | $\gamma_{T_x}[mm]$ |
|---|---|---|---|---|
| 500 | 0.0192 | 0.00519 | 23.1 | 5.1 |
| 750 | -0.0355 | 0.00538 | 19.7 | 11.2 |
| 1170 | -0.0422 | 0.00411 | 18.3 | 9.8 |
| 1650 | 0.0632 | 0.00215 | 38.1 | 11.1 |
| 2150 | -0.0571 | 0.00118 | 0.65 | 27.1 |
| 3100 | -0.0355 | 0.00183 | 17.6 | 154.0 |
| 3642 | -0.0141 | 0.00155 | 24.3 | $\infty$ |
| 6135 | -0.0154 | 0.00092 | 27.3 | $\infty$ |
| 9700 | -0.0290 | 0.00468 | 26.1 | $\infty$ |

Table 2. Numerical parameter values of temperature profiles in Figure 6.

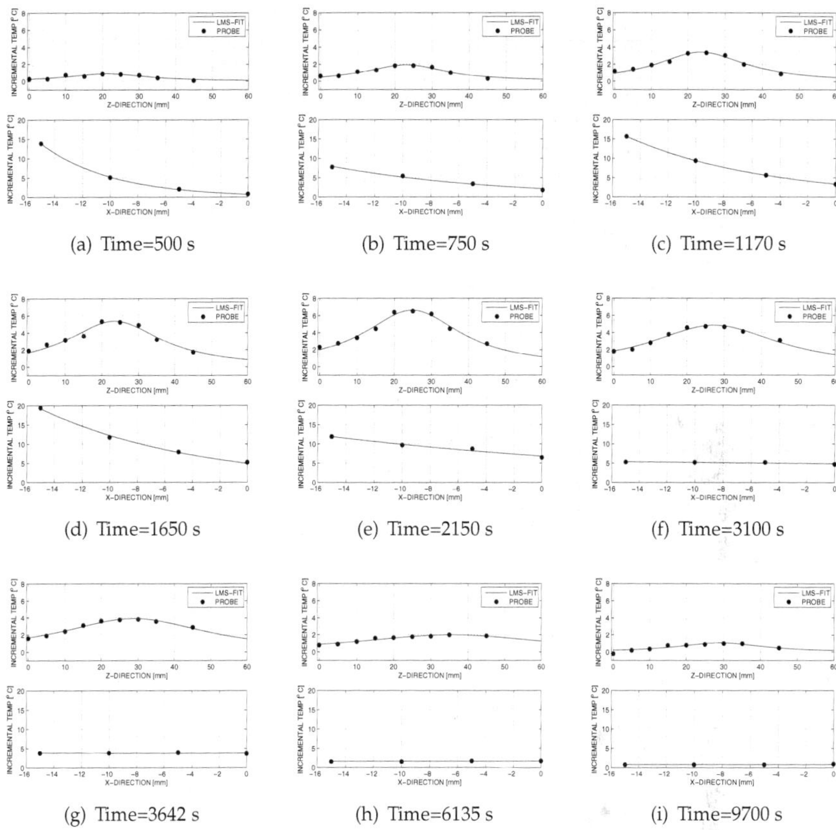


(a) Time=500 s (b) Time=750 s (c) Time=1170 s

(d) Time=1650 s (e) Time=2150 s (f) Time=3100 s

(g) Time=3642 s (h) Time=6135 s (i) Time=9700 s

Fig. 6. Fiberoptic temperature profiles in phantom along two orthogonal axes for different times. Solid line: Fitted model in Equations (4) and (6).

in the order of centimeters. This limitation, combined with the necessity of proper shielding to avoid interfering signals from other competing EM sources, put constraints on the realization of this observation technique.

Since medical microwave radiometry only offers sparse temperature data sets, either *a priori* experimental knowledge (Mizushina et al., 1993) or complementary thermal modeling (Bardati et al., 1993) have been proposed to help solve the underdetermined inverse problem of characterizing temperature distributions at depth in living tissue.

In the present study we evaluate the robustness of a proposed temperature estimator scheme based on single-band radiometric scanning. A bistatic antenna pitch-catch setup on a glycol based phantom was used to test the performance of a single-band miniature microwave radiometer. One large spiral antenna was used to heat the volume under investigation using the medical 915 MHz frequency. A similar, but smaller, spiral antenna was also put flush to the phantom to observe thermal radiation. Since no blood perfusion is present in the phantom, only 9 W was required to heat the volume up to therapeutic hyperthermia temperatures

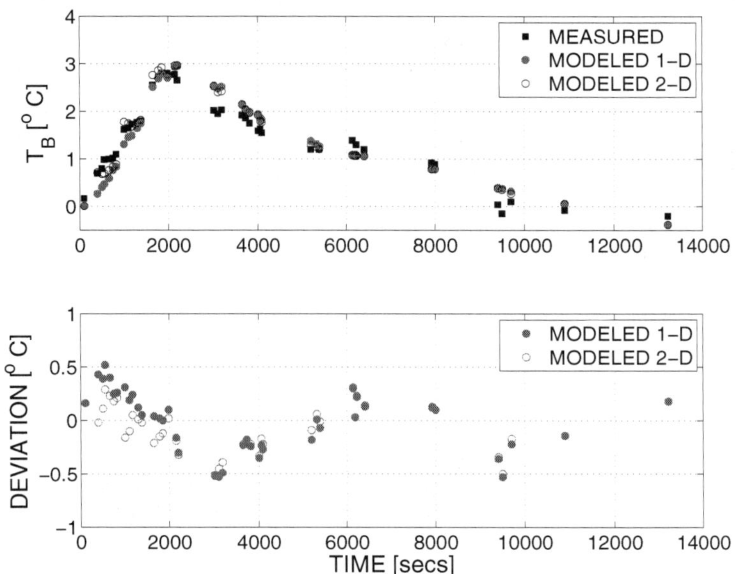


Fig. 7. Differential radiometric brightness temperature versus time. Measured and 1-D & 2-D models (upper panel), deviation between models and measurements (lower panel).

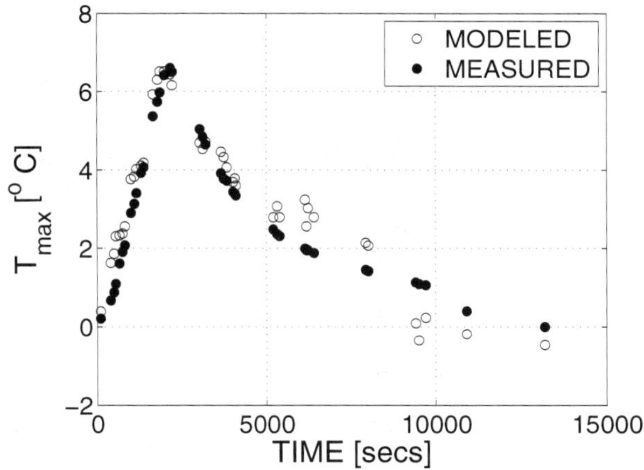


Fig. 8. Measured and radiometrically modeled ($\nu$=0.65 in Equation (13)) maximum temperature within heated phantom vs time.

(>40$^o$C). The heating was conducted in several burst to mimic the clinical scenario of interspersed heating and radiometric temperature reading. In order to investigate long term stability of the radiometric instrumentation, temperature readings were also performed in the cooling phase, in which the phantom interchanged energy with the surroundings (ambient temperature of 21°C) back to the initial pre-heating state.

Overall, the concordance between the modeled and measured brightness temperatures was satisfactory, with deviations typically less than $\pm 0.3^oC$ of the maximum $T_B$ temperature elevation of 3.0°C. Since the proportionality factor $\nu < 1$ in Equation (13) , the maximum temperature elevation in the phantom is somewhat higher than the maximum observed brightness temperature. According to Figure 8, the maximum increase along the radiometric observation axis is about 6.5°C occuring at z=25 mm (see Figure 6) 2150 secs into the experiment.

The pitch-catch experimental setup produces temperature gradients in two directions. Hence, the radiometric weighting function is generally 2-D in cases where temperature equilibrium is not obtained along the transversal x-axis. Thus, one pertinent question is whether the proposed 2-D model performs better than the simplified 1-D model in such cases. This is indeed the case as can be observed in the heating phase of the experiment (Figure 7) displaying an improvement of typically 0.2°C in accuracy. However, when the transversal temperature gradient can be neglected, the 1-D and 2-D models perform equally well.

Finally, the ability to predict the maximum temperature within the heated volume given a single-band radiometric reading was investigated. In the heating phase, a high degree of concordance was observed between measured and estimated maximum temperature. As the cooling process started, larger deviations were observed as a consequence of more extreme values of the model parameters (see Figure 8). Fortunately, the cooling phase is less realistic with respect to a clinical hyperthermia scenario, as more steady state conditions with temperatures above 40°C are expected in such cases.

## 5. Conclusion

The accuracy of using mono-frequency microwave radiometry to retrieve maximum bulk temperatures of a heated glycol-based phantom was experimentally studied using a miniature radiometer operating at 3.5 GHz. A computationally fast retrieval scheme, based on a single brightness temperature, was implemented to predict the dynamic brightness temperature during both the heating and cooling part of the experiment. In cases of highest degree of temperature inhomogeneity within the phantom, more advanced temperature distribution models were needed to predict the observed variations. As the temperature distribution became more homogeneous towards the end of the experiment, 1-D temperature profiles were sufficient to adequately model the observations.

## 6. References

Al-Alousi, L. M., Anderson, R. A. & Land, D. V. (1994). A non invasive method for postmortem temperature measurement using a microwave probe, *Forensic Science Int.* 64: 35–46.

Bardati, F., Brown, V. J. & Bernardo, G. D. (1991). Multi-Frequency Microwave Radiometry for Retrieval of Temperature Distributions in the Human Neck, *The Journal of Photographic Science* 39: 157–160.

Bardati, F., Brown, V. J. & Tognolatti, P. (1993). Temperature Reconstruction in a Dielectric Cylinder by Multi-Frequency Microwave Radiometry, *Journal Electromagnetic Waves and Applications* 7(1): 1549–1571.

Calderwood, S. K., Theriault, J. R. & Gong, J. (2005). How is the immune response affected by hyperthermia and heat shock proteins?, *International Journal Hyperthermia* 21: 713–716.

Camart, J. C., Despretz, D., Prevost, B., Sozanski, J. P., Chivé, M., & Pribetich, J. (2000). New 434 MHz interstitial hyperthermia system monitored by microwave radiometry: theoretical and experimental results, *Int Journal of Hyperthermia* 16: 95–111.

Carlier, J., Camart, J. C., Dubois, L. & Priebetich, J. (2002). Modeling of planar applicators for microwave thermotherapy, *IEEE Transactions Microwave Theory Techniques* 50(12): 3036–3042.

Chang, J. T., Paulsen, K., Meaney, P. & Fanning, M. (1998). Non-invasive thermal assessment of tissue phantoms using an active near field microwave imaging technique, *International Journal Hyperthermia* 14(6): 513–534.

Corry, P. & Dewhirst, M. (2005). Special Issue: Thermal Medicine, Heat Shock Proteins and Cancer, *International Jornal of Hyperthermia* 8: 675–790.

der Zee, J. V., Holt, B. V. D., Rietveld, P. J. M. & et al. (1999). Reirradiation combined with hyperthermia in recurrent breast cancer results in worthwhile local palliation, *British Journal Cancer* 99: 483–490.

Dewhirst, M., Sapareto, S. & Connor, W. G. (1984). Importance of minimum tumor temperature in determining early and long term responses of spontaneous canine and feline tumors to heat and radiation, *Cancer Research* 44: 43–50.

Dewhirst, M. W., Vujaskovic, Z., Jones, E. & Thrall, D. (2005). *International Journal of Hyperthermia* 8: 779–790.

Edrich, J. & Hardee, P. C. (1974). Thermography at millimeter wavelength, *Proceedings of IEEE* 62: 1391–1392.

Enander, B. & Larson, G. (1974). Microwave radiometric measurements of the temperature inside a body, *Electronic Letters* 10: 317–318.

Gabrielyan, E. S., Khachatryan, L. A., Nalbaudyan, S. G., & Gregorian, F. A. (1992). Microwave method for determining cerebral blood flow", *Methods (New York: Plenum)* pp. 713–715.

Gopal, M. K., Hand, J. W., Lumiori, M. L. D., Alkhairi, S., Paulsen, K. D. & Cetas, T. C. (1992). Current sheet applicator arrays for superficial hyperthermia of chestwall lesions, *International Journal of Hyperthermia* 8(2): 227–240.

Gradshteyn (1980). *Table of Integrals, Series, and Products*, Academic Press, San Diego.

Hand, J. W., Leeuwen, G. M. J. V., Mizushina, S., de Kamer, J. B. V., Maruyama, K., Sugiura, T., Azzopardi, D. V., & Edwards, A. D. (2001). Monitoring of Deep Brain Temperature in Infants Using Multi-Frequency Microwave Radiometry and Thermal Modeling, *Physics in Medicine and Biology* 46: 1885–1903.

Iskander, M. F., Durney, C. H., Grange, T., & Smith, C. S. (1984). Radiometric Technique for Measuring Changes in Lung Water, *IEEE Trans. Microwave Theory Techiques* MTT-32: 554–556.

J. Schaeffer, A. El Mahdi, A. H. & Carr, K. (1986). Detection of extravasation of antineoplastic drugs by microwave radiometry, *Cancer Letters* 31: 285–291.

Jacobsen, S. & Stauffer, P. (2003). Nonparametric 1-D Temperature Restorartion in Lossy Media Using Tikhonov Regularization on Sparse Radiometry Data, *IEEE Transactions Biomedical Engineering* 50(2): 178–188.

Jacobsen, S. & Stauffer, P. R. (2007). Can we settle with single-band radiometric temperature monitoring during hyperthermia teatment of chestwall recurrence of breast cancer using a dual-mode transceiving applicator, *Physics in Medicine and Biology* 52(4): 911–928.

Jones, E. L., Oleson, J. R., Prosnitz, L., Samulski, T., Vujaskovicl, Z., Yu, D., Sanders, L. & Dewhirst, M. (2005). Randomized trial of hyperthermia and radiation for superficial tumors, *Journal of Clinical Oncology* 23: 3079–3085.

Kapp, D. S. & Cox, R. S. (1993). Cumulative minutes of isoeffective hyperthermia with T90=43$^o$C is the treatment parameter most predictive for outcome in patients with metastatic adenocarcinoma of the breast and single tumor nodules per treatment field, *International Journal of Radiation Oncology Biology Physics* 27: 315.

Klemetsen, O., Birkelund, Y., Jacobsen, S. K., Maccarini, P. & Stauffer, P. R. (2011). Design of Medical Radiometer Front-End for Improved Performance, *Progress In Electromagnetic Research B* 27: 289–306.

Lee, E. R. (1995). Electromagnetic superficial heating technology, *in* P. Fessenden & C. C. Vernon (eds), *Thermoradiotherapy and Thermochemotherapy*, Springer-Verlag, pp. 193–217.

Lee, W. M., Gelvich, E. A., der Baan, P. V., Mazokhin, V. N. & Rhoon, G. C. V. (2004). Assessment of the performance characteristics of a prototype 12-element capacitive contact flexible microstrip applicator (CFMA-12) for superficial hyperthermia, *International Journal of Hyperthermia* 20(6): 607–624.

MacDonald, A. G., Land, D. V. & Sturrock, R. D. (1994). Microwave thermography as a non-invasive assessment of disease activity in inflammatory arthritis, *Clinical Rheumatology* 13: 589–592.

Maruyama, K., Mizushina, S., Sugiura, T., van Leeuwen, G. M. J., Hand, J. W., Marrocco, G., Bardati, F., Edwards, A., Azzopradi, D. & Land, D. (2000). Feasibility of Noninvasive Measurement of Deep Brain Temperature in Newborn Infants by Multifrequency Microwave Radiometry, *IEEE Trans Microwave Theory Techniques* 48: 2141–2147.

Mizushina, S., Shimizu, T., Suzuki, K., Kinomura, M., Ohba, H. & Sugiura, T. (1993). Retrieval of Temperature-Depth Profiles in Biological Objects from Multi-Frequency Microwave Radiometric Data, *Journal Electromagnetic Waves and Applications* 7(11): 1515–1547.

Moskowitz, M. J., Ryan, T. P., Paulsen, K. D. & Mitchell, S. E. (1995). Clinical implementation of electrical-impedance tomography with hyperthermia, *International Journal Hyperthermia* 11(2): 141–149.

Mouty, S., Bocquet, B., Ringot, R., Rocourt, N. & Devos, P. (2000). Microwave radiometric imaging (MWI) for the characterization of breast tumors, *The European Journal of Applied Physics* AP 10: 73–78.

Ohba, H., Kinomura, M., Ito, M., Sugiura, T. & Mizushina, S. (1995). Multi-frequency microwave radiometry for noninvasive thermometry using a new temperature profile model function, *Trans. IEICE Electron.* E-78: 1071–1081.

Rhoon, G. C. V. & Wust, P. (2005). Special Issue: Non Invasive thermometry for thermotherapy, *International Journal of Hyperthermia* 21(6): 489–600.

Rosen, A., Stuchly, M. A. & Vorst, A. V. (2002). Applications of RF/Microwaves in Medicine, *IEEE Trans. Microwave Theory Techniques* 50: 963–974.

Rossetto, F. & Stauffer, P. R. (2001). Theoretical characterization of dual concentric conductor microwave applicators for hyperthermia at 433 MHz, *International Journal Hyperthermia* 17(3): 258–270.

Rutt, B. K., Fike, J. R. & Stauffer, P. R. (1986). *In: 34th Meeting of Radiation Research Society; 1986; Las Vegas* p. 35.

Samulski, T. V., Clegg, S. T., Das, S., Macfall, J. & Prescott, D. M. (1994). Application of a new techology in clinical hyperthermia, *International Journal Hyperthermia* 10(3): 389–394.

Sapareto, S. A. & Dewey, W. C. (1984). Thermal dose determination in cancer therapy, *International Journal of radiation Oncology, Biology & Physics* 10: 787–800.

Seip, R. & Ebbini, E. S. (1995). Noninvasive Estimation of Tissue Temperature Response to Heating Fields Using Diagnostic Ultrasound, *IEEE Transactions on Biomedical Engineering* 42(8): 828–839.

Stauffer, P. (2005). Evolving technology for thermal therapy of cancer, *Internation Journal of Hyperthermia* 21(8): 731–744.

Tennant, A. & Anderson, A. P. (1990). A robot-controlled microwave antenna system for uniform hyperthermia treatment of superficial tumours with arbitrary shape, *International Journal of Hyeprthermia* 6(1): 193–202.

Thrall, D. E., LaRue, S. M., Yu, D., Samulski, T., Sanders, L., Case, B., Rosner, G., Azuma, C., Poulsen, J., Pruitt, A. F., Stanley, W., Hauck, M. L., Williams, L., Hess, P. & Dewhirst, M. (2005). Thermal dose is related to duration of local control in canine sarcomas treated with thermoradiotherapy, *Clinical Cancer Research* 11: 5206–5214.

Vernon, C. C., Hand, J. W., Field, S. B., Machin, D., Whaley, J. B., van der Zee, J., van Putten, W. L. J., van Rhoon, G. C., van Dijk, J. D. P., Gonzalez-Gonzalez, D., Liu, F. F., Goodman, P. & Sherar, M. (1996). Radiotherapy with or without hyperthermia in the treatment of superficial localized breast cancer: Results from five randomized controlled trials, *International Journal Radiation Oncology Biology Physics* 35(4): 731–744.

3

# Role of Microwave Heating in Antigen Retrieval in Cryosections of Formalin-Fixed Tissues

Hrvoje Brzica, Davorka Breljak, Ivana Vrhovac and Ivan Sabolić
*Molecular Toxicology, Institute for Medical Research and Occupational Health, Zagreb*
*Croatia*

## 1. Introduction

Immunocytochemistry is a common method for studying localization of proteins in the mammalian tissues and cultured cells. Tissues and cells to be used in immunocytochemical experiments are often fixed and preserved with 2%-10% *p*-formaldehyde (PFA, formaldehyde, formalin), embedded in paraffin or frozen, and sectioned. In histological research, PFA has been used as a fixative preferable to alcohols since the end of 19$^{th}$ century. It is praised for its effectiveness in a broad spectrum of concentrations and conditions without producing "overfixation". Moreover, in comparison with the alcohol-fixed tissues, the PFA-fixed tissues show minor tissue distortion and better staining with the "classical dyes", such as hematoxylin and aniline dyes. The effects of PFA fixation were shown to be pH-, temperature-, and protein type-dependent (Fox et al., 1985).

In contrast to excellent results of PFA fixation when used on tissues to be stained with the afore mentioned "classical dyes", in immunocytochemical studies the PFA-fixed tissues often do not yield acceptable results; fixation with PFA denatures the proteins while preserving antigenicity, but it can affect the tissue and cell integrity by damaging the structure and it can "mask/hide" the antibody binding sites (epitopes) (McNicol & Richmond, 1998). Sometimes the severity of this phenomenon affects the availability of epitopes, thus making it harder or even impossible for antibodies to react with their binding sites. Although the problem of epitope masking has been recognized a long time ago, it is still unclear how PFA changes proteins in tissues. In general, the epitope masking occurs due to PFA-induced cross-linking of the reactive sites on the same protein and/or among adjacent proteins *via* bridges that can be formed between methylene, amino, imino, aromatic, and other reactive groups (Fraenkel-Conrat et al., 1948, 1949). However, this masking does not happen at the same rate for all kinds of proteins; in tissues fixed for a prolonged time period, different proteins react differently, and the consecutive staining efficiency is protein-dependent. For example, in the absence of any unmasking technique, the proteins vimetin and desmin failed to be stained with the respective monoclonal antibodies after one day of tissue fixation, the lymphocyte antigens LN1, LN2, and LN3 were stained weakly with polyclonal antibodies in the cells fixed in formalin for three days, while thyroglobulin and carcinoembryonic antigen showed good staining with polyclonal antibodies in the tissues even after 14 days of formalin fixation (Leong & Gilham, 1989). These phenomena can be only partially explained by the antigen affinity for an antibody, where monoclonal antibodies, being more specific in nature then the polyclonal antibodies, have fewer reactive binding sites (McNicol & Richmond, 1998). Some additional explanations of the epitope masking problem have been offered by Seshi et al., 2004, who used crafted peptides to mimick the antibody binding sites on the targeted proteins; the authors were able to divide peptides in three distinct groups based on their sensitivity to PFA: (a) PFA-sensitive peptides, that contained tyrosine and arginine, (b) peptides without tyrosine and arginine, which were refractory to PFA, and (c) tyrosine-containing peptides, which were sensitive to PFA only if an another, arginine-containing peptide was in the vicinity. Other data have shown that the PFA treatment can change peptides depending on their amino acid sequence (Metz et al., 2004), and that calcium and some other divalent cations could affect the epitope masking by forming tight complexes with proteins during fixation (Morgan et al., 1994). Recent studies have shown that not only the reactive sites (amino groups, tryptophan, tyrosine, lysine, etc...), but also their position in the three dimensional protein structure may be important for masking the epitopes during the PFA fixation (Toews et al., 2008). The resulting changes in protein conformation can make epitopes inaccessible to antibodies in immunocytochemical studies (McNicol & Richmond, 1998).

In order to overcome the problem of masked (hidden) epitopes caused by PFA fixation, one has to unmask/recover these epitopes by various protocols of antigen retrieval before applying specific antibodies. These protocols are normally used with sections of the paraffin-embedded tissues, and include heating at high temperatures, treatment with various alcohols, detergents or high pressure, or various combinations of these procedures. Antigen retrieval is commonly described as a "high temperature heating method to recover the antigenicity of tissue sections that had been masked by formalin fixation" (Shi et al., 2001), but in some cases, the hidden epitopes can be recovered only with the detergent treatment, without heating (Brown et al., 1996). Good antigen retrieval protocols can significantly increase the staining even in the archival tissues that have been stored in formaline for months (Hann et al., 2001).

Although the method of antigen retrieval has been in use for a few decades, possible mechanisms behind it are still unclear. A heat, combined with buffers of various pH values, seems to be a primary factor in most antigen retrieval protocols, and supposedly breaks down protein cross-links formed during fixation. In some cases, chelating agents can also help by breaking the fixation-generated calcium-protein complexes (Gown, 2004; Leong et al., 2007). In paraffin-embedded tissue sections, a microwave oven was found to be preferable to conventional oven as a heat source (Shi et al., 1991), whereas in a study by Cuevas et al., 1994, with 80 different antibodies, microwave heating of the paraffin sections significantly enhanced the staining with 41 antibodies, thus confirming microwave as an important tool in immunocytochemistry. In addition to conventional and microwave ovens, other sources of heat can be used, such as the pressure cooker (Norton et al., 1994) and hotplate for heating tissue sections (Hann et al., 2001) or whole tissues before sectioning (Ino, 2003). Besides heating, some antigen retrieval protocols include treatment with high vapour pressure, various detergents, buffers of different pH and concentration, and various metal ions, that can be used alone or in various combinations without or with heating (Shi et al., 2001).

A harsh antigen retrieval protocols, that include microwave heating and alcohol treatment, are common steps for unmasking epitopes in sections of the paraffin-embedded, PFA-fixed tissues (Cattoretti et al., 1993; Cuevas et al., 1994; Hoetelmans et al., 2002; Leong et al., 2002), whereas the use of such protocols with cryosections of the PFA-fixed tissues is less common. It is assumed that the availability of epitopes in tissue cryosections is a lesser problem than in the sections of paraffin-embedded tissues. However, the treatment of tissue cryosections with sodium dodecyl sulfate (SDS), without heating, has been efficiently used to enhance immunostaining with some antibodies (Brown et al., 1996; Sabolić et al., 1999), thus indicating that unmasking techniques may be beneficial for revealing the antibody binding epitopes also in cryosections. Furthermore, we have recently described that heating cryosections of the PFA-fixed rat and mouse organs in a microwave oven can be used to enhance immunostaining with specific antibodies for various transporters of organic anions (Bahn et al., 2005; Breljak et al., 2010; Brzica et al., 2009b; Ljubojević et al., 2004 & 2007; Yokoyama et al., 2008), glucose (Balen et al., 2008; Sabolić et al., 2006), and sulfate (Brzica et al., 2009a).

In order to demonstrate the importance of various unmasking protocols for immunocytochemical studies in tissue cryosections, we have used cryosections of the rat kidney and liver tissues that had been fixed with 4% PFA *in vivo*, treated them with various antigen retrieval protocols without and with the use of microwave heating, and tested for the intensity of immunostaining of several representative proteins known to reside in the cell membrane (cell adhesion molecule 105 (CAM105), megalin ($GP_{330}$), Na/K-ATPase, aquaporin 1 (AQP1)), cytoplasm (metallothionein), cell membrane and intracellular organelles (vacuolar $H^+$-ATPase (V-ATPase)), and cytoskeleton (actin, tubulin). Using the V-ATPase as an example, we have also tested effects of heating time and different power settings of the microwave oven on the intensity and distribution of antibody staining. The immunostaining was studied by the method of indirect immunofluorescence, where specific epitopes on the proteins were first labeled with the relevant primary polyclonal or monoclonal antibodies, followed by labeling with the fluorescent molecule-conjugated secondary antibodies and fluorescence microscopy.

## 2. Material and methods

### 2.1 Antibodies and other material

Primary antibodies used in these studies were monoclonal or polyclonal affinity purified antibodies raised against the holoproteins or defined, protein-specific peptide sequences. Some of them were purchased commercially, and some were non-commercial, but their use in immunocytochemistry has been described in our previous publications (down indicated in brackets). The commercial antibodies against the following proteins were used: Na/K-ATPase (monoclonal anti-peptide antibody against the $\alpha 1$ subunit of the human protein, Santa Cruz Biotechnology, CA, USA), metallothionein (monoclonal antibody against the polymerized MT1 and MT2 horse holoproteins; Dako North America, CA, USA), actin (monoclonal anti-peptide antibody; Millipore, MA, USA), and $\alpha$-tubulin (monoclonal antibody against the see urchin filament; Sigma, MO, USA). The non-commercial antibodies against the following proteins were used: megalin (polyclonal antibody against the rat holoprotein; characterized in Abbate et al., 1994 and Sabolić et al., 2002), CAM105 (polyclonal antibody against the rat holoprotein; characterized in Sabolić et al., 1992a), AQP1; polyclonal antibody against the rat holoprotein; characterized in Sabolić et al., 1992b), and V-ATPase (polyclonal anti-peptide antibody against the 31 kDa ("E") subunit;

characterized in Herak-Kramberger et al., 2001). Cryosections were incubated in an optimal concentration of the antibody, as defined in preliminary experiments (data not shown), in a refrigerator over night (ON; 12-14 hours).

Secondary antibodies were the CY3-labeled goat anti-rabbit (GARCY3) or donkey anti-mouse IgG (DAMCY3), purchased from Jackson ImmunoResearch (West Grove, PA, USA), and FITC-labeled goat anti-rabbit (GARF) or goat anti-mouse (GAMF) IgG, purchased from Kirkergaard & Perry Laboratories (Gaithersburg, MD, USA). These antibodies were applied in the following concentrations: GARCY3 - 1.5 μg/mL , DAMCY3 - 1.25 μg/mL, GARF - 2.5 μg/mL, and GAMF - 5 μg/mL, at room temperature (RT) for 1 hour. All incubations with the antibodies were performed in the moist chamber.

The following noncommercial solutions have been prepared and used:

PBS (phosphate buffered saline (in mM): 137 NaCl, 2.7 KCl, 8 $Na_2HPO_4$, 2 $K_2PO_4$, pH 7.4)

PFA (4% w/v *p*-formaldehyde in PBS)

HS-PBS (high salt PBS; PBS containing a triple (411 mM) concentration of NaCl)

BSA (bovine serum albumin; 1% w/v in PBS)

Citrate buffer (10 mM citrate buffer, pH 3, pH 6, and pH 8)

Sucrose solution (30% w/v sucrose in PBS)

SDS (detergent sodium dodecyl-sulphate; 1% w/v in PBS)

T-X-100 (Triton-X-100; 0.1%, 0.5%, and 2% w/v in PBS)

Alcohols xylol and propanol were used undiluted (100%), whereas ethanol was used in concentrations of 60%, 70%, and 96%, prepared v/v in distilled water.

The ingredients in these solutions, and other chemicals used in this study were of the highest purity, and were purchased from Sigma (St. Louis, MO, USA). The SuperFrost-plus microscope slides were purchased from Thermo Scientific (Menzel-Gläser, Braunschweig, Germany). The fluorescence fading retardant (Vectashield) was purchased from Vector Laboratories Inc., CA, USA.

### 2.2 Animals, tissue fixation, cryosections

Adult male Wistar strain rats were anaesthetized and their circulatory system was perfused *in vivo* using the Masterflex pump (Cole-Parmer, IL, USA) *via* the left ventricle of the heart, first with aerated (95% $O_2$/5% $CO_2$) and temperature equilibrated (37 $^0$C) PBS for 2-3 min to remove blood, and then with 150 mL of fixative (4% PFA) for 4-5 min. Kidneys and the largest liver lobe were removed, cut in ~1 mm thick slices, postfixed in the same fixative at 4 $^0$C for 24 hours, and then washed with four abundant volumes of PBS to remove PFA. The tissues were then rinsed with, and stored refrigerated in PBS containing 0.02% $NaN_3$ until use. For cryosectioning, tissues were soaked in 30% sucrose solution at 4 $^0$C overnight, embedded in OCT-medium (Tissue-Tek, Sakura, Japan), and frozen at -25 $^0$C. Four μm-thick cryosections were sliced in a Leica CM 1850 cryostat (Leica instruments GmbH, Nussloch, Germany), collected on SuperFrost-plus microscope slides, dried at room temperature for 2-3 hours, and stored refrigerated until use.

### 2.3 Protocols of antigen retrieval

In order to reveal optimal protocols for unmasking the antibody-binding epitopes, tissue cryosections were treated with different protocols and then subjected to immunostaining procedure with specific antibodies. The immunostaining in the absence of any antigen retrieval (Protocol A) was compared with that following treatment with SDS (Protocol B), following heating in a microwave oven in strongly acidic (pH 3), weakly acidic (pH 6), and alkaline (pH 8) citrate buffers (Protocols C, D, and E, respectively), and following treatment in alcohols and consecutive heating in a microwave oven in the same citrate buffers (Protocols F, G, and H, respectively). The incubations were performed either at room temperature (RT) or in a refrigerator (4 $^0$C). Heating of cryosections was performed in a conventional microwave oven with the variable power settings (max. 800 W). Therefore, in these studies the following protocols were applied:

**Protocol A,** untreated cryosections (steps): Rehydration in PBS (15 min/RT) - incubation in BSA (30 min/RT) - incubation in primary antibody (refrigerator/ON) - rinsing in HS PBS (2x5 min/RT) - rinsing in PBS (2x5 min/RT) - incubation in secondary antibody (60 min/RT) - rinsing in HS PBS (2x5 min/RT) - rinsing in PBS (2x5 min/RT) – covering with Vectashield and the cover glass, and sealing lateral openings with nail polish.

**Protocol B**, SDS-treatment (steps): Rehydration in PBS (15 min/RT) - incubation in SDS (5 min/RT) – rinsing in PBS (4x5 min/RT) - incubation in BSA (30 min/RT) - incubation in primary antibody (refrigerator/ON) – other steps are the same as listed in protocol A.

**Protocol C**, microwave heating (steps): Rehydration in PBS (15 min/RT) – immersion in citrate buffer, pH 3, and heating in a microwave oven (4x5 min/800 W, with intermittent addition of the evaporated buffer; cryosections should be fully immersed in the buffer all the time) - cooling down of cryosections in the buffer (20 min/RT) – rinsing in PBS (3x5 min/RT) – incubation in 0,5% T-X-100 (15 min/RT) – incubation in 2% T-X-100 (30 min/RT), - rinsing in PBS (2x5 min/RT) – incubation in BSA (30 min/RT) - incubation in primary antibody (refrigerator/ON) - incubation in 0.1% T-X-100 (10 min/RT) – rinsing in PBS (2x5 min/RT) - incubation in secondary antibody (1 hour/RT) – incubation in 0.1% T-X-100 (10 min/RT) – rinsing in PBS (2x5 min/RT) covering with Vectashield and the cover glass, and sealing with nail polish.

**Protocol D**, microwave heating (steps): The same as protocol C, except the citrate buffer had pH 6.

**Protocol E**, microwave heating (steps): The same as protocol C, except the citrate buffer had pH 8.

**Protocol F**, alcohol treatment and microwave heating (steps): Immersion (without rehydration) and incubation in 100% xylol (30 min/RT) - incubation in 100% propanol (5 min/RT) – incubation in 96% ethanol (5 min/RT) – incubation in 70% ethanol (5 min/RT) – incubation in 60% ethanol (5 min/RT) – rinsing in destilled $H_2O$ (5 min/RT) - rehydration in PBS (15 min/RT) - immersion in citrate buffer, pH 3, and heating in a microwave oven (4x5 min/800 W, with intermittent addition of the evaporated buffer; cryosections should be fully immersed in the buffer all the time) – the consecutive steps are the same as listed in protocol C.

**Protocol G**, alcohol-treatment and microwave heating (steps): The same as the protocol F, except the citrate buffer had pH 6.

**Protocol H**, alcohol-treatment and microwave heating (steps): The same as the protocol F, except the citrate buffer had pH 8.

In an additional set of experiments, the antigen retrieval-dependent immunostaining with the anti-V-ATPase antibody in cryosections of the PFA-fixed rat kidney cortex was used to test (a) the heating time-dependent effects at the fixed power, and (b) the power-dependent effects at the fixed heating time. Accordingly, to test the time-dependency, cryosections of the renal cortex were heated in a microwave oven at 800 W for 5 min, 2x5 min, 3x5 min and 4x5 min, using other conditions defined in the optimal protocol. To demonstrate the power-dependent effects, cryosections of the renal cortex were heated in a microwave oven with 4x5 min cycles at 80 W, 160 W, 400 W, 560 W and 800 W, using other conditions defined in the optimal protocol. In both cases, the control cryosections were processed similarly, but without being heated.

The fluorescence was examined with an Opton III RS fluorescence microscope (Opton Feintechnik, Oberkochen, Germany) and photographed using the Spot RT slider camera and software (Diagnostic Instruments, Sterling Heights, MI, USA). Each set of images, related to the retrieval protocol for a specific antibody/antigen pair, was photographed under the same recording parameters. The images were imported into Adobe Photoshop 6.0 for processing, assembling, and labeling. In some cases, the CY3- or FITC-related fluorescence was converted into black and white mode using the same software.

## 3. Results

### 3.1 Antigen retrieval of epitopes specific for proteins located in cell membrane domains

In these experiments, cryosections of the PFA-fixed rat kidney or liver tissues underwent the above-listed unmasking protocols, followed by immunostaining of proteins known to be largely localized in the specific cell membrane domains of the nephron epithelium and hepatocytes: cell adhesion molecule CAM105, an ectoprotein expressed in the luminal (apical/brush-border (BBM)) membrane of the proximal tubules and the plasma membrane of peritubular capillaries (Sabolić et al., 1992a), megalin, a glycoprotein expressed in the BBM and subapical vesicles of the proximal tubules (Abbate et al., 1994; Kerjaschki & Farquhar, 1983; Sabolić et al., 2002), Na/K-ATPase, an integral membrane protein expressed in the contraluminal (basolateral (BLM)) cell membrane of the proximal and other tubules (Kashgarian et al., 1985; Sabolić et al., 1999) and in the hepatocyte sinusoidal membrane, and AQP1, a transmembrane protein expressed in both luminal and contraluminal membranes of the proximal tubules and thin descending limbs (Nielsen et al., 1993; Sabolić et al., 1992b).

#### 3.1.1 CAM105

Previous studies showed localization of CAM105 in the BBM of S1 and S3, but not S2 segments of the rat kidney proximal tubules, and also in the plasma membrane of peritubular capillaries (Sabolić et al., 1992a). The present study was performed on cryosection of the kidney cortex, thus showing the protein expression in proximal tubule S1 segments and peritubular capillaries (Fig. 1). As shown in Fig. 1A, in the absence of any antigen retrieval conditions, a limited staining was observed in the BBM of S1 and in peritubular capillaries (arrowheads), but not in S2. In the SDS-treated cryosections (B), the staining distribution was similar to, but the intensity was weaker than that in (A). In cryosections treated with heating in a microwave oven (C-E), the staining intensity dependent on buffer pH, being weak at pH 3 (C), intermediate at pH 6 (D), and strong at pH 8 (E). Localization of the staining in the BBM of S1 and in the plasma membrane of peritubular capillaries, remained unchanged. Alcohol treatment plus microwave heating resulted in strong reduction of the staining intensity, to almost none at pH 3 (F) and pH 6 (G), while peritubular capillaries remained stained. However, at pH 8 (H), the staining in both location was present, but the intensity was weaker than in E. In summary, the CAM105-related staining of peritubular capillaries was present in all conditions, while the staining of BBM was dependent on the applied protocol. The best staining of CAM105 was achieved after using protocol E.

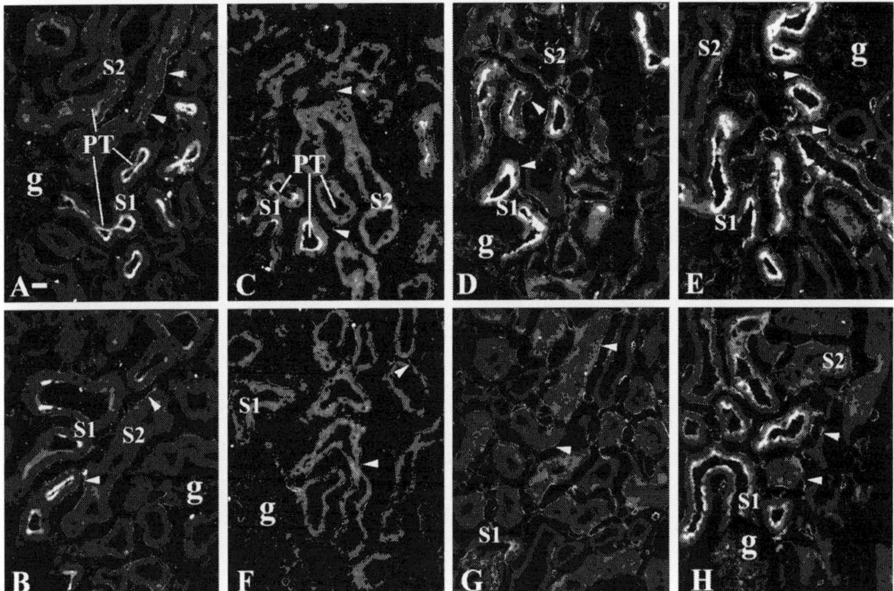


Fig. 1. Distribution and intensity of immunostaining of CAM105 in cryosections of the PFA-fixed rat kidney cortex tissue that underwent various antigen retrieval protocols listed in the section 2.3. g, glomeruli; PT, proximal tubules (S1, initial segment; S2, convoluted segment); arrowheads, peritubular capillaries. Bar, 20 μm.

### 3.1.2 Megalin

Immunostaining of megalin in the kidney cortex in cryosections treated with different antigen retrieval protocols is demonstrated in Fig. 2, A-H. A relatively strong staining of the proximal tubule apical domain (BBM and subapical vesicles) was present already in untreated cryosection (A), while SDS-treatment resulted in an additional intensity (B). Protocols with the microwave heating (C-E) showed the pH-dependent staining intensity with a sequence: pH 3 > pH 6 > pH 8, where the staining pattern and intensity at pH 3 (C) were similar to those in control conditions (c.f. A). Protocols with the alcohol treatment followed by heating (F-H) caused overall weaker staining when compared to protocols C-E, while displaying the same pH-dependency. Therefore, megalin immunostaining was not improved with heating cryosections at any pH without or with alcohol pretreatment, and the protocol with SDS-treatment alone (B) appears to be the best antigen-revealing method for megalin epitopes in the proximal tubules.

### 3.1.3 Na/K-ATPase

Fig. 3 (A-H) shows effects of different unmasking protocols on immunostaining of Na/K-ATPase in cryosections of the rat kidney cortex. In control conditions, Na/K-ATPase was stained weakly in the BLM of proximal tubules and slightly stronger in distal tubules and cortical collecting ducts (A). Following the SDS-treatment, an enhanced staining was observed in the distal tubules and collecting ducts, but the staining in proximal tubules remained unchanged (B). Protocols with the microwave heating at pH 3 (C) and pH 6 (D) led to an enhancement in staining in all tubule profiles, particularly in the proximal and distal tubules (pH 3 < pH 6), whereas at pH 8 (E), the staining was weak in all tubules. In alcohol+heating-treated cryosections (F-H), the overall staining intensity was weaker than, but the staining pattern was similar to that in C-E. Therefore, protocol D represents the optimal antigen revealing approach for immunostaining of Na/K-ATPase along the rat nephron.

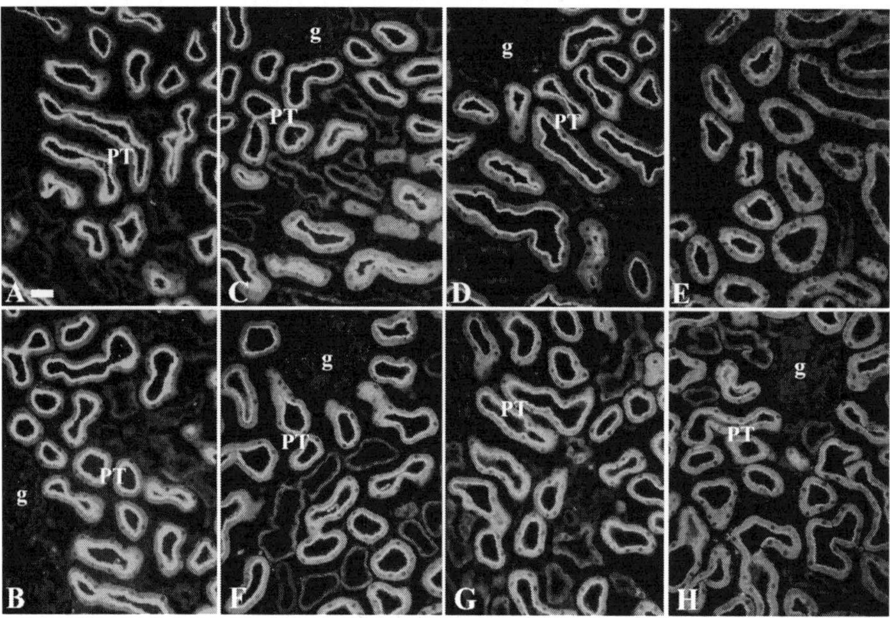


Fig. 2. Distribution and intensity of immunostaining of megalin in cryosections of the PFA-fixed rat kidney cortex tissue that underwent various antigen retrieval protocols listed in the section 2.3. g, glomeruli; PT, proximal tubules. Bar, 20 μm.

In order to check if the same protocol is valid for unmasking the epitopes of Na/K-ATPase in hepatocytes, cryosections of the PFA-fixed liver tissue underwent treatment with various protocols (Fig. 4, A-H). In control cryosections (A), and in cryosections treated with SDS (B), no significant staining of Na/K-ATPase was observed. In the microwave heating-treated cryosections (C-E), protocols C and E were only slightly unmasking (C < E), whereas the protocol D was better, exhibiting a fair staining of the hepatocyte sinusoidal membrane. Following alcohol treatment+microwave heating (F-H), protocols F and H were also weak, whereas protocol G was strongly unmasking, showing the strongest staining intensity in the hepatocyte sinusoidal membrane (arrows), whereas the bile canaliculi remained unstained (arrowheads). Therefore, unlike the situation in kidneys, where the best antigen revealing conditions were in protocol D, the best conditions for unmasking Na/K-ATPase in the liver cells were in protocol G. This experiment indicates that the unmasking conditions of a common antigen in tissue cryosection of different organs can be different.

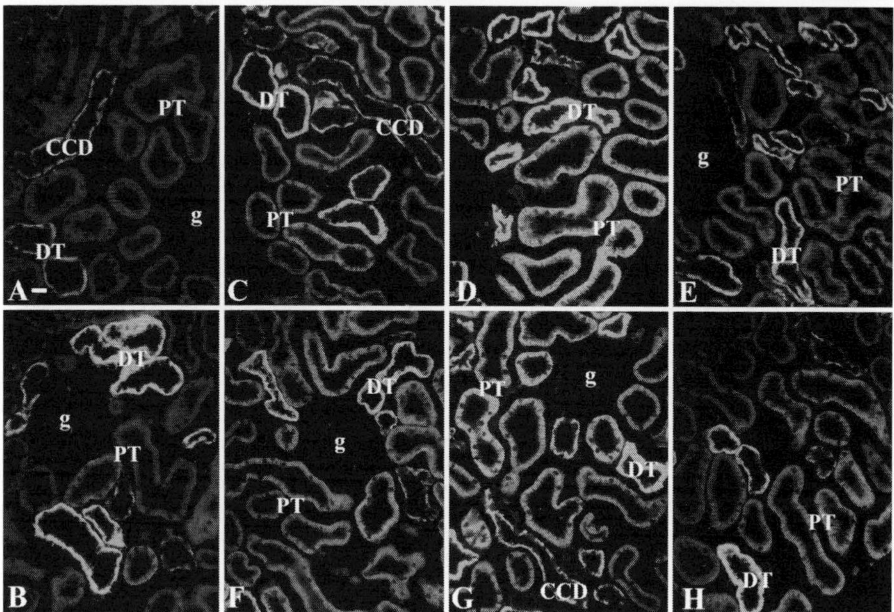


Fig. 3. Distribution and intensity of immunostaining of Na/K-ATPase in cryosections of the PFA-fixed rat kidney cortex tissue that underwent various antigen retrieval protocols listed in the section 2.3. g, glomeruli; PT, proximal tubules; DT, distal tubules; CCD, cortical collecting ducts. Bar, 20 μm.

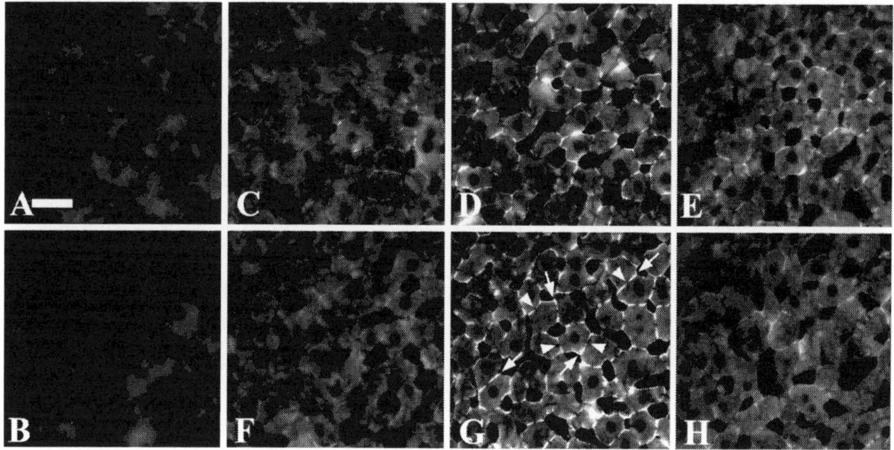


Fig. 4. Distribution and intensity of immunostaining of Na/K-ATPase in cryosections of the PFA-fixed rat liver tissue that underwent various antigen retrieval protocols listed in the section 2.3. Arrows, the stained sinusoidal membrane; arrowheads, unstained bile canaliculi. Bar, 20 μm.

### 3.1.4 AQP1

AQP1 was previously localized to the luminal and contraluminal cell membrane domains of the proximal tubules and thin descending limbs (Nielsen et al., 1993; Sabolić et al., 1992b). Fig. 5. shows effects of various antigen retrieval protocols on immunostaining of AQP1 in cryosections of the rat kidney cortex, where only the proximal tubules are expected to be positive for this water channel. In cryosection treated with protocol A, a fair staining of AQP1 was observed in both BBM (arrowheads) and BLM (arrows) in the proximal tubules (A), whereas in the SDS-treated cryosections, the overall staining was much weaker (B). Following microwave heating (C-E), the staining was pH-dependent, being in both membrane domains strongest at pH 3 (C), weaker and present largely in the BBM at pH 6 (D), and negligible in both membrane domains at pH 8 (E). Treatment of cryosections with alcohols+microwave heating (F-H) resulted in distinct staining pattern; at pH 3, the overall staining was relatively weak, and present largely in the BBM (F), at pH 6, the staining was hardy visible in both membrane domains (G), whereas at pH 8, the staining in the BLM was stronger than in the BBM (H). Therefore, this experiment indicates that protocol C is the best for unmasking the AQP1 epitopes in both BBM and BLM of the proximal tubules, protocols D and F preferably unmask these epitopes in the BBM, whereas protocol H preferably unmasks the same epitopes in the BLM.

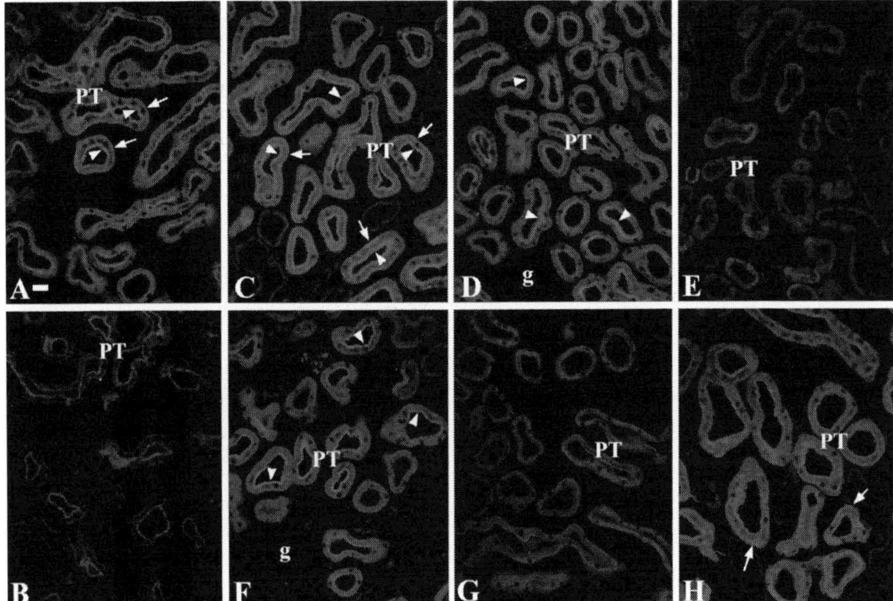


Fig. 5. Distribution and intensity of immunostaining of AQP1 in cryosections of the PFA-fixed rat kidney cortex tissue that underwent various antigen retrieval protocols listed in the section 2.3. g, glomeruli; PT, proximal tubules; arrows, BLM; arrowheads, BBM. Bar, 20 μm.

### 3.1.5 V-ATPase

V-ATPase is localized in the BBM, subapical vesicles, and in various intracellular acidic organelles in the proximal tubule cells and in other cell types along the rat nephron

(Brown et al., 1988). As shown in Fig. 6, a negligible staining of V-ATPase was recorded in control (A) and in the SDS-treated (B) cryosections of the kidney cortex. However, the microwave heating at pH 3 caused a strong unmasking effect, resulting in bright staining of the proximal tubule apical domain and intercalated cells in the cortical collecting ducts (C).

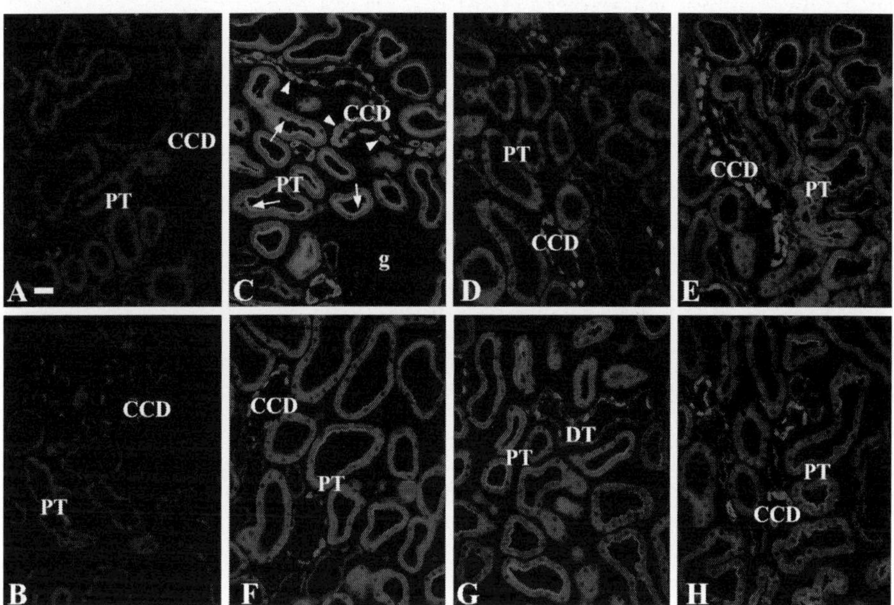


Fig. 6. Distribution and intensity of immunostaining of V-ATPase in cryosections of the PFA-fixed rat kidney cortex that underwent various antigen retrieval protocols listed in the section 2.3. g, glomeruli; PT, proximal tubules; DT, distal tubules; CCD, cortical collecting ducts; arrows, apical cell domain (BBM and subapical organelles); arrowheads, intercalated cells in CCD. Bar, 20 μm.

In cryosections treated with protocols D and E, the overall staining was much weaker than in those treated with protocol C, but intercalated cells in the collecting ducts were generally stained better than the proximal tubules. Compared with the staining after protocol C, treatment of cryosections with alcohols combined with microwave heating (F-H) resulted in overall weaker staining in all tubule profiles and in all conditions, with the pattern pH 3 $\leq$ pH 6 $\leq$ pH 8. Therefore, protocol C appeared to be an optimal unmasking approach for immunocytochemical presentation of V-ATPase in cryosections of the PFA-fixed kidney tissue.

### 3.2 Antigen retrieval of epitopes specific for cytoplasmic protein metallothionein

In the rat kidney, metallothionein is expressed in the cytoplasm of various cells localized along the nephron. In the kidney cortex, a variable expression of this protein has been detected in the proximal tubule cells (Sabolić et al., 2010, and unpublished data).

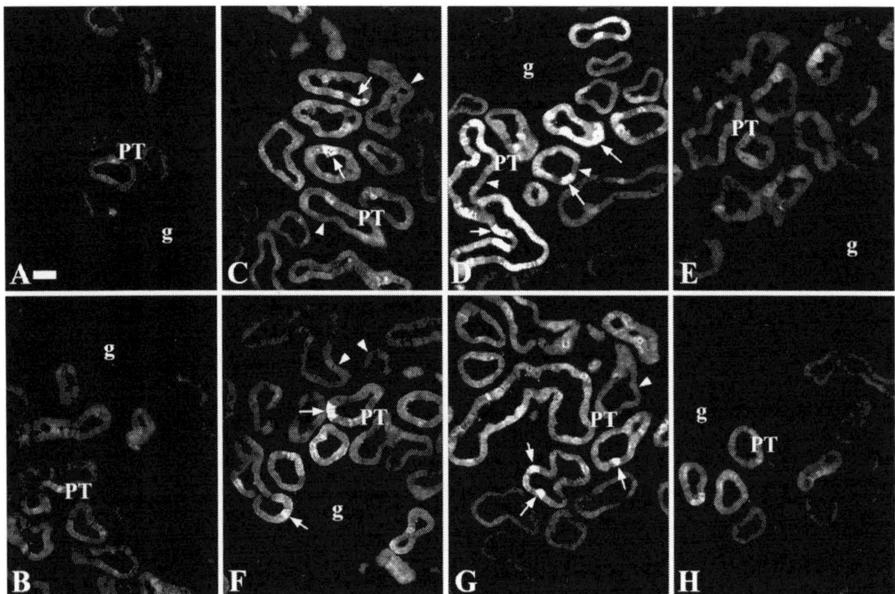


Fig. 7. Distribution and intensity of immunostaining of metallothionein in cryosections of the PFA-fixed rat kidney cortex that underwent various antigen retrieval protocols listed in the section 2.3. g, glomeruli; PT, proximal tubules; arrows, strongly-stained cells; arrowheads, weakly-stained cells. Bar, 20 μm.

In cryosections processed with protocol A, only a weak staining was observed in the cell cytoplasm of some proximal tubules (A). The SDS-treatment increased the number of metallothionein-positive tubules, and slightly increased the staining intensity (B). Following heating in the microwave oven at pH 3, the staining was further increased, exhibiting the cells with a variable staining intensity from weak (arrowheads) to strong (arrows) (C). The intensity and the heterogeneity in the staining intensity was additionally enhanced in cryosections treated with the microwave heating at pH 6 (D), but significantly diminished after heating at pH 8 (E). In cryosections treated with protocols F-H, in all these conditions the overall staining, particularly that in H, was weaker than in D. In conclusion, protocol D appears to be optimal for immunolocalization of metallothionein in cryosections of the PFA-fixed rat renal cortex.

### 3.3 Antigen retrieval of epitopes specific for cytoskeletal proteins

#### 3.3.1 Actin

Fig. 8, A-H shows effects of various protocols of antigen retrieval on actin-related immunostaining in cryosections of the PFA-fixed rat kidney cortex. Control (A) and the SDS-treated cryosections (B) showed no significant staining of actin in any cortical structure. Following heating in the microwave oven (C-E), the staining intensity gradually increased with increasing pH (pH 3 < pH 6 < pH 8) in the glomeruli, proximal tubule BBM, and in individual cells (probably intercalated cells) of the cortical collecting duct.

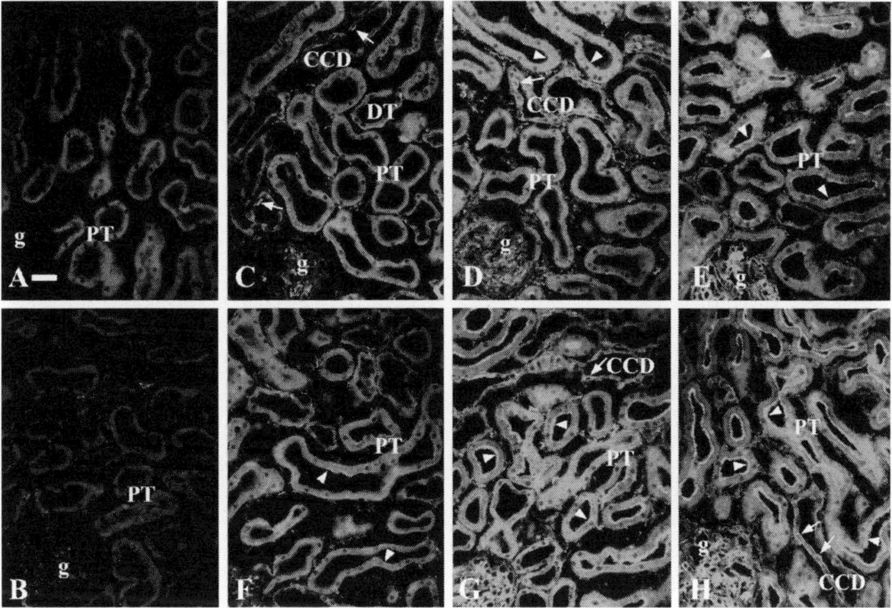


Fig. 8. Distribution and intensity of immunostaining of actin in cryosections of the PFA-fixed rat kidney cortex that underwent various antigen retrieval protocols listed in the section 2.3. g, glomeruli; PT, proximal tubules; DT, distal tubules; CCD, cortical collecting duct; arrows, the actin-positive individual (probably intercalated) cells in CCD; arrowheads, the actin-positive BBM in proximal tubules. Bar, 20 μm.

Cryosections treated with protocols that included treatment with alcohols and microwave heating (F-H) exhibited the same, pH-dependent staining pattern, but the staining intensity was higher in all conditions. Therefore, pretreatment with alcohols, followed by microwave heating in alkaline condition, as present in protocol H, enabled the strongest immunostaining of actin in cryosections of the PFA-fixed rat kidney cortex.

### 3.3.2 α-Tubulin

Effects of different antigen retrieval protocols on immunostaining with the anti-α-tubulin antibody was performed in cryosections of the PFA-fixed rat kidney outer stripe (Fig. 9, A-H). As shown in our previous publication (Sabolić et al., 2002), in the proximal tubule cells α-tubulin forms the apicobasally-oriented bundles of microtubules, which are particularly rich in the cells of proximal tubule straight segments (S3) in the outer stripe zone. A weak staining of α-tubulin in the S3 segments was evident in cryosections that had been processed with protocol A. The staining was somewhat stronger in the SDS-treated cryosections (B). Heating in the microwave oven did not significantly affect the staining intensity at pH 3 (C), heating at pH 6 resulted in a bright staining of microtubule bundles (D), whereas after heating at pH 8 (E), the staining was weaker than in D.

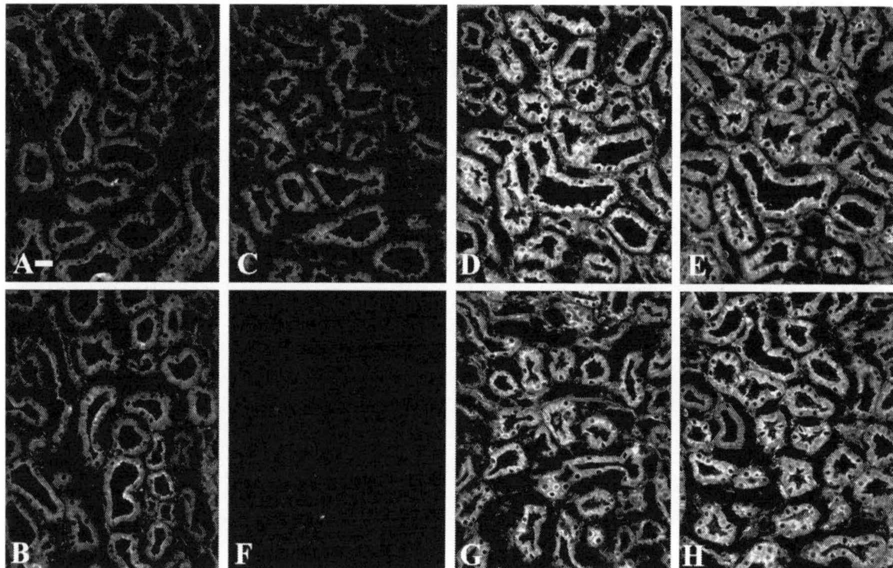


Fig. 9. Immunostaining of α-tubulin in cryosections of the PFA-fixed rat kidney outer stripe following various antigen retrieval protocols listed in the section 2.3. The staining is localized in intracellular bundles of microtubules largely in the proximal tubule S3 segments. Bar, 20 μm.

Alcohol treatment combined with heating at pH 3 unmasked virtually nothing (F), whereas similar protocols at pH 6 (G) and pH 8 (H) resulted in a relatively strong and similar staining, yet weaker then that elicited with protocol D. Therefore, protocol D was found to be optimal for unmasking α-tubulin epitopes in cryosections of the PFA-fixed rat kidney tissue.

Various antigen retrieval protocols were also applied on cryosections of the PFA-fixed liver tissue (Fig. 10, A-H) in order to compare the staining pattern with that in the kidney (c.f., Fig. 9). A very weak staining was observed in control conditions (A), and the SDS-treatment resulted in only limited enhancement of staining (B). After heating cryosections in a microwave oven at pH 3, no significant staining was obtained (C). However, heating at pH 6 resulted in strong staining of microtubules in hepatocytes (D), whereas at pH 8 (E), the staining was weaker than at pH 6. Similar pattern, but at the lower level of staining intensity was observed in cryosections that had been treated with protocols F-H. Therefore, unlike the situation with Na/K-ATPase (c.f., Fig. 3 and Fig. 4), where the optimal unmasking protocols in cryosection of the kidney and liver tissues were different, the optimal protocols for unmasking α-tubulin in cryosections of both organs were similar.

### 3.4 Effects of heating power for a fixed time and of time-dependent exposure to fixed heating power

The preceding experiments showed that for most antigens to be localized by their antibodies, the optimal unmasking effect was achieved with one of the protocols that included heating of cryosections in a microwave oven at the power of 800 W for 20 min. Megalin was the only antigen that did not need heating for optimal presentation. These data thus indicate an indispensable role of microwave heating in unmasking epitopes for immunocytochemical presentation of a variety of antigens in cryosections of the PFA-fixed tissues. In the following two experiments we tested the importance of (a) microwave heating power at the fixed time, and (b) duration of the microwave heating at the fixed power, for unmasking efficiency using V-ATPase as the representative antigen and general conditions in protocol C. The conditions in this protocol were found to be optimal for unmasking epitopes of V-ATPase in cryosection of the PFA-fixed rat kidney cortex (c.f. Fig. 6).

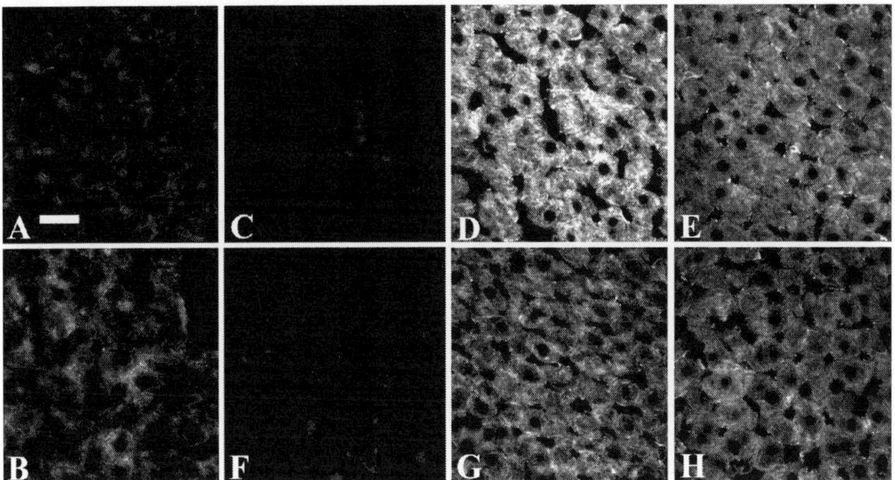


Fig. 10. Immunostaining of α-tubulin in cryosections of the PFA-fixed rat liver following various antigen retrieval protocols listed in the section 2.3. Bar, 20 μm.

Fig. 11. shows the effects of microwave heating of cryosections in citrate buffer, pH 3, at different power settings (0-800 W) for 20 min on the V-ATPase immunostaining. Incubation of cryosections for 20 min without heating resulted in a very weak staining of the cells in collecting ducts (a). After heating at 80 W (b) and 160 W (c), the staining intensity of the collecting duct cells was slightly increased, while other tubule profiles remained unstained. At 400 W (d), the staining intensity of the individual collecting duct cells was stronger, and a weak staining of the proximal tubule BBM was also observed. The overall staining in both collecting duct cells and proximal tubule BBM further increased at 560 W (e), and the brightest staining was recorded after heating at 800 W (f).

Effects of heating duration (0-20 min) at the fixed power (800 W) are shown in Fig. 12. Heating was applied in cycles, 5 min each. An incubation of cryosections at room temperature (heating duration 0 min) resulted in weak staining of collecting ducts (a). Heating for 5 min (one cycle) only slightly increased the staining intensity of the cells in collecting ducts (b). After heating for 10 min (two cycles), a limited staining was observed in the proximal tubule BBM, while the staining intensity of the collecting duct cells further increased (c). Three cycles of heating (15 min) revealed a strong staining in both nephron segments (d), whereas four heating cycles (20 min) resulted in an additional, but slight increase in the staining intensity (e). From the data in these two experiments we conclude that both the heating power and duration of heating in a microwave oven are crucial for an optimal retrieval of hidden antibody-binding epitopes in cryosections of the PFA-fixed tissues.

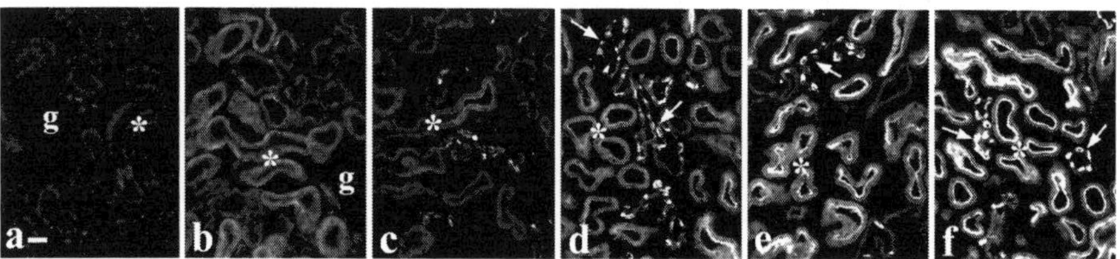


Fig. 11. Effect of power setting of microwave oven on the staining pattern and intensity of V-ATPase in cryosections of the PFA-fixed rat kidney cortex. (a) Without heating, and heating at (b) 80 W, (c) 160 W, (d) 400 W, (e) 560 W, and (f) 800 W for 20 min, other conditions being in protocol C. g, glomeruli; asterisks, proximal tubules; arrows, cortical collecting ducts. Bar, 20 μm.

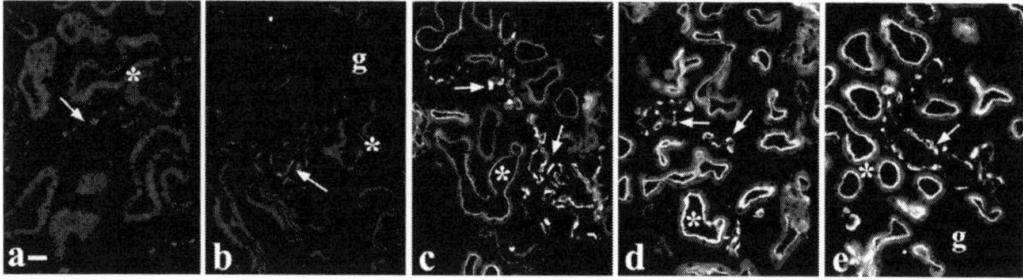


Fig. 12. Effect of duration of microwave heating at the fixed power setting (800 W) on the staining pattern and intensity of V-ATPase in cryosections of the PFA-fixed rat kidney cortex. (a) Without heating, and heating for (b) 5 min, (c) 10 min, (d) 15 min, and (e) 20 min, other conditions being in protocol C. g, glomeruli; asterisks, proximal tubules; arrows, cortical collecting ducts. Bar, 20 μm.

## 4. Discussion

A good immunocytochemistry is largely based on experience collected during the experimental search for optimal conditions of presenting specific antibody-binding epitopes. Our experience indicates that these conditions can not be predicted with certainty for any antibody and tissue or cells, and various protocols of antigen retrieval should be tested with each new antibody and tissue before starting detailed immunocytochemical studies.

The SDS-treatment of cryosections of the PFA-fixed tissues and cultured cells was previously demonstrated as a beneficial technique for unmasking epitopes of several membrane-bound proteins (Alper et al., 1997; Brown et al., 1996; Sabolić et al., 1999). However, using similar cryosections of the rat and mouse tissues, we have recently applied a few harsh antigen retrieval protocols that are commonly used to unmask hidden epitopes in the PFA-fixed, paraffin-embedded sections, and efficiently demonstrated the expression of various transporters in the respective organs (Bahn et al., 2005; Balen et al., 2008; Breljak et al., 2010; Brzica et al., 2009a & 2009b; Ljubojević et al., 2004 & 2007; Yokoyama et al., 2008). In these studies, in preliminary experiments we have always tested a battery of antigen-revealing protocols with each antibody and tissue cryosections until one, a protocol that resulted in the highest immunostaining intensity, was defined and further used. In most cases, the SDS-treatement had only a limited potency in unmasking the tested antigens, and one of the revealing protocols with microwave heating without or with alcohol pretreatment usually proved to be more efficient in revealing cellular localization of the protein-specific epitopes and the intensity of their staining. In the present study, we describe our experience in using various antigen revealing protocols, and the importance of microwave heating in these protocols in order to compare their efficiency in unmasking epitopes on the proteins that reside in specific cellular locations. The major data from these experiments are summarized in Table 1.

As shown in Table 1, except for megalin in the proximal tubule BBM, whose staining was slightly enhanced only following treatment of cryosections with SDS (protocol B), all other antigens were better stained after being treated with one of the protocols that included microwave heating, without (protocols C-E) or with pretreatment with alcohols (protocols F-H). Each tested antigen needed one distinct unmasking protocol to be optimally exposed, with an unpredictable preference for the buffer pH and presence of alcohols.

| Antigen | Cellular localization | SDS treatment | Microwave heating | | | | | |
|---|---|---|---|---|---|---|---|---|
| | | | - Alcohol treatment | | | +Alcohol treatment | | |
| | | | pH 3 | pH 6 | pH 8 | pH 3 | pH 6 | pH 8 |
| CAM105 | PT - BBM | -1 | -1 | +2 | **+3** | -2 | -2 | +2 |
| Megalin | PT - BBM & SAV | **+1** | 0 | -1 | -2 | -1 | -1 | -2 |
| Na/K-ATPase | | | | | | | | |
| Kidney | BLM | +1 | +1 | **+3** | +1 | +1 | +2 | +1 |
| Liver | SM | 0 | +1 | +2 | +1 | 0/+1 | **+3** | 0/+1 |
| AQP1 | BBM & BLM | -2 | **+2** | 0/-1 | -3 | 0/-1 | -3 | 0/-1 |
| V-ATPase | BBM & SAV & IO | 0 | **+3** | +1 | +1 | +1 | +1 | +1 |
| Metallothionein | PT - Cytoplasm | +1 | +2 | **+3** | +1 | +2 | +2 | +1 |
| Actin | Cytoskeleton | 0/-1 | +1 | +1 | +2 | +1 | +2 | **+3** |
| Tubulin | | | | | | | | |
| Kidney | Cytoskeleton | +1 | 0 | **+3** | +2 | -3 | +2 | +2 |
| Liver | Cytoskeleton | +1 | -1 | **+3** | +2 | -1 | +2 | +1/+2 |

Table 1. Relative efficiency of various antigen retrieval protocols B-H in unmasking hidden epitopes in cryosections of the PFA-fixed tissues, as observed in the present study. The efficiency was based on visual estimation of the immunostaining intensity, relative to that obtained for the respective antigen processed with protocol A. PT, proximal tubule; BBM, brush-border membrane; SAV, subapical vesicles; BLM, basolateral membrane; SM, sinusoidal membrane; IO, intracellular organelles. (-) Relative decrease in staining intensity, (0) similar staining intensity, and (+) relative increase in staining intensity, when compared to the intensity in untreated (control) cryosections. Conditions that resulted in relatively strongest unmasking effect and staining intensity were indicated in bold.

The beneficial effect of the buffer pH combined with heating in unmasking hidden antigens has been previously well documented with paraffin sections (Evers & Uylings, 1994; Grossfeld et al., 1996; Shi et al., 1995; Shi et al., 1997). Furthermore, also unlike megalin, whose epitopes were temperature sensitive and were poorly exposed and stained following microwave heating, all other antigens required heating in a microwave oven either without or with alcohol pretreatment for best unmasking effects. Since the benefits of microwave heating for antigen retrieval in paraffin sections have been described (Shi et al., 1991), various time-dependent and power-related protocols have been used to unmask the hidden epitopes for different antibodies. Most of them played with duration of heating at the boiling point of the buffers at the maximum power settings (Evers & Uylings, 1994; Hoetelmans et al., 2002; Munakata & Hendricks, 1993; Shi et al., 1991). Our data obtained with an optimal protocol for unmasking the V-ATPase epitopes following time-dependent and power-dependent heating in a microwave oven indicate the importance of such studies for defining optimal antigen retrieval conditions also with cryosections.

Our data indicate that the optimal protocols for unmasking specific epitopes can be different in cryosections of different tissues. Thus, the optimal conditions for revealing the Na/K-ATPase epitopes in cryosections of the kidney cortex were found in protocol D (microwave heating in buffer pH 6, without alcohol pretreatment), whereas the same epitopes in cryosections of the liver tissue were best exposed in using protocol F (microwave heating in buffer pH 6, with alcohol pretreatment). Similar data were recently found in the studies of sulfate anion transporter sat-1 in cryosections of the rat kidney and liver with a monoclonal antibody; the SDS-treatment was best in unmasking the sat-1 epitopes in cryosections of the kidney cortex, whereas in cryosections of the liver tissue, protocol D (microwave heating in buffer pH 6) was needed for optimal immunostaining (Brzica et al., 2009a). However, some epitopes were optimally retrieved in different tissue with the same protocols, as shown for tubulin epitopes, which were best exposed with the same protocol D in cryosections of both kidney and liver tissues. All these data support the previously reported findings that in *in vivo* conditions, the microenvironment in each tissue can play a role in antigen masking in addition to reactive sites of the protein and/or their three dimensional structures (Toews et al., 2008).

## 5. Conclusion

Based on the data obtained in the preceding experiments, we conclude that: (a) the antigen retrieval protocols, that are usually used with sections of the formalin-fixed, paraffin-embedded tissues, can be efficiently used with cryosections of the PFA-fixed tissues, (b) in order to unmask the hidden epitopes in tissue cryosection, in most, but not all cases the retrieval protocols that include microwave heating give the best unmasking effect and immunostaining intensity, (c) the unmasking effect with a defined protocol is epitope-related and dependent on the buffer pH, d) the unmasking efficiency is dependent on heating power (more power - better unmasking effect) and heating time in a microwave oven (longer heating - better unmasking effect); however, the heating power of 800 W, applied for 20 min, is enough to obtain the maximal unmasking effect and immunostaining intensity with a variety of antibodies, and (d) cryosections of different organs can respond differently to the antigen retrieval conditions required for the same epitope. The later observation indicates the necessity of using various antigen revealing protocols with cryosections of each tissue separately and for each pair of antigen-antibody before starting the immunocytochemical studies rutinely. This way, not only these studies can yield in an optimal staining intensity and distribution, but can also significantly diminish the amount of primary antibodies.

## 6. Acknowledgements

We thank Mrs. E. Heršak for excellent technical assistance, and Dr. D. Brown and Dr. Sylvie Breton from Massachusetts General Hospital/Harvard Medical School in Boston, MA (USA) for generous donation of the anti-V-ATPase, anti-AQP1, and anti-megalin antibodies, and Dr. Sue-Hwa Lin (The M. D. Anderson Cancer Center, University of Texas, Houston, Texas, USA) for donation of the anti-CAM105 antibody. This work was supported by grant No. 0022-0222148-2146 from the Ministry of Science, Education and Sports, Republic of Croatia (I.S.).

## 7. References

Abbate, M., Bachinsky, D. R., McCluskey, R. T. & Brown, D. (1994). Expression of gp330 and gp330/α2-macroglobulin receptor-associated protein in renal tubular differentiation, *Journal of the American Society of Nephrology* 4 (12): 2003-2015.

Alper, S., Stuart-Tilley, A. K., Biemesderfer, D., Shmukler, B. E. & Brown, D. (1997). Immunolocalization of AE2 anion exchanger in rat kidney, *American Journal of Physiology Renal Physiology* 273 (4): F601-F614.

Bahn, A., Ljubojević, M., Lorenz, H., Schultz, C., Ghebremedhin, E., Ugele, B., Sabolić, I., Burckhardt, G. & Hagos, Y. (2005). Murine renal organic anion transporters mOAT1 and mOAT3 facilitate the transport of neuroactive tryptophan metabolites, *American Journal of Physiology Cell Physiology* 289 (5): C1075–C1084.

Balen, D.,Ljubojević, M., Breljak, D., Brzica, H., Žlender, V., Koepsell, H. & Sabolić, I. (2008). Revised immunolocalization of the $Na^+$-D-glucose cotransporter SGLT1 in rat organs with an improved antibody, *American Journal of Physiology Cell Physiology* 295 (2): C475–C489.

Breljak, D., Ljubojević, M., Balen, D., Žlender, V., Brzica, H., Micek, V., Kušan, M., Anzai, N. & Sabolić, I. (2010). Renal expression of organic anion transporter Oat5 in rats and mice exhibits the female-dominant sex differences, *Histology & Histopathology* 25 (11): 1385-1402.

Brown, D., Hirsch, S. & Gluck, S. (1988). Localization of proton-pumping ATPase in rat kidney, *Journal of Clinical Investigation* 82 (6): 2114-2126.

Brown, D., Lydon, J., McLaughlin, M., Stuart-Tilley, A., Tyszkowski, R. & Alper, S. (1996). Antigen retrieval in cryostat tissue sections and cultured cells by treatment with sodium dodecyl sulfate (SDS), *Histochemistry & Cell Biology* 105 (4): 261-267.

Brzica, H., Breljak, D., Krick, W., Lovrić, M., Burckhardt, G., Burckhardt B. C. & Sabolić, I. (2009a). The liver and kidney expression of sulfate anion transporter sat-1 in rats exhibits male-dominant gender differences, *Pflügers Archiv - European Journal of Physiology* 457 (6): 1381-1392.

Brzica, H., Breljak, D., Ljubojević, M., Balen, D., Micek, V., Anzai, N. & Sabolić, I. (2009b). Optimal methods of antigen retrieval for organic anion transporters in cryosections of the rat kidney, *Archives of Industrial Hygiene and Toxicology* 60 (1): 7-17.

Cattoretti, G., Piler, S., Parravicini, C., Becker, H. G., Poggi, S., Bifulco, Ckey, G., D'Amato, L., Sabattini, E., Feudale, E., Reynolds, F., Gerdes, J. & Rilke, F. (1993). Antigen unmasking on formalin-fixed, paraffin embeded tissue sections, *Journal of Pathology* 171 (2): 83-98.

Cuevas, E. C., Bateman, A. C., Wilkins, B, S., Johnson, P. A., Williams, J. H., Lee, A. H., Jones, D. B. & Wright, D. H. (1994). Microwave antigen retrieval in immunocytochemistry: a study of 80 antibodies, *Journal of Clinical Pathology* 47 (5): 448-452.

Evers, P. & Uylings, H. B. M. (1994). Microwave-stimulated antigen retrieval is pH and temperature dependent, *Journal of Histochemistry & Cytochemistry* 42 (12): 1555-1563.

Fox, C. H., Johnson, F. B., Whiting, J. & Roller P. P. (1985). Formaldehyde fixation, *Journal of Histochemistry and Cytochemistry* 33 (8): 845–853.

Fraenkel-Conrat, H. & Olcott, H. S. (1948). Reaction of proteins with formaldehyde VI. Cross-linking of amino groups with phenol, imidazole or indole groups. *Journal of Biological Chemistry* 174 (3): 827-843.

Fraenkel-Conrat, H., Brandon, B. A. & Olcott, H. S. (1949). The reaction of formaldehyde with proteins VII. Demonstration of intermolecular cross-linking by means of osmotic pressure measurements, *Journal of Biological Chemistry* 177 (1): 477-486.

Gown, A. M. (2004). Unmasking the mysteries of antigen or epitope retrieval and formalin fixation, *American Journal of Clinical Pathology* 121 (2): 172-174.

Grossfeld, G. D., Shi, S. R., Ginsberg, D. A., Rich, K. A., Skinner, D. G., Taylor, C. R. & Cote, R. J. (1996). Immunohistochemical detection of thrombospondin-1 in formalin-fixed, paraffin-embedded tissue, *Journal of Histochemistry & Cytochemistry* 44 (7): 761-766.

Hann, C. R., Springett, M. J. & Johnson, D. H. (2001). Antigen retrieval of basement membrane proteins from archival eye tissues, *Journal of Histochemistry & Cytochemistry* 49 (4): 475-482.

Herak-Kramberger, C. M., Breton, S., Brown, D., Kraus, O. & Sabolić, I. (2001). Distribution of the vacuolar $H^+$ATPase along the rat and human male reproductive tract, *Biology of Reproduction* 64 (6): 1699–1707.

Hoetelmans, R. W. M., van Slooten, H. J., Keijzer, R., van de Velde, C. J. H. & van Dierendonck, J. H. (2002). Comparison of the effects of microwave heating and high pressure cooking for antigen retrieval of human and rat Bcl-2 protein in formaldehyde- fixed, paraffin embedded sections, *Biotechnic & Histochemisty* 77 (7): 137-144.

Ino, H. (2003). Antigen retrieval by heating en bloc for pre-fixed frozen material, *Journal of Histochemistry & Cytochemistry* 51 (8): 995-1003.

Kashgarian, M., Biemsderfer, D., Caplan, M. & Forbush, B. III (1985). Monoclonal antibody to Na/K-ATPase: Immunocytochemical localization along nephron segments, *Kidney International* 28 (6): 899-913.

Kerjaschki, D. & Farquhar, M. G. (1983). Immunocytochemical localization of the heymann nephritis antigen (gp330) in glomerular epithelial cells of normal lewis rats, *Journal of Experimental Medicine* 157 (2): 667-686.

Leong, A. S. & Gilhmam, P. N. (1989). The effect of progressive formaldehyde fixation on the preservation of tissue antigens, *Pathology* 21 (4): 266-268.

Leong, A. S., Lee, E. S., Yin, H., Kear, M., Haffajee, Z. & Pepperall, D. (2002). Superheating antigen retrieval, *Applied Immunohistochemistry & Molecular Morphology* 10 (3): 263-268.

Leong, T. Y. & Leong A. S. (2007). How does antigen retrieval work, *Advances in Anatomic Pathology* 14 (2): 129-131.

Ljubojević, M., Balen, D., Breljak, D., Kušan, M., Anzai, N., Bahn, A., Burckhardt, G. & Sabolić, I. (2007). Renal expression of organic anion transporter OAT2 in rats and mice is regulated by sex hormones, *American Journal of Physiology Renal Physiology* 292 (1): F361-F372.

Ljubojević, M., Herak-Kramberger, C. M., Hagos, Y., Bahn, A., Endou, H., Burckhardt, G. and Sabolić, I. (2004). Rat renal cortical OAT1 and OAT3 exhibit gender differences determined by both androgen stimulation and estrogen inhibition, *American Journal of Physiology Renal Physiology* 287 (1): F124-F138.

McNicol, A. M. & Richmond, J. A. (1998). Optimizing immunohistochemistry: antigen retrieval and signal amplification, *Histopathology* 32 (2): 97-103.

Metz, B., Kersten, G. F. A., Hoogerhout, P., Brugghe, H. F., Timmermans, H. A. M., de Jong, A., Meiring, H., ten Hove, J., Hennink, W. E., Crommelin D. J. A. & Jiskoot, W. (2004). Identification of formaldehyde-induced modifications in proteins (reaction with model peptides), *Journal of Biological Chemistry* 279 (8): 6235-6243.

Morgan, J. M., Navabi, H., Schmid, K. W. & Jasani, B. (1994). Possible role of tissue bound calcium ions in citrate-mediated high-temperature antigen retrieval, *Journal of Pathology* 174 (4): 301-307.

Munakata, S. & Hendricks, J. B. (1993). Effect of fixation time and microwave oven heating time on retrieval of the Ki-67 antigen from paraffin embedded tissue, *Journal of Histochemistry & Cytochemistry* 41 (8): 1241-1246.

Nielsen, S., Smith, B. L., Christensen, E. I., Knepper, M. A. & Agre P. (1993). CHIP28 water channels are localized in constitutively water-permeable segments of the nephron, *Journal of Cell Biology* 120 (2): 371–383.

Norton, A. J., Jordan, S. & Yeomans, P. (1994). Brief, high-temperature heat denaturation (pressure cooking): A simple and effective method of antigen retrieval for routinely processed tissues, *Journal of Pathology* 174 (4): 371-379.

Sabolić, I., Breljak, D., Škarica, M. & Herak-Krambeger, C. M. (2010). Role of metallothionein in cadmium traffic and toxicity in kidneys and other mammalian organs, *Biometals* 23 (5): 897-926.

Sabolić, I., Čulić, O., Lin, H. S. & Brown, D. (1992a). Localization of ecto-ATPase in rat kidney and isolated renal cortical membrane vesicles, *American Journal of Physiology,* 262 (2): F217-F228.

Sabolić, I., Herak-Krambeger, C. M., Breton, S. & Brown, D. (1999). Na/K-ATPase in Intercalated cells along the rat nephron revealed by antigen retrieval, *Journal of the American Society of Nephrology* 10 (5): 913-922.

Sabolić, I., Ljubojević, M., Herak-Kramberger, C. M., & Brown, D. (2002). Cd-MT causes endocytosis of brush-border transporters in rat renal proximal tubules, *American Journal of Physiology Renal Physiology* 283 (6): F1389–F1402.

Sabolić, I., Valenti, G., Verbavatz, J. M., Van Hoek, A. N., Verkman, A. S., Ausiello, D. A. & Brown, D. (1992b). Localization of the CHIP28 water channel in rat kidney, *American Journal of Physiology Cell Physiology* 263 (6): C1225–C1233.

Seshi, R. S., Vani, K., Messana, E. & Bogen S. A. (2004). A molecular mechanism of formalin fixation and antigen retrieval, *American Journal of Clinical Pathology* 121 (2): 190-199.

Shi, S.-R., Key, M. E. & Kalra, K. L. (1991). Antigen retrieval in formalin-fixed, paraffin-embedded tissues: an enhancement method for immunohistochemical staining based on microwave oven heating of tissue sections, *Journal of Histochemistry & Cytochemistry* 39 (6): 741-748.

Shi, S.-R., Imam, A., Young, L., Cote, R. J. & Taylor, C. R. (1995). Antigen retrieval immunohistochemistry under the influence of pH using monoclonal antibodies, *Journal of Histochemistry & Cytochemistry* 43 (2): 193-201.

Shi, S.-R., Cote, R. J. & Taylor, C. R. (1997). Antigen retrieval immunohistochemistry: past, present, and future, *Journal of Histochemistry & Cytochemistry* 45 (3): 327-343.

Shi, S.-R., Cote, R. J & Taylor, C. R. (2001). Antigen retrieval techniques: current perspectives, *Journal of Histochemistry & Cytochemistry* 49 (8): 931-937.

Toews, J., Rogalski, J. C., Clark, T. J. & Kast, J. (2008). Mass spectrometric identification of formaldehyde-induced peptide modifications under *in vivo* protein cross-linking conditions, *Analytica Chemica Acta* 618 (2): 168-183.

Yokoyama, H., Anzai, N. Ljubojević, M., Ohtsu, N., Sakata, T., Miyazaki, H., Nonoguchi, H., Islam, R., Onozato, M. L., Tojo, A., Tomita, K., Kanai, Y., Igarashi, T., Sabolić, I. & Endou, H. (2008). Functional and immunochemical characterization of a novel organic anion transporter Oat8 (Slc22a9) in rat renal collecting duct, *Cellular Physiology & Biochemistry* 21 (4): 269-278.

# 4

# Microwave Assisted Disinfection Method in Dentistry

Carlos Eduardo Vergani[1], Daniela Garcia Ribeiro[2], Lívia Nordi Dovigo[1], Paula Volpato Sanitá[1] and Ana Claudia Pavarina[1]

*[1]Araraquara Dental School, UNESP- Univ Estadual Paulista, Department of Dental Materials and Prosthodontics, Araraquara, SP,*
*[2]Department of Dentistry, Ponta Grossa State University, Ponta Grossa, PR, Brazil*

## 1. Introduction

Control of cross-infection has been a subject of interest to the dentistry area over the last few decades, due to the concern about the transmission of infectious-contagious diseases, such as AIDS, hepatitis, tuberculosis, pneumonia, and herpes, between the dental patients and dental personnel and the dental office and dental prosthesis laboratory (Codino & Marshall, 1976; Infection control in the dental office, Council on Dental Materials and Devices, 1978; Sande et al., 1975). This concern began in the 1970s, when the microorganism *Mycobacterium tuberculosis* was isolated in patients' molds (Ray & Fuller, 1963) and it was found that *Mycoplasma pneumoniae* was transmitted to laboratory technicians when they performed wear on contaminated dentures (Sande et al., 1975). Several items, such as instruments (Tarantino et al., 1997), dental burs (Rizzo, 1993), impressions (Abdelaziz et al., 2004), gypsum casts (Berg et al., 2005; Davis et al., 1989), dental appliances (Rohrer & Bulard, 1985), and others are often heavily contaminated with microorganisms from saliva and blood. Powell et al. (1990) also observed that 67% of all the materials received in the laboratories, among them crowns, molds wax records, and dentures, were contaminated with pathogenic microorganisms. The constant exposure of dental personnel and instruments to saliva and blood in virtually every procedure is an ever present hazard and potential contributor to the transmission of infection. Therefore, the use of an inadequate disinfection procedure in the handling of dental materials not only places the unwary staff at risk, but also results in a high level of avoidable cross-contamination.

Dental practitioner has a legal and ethical responsibility to prevent infections in patients and staff members and an interest in protecting her-himself from contracting a disease from a patient. The Centers for Disease Control and Prevention of the United States (Kohn et al., 2004) have recommended sterilization and disinfection procedures to guarantee the safety of dental treatment in the dental health care settings. Typical methods of sterilization and disinfection include dry heat at 160 to 180°C for two hours, wet steam under pressure at 121°C for at least 15 minutes (autoclaving), immersion in chemical solutions, gamma radiation, and ethylene dioxide gas. However, these methods can be time-consuming and some of them require special and costly equipment, knowledgeable operative personnel, and constant surveillance. In order to overcome these limitations, microwave irradiation has been recommended as a practical physical sterilization method (Cottone et al., 1991) that is as effective as autoclaving (Tate et al., 1995). The lethal effects of a high frequency electric field on microorganisms was first described in 1925 (Kahler, 1929) and the destructive effects of strong radio-frequency fields on microorganisms were investigated in 1954 (Brown & Morrison, 1954). In fact, there are studies that have pointed out microwaves as a method for disinfecting food (Culkin & Fung, 1975), contact lenses (Hiti et al., 2001; Rohrer et al., 1986), laboratory microbiologic materials (Border & Rice-Spearman, 1999; Latimer, 1977), items of intimate clothing contaminated with *Candida albicans* (Friedrich & Phillips, 1988), hospital garbage (Hoffman & Hanley, 1994), coloring matter used in the cosmetic industry (Jasnow, 1975), and instruments used in medicine (Rosaspina et al., 1994). Hence, interest in this area has been maintained and in 1985 the technology was applied to the sterilization of dental appliances (Rohrer & Bulard, 1985).

In dentistry, microwave irradiation has been used for several purposes, including disinfection of toothbrushes (Nelson-Filho et al., 2011; Spolidorio et al., 2011), tongue scrapers (Spolidorio et al., 2011), instruments (Tarantino, 1997), contaminated gauze (Border & Rice-Spearman, 1999; Cardoso et al., 2007), dental burs (Fais et al., 2009; Rizzo, 1993), composite polishing instruments (Tate et al., 1995, 1996), molds made from elastomers (Abdelaziz et al., 2004), and gypsum casts (Berg et al., 2005; Davis et al., 1989). Furthermore, microwave irradiation has been widely accepted for polymerizing acrylic resin (Ilbay et al., 1994), drying gypsum products and investment materials (Hersek et al., 2002; Luebke & Chan, 1985; Luebke & Schneider, 1985; Tuncer et al., 1993), and as a post-polymerization treatment for reducing the residual monomer contents of polymerized acrylic resins and its cytotocixity (Jorge et al., 2009; Nunes de Mello et al., 2003). Besides these purposes, some relevant applications of microwave energy in the dentistry field are the disinfection of removable dentures (Dovigo et al., 2009; D.G. Ribeiro et al., 2009; Rohrer & Bulard, 1985; Sanitá et al., 2009; Silva et al., 2006) and the use of this disinfection method to treat patients with oral candidiasis (Banting & Hill, 2001; Neppelenbroek et al., 2008; Webb et al., 2005).

Although the lethal action of microwaves on microorganisms is well established in the literature, the mechanism of destruction of microwaves is not completely understood. While some studies sustain that the effect of microwave irradiation on microorganisms is directly of thermal character (Fitzpatrick et al., 1978; Jeng et al., 1987; Yeo et al., 1999), others claim that the killing of the organisms probably also results from the non-thermal effects of microwaves (Carrol & Lopez, 1969; Culkin & Fung, 1975; Olsen, 1965; Rohrer et al., 1986). In order to attempt these mechanisms of destruction of microwaves, many different microwave regimes have been tested and advocated (Banting & Hill, 2001; Neppelenbroek et al., 2008; Rohrer & Bulard, 2001; Sanitá et al., 2009). Efficacy of microwave irradiation seems to be associated with the vehicle in which the dentures are immersed, the time of exposure, the level of power of the microwave oven, and the type of microorganisms. In addition, when selecting a disinfection procedure, its effect on the physical and mechanical properties of the irradiated materials must be carefully considered. Thus, the establishment of different protocols must be essential to each particular case, with the goal of achieving consistent sterilization without harming dental materials.

Based on the given information, the purposes of this chapter are: 1. to explain and describe the range of applications of microwave irradiation in dentistry field; 2. to review the microbiological effectiveness and recommended protocols of microwave irradiation; 3. to show the mechanisms of action of the microwaves on the microorganisms; and 4. to discuss the effects of microwave irratiation on the properties of dental materials and appliances.

## 2. Applications in dentistry field

The recognition of the potential for transmission of numerous infectious microorganisms during dental procedures has led to an increased concern for infection control in dental practice. Approaches to the clinical use of microwaves for preventing cross-infection have shown relevant results. Devices and instruments used in dental offices have been identified as a source of cross-contamination among patients and from patients to dental personnel. With this in mind, investigations were undertaken to explore the efficacy of microwave irradiation in disinfecting dental mirrors (Tarantino et al., 1997) and handpieces (Rohrer & Bulard, 1985). In addition, dental burs, which may become heavily contaminated with necrotic tissues, saliva, blood, and potential pathogens during use, can also be sterilized by microwave irradiation (Fais et al., 2009; Rizzo, 1993; Rohrer & Bulard, 1985). In order to prevent cross-infection, microwave energy can also be used to the disinfection of finishing and polishing instruments (Tate et al., 1995, 1996). As any another device used in dental offices, finishing and polishing instruments routinely come into contact with patient's saliva and blood and may also act as a source of cross-contamination. In accordance with the studies of Tate et al. (Tate et al., 1995, 1996), these dental devices can be effectively sterilized by microwave irradiation.

Another common dental procedure that may cause cross-infection, especially between patients and dental laboratory personnel, is the making of impressions. Previous studies showed that the majority of impressions arriving at a dental laboratory were contaminated with bacteria and other microorganisms (Egusa et al., 2008; Powel et al., 1990; Ray & Fuller, 1963; Sande et al., 1975; Sofou et al., 2002), irrespective of whether they had been exposed to a disinfectant procedure or merely rinsed with tap water (Sofou et al., 2002). The study of Egusa et al. (2008) showed an extensive contamination of alginate impressions with oral streptococci, staphylococci, *Candida* spp., methicilin resistant *Staphylococcus aureus* (MRSA), and *Pseudomonas aeruginosa*. These results indicate that a large number of microbes are retained on impression materials and are viably transferred onto the surface of stone casts. In fact, it has been shown that microorganisms can be transferred from impressions to gypsum models (Egusa et al., 2008; Leung & Schonfeld, 1983) and that the dental casts are potential sources of microbial transmission (Egusa et al., 2008; Leung & Schonfeld, 1983; Sofou et al., 2002). Also, even a cast from a properly disinfected impression may subsequently become contaminated by a technician or clinician. Considering all these information, microwave irradiation has been suggested to the disinfection of both impressions (Abdelaziz et al., 2004) and dental casts (Berg et al., 2005; Davis et al., 1989). The clinical relevance of impressions and dental casts microwave disinfection is that this procedure can be performed quickly and repeatedly, without the use of toxic, pungent, or allergenic chemicals. However, the disinfection of impression materials hinders possible cross-contamination only at the time the cast is poured. Because casts become contaminated after the intra-oral adjustments of dental appliances, they must be regarded as the major vehicle for cross-contamination and should be disinfected using microwave energy throughout all phases of the dental treatment.

Another application of microwave irradiation in preventing cross-infection is the disinfection of disposable materials. Although there is little scientific support for this purpose, some reports documented the disinfection of contaminated gauze and swabs (Border & Rice-Spearman, 1999; Cardoso et al., 2007) with the use of microwave irradiation. Cardoso et al. (2007) and Border & Rice-Spearman (1999) demonstrated that a short period of exposure of contaminated gauze pieces and swabs to microwave energy (30 seconds) inhibit the growth of pathogenic microorganisms. The authors suggested that microwave oven could be used instead of an autoclave in a variety of clinical and research settings because the procedure is rapid and the equipment cost is minimal.

Among the purposes of microwave energy in cross-infection prevention, one of the most important is the disinfection of removable dentures. In addition to its contamination by the oral microorganisms, it has been reported that dentures are contaminated at various stages during their fabrication (Verran et al., 1996; Wakefield, 1980; Williams et al., 1985) and are capable of transmitting microorganisms to other materials, dental equipments, laboratory, and technicians (Kahn et al., 1982). Microscopic studies have also demonstrated that a biofilm similar to that formed on natural teeth is present on dentures (Budtz-Jorgensen & Theilade, 1983). Large quantity of *Candida* spp. (Budtz-Jorgensen & Theilade, 1983) and some bacterial species associated with systemic diseases have been found in removable dentures, with predominance of gram-positive bacteria, as *Staphylococcus* spp., *Streptococcus* spp., and *Actinomyces* spp. (Chau et al., 1995; Glass et al., 2001a; Monsenego, 2000). Gram-negative species, such as *Neisseria perfava, P. aeruginosa, Klebsiella pneumoniae,* and *Enterobacter cloacae,* have also been identified (Henderson et al., 1987; Latimer, 1977). In fact, in vivo studies (D.G. Ribeiro et al., 2009; Rossi et al., 1996) found *C. albicans, S. aureus, Streptococcus mutans,* and MRSA on the surfaces of dentures from patients, with *C. albicans* having the highest prevalence in these biofilms. Therefore, the denture can function as a reservoir of microorganisms, enabling the transmission of diseases in the dental office and from it to the prosthetic laboratories. Manipulation of dentures in the different dental procedures may also disseminate the microorganisms throughout the environment in the form of aerosols (Clifford & Burnett, 1995). These pathogens may be inhaled by the dentist, assistants, and laboratory technicians, resulting in cross-infection between patients and dental personnel. In the context of denture microwave disinfection, the first studies were performed in order to demonstrate the effectiveness of microwave irradiation in inactivating microorganisms adhered to complete dentures (Thomas & Webb, 1995; Rohrer & Bulard, 1985; Webb et al., 1998). Using several protocols, positive results were obtained by in vitro and in vivo studies, proving that microwave energy can be an effective method in the disinfection of dentures (Dovigo et al., 2009; Glass et al., 2001b; D.G. Ribeiro et al., 2009; Sanitá et al., 2009; Silva et al., 2006). Thus, performing microwave disinfection of a removable denture before it is transferred to a dental laboratory, and immediately before it is returned to the patient, provides a measure of infection control for all parties.

In the course of time, the fit of dentures progressively declines as a result of time-dependent changes in the supporting tissue. In this context, hard chairside reline acrylic resins or soft denture liners are proposed for permanent or temporary improvement of denture fit. These auto-polymerizing acrylic resins allow the clinician to reline a removable prosthesis directly in the mouth in intimate contact with a large area of oral mucosa. Although such liners improve the denture fit and offer comfort to some edentulous and partially edentulous patients, these materials may present a source of other problems. One of their greatest disadvantages is the ongoing task of hygiene maintenance. Denture reline materials, especially the soft liners, have been found to be more prone to microbial adhesion than acrylic resin base materials because of their surface texture and higher porosity (Nikawa et al., 1992). The rougher surface of a prosthesis with a soft liner exhibits greater colonization by *C. albicans* when compared with a conventional acrylic resin denture (Wright et al., 1985). As a result, the oral mucosa is more susceptible to infections. A study evaluated the effectiveness of microwave disinfection on three hard chairside reline resins and observed a consistent sterilization after microwave exposure (Neppelenbroek et al., 2003). In other investigations (Dixon et al., 1999; Mima et al., 2008), microwave irradiation resulted in sterilization of a hard chairside reline resin and soft denture liners contaminated by four pathogenic microorganisms. It is also important to emphasize that disinfection by microwaves promotes inactivation of the pathogenic microorganisms present on both the surface and inside the pores of the acrylic materials (Chau et al., 1995; Dixon et al., 1999).

By being a reservoir of pathogens, the tissue surface of the acrylic resin denture enhances the infective potential of microorganisms and favors the appearance of oral infections (Budtz-Jorgensen, 1990). Oral candidiasis, represented by denture stomatitis in denture wearers, is one of the most common manifestations of disease associated with the use of removable dentures. Denture stomatitis is mainly caused by microorganisms of the *Candida* genus and normally affects the palate of approximately 65% of denture wearer patients (Chandra et al., 2001). This infection is characterized by the presence of multiple hyperemic points in the mucosa subjacent to the removable dentures of patients and, in more advanced cases, diffused erthymatous areas and papillary hyperplasia of the palate may also be observed (Newton, 1962). Considering that microbial adherence and colonization on dental prostheses favor the appearance of denture stomatitis, the cleansing and disinfection of dentures is fundamental to prevent this disease. Recent studies have suggested denture disinfection by microwave irradiation for the treatment of denture stomatitis. The microwave regimes established in laboratorial studies provided the baseline for subsequent clinical trials. Banting & Hill (2001) conducted the first study that evaluated the effectiveness of microwave energy for denture disinfection as a co-adjuvant in the treatment of denture stomatitis. The authors observed that the method was effective in the reduction of the clinical signs of infection. These findings are in agreement with those found by Webb et al. (2005) a few years later. A more recent study conducted by Neppelenbroek et al. (2008) also evaluated the effectiveness of complete denture disinfection by microwave energy in the treatment of patients with denture stomatitis. In agreement to Banting & Hill (2001) and Webb et al. (2005), it was observed that disinfection of the dentures by microwaves was effective for the treatment of denture stomatitis (Neppelenbroek et al., 2008). Investigations are still being carried out to evaluate the effectiveness of microwave irradiation in the treatment of denture stomatitis (Silva et al., 2008; Vergani et al., 2008). Further tests have been performed in order to evaluate the effectiveness of denture irradiation in the treatment of diabetic denture wearer patients with denture stomatitis, showing promising outcomes (Sanitá et al., 2010, 2011). Based on the above mentioned clinical studies, the use of microwaves has shown important results for the treatment of denture stomatitis. This disease is one of the most frequent opportunistic infections found in denture wearers, including the diabetic patients, and it may extend regionally and result in a systemic infection that is associated with high mortality rates (Colombo et al., 1999; Meunier-Carpentier et al., 1981). Hence, the prevention of colonization of the oropharynx is critically important in preventing systemic infections due to *Candida*.

A recent study has shown that pathogenic microorganisms can adhere in toothbrushes and tongue scrapers made from stainless steel- and polystyrene-based injection-moulded plastic (Spolidorio et al., 2011). Thus, toothbrushes and tongue scrapers become contaminated after use and, if not properly disinfected, may be a reservoir of microorganisms that maintain their viability for a significant amount of time, ranging 24 hours to 7 days (Nelson-Filho et al., 2006). Microbial survival promotes reintroduction of potential pathogens in the oral cavity or dissemination to other individuals when cleaning devices are stored together or shared (Ankola et al., 2009). Hence, these cleaning devices should be regularly disinfected. With this in mind, investigations have demonstrated the efficacy of microwave irradiation for disinfection of toothbrushes and tongue cleaners (Nelson-Filho et al., 2011; Spolidorio et al., 2011), suggesting that this may be a practical and low-cost alternative method of disinfection that can be easily used in the oral hygiene care practices.

This section of the chapter describes the several applications of microwave energy in dentistry field. The use of microwave irradiation for rapid disinfection of different dental materials and appliances may be an important tool in the context of the prevention of cross-infection in dental practice. In addition, there is sufficient scientific evidence that the use of this physical method of denture disinfection is effective in the treatment of denture stomatitis.

## 3. Microbiological effectiveness and recommended protocols

The microbiological effectiveness of microwave irradiation has been documented in the literature and the effectiveness seems to be directly related to the protocol adopted. When defining a microwave irradiation protocol, the parameters to be considered are: the time of exposure; the level of power of the microwave oven; the material to be irradiated; the vehicle in which the material is immersed, and the type of microorganisms that colonize the material. Considering these parameters and the several applications of microwave energy in dentistry field, different protocols of microwave irradiation are available and have been tested.

There are some studies that evaluate microwave irradiation for the disinfection of dental air turbine handpieces (Rohrer & Bulard, 1985), mirrors (Tarantino et al., 1997), burs (Fais et al., 2009; Rizzo, 1993; Rohrer & Bulard, 1985), and finishing and polishing instruments (Tate et al., 1995, 1996). In the study of Rohrer & Bulard (1985), handpieces contaminated with a mixture of four aerobic bacteria (*Staphylococcus epidermis, S. aureus, K. pneumonia, Bacillus subtilis, Clostridium histolyticum*) and *C. albicans* were consistently sterilized with an exposure of 10 minutes at 720W to microwave irradiation when the materials were attached to a three-dimensional rotating device. These authors also observed that handpieces contaminated by both polio type 1 and herpes simplex type 1 viruses were consistently sterilized after an exposure to microwave irradiation. Another in vitro investigation proved that dental mirrors contaminated with *S. aureus, B. subtilis*, and *Bacillus stearothermophilus* can be sterilized by microwave irradiation at 600W for 4 minutes, with the instruments immersed in an aldehyde solution (Tarantino et al., 1997). A microwave regime of 10 minutes at 720W is sufficient to sterilize carbide and diamond burs contaminated with a mixture of *S. aureus, K. pneuminiae*, and *B. subtilis* (Rohrer & Bulard, 1985). In another effective protocol , carbide burs are individually placed in a loosely capped test tube with 10 mL of distilled water, transferred to the right lateral position inside the microwave oven, and then exposed to 5 minutes at 600W (Fais et al., 2009). A similar protocol consisting of 6 minutes of irradiation at 750W (Tate et al., 1995, 1996) can be adopted when disinfecting finishing and polishing instruments (Enhance finishing cups, L.D. Caulk Co., Milford, DE, USA and Min-Identoflex fine cups, Centrix Inc., Shelton, CT, USA).

Likewise, 10 minutes of irradiation at high power can sterilize customized impressions made from both vinyl polysiloxane (Cinch Platinum, Parkell, Farmingdale, NY, USA) and polyether (Impregum F, 3M ESPE AG Dental Products, Seefeld, Germany) rubber impression materials (Abdelaziz et al., 2004). Microwave irradiation can also be used for gypsum casts disinfection (Berg et al., 2005; Davis et al., 1989). In the studies of Davis et al. (1989) and Berg et al. (2005), molds were contaminated with pathogenic microorganisms and the stone casts obtained were submitted to microwave irradiation. While *Serratia marcescens* cells on casts were inactivated by microwave irradiation at 900W for 1, 5 or 20 minutes (Davis et al., 1989), *S. aureus* and *P. aeruginosa* were killed after 5 minutes of irradiation (Berg et al., 2005). Microwave irradiation was ineffective in killing *B. subtilis* transferred to stone casts (Davis et al., 1989). Although a complete inactivation was not obtained, the gypsum casts submitted to microwaves for 20 minutes exhibited less growth than the samples irradiated for shorter times (Davis et al., 1989).

There are also two available protocols for the microwave disinfection of disposable materials, such as swabs and gauze. Cardoso et al. (2007) used stock cultures of *Escherichia coli, S. aureus, S. epidermidis, P. aeruginosa,* and *C. albicans* to contaminate gauze and swabs. Thereafter, the materials were placed into autoclave bags, sealed, and irradiated for 30 seconds at 1000W. This time/power regime was efficient in the sterilization of the disposable materials tested. When a lower power was used (650W), 30 seconds of microwave irradiation were also sufficient to sterilize gauze contaminated with the same pathogenic bacteria and fungi (Border & Rice-Spearman, 1999). With the regard of oral cleaning devices, disinfection of toothbrushes contaminated with a suspension containing *S. mutans* was obtained after exposure to microwaves for 7 minutes at 1100W (Nelson-Filho et al., 2011). However, the results from another study demonstrated that toothbrushes and tongue scrapers contaminated with *C. albicans, S. aureus,* and *S. mutans* were effectively disinfected after 1 minute of microwaving at 650W (Spolidorio et al., 2011).

Given the efficacy of microwave disinfection, much attention has been focused on the use of this method for denture decontamination. Various studies have been conducted to determine the most suitable time/power protocol when using microwave irradiation to disinfect dentures. Some studies have used home microwave ovens for the inactivation of pathogenic microorganisms, such as those recommended by the *Handbook of Disinfectants and Antiseptics* (Cole & Robison, 1996). Among these microorganisms, there are those considered indicators of sterilization, such as *S. aureus* (gram-positive bacteria), *P. aeruginosa* (gram-negative bacteria), and *B. subtilis* (sporulated aerobic microorganisms). In this context, Rohrer & Bulard (1985) investigated the possibility of using microwave irradiation to sterilize dentures and concluded that 8 minutes of irradiation at 720W were sufficient to sterilize the dentures contaminated with a mixture of five bacteria and a fungus. To obtain these results, the authors attached the dentures to a three dimensional rotating device. However, such device is not commercially available or practical for use by dentists or health care facilities. Ten years later, Thomas & Webb (1995) observed that an unmodified domestic microwave oven could be used in the disinfection of dentures. They also demonstrated that sterilization of dentures inoculated with *C. albicans* and *Streptococcus gordonii* could be achieved at 2, 4, 6, 8, and 10 minutes exposure times at high setting (650W) (Webb et al., 1998). Likewise, using an unmodified domestic microwave oven with a rotating table, Baysan et al. (1998) observed that microwave irradiation at 650W for 5 minutes promoted a reduction in the quantity of *C. albicans* and *S. aureus* present in resilient relining materials. The study of Meşe & Meşe (2007) also investigated the effect of microwave energy on the growth of *C. albicans* in resilient relining material and verified the reduction in the colony counts of the fungus after dry exposure to microwaves for 5 minutes (650W). The results obtained with the aforementioned microbiological studies, in which the test specimens or dentures were irradiated in a dry state, are variable with regard to the effectiveness of disinfection by microwaves. Since irradiation in water provides uniform heating of the materials, Dixon et al. (1999) suggested immersing the samples in water during exposure to microwaves. This procedure was considered adequate for eliminating microorganisms, including those located inside the pores. The authors inoculated *C. albicans* in resilient reliners and a heat-polymerized resin and showed that the specimens immersed in water during irradiation were completely disinfected after 5 minutes at maximum power, while those not immersed in water were only partially disinfected after the same time of exposure. Bearing in mind these results, Neppelenbroek et al. (2003) evaluated the effectiveness of a home microwave oven for the inactivation of *S. aureus, P. aeruginosa, B. subtilis*, and *C. albicans* present in three reline resins. The contaminated samples were immersed in 200 mL of sterile distilled water and irradiated for 6 minutes at 650W. It was observed that all the test specimens were sterilized after irradiation, as no microbiological growth was noticed after the irradiated specimens had remained incubated for 7 days. Silva et al. (2006) evaluated the same protocol for disinfecting simulated complete dentures and observed that the lethal action varied according to the microorganisms tested. While complete disinfection was achieved for the dentures contaminated with *S. aureus* and *C. albicans*, those contaminated with *P. aeruginosa* and *B. subtilis* showed little, but detectable, microbial growth. The different results from those of Neppelenbroek et al. (2003) are probably related to the larger surface area of the complete dentures, given that the number of microorganism colonies on the acrylic resin surface is proportional to the total area involved. In addition, this difference could be related to the microorganisms tested. A greater resistance of *B. subtilis* to microwaves has been reported (Davis et al., 1989; Najdovski et al., 1991) and these results are probably related to the sporulation capacity of *B. subtilis*. Bacterial spores are metabolically inactive and particularly resistant to situations of stress, such as heating and radiation. Microwaves promote heating of the test specimens and water in which they are immersed, thus there is the possibility of spore formation during this procedure (Najdovski et al., 1991).

A more recent study (Mima et al., 2008) showed that the experimental protocols advocated by Dixon et al. (1999), Neppelenbroek et al. (2003), and Silva et al. (2006) could be used with lower exposure times. Test specimens contaminated with four microorganisms (*S. aureus, P. aeruginosa, B. subtilis*, and *C. albicans*) were immersed in water and submitted to microwave irradiation (650W) at different exposure times (5, 4, 3, 2, and 1 minutes). It was observed that all the test specimens irradiated for 3, 4, and 5 minutes were completely disinfected after microwave irradiation. When the time of irradiation was reduced to 2 minutes, the samples contaminated with *C. albicans* were completely disinfected while those inoculated with *S. aureus, P. aeruginosa*, and *B. subtilis* demonstrated microbial growth. When submitted to sterilization by humid heat, bacterial cells are inactivated at higher temperatures than fungal cells (Pelczar et al., 1993). Therefore, irradiation for at least 2 minutes promoted sufficient water heating to inactivate *C. albicans* but not the bacteria. Moreover, the yeast cells are larger than those of the bacteria (Verran & Maryan, 1997). Therefore, one could suppose that the *C. albicans* cells contained more water in their composition than did the other microorganisms tested, and therefore, they had been more susceptible to microwave irradiation. In spite of this in vitro study has evaluated small size test specimens that had surfaces with vitreous characteristics, differently from those observed clinically, its results indicated that the protocol should be evaluated in other experimental conditions. Therefore, laboratorial investigations were performed to evaluate the effectiveness of this protocol in the disinfection of dentures contaminated by several microorganisms. Sanitá et al. (2009) demonstrated that simulated complete dentures inoculated with different *Candida* spp. (*C. albicans*, *Candida tropicalis*, and the intrinsically resistant *Candida glabrata*, *Candida dubliniensis*, and *Candida krusei*) were completely disinfected by microwave irradiation for 3 minutes at 650W. Similar results have been reported for complete dentures contaminated with *S. aureus* and *P. aeruginosa* (Dovigo et al., 2009).

Given the promising results, some protocols were tested in clinical trials to evaluate the effectiveness of microwave irradiation in disinfecting patients'dentures (Glass et al., 2001b; D.G. Ribeiro et al., 2009). In one study, fragments of dentures that had been worn for periods ranging from 12 days to 48 years were immersed in a chemical solution (MicroDent Sanitizing and Cleaning System®) and exposed to microwaves for 2 minutes (Glass et al., 2001b). This method showed positive results for denture decontamination, considering that no microbial growth was observed on the fragments. In addition, the study of D.G. Ribeiro et al. (2009) showed that 3 minutes of irradiation at 650W resulted in complete inactivation of the biofilm present on dentures of 30 patients. It emerged also from this study that a lower reduction in the count of microorganisms (*C. albicans*, *S. aureus*, and *S. mutans*) was observed when a lower time of exposure was used (2 minutes).

In terms of treating denture stomatitis, Banting & Hill (2001) conducted the first clinical study that evaluated the effectiveness of microwave energy for denture disinfection as a co-adjuvant treatment. This study was performed in 2001, when the effective protocol of 3 minutes at 650W had not yet established. Patients received topical antifungal medication (nystatin/ three times a day) for 14 days and had their dentures irradiated (850W for 1 minute) on three different days (1st, 5th and 10th day). The authors observed that the method was effective in the reduction of the clinical signs of infection and of the invasive forms of *C. albicans* (pseudohyphas) adhered on the surfaces of the dentures. In another clinical study, microwave disinfection of dentures (10 minutes at 350W) in a daily basis during one week reduced the palatal inflammation and the numbers of *Candida* on cultures from the palates and dentures of patients (Webb et al., 2005). A more recent study (Neppelenbroek et al., 2008) also evaluated the effectiveness of denture microwave disinfection in the treatment of patients with denture stomatitis. The microwave treatment protocol adopted was 6 minutes of irradiation of the complete dentures at 650W, three times a week for 30 days. It was verified that disinfection of the dentures by microwaves was effective for the treatment of denture stomatitis and for the reduction of the mycelial forms of *Candida* spp. In agreement to Banting & Hill (2001), these authors also observed that the levels of recurrence of *C. albicans* on the internal surfaces of the dentures and the supporting mucosa were significantly reduced in the patients whose dentures were microwaved (Neppelenbroek et al., 2008). Other clinical investigations have been carried out to evaluate the effectiveness of microwave irradiation in the treatment of denture stomatitis. A modification on the protocol proposed by Neppelenbroek et al. (2008) was evaluated by Vergani et al. (2008) and Silva et al. (2008). These authors demonstrated that denture microwave irradiation for 3 minutes at 650W, three times weekly for 14 days, is able to reduce the clinical signs of denture stomatitis on the palatal mucosa and the *Candida* colonization on complete dentures. Further clinical evaluations have also been conducted in order to evaluate the effectiveness of this protocol in the treatment of diabetic denture wearer patients with denture stomatitis (Sanitá et al., 2010, 2011). It was observed that microwave disinfection of complete dentures, by itself, was as effective as nystatin, the more conventional topical antifungal medication, in reducing the *Candida* counts and the clinical signs of denture stomatitis infection in patients with diabetes mellitus.

Based on the aforementioned studies, it can be seen that several regimes of microwave irradiation in relation to time/power are available. The protocol must be selected in accordance to the specific application of the microwave energy. Regardless these parameters, microwave irradiation is a potentially effective method for inactivating various microbial species present on dental materials, many of which are related to oral pathologies.

## 4. Mechanisms of action on the microorganisms

While the inhibitory effect of microwave irradiation on microorganisms is being researched extensively, how microwave brings about this effect has been a matter of discussion. Some authors believe that microorganism inactivation by microwave irradiation is explained by a thermal effect (Fitzpatrick et al., 1978; Jeng et al., 1987; Yeo et al., 1999). Nevertheless, several studies suggest that, in addition to the heat generated around the microorganisms, there are other mechanisms resulting directly from the interaction of the electromagnetic field (Carrol & Lopez, 1969; Culkin & Fung, 1975; Olsen, 1965; Rohrer et al., 1986). Various mechanisms have been suggested to explain the nature of the so called non-thermal theory. Depending on the chemical composition of the microorganisms and the surrounding medium, the microbial cells may be selectively heated by the microwaves (Carrol & Lopez, 1969; Hiti et al., 2001; Yeo et al., 1999). Therefore, a certain frequency of microwave energy may be absorbed by certain fundamental biological molecules, such as the nucleic acids (Rohrer et al., 1986). Moreover, the level of molecular response from the biological system to the quantity of thermal energy may also explain the non-thermal effect of microwaves (Carrol & Lopez, 1969). The structural changes in the more peripheral layer around the biological macromolecules may alter their stability and function, and, consequently, these molecules may be denatured in an irreversible manner (Culkin & Fung, 1975). Studies have also demonstrated that the exposition of bacterial suspensions to microwave irradiation caused reduction on viable cell counts and increased the leaching of DNA and protein (Woo et al., 2000). These results suggest that microwaves caused changes in structural integrity and permeability of cell membrane and cell wall that have detrimentally affected the cell metabolism and lead to cell death (Campanha et al., 2007; Carrol & Lopez, 1969; Culkin & Fung, 1975; Olsen, 1965). Campanha et al. (2007) verified that leveduriform suspensions submitted to microwave irradiation at 650W for 6 minutes presented significantly lower cell count values and a larger number of substances released in comparison with the non-irradiated suspensions. The distinction between integral and non-integral cells was made based on the entry of methylene blue coloring into the cells, which is an indirect form of evaluating the cell membrane and wall integrity. Disintegrated cells were found in the irradiated suspension, indicating an alteration in the permeability or integrity of these structures. Moreover, the cells of this suspension lost their characteristic refringence, in spite of preserving their ellipsoidal leveduriform morphology. It was also demonstrated that after irradiation by microwaves, the release of electrolytes ($K^+$, $Ca^{++}$) and nucleic acids was significantly higher in the irradiated suspensions than that from the non-irradiated (Campanha et al., 2007). However, despite of cell inactivation, the optic density and cell concentration were not altered in comparison with those of the control suspensions, indicating that cells were not completely lysed. Irrespective of the mechanism of microwaves on pathogenic microorganisms being thermal or non-thermal, it is known that the effect of inactivation occurs mainly in the presence of water, this being an important factor for sterilization in microwave ovens. Freeze-dried or dry organisms are unlikely to be affected, even when submitted to prolonged exposures, indicating that humidity plays an important role in microwave energy absorption (Dixon et al., 1999; Watanabe et al., 2000). The water molecules present in the medium or inside the cells, being diploid, interact with the electromagnetic field of the microwaves. Consequently, numerous intermolecular collisions may occur and this vibration produces heating (Najdovski et al., 1991). This increase in temperature may cause protein and DNA denaturation (Ponne & Bartels, 1995). The consistent results from several studies, in which strains were completely inactivated when microwave irradiation was carried out with the specimens immersed in water, confirm this hypothesis.

According to scanning electron microscopic (SEM) studies (Neppelenbroek et al., 2003; Mima et al., 2008), microwave irradiation not only inactivated C. *albicans*, but also removed the nonviable yeast cells from resin surface. In this case, the irradiation was performed with the resin specimens immersed in water and, since water started to boil after approximately 1.5 minutes of irradiation, the movement of the water particles probably removed microbial cells from the resins. Verran & Maryan (1997) reported that the larger yeasts cells are more easily dislodged from acrylic resin surfaces compared with smaller bacteria. Considering the information discussed above, it can be concluded that, the nature of the lethality of the microwave irradiation for microorganisms may be a combination of thermal and non-thermal effects.

## 5. Effects on physical and mechanical properties of dental materials and appliances

Several investigations have focused on finding the adequate microwave parameters for microbial inactivation and cross-infection control. Different irradiation protocols have proved to be remarkably effective for disinfection of dental prostheses and other materials frequently used in dental practice. However, to enable this method of disinfection to be safely recommended, it is important to clearly demonstrate that it does not exert deleterious effects on the physical and mechanical properties of the materials submitted to microwaves. For this reason, studies have been conducted to examine the effect of microwave disinfection on dental instruments and burs, impressions, gypsum products, acrylic resins, denture lining materials, and artificial teeth.

Rohrer & Bulard (1985) exposed dental air turbine handpieces to microwaves for 2, 4, 6, 8, 10, and 15 minutes (720W). After 25 cycles of 10 minutes, the dental handpiece tested showed no decrease in the pressure reading and no apparent alteration in sound or cutting power. Another study (Tate et al., 1996) evaluated instrument performance of two composite finishing and polishing systems before and after three cycles of microwave irradiation (6 minutes at 750W). The sample surfaces were examined with a profilometer after the finishing procedure and the results demonstrated that the systems tested can be submitted to microwave irradiation at least three times without affecting performance. Questions have also been raised about the effects of microwave regimens on the microscopic characteristics, durability, and strength of dental burs, which can have their sharpness and ability to effectively cut tooth structure altered. The effect of sterilization with microwaves on diamond burs was evaluated by Rizzo (1993). The author evaluated the dental burs by viewing them under stereomicroscope before and after sterilization cycles. It was found that no damage was present after 15 cycles. The possible influence of microwave irradiation on the cutting capacity of carbide burs was also investigated (Fais et al., 2009). The burs were used to cut glass plates in a cutting machine set for 12 cycles of 2.5 minutes each and, after each cycle, they were exposed to microwave irradiation for 5 minutes at 600W. The cutting capacity of the burs was determined by a weight-loss method. Compared to the control conditions, the microwaved burs showed a statistically significant decrease in their cutting capacity. Thus, microwave irradiation requires further investigations before final recommendations can be made for disinfection of carbide burs.

Microwave disinfection of rubber impressions was also suggested by some authors as an alternative approach of controlling microbial transmission. In this context, the reproducibility of rubber impressions after microwave irradiation (10 minutes/720W) has been evaluated and the results compared with other disinfection methods (Abdelaziz et al., 2004). Microwave sterilization had a small effect on accuracy of impressions and this procedure has been recommended as a suitable technique for sterilizing rubber impressions.

Another application of microwaves in dentistry is the disinfection of gypsum casts. Although there are no studies evaluating the disinfection protocols on the properties of gypsum materials, the effect of drying casts by microwaves has been described. In this context, microwave irradiation of gypsum casts has been tested as to its effect on the resistance to fracture (Hersek et al., 2002; Luebke & Schneider, 1985; Tuncer et al., 1993) and hardness (Luebke & Chan, 1985). In general, the results indicated an improvement in these properties. However, there was some concern that a decrease in the compressive strength and the appearance of cracks or porosities in the surface might occur when gypsum casts were exposed to irradiation with a very high power (1450W). Other physical and mechanical properties, such as abrasion resistance and dimensional stability, should be performed to confirm the clinical applicability of this procedure.

One of the main applications of microwaves in dentistry is to disinfect dental prosthesis. A large number of investigations have been published in the past and recent years concerning its effectiveness and limitations. Laboratorial investigations aimed at identifying if microwave exposure affects the surface hardness of the denture base acrylic resins, relining materials, and artificial denture teeth. The hardness of a material is the result of the interaction of several properties, such as ductility, malleability, and resistance to cutting, and, therefore, hardness tests may be used as an indicator of these properties (Anusavice, 1996). Also, the hardness of materials is related to its resistance to local plastic deformation. Two universal types of microhardness test, Vickers and Knoop, are standard methods for measuring hard surfaces, while the Shore A measures hardness in terms of the elasticity of the material. It has been demonstrated that the Vickers hardness of a heat-polymerizable acrylic resin was not changed after different times of exposure to microwaves (6, 5, 4, 3, 2, and 1 minutes) at 650W (Machado et al., 2009; D.G. Ribeiro et al., 2008). The Knoop hardness values of a denture base resin were also not changed after two cycles of microwave disinfection for 6 minutes at 690W (Sartori et al., 2008). In addition, no alterations in the Shore A hardness values were detected after microwave irradiation (3 minutes/500W) of resilient relining materials (Pavan et al., 2007). Other studies showed, however, increased hardness of denture base materials associated with microwave irradiation. Polyzois et al. (1995) evaluated two microwave disinfection protocols on the hardness of test specimens of a heat-polymerizable resin (3 or 15 minutes at 500W). Both protocols provided an increase in microhardness values. Similar findings were described by Machado et al. (2005) for two resilient relining materials submitted to seven irradiation cycles for 6 minutes (650W). D.G. Ribeiro et al. (2008) evaluated the effect of different times of exposure to microwaves (5, 4, 3, 2, and 1 minutes/ 650W) on the hardness of four reline materials and the findings suggested that the disinfectant method promoted an increase in the hardness of the reline resins. The increase in microhardness values after microwave disinfection may be related to the increase in temperature during the irradiation procedure. Arab et al. (1989) reported an increase in hardness values when heat-polymerizable resins were immersed in water heated to 100°C. Similarly, an increase in Vickers hardness of a reline resin after heat treatment in a water bath at 55°C for 10 minutes has been reported (Seó et al., 2007a). These results may be attributed to the reduction in the level of residual monomer, as a result of the complementary processes of polymerization (Lamb et al., 1983; Sideridou et al., 2004) and diffusion of residual monomer, both favored by the increase in water temperature during microwave irradiation.

Besides the acrylic denture base resins, dentures also comprise artificial teeth. For this reason, the effect of microwave disinfection on the surface hardness of artificial teeth commonly used for denture construction was also evaluated (Campanha et al., 2005). Two microwave disinfection cycles of 6 minutes each (650W), preceded or not by immersion in distilled water for 90 days, were tested. From these experiments, a reduction in surface hardness of the artificial denture teeth was observed after microwave irradiation. It seems that the high temperature associated with the movements of the molecules probably caused an increase in the speed of diffusion of the water molecules into the polymer, facilitating the movement of the polymeric chains during performance of the hardness (Takahashi et al., 1998). Thus, it is feasible that the reduction in hardness after irradiation is related to water sorption rather than to microwave irradiation. In fact, the teeth from the group that was microwaved after 90 days of water saturation presented no significant alteration in hardness after disinfection. Since microwave disinfection involves the exposition of dentures to water at high temperature, it has been hypothesized that this may affect the bond strength between the artificial teeth and the acrylic resin from which dentures are made. Results from a study evaluating the effect of microwave disinfection (6 minutes/650W) on the bond strength of two types of denture teeth to three acrylic resins showed that microwave disinfection did not adversely affect the bond strength of all tested materials, with the exception of one tooth/resin combination (R.C. Ribeiro et al., 2008). In another study microwave irradiation for 3 minutes at 650W promoted a reduction in the impact strength of the tooth/acrylic resin interface, which could be explained by the increase in the degree of conversion of self-polymerizing resin, reducing the cohesion at the interface of the samples (Consani et al., 2008a).

The effects of microwave irradiation on other surface properties, such as roughness and porosity, have also been investigated. According to Allison & Douglas (1973), smoother surface retain a smaller quantity of biofilm, thus avoiding the proliferation of microorganisms on the acrylic surface of dentures. Surface roughness is an important characteristic of dental materials and, therefore, there is a direct correlation between the values of roughness and bacterial adherence. Moreover, according to Yannikakis et al. (2002), the presence of pores may reduce the mechanical properties of acrylic resin, as well as interfere in denture hygiene. It seems that the effect of microwave irradiation on a denture's surface roughness and porosity depends on the time of exposure, number of cycles, and the type of denture resin used. Novais et al. (2009) investigated the occurrence of porosity on the surface of four self-polymerizable acrylic resins and one heat-polymerizable resin, after two or seven cycles of microwave disinfection (6 minutes/ 650W). The number of pores found in two out of five resins remained similar after microwaving, while a reduction in porosity was observed in two resins after seven disinfection cycles. In these cases, it was suggested that the high temperatures may have been attained during exposure to microwaves, which led to a greater degree of conversion and continuation of polymerization. Seven cycles of microwave disinfection increased the number of pores in one material. According to the authors, this material presented a high level of residual monomer and, therefore, a probable explanation for the increase in the number of pores in this resin was related to monomer vaporization. Another investigation also showed that the use of microwave energy can modify the surface texture of acrylic resins (Sartori et al., 2006). It was reported that two microwave disinfection cycles (6 minutes at 690W) promoted an increase in surface roughness of an acrylic resin. However, only one material was evaluated and the effect of reduced exposure times on surface roughness was not investigated.

From the literature, it seems that microwave disinfection can play a role in promoting changes in denture materials. The flexural strength of five chairside reline resins and one denture base resin were evaluated after two and seven cycles of microwave disinfection at 650W for 6 minutes (Pavarina et al., 2005). The flexure strength of three resins presented significant increase in strength values. In contrast, two reline resins presented reduced flexure values after microwave irradiation. The heating provided by each of the seven cycles of microwave disinfection may have increased the absorption of water of some of the evaluated materials, resulting in a reduction in the flexural strength values. In view of these results, shorter times of exposure to microwaves, and their effects on the flexural strength of acrylic resins, were investigated and the findings showed that the flexural strength of four reline materials and one heat-polymerizable resin was not detrimentally affected after 5, 4, 3, 2, and 1 minutes of microwave irradiation at 650W (D.G. Ribeiro et al., 2008). Indeed, the disinfection method was capable of significantly increase the flexural strength of one reline resin. A similar result was described elsewhere after a disinfection cycle for 3 minutes at 650W (Consani et al., 2008b). In addition, another investigation (Polizois et al., 1995) observed that the flexural properties of a heat-polymerizable resin remained unaltered after the use of a low power (500W) associated with a long exposure time (15 minutes). However, in this study the samples were irradiated in a dry condition, a procedure that has been shown to be less effective for microbiological inactivation. In a more clinically relevant approach, the effect of denture microwave disinfection on the maximum fracture load, deflection at fracture, and fracture energy of intact and relined denture bases was evaluated. After exposed to microwave irradiation for 7 days (6 minutes/650W), the strength of the denture bases was similar to the strength of those immersed in water for 7 days (Seó et al., 2008).

An aspect that has also been investigated is the influence of microwave disinfection on bond strength between resilient liners and denture base acrylic resin. Test specimens made of resilient resins bonded to a denture base resin were submitted to microwave disinfection for two and seven irradiation cycles of 6 minutes at 650W (Machado et al., 2005). Microwave disinfection did not compromise the adhesion of resilient liners to the denture base resin. In a subsequent study, seven microwave disinfection cycles (6 minutes/650W) did not decrease the torsional bond strength between two hard reline resins and a denture base resin (Machado et al., 2006). Therefore, in general, the use of microwaves for denture disinfection does not appear to have any negative effect on the bond strength of reline materials frequently used in dental practice. Recently, a clinical study (R.C. Ribeiro et al., 2009) also reported the color stability of a hard chairside reline resin after microwave disinfection.

Another important aspect that should be considered prior to the selection of a disinfection procedure is the maintenance of adequate adaptation between the denture base and residual ridge. Clinically, this characteristic is fundamental both for denture retention and the preservation of the supporting tissues. Changes in denture base adaptation could act as one of the causes of alveolar bone loss and be indirectly responsible for decreasing the retention and stability of the denture (De Gee et al., 1979). A denture that exactly reproduces the supporting tissue may assure a uniform distribution of forces on the largest possible area of surface. Thus, analyses of the effect of microwave disinfection on the dimensional stability of denture bases resins have been conducted. Burns et al. (1990) aimed to determine the possible influence of microwaving on the dimensional stability of heat-polymerizable, self-polymerizable, and light-polymerizable resins. Test specimens were fabricated and submitted to measurements of mass and length before and after microwave irradiation (15 minutes/ 650W). The results showed that all the materials maintained dimensional stability after the disinfection procedure. In another study, microwave disinfection (3 or 5 minutes at 650W) promoted small, clinically insignificant dimensional changes on test specimens of a heat-polymerizable resin (Polyzois et al., 1995). Contrasting results were obtained in the study of Gonçalves et al. (2006), who evaluated the effect of two or seven cycles of microwave disinfection (6 minutes/ 650W) on the linear dimensional change of four reline resins and one denture base resin. Three of the evaluated resins presented significant alteration in the linear dimensional after disinfection. In spite of the positive and negative findings found by these different studies, all of the presented results cannot be extrapolated directly to a clinical situation, since the test specimens used did not simulate the dimensions and shape of denture bases. Standardized dentures were fabricated in the study of Thomas & Webb (1995) in order to evaluate the effect of microwaves on their dimensional stability. After 10 minutes of exposure to microwaves at 604W, some of the areas measured in the dentures presented significant shrinkage or expansion. Similarly, Sartori et al. (2008) observed that the disinfection cycles of 6 minutes (690W) could compromise the dimensional stability of denture base resins. In another study, however, a lower power setting (331W) decreased the occurrence of dimensional changes when the dentures were microwaved for 6 minutes (Thomas & Webb, 1995). The protocol of 6 minutes/ 650W of microwave irradiation was also tested for the dimensional stability of denture bases (Seó et al., 2007b). Repeated disinfection cycles were performed (two and seven) and an increase in shrinkage was observed both in intact bases and bases relined with heat-polymerizable resins. As the bases were immersed in water during irradiation, the results could be related to the possible occurrence of complementary polymerization as a result of water heating. In another investigation, the use of the same microwave power (650W), but with irradiation for 3 minutes, promoted no deleterious effects on the adaptation of denture bases (Consani et al., 2007). In fact, the authors observed that microwave irradiation improved the adaptation of bases in some experimental conditions. Pavan et al. (2005) also suggested that use of shorter irradiation times preserves the dimensional stability of dentures. The authors fabricated 30 denture bases that were submitted to irradiation for 3 minutes/500W or 10 minutes/ 604W. No dimensional alteration was observed in the bases irradiated for 3 minutes. Taken together, the results from all the cited studies suggest that short microwave irradiation times should be used, so that the adaptation of denture bases to subjacent tissue is not changed. Recently, Basso et al. (2010) performed a clinical evaluation of the effect of 3 minutes of microwave irradiation at 650W once or three times a week on the linear dimensional stability of complete dentures. Measurements were taken before the first microwave disinfection (baseline) and after each week of disinfection. Furthermore, the dentures were monitored clinically. Three times weekly irradiation showed significantly higher shrinkage in all evaluated weeks. This result could be attributed to the heating generated by microwave irradiation in an already polymerized material. Even though three microwave disinfections showed statistically significant greater shrinkage, the clinical evaluation did not reveal any change. Therefore, the authors suggested that microwave irradiation can be used clinically for the disinfection of dentures and treatment of denture stomatitis.

Given the information above, it is clear that discordant results have been published regarding the risks of denture microwave disinfection. It seems that the occurrence of negative effects on physical and mechanical properties of dentures depends on the microwave protocol tested and type of material evaluated. It also seems that the use of short exposure times could minimize the occurrence of harmful effects on the dental prostheses. Taking all the results into account it would seem that the microwave regime of 3 minutes at 650W is adequate for denture disinfection without causing significant detrimental effects on the denture materials.

## 6. Conclusion

According to the information discussed in this chapter, there is scientific evidence to support the efficacy of microwave irradiation in preventing cross-infection and treating denture stomatitis. Several protocols of irradiation were described and discussed, and the microbiology effectiveness of microwave energy was clearly demonstrated. Regardless of all the parameters used, we can concluded that microwave disinfection is an effective, quick, easy, and inexpensive versatile tool that can be performed by dentists, assistants, technicians, patients and/or their caregivers to inactivate microorganisms. In addition, the use of a microwave oven does not require special storage and does not induce resistance for fungi or other microorganisms. Thus this method may have an important potential use in dental offices, dental laboratories, and institutions and hospitals in which patients are treated, especially those wearing removable dentures.

## 7. References

Abdelaziz, K.M.; Hassan, A.M. & Hodges JS. (2004). Reproducibility of sterilized rubber impressions. *Brazilian Dental Journal*, Vol.15, No.3, (December 2004), pp. 209-213, ISSN 0103-6440

Allison, R.J. & Douglas, W.H. (1973). Microcolonization of the denture-ftting surface by *Candida albicans*. *Journal of Dentistry*, Vol.1, No.5, (June 1973), pp. 198-201, ISSN 0300-5712

Ankola, A.V.; Hebbal, M. & Eshwar, S. (2009). How clean is the toothbrush that cleans your tooth? *International Journal of Dental Hygiene*, Vol.7, No.4, (November 2009), pp. 237–240, ISSN 1601-5029

Anusavice, K.J. (1996). *Phillips' science of dental materials*. (10th Ed), W.B. Saunders, ISBN 0721657419, Philadelphia, United States

Arab, J.; Newton, J.P. & Lloyd, C.H. (1989). The effect of an elevated level of residual monomer on the whitening of denture base and its physical properties. *Journal of Dentistry*, Vol.17, No.4, (August 1989), pp. 189-194, ISSN 0300-5712

Banting, D.W. & Hill, S.A. (2001). Microwave disinfection of dentures for the treatment of oral candidiasis. *Special Care in Dentistry*, Vol.21, No.1, (January 2001), pp. 4-8, ISSN 0275-1879

Basso, M.F.; Giampaolo, E.T.; Vergani, C.E.; Machado, A.L.; Pavarina, A.C. & Compagnoni, M.A. (2010). Influence of microwave disinfection on the linear dimensional stability of complete dentures: a clinical study. *The International Journal of Prosthodontics*, Vol.23, No.4, (August 2010), pp. 318-320, ISSN 0893-2174

Baysan, A.; Whiley, R. & Wright, P.S. (1998). Use of microwave energy a long-term soft lining material contaminated with *Candida albicans* or *Staphylococcus aureus*. *The Journal of Prosthetic Dentistry*, Vol.79, No.4, (April 1998), pp. 454-458, ISSN 0022-3913

Berg, E.; Nielsen, O. & Skaug, N. (2005). High-level microwave disinfection of dental gypsum casts. *The International Journal of Prosthodontics*, Vol.18, No.6, (December 2005), pp. 520-525, ISSN 0893-2174

Border, B.G. & Rice-Spearman, L. (1999). Microwaves in the laboratory: effective decontamination. *Clinical Laboratory Science*, Vol.12, No.3, (June 1999), pp. 156-160, ISSN 0894-959X

Brown, G.H & Morrison, W.C. (1954). An exploration of the effects of strong radio-frequency fields on micro-organism in aqueous solutions. *Food Technology*, Vol.8, pp. 361-366, ISSN: 0015-6639

Budtz-Jörgensen, E. & Theilade, E. (1983). Regional variations in viable bacterial and yeast counts of 1-week-old denture plaque in denture-induced stomatitis. *Scandinavian Journal of Dental Research*, Vol.91, No.4, (August 1983), pp.288-295, ISSN 0029-845X

Budtz-Jörgensen, E. (1990). Etiology, pathogenesis, therapy, and prophylaxis of oral yeast infections. *Acta odontologica Scandinavica*, Vol.48, No.1, (February 1990), pp. 61-69, ISSN 0001-6357

Burns, D.R.; Kazanoglu, A.; Moon, P.C. & Gunsolley, J.C. (1990). Dimensional stability of acrylic resin materials after microwave sterilization. *The International Journal of Prosthodontics*, Vol.3, No.5, (October 1990), pp. 489-493, ISSN 0893-2174

Campanha, N.H.; Pavarina, A.C.; Vergani, C.E. & Machado, AL. (2005). Effect of microwave sterilization and water storage on the Vickers hardness of acrylic resin denture teeth. *The Journal of Prosthetic Dentistry*, Vol.93, No.5, (May 2005), pp. 483-487, ISSN 0022-3913

Campanha, N.H.; Pavarina, A.C.; Brunetti, I.L.; Vergani, C.E.; Machado, A.L. & Spolidorio, D.M.P. (2007). *Candida albicans* inactivation and cell membrane integrity damage by microwave irradiation. *Mycoses*, Vol.50, No.2, (March 2007), pp. 140-147, ISSN 1439-0507

Cardoso, V.H.; Goncalves, D.L.; Angioletto, E.; Dal-Pizzol, F. & Streck, E.L. (2007). Microwave disinfection of gauze contaminated with bacteria and fungi. *Indian Journal of Medical Microbiology*, Vol.25, No.4, (October 2007), pp. 428-429, ISSN 0255-0857

Carrol, D.E. & Lopez, A. (1969). Lethality of radio-frequency energy upon microorganisms in liquid, buffered, and alcoholic food systems. *Journal of Food Science*, Vol.34, No.4, (July 1969), pp. 320-324, ISSN 1750-3841

Chandra, J.; Mukherjee, P.K.; Leidich, S.D.; Faddoul, F.F.; Hoyer, L.L.; Douglas, L.J. & Ghannoum, M.A. (2001). Antifungal resistance of candidal biofilms formed on denture acrylic in vitro. *Journal of Dental Research*, Vol.80, No.3, (March 2001), pp. 903-908, ISSN 0022-3913

Chau, V.B.; Saunders, T.R.; Pimsler, M. & Elfring, D.R. (1995). In-depth disinfection of acrylic resins. *The Journal of Prosthetic Dentistry*, Vol.74, No.3, (September 1995), pp. 309-313, ISSN 0022-3913

Clifford, T.J. & Burnett, C.A. (1995). The practice of consultants in restorative dentistry (UK) in routine infection control for impressions and laboratory work. *The European Journal of Prosthodontics and Restorative Dentistry*, Vol.3, No.4, (June 1995), pp. 175-177, ISSN 0965-7452

Codino, R.J. & Marshall, W.E. (1976). Control of infection in the dental operatory. *Dental Survey*, Vol.52, No.5, (May 1976), pp. 42-50, ISSN 0011-8788

Cole, E.C. & Robison, R. (1996). Test methodology for evaluation of germicides. In: *Handbook of Disinfectants and Antiseptics*. J.M. Ascenzi, 1-13, Marcel Dekker, ISBN 0-8247-9524-5, New York, United States

Colombo, A.L.; Nucci, M.; Salomão, R.; Branchini, M.L.; Richtmann, R.; Derossi, A. & Wey, S.B. (1999). High rate of non-*albicans* candidemia in Brazilian tertiary care hospitals. *Diagnostic Microbiology and Infectious Disease*, Vol.34, No.4, (August 1999), pp. 281-286, ISSN 0732-8893

Consani, R.L.; Mesquita, M.F.; de Arruda Nobilo, M.A. & Henriques, G.E. (2007). Influence of simulated microwave disinfection on complete denture base adaptation using different flask closure methods. *The Journal of Prosthetic Dentistry*, Vol.97, No.3, (March 2007), pp. 173-178, ISSN 0022-3913

Consani, R.L.; Mesquita, M.F.; Zampieri, M.H.; Mendes, W.B. & Consani, S. (2008). Effect of the simulated disinfection by microwave energy on the impact strength of the tooth/acrylic resin adhesion. *The Open Dentistry Journal*, Vol.2, (January 2008), pp. 3-7, ISSN 1874-2106

Consani, R.L.; Vieira, E.B.; Mesquita, M.F.; Mendes, W.B. & Arioli-Filho, J.N. (2008). Effect of microwave disinfection on physical and mechanical properties of acrylic resins. *Brazilian Dental Journal*, Vol.19, No.4, pp. 348-353, ISSN 0103-6440

Cottone, J.A.; Tererhalmy, G.T. & Molinari, J.A. (1991). *Practical Infection Control in Dentistry* (1st Edition), Lea & Febiger, ISBN 0683021389, Malvern, PA

Culkin, K.A. & Fung, D.Y.C. (1975). Destruction of *Escherichia coli* and *Salmonella typhimurium* in microwave-cooked soups. *Journal of Milk and Food Technology*, Vol.38, No.1, pp. 8-15, ISSN 0022-2747

Davis, D.R.; Curtis, D.A. & White, J.M. (1989). Microwave irradiation of contaminated dental casts. *Quintessence International*, Vol.20, No.8, (August 1989), pp. 583-585, ISSN 0033-6572

De Gee, A.J.; Ten Harkel, E.C. & Davidson, C.L. (1979). Measuring procedure for the determination of the three-dimensional shape of dentures. *The Journal of Prosthetic Dentistry*, Vol.42, No.2, (August 1979), pp. 149-53, ISSN 0022-3913

Dixon, D.L.; Breeding, L.C. & Faler, T.A. (1999). Microwave disinfection of denture base materials colonized with *Candida albicans*. *The Journal of Prosthetic Dentistry*, Vol.81, No.2, (February 1999), pp. 207-214, ISSN 0022-3913

Dovigo, L.N.; Pavarina, A.C.; Ribeiro, D.G.; de Oliveira, J.A.; Vergani, C.E. & Machado, A.L. (2009). Microwave disinfection of complete dentures contaminated in vitro with selected bacteria. *Journal of Prosthodontics*, Vol.18, No.7, (October 2009), pp. 611-617, ISSN 1059-941X

Egusa, H.; Watamoto, T.; Abe, K.; Kobayashi, M.; Kaneda, Y.; Ashida, S.; Matsumoto, T. & Yatani, H. (2008). An analysis of the persistent presence of opportunistic pathogens on patient-derived dental impressions and gypsum casts. *The International Journal of Prosthodontics*, Vol.21, No.1, (February 2008), pp. 62-68, ISSN 0893-2174

Fais, L.M.; Pinelli, L.A.; Adabo, G.L.; Silva, R.H.; Marcelo, C.C. & Guaglianoni, D.G. (2009). Influence of microwave sterilization on the cutting capacity of carbide burs. *Journal of Applied Oral Science*, Vol.17, No.6, (December 2009), pp. 584-589, ISSN 1678-7757

Fitzpatrick, J.A.; Kwao-Paul, J. & Massey, J. (1978). Sterilization of bacteria by means of microwave heating. *Journal of Clinical Engineering*, Vol.3, No.1, (March 1978), pp. 44-47, ISSN 0363-8855

Friedrich Jr, E.G. & Phillips, L.E. (1988). Microwave sterilization of *Candida* on underwear fabric. A preliminary report. *The Journal of Reproductive Medicine*, Vol.33, No.5, (May 1988), pp. 421-422, ISSN 0024-7758

Glass, R.T.; Bullard, J.W.; Hadley, C.S.; Mix, E.W. & Conrad, R.S. (2001a). Partial spectrum of microorganisms found in dentures and possible disease implications. *The Journal of the American Osteopathic Association*, Vol.101, No.2, (February 2001), pp. 92-94, ISSN 1945-1997

Glass, R.T.; Goodson, L.B.; Bullard, J.W. & Conrad, R.S. (2001b). Comparison of the effectiveness of several denture sanitizing systems: a clinical study. *Compendium of Continuing Education in Dentistry*, Vol.22, No.12, (December 2001), pp. 1093-1096, 1098, 1100-2 passim; quiz 1108, ISSN 1548-8578

Gonçalves, A.R.; Machado, A.L.; Giampaolo, E.T.; Pavarina, A.C. & Vergani, C.E. (2006). Linear dimensional changes of denture base and hard chair-side reline resins after disinfection. *Journal of Applied Polymer Science*, Vol.102, No.2, (October 2006), pp. 1821-1826, ISSN 0021-8995

Henderson, C.W.; Schwartz, R.S.; Herbold, E.T. & Mayhew R.B. (1987). Evaluation of the barrier system, an infection control system for the dental laboratory. *The Journal of prosthetic dentistry*, Vol.58, No.4, (October 1987), p. 517-521, ISSN 1097-6841

Hersek, N.; Canay, S.; Akça, K. & Ciftçi, Y. (2002). Tensile strength of type IV dental stones dried in a microwave oven. *The Journal of Prosthetic Dentistry*, Vol.87, No.5, (May 2002), pp. 499-502, ISSN 1097-6841

Hiti, K.; Walochnik, J.; Faschinger, C.; Haller-Schober, E.M. & Aspock, H. (2001). Microwave treatment of contact lens cases contaminated with *Acanthamoeba*. *Cornea*, Vol.20, No.5, (July 2001), pp. 467-470, ISSN 1536-4798

Hoffman, P.N. & Hanley, M.J. (1994). Assessment of a microwave-based clinical waste decontamination unit. *The Journal of Applied Bacteriology*, Vol.77, No.6, (December 1994), pp. 607-612, ISSN 0021-8847

Ilbay, S.G.; Güvener, S. & Alkumru, H.N. (1994). Processing dentures using a microwave technique. *Journal of Oral Rehabilitation*, Vol.21, No.1, (January 1994), pp.103-109, ISSN 1365-2842

Infection control in the dental office. Council on Dental Materials and Devices. Council on Dental Therapeutics. (1978). *Journal of the American Dental Association*, Vol.97, No.4, (October 1978), pp. 673-677, ISSN 0002-8177

Jasnow, S.B. & Smith, J.L. (1975). Microwave sanitization of color additives used cosmetics: feasibility study. *Applied Microbiology*, Vol.30, No.2, (August 1975), pp. 205-211, ISSN 0003-6919

Jeng, D.K.; Kaczmarek, K.A.; Woodworth, A.G. & Balasky, G. (1987). Mechanism of microwave sterilization in the dry state. *Applied and Environmental Microbiology*, Vol.53, No.9, (September 1987), pp. 2133-2137, ISSN 1098-5336

Jorge, J.H.; Giampaolo, E.T.; Vergani, C.E.; Pavarina, A.C.; Machado, A.L. & Carlos, I.Z. (2009) Effect of microwave postpolymerization treatment and of storage time in water on the cytotoxicity of denture base and reline acrylic resins. *Quintessence International*, Vol.40, No.10, (November-December 2009), pp. e93-100, ISSN 0033-6572

Kahler, H. (1929). The nature of the effect of a high-frequency electric field upon *Paramecium*. *Public Health Reports*, Vol.44, No.7, (February 1929), pp. 339-347, ISSN 0033-3549

Kahn, R.C.; Lancaster, M.V. & Kate, Jr.W. (1982). The microbiologic crosscontamination of dental prostheses. *The Journal of Prosthetic Dentistry*, Vol.47, No.5, (May 1982), pp. 556-559, ISSN 1097-6841

Kohn, W.G.; Harte, J.A.; Malvitz, D.M.; Collins, A.S.; Cleveland, J.L. & Eklund, K.J. (2004). Centers for Disease Control and Prevention. Guidelines for infection control in dental health care settings-2003. *The Journal of the American Dental Association*, Vol.135, No.1, (January 2004), pp. 33-47, ISSN 0002-8177

Lamb, D.J.; Ellis, B. & Priestley, D. (1983). The effects of process variables on levels of residual monomer in autopolymerizing dental acrylic resin. *Journal of Dentistry*, Vol.11, No.1, (March 1983), pp. 80-88, ISSN 0300-5712

Latimer, J.M. (1977). Microwave oven irradiation as a method for bacterial decontamination in a clinical microbiology laboratory. *Journal of Clinical Microbiology*, Vol.6, No.4, (October 1977), pp. 340-342, ISSN 0095-1137

Leung, R.L. & Schonfeld, S.E. (1983). Gypsum casts as a potential source of microbial cross-contamination. *The Journal of Prosthetic Dentistry*, Vol.49, No.2, (February 1983), pp. 210-211, ISSN 1097-6841

Luebke, R.J. & Schneider, R.L. (1985). Microwave oven drying of artificial stone. *The Journal of Prosthetic Dentistry*, Vol.53, No.2, (February 1985), pp. 261-265, ISSN 1097-6841

Luebke, R.J. & Chan, K.C. (1985). Effect of microwave oven drying on surface hardness of dental gypsum products. *The Journal of Prosthetic Dentistry*, Vol.54, No.3, (September 1985), pp. 431-435, ISSN 1097-6841

Machado, A.L.; Breeding, L.C. & Puckett, A.D. (2005). Effect of microwave disinfection on the hardness and adhesion of two resilient liners. *The Journal of Prosthetic Dentistry*, Vol.94, No.2, (August 2005), pp. 183-189, ISSN 1097-6841

Machado, A.L.; Breeding, L.C. & Puckett, A.D. (2006). Effect of microwave disinfection procedures on torsional bond strengths of two hard chairside denture reline materials. *Journal of Prosthodontics,* Vol.15, No.6, (November-December 2006), pp. 337-344, ISSN 1532-849X

Machado, A.L.; Breeding, L.C.; Vergani, C.E. & da Cruz Perez, L.E. (2009). Hardness and surface roughness of reline and denture base acrylic resins after repeated disinfection procedures. *The Journal of Prosthetic Dentistry,* Vol.102, No.2, (August 2009), pp. 115-122, ISSN 1097-6841

Meşe, A. & Meşe, S. (2007). Effect of microwave energy on fungal growth of resilient denture liner material. *Biotechnology & Biotechnological Equipment,* Vol.21, No.1, (January 2007), pp. 91-93, ISSN 1310-2818

Meunier-Carpentier, F.; Kiehn, T.E. & Armstrong, D. (1981). Fungemia in the immunocompromised host. *The American Journal of Medicine,* Vol.71, No.3, (September 1981), pp. 363-370, ISSN 0002-9343

Mima, E.G.; Pavarina, A.C.; Neppelenbroek, K.H.; Vergani, C.E.; Spolidorio, D.M. & Machado, A.L. (2008). Effect of different exposure times of microwave irradiation on the disinfection of a hard chairside reline resin. *Journal of Prosthodontics,* Vol.17, No.4, (June 2008), pp. 312-317, ISSN 1532-849X

Monsenego, P. (2000). Presence of microorganisms on the fitting denture complete surface: study in vivo. *Journal of Oral Rehabilitation,* Vol.27, No.8, (August 2000), pp. 708-713, ISSN 1365-2842

Najdovski, L.; Dragas, A.Z. & Kotnik, V. (1991). The killing activity of microwaves on some non-sporogenic and sporogenic medically important bacterial strains. *The Journal of hospital infection,* Vol.19, No.4, (December 1991), pp. 239-247, ISSN 0195-6701

Nelson-Filho, P.; Faria, G.; da Silva, R.A.; Rossi, M.A. & Ito IY. (2006). Evaluation of the contamination and disinfection methods of toothbrushes used by 24- to 48-month-old children. *Journal of Dentistry for Children,* Vol.73, No.3, (September-December 2006), pp. 152-158, ISSN 1935-5068

Nelson-Filho, P.; da Silva, L.A.; da Silva, R.A.; da Silva, L.L.; Ferreira, P.D. & Ito, I.Y. (2011). Efficacy of microwaves and chlorhexidine on the disinfection of pacifiers and toothbrushes: an in vitro study. *Pediatric Dentistry,* Vol.33, No.1, (January-February 2011), pp. 10-13, ISSN 0164-1263

Neppelenbroek, K.H.; Pavarina, A.C.; Spolidorio, D.M.; Vergani, C.E.; Mima, E.G. & Machado, A.L. (2003). Effectiveness of microwave sterilization on three hard chairside reline resins. *The International Journal of Prosthodontics,* Vol.16, No.6, (November-December 2003), pp. 616-620, ISSN 0893-2174

Neppelenbroek, K.H.; Pavarina, A.C.; Palomari Spolidorio, D.M.; Sgavioli Massucato, E.M.; Spolidorio, L.C. & Vergani, C.E. (2008). Effectiveness of microwave disinfection of complete dentures on the treatment of *Candida*-related denture stomatitis. *Journal of Oral Rehabilitation,* Vol.35, No.11, (November 2008), pp. 836-846, ISSN 1365-2842

Newton, A.V. (1962). Denture sore mouth. A possible etiology. *British Dental Journal,* Vol.112, No.1, (May 1962), pp. 357-360, ISSN 0007-0610

Nikawa, H.; Iwanaga, H.; Kameda, M. & Hamada T. (1992). In vitro evaluation of *C. albicans* adherence to soft denture-lining materials. *The Journal of Prosthetic Dentistry,* Vol.68, No.5, (November 1992), pp. 804-808, ISSN 1097-6841

Novais, P.M.; Giampaolo, E.T.; Vergani, C.E.; Machado, A.L.; Pavarina, A.C. & Jorge, J.H. (2009). The occurrence of porosity in reline acrylic resins. Effect of microwave disinfection. *Gerodontology,* Vol.26, No.1, (March 2009), pp. 65-71, ISSN 1741-2358

Nunes de Mello, J.A.; Braun, K.O.; Rached, R.N. & Del Bel Cury, A.A. (2003). Reducing the negative effects of chemical polishing in acrylic resins by use of an additional cycle of polymerization. *The Journal of Prosthetic Dentistry,* Vol.89, No.6, (June 2003), pp. 598-602, ISSN 1097-6841

Olsen, C.M. (1965). Microwaves inhibit bread mold. *Food Engineering,* Vol.37, No.7, (1965), pp. 51-53, ISSN 0260-8774

Pavan, S.; Arioli Filho, J.N.; Dos Santos, P.H. & Mollo, F.deA.Jr. (2005). Effect of microwave treatments on dimensional accuracy of maxillary acrylic resin denture base. *Brazilian Dental Journal,* Vol.16, No.2, (May-August 2005), pp. 119-123, ISSN 1806-4760

Pavan, S.; Arioli Filho, J.N.; Dos Santos, P.H.; Nogueira, S.S. & Batista A.U. (2007). Effect of disinfection treatments on the hardness of soft denture liner materials. *Journal of prosthodontics,* Vol.16, No.2, (March-April 2007), pp. 101-106, ISSN 1532-849X

Pavarina, A.C.; Neppelenbroek, K.H.; Guinesi, A.S.; Vergani, C.E.; Machado, A.L. & Giampaolo, E.T. (2005). Effect of microwave disinfection on the flexural strength of hard chairside reline resins. *Journal of Dentistry,* Vol.33, No.9, (October 2005), pp. 741-748, ISSN 1879-176X

Pelczar, Jr.M.J.; Chan, E.C.S. & Krieg, N.R. (1993). Control of 1 microorganisms: principles 2 and physical agents. In: *Microbiology: concepts and applications,* McGraw-Hill, pp. 200–220, ISBN 0071129146, New York

Polyzois, G.L.; Zissis, A.J. & Yannikakis, S.A. (1995). The effect of glutaraldehyde and microwave disinfection on some properties of acrylic denture resin. *The International Journal of Prosthodontics*, Vol.8, No.2, (March-April 1995), pp. 150-154, ISSN 0893-2174

Ponne, C.T. & Bartels, P.V. (1995). Interaction of electromagnetic energy with biological material - relation to food processing. *Radiation Physics and Chemistry,* Vol.45, No.4, (April 1995), pp. 591-607, ISSN 0969-806X

Powell, G.L.; Runnells, R.D.; Saxon, B.A. & Whisenant, B.K. (1990). The presence and identification of organisms transmitted to dental laboratories. *The Journal of Prosthetic Dentistry*, Vol.64, No.2, (August 1990), pp. 235-237, ISSN 0022-3913

Ray, K.C. & Fuller, M.L. (1970). Isolation of *Mycobacterium* from dental impression material. *The Journal of Prosthetic Dentistry*, Vol.24, No.3, (September 1970), pp. 335-338, ISSN 0022-3913

Ribeiro, D.G.; Pavarina, A.C.; Machado, A.L.; Giampaolo, E.T. & Vergani, C.E. (2008). Flexural strength and hardness of reline and denture base acrylic resins after different exposure times of microwave disinfection. *Quintessence International,* Vol.39, No.10, (November 2008), pp. 833-840, ISSN 0033-6572

Ribeiro, D.G.; Pavarina, A.C.; Dovigo, L.N.; Palomari Spolidorio, D.M.; Giampaolo, E.T. & Vergani, C.E. (2009). Denture disinfection by microwave irradiation: a randomized clinical study. *Journal of Dentistry*, Vol.37, No.9, (September 2009), pp. 666-672, ISSN 0300-5712

Ribeiro, R.C.; Giampaolo, E.T.; Machado, A.L.; Vergani, C.E. & Pavarina, A.C. (2008). Effect of microwave disinfection on the bond strength of denture teeth to acrylic resins. *The International Journal of Adhesion and Adhesives*, Vol.28, No.6, (September 2008), pp. 296-301, ISSN 0143-7496

Ribeiro, R.C; Izumida, F.E.; Moffa, E.B.; Basso, M.F.M.; Giampaolo, E.T.; & Vergani, C.E. (2009). Color stability of reline resin after microwave disinfection: clinical evaluation. *Journal of Dental Research*, Vol.88, Special Issue Letter A, Abstract number 1755, ISSN 0022-0345.

Rizzo, R. (1993). The effects of sterilization with microwaves on diamond burs. *Minerva Stomatologica*, Vol.42, No.3, (March 1993), pp. 93-96, ISSN 0026-4970

Rohrer, M.D.; Bulard, R.A. (1985). Microwave sterilization. *Journal of the American Dental Association*, Vol.110, No.2, (February 1985), p.194-198, ISSN 0002-8177

Rohrer, M.D.; Terry, M.A.; Bulard, R.A.; Graves, D.C. & Taylor, E.M. (1986). Microwave sterilization of hydrophilic contact lenses. *American Journal of Ophthalmology*, Vol.101, No.5, (January 1986), pp. 49-57, ISSN 0002-9394

Rosaspina, S.; Salvatorelli, G.; Anzanel, D. & Bovolenta, R. (1994). Effect of microwave radiation on *Candida albicans*. *Microbios*, Vol.78, No.314, pp. 55-59, ISSN 0026-2633

Rossi, T.; Peltonen, R.; Laine, J.; Eerola, E.; Vuopio-Varkila, J. & Kotilainen, P. (1996). Eradication of the long-term carriage of methicillin-resistant *Staphylococcus aureus* in patients wearing dentures: a follow-up of 10 patients. *The Journal of Hospital Infection*, Vol.34, No.4, (December 1996), pp. 311-320, ISSN 0195-6701

Sande, M.A.; Gadot, F. & Wenzel, R.P. (1975). Point source epidemic of *Mycoplasma pneumoniae* infection in a prosthodontics laboratory. *The American Review of Respiratory Disease*, Vol.112, No.2, (August 1975), pp. 213-217, ISSN 0003-0805

Sanitá, P.V.; Vergani, C.E.; Giampaolo, E.T.; Pavarina, A.C. & Machado, A.L. (2009). Growth of *Candida* species on complete dentures: effect of microwave disinfection. *Mycoses*, Vol.52, No.2, (March 2009), pp. 154-160, ISSN 0933-7407

Sanitá, P.V.; Vergani, C.E.; Machado, A.L.; Giampaolo, E.T. & Pavarina, A.C. (2010). Denture microwave disinfection for reducing *Candida* infection in diabetic patients. *Journal of Dental Research*, Vol.89, Special Issue Letter B, Abstract number 1798, ISSN 0022-0345

Sanitá, P.V.; Machado, A.L.; Dovigo, L.N.; Giampaolo, E.T.; Pavarina, A.C. & Vergani, C.E. (2011). Denture microwave disinfection in treating diabetics with denture stomatitis. *Journal of Dental Research*, Vol.90, Special Issue Letter A, Abstract number 2416, ISSN 0022-0345

Sartori, E.A.; Schmidt, C.B.; Walber, L.F. & Shinkai, R.S. (2006). Effect of microwave disinfection on denture base adaptation and resin surface roughness. *Brazilian Dental Journal*, Vol.17, No.3, (2006), pp. 195-200, ISSN 0103-6440

Sartori, E.A.; Schmidt, C.B.; Mota, E.G.; Hirakata, L.M. & Shinkai, R.S. (2008). Cumulative effect of disinfection procedures on microhardness and tridimensional stability of a poly(methyl methacrylate) denture base resin. *Journal of Biomedical Materials Research. Part B, Applied Biomaterials*, Vol. 86B, No.2, (August 2008), pp. 360-364, ISSN 1552-4973

Seó, R.S.; Vergani, C.E.; Giampaolo, E.T.; Pavarina, A.C. & Machado, A.L. (2007a). Effect of a post-polymerization treatment on the flexural strength and Vickers hardness of reline and acrylic denture base resins. *Journal of Applied Oral Science*, Vol.15, No.6, (December 2007), pp. 506-511, ISSN 1678-7757

Seó, R.S.; Vergani, C.E.; Pavarina, A.C.; Compagnoni, M.A. & Machado, A.L. (2007b). Influence of microwave disinfection on the dimensional stability of intact and relined acrylic resin denture bases. *The Journal of Prosthetic Dentistry*, Vol.98, No.3, (September 2007), pp. 216-223, ISSN 0022-3913

Seó, R.S.; Vergani, C.E.; Giampaolo, E.T.; Pavarina, A.C.; dos Santos Nunes Reis, J.M. & Machado, A.L. (2008). Effect of disinfection by microwave irradiation on the strength of intact and relined denture bases and the water sorption and solubility of denture base and reline materials. *Journal of Applied Polymer Science*, Vol.107, No.1, (January 2008), pp. 300-308, ISSN 1097-4628

Sideridou, I.; Achilias, DS. & Kyrikou, E. (2004). Thermal expansion characteristics of light-cured dental resins and resin composites. *Biomaterials*, Vol.25, No.15, (July 2004), pp. 3087-3097, ISSN 0142-9612

Silva, M.M.; Vergani, C.E.; Giampaolo, E.T.; Neppelenbroek, K.H.; Spolidorio, D.M.P. & Machado, A.L. (2006). Effectiviness of microwave irradiation on the disinfection of complete dentures. *The International Journal of Prosthodontics*, Vol.19, No.3, (May-June 2006), pp. 151-156, ISSN 0893-2174

Silva, M.M.; Vergani, C.E.; Mima, E.G.O.; Pavarina, A.C.; Machado, A.L. & Giampaolo, E.T. (2008). Effect of microwave disinfection in the treatment of denture stomatitis. *Journal of Dental Research*, Vol.87, Special Issue Letter B, Abstract number 1276, ISSN 0022-0345

Sofou, A.; Larsen, T.; Fiehn, N.E. & Owall, B. (2002). Contamination level of alginate impressions arriving at a dental laboratory. *Clinical Oral Investigations*, Vol.6, No.3, (September 2002), pp. 161–165, ISSN 1432-6981

Spolidorio, D.; Tardivo, T.; dos Reis Derceli, J.; Neppelenbroek, K.; Duque, C.; Spolidorio, L. & Pires, J. (2011). Evaluation of two alternative methods for disinfection of toothbrushes and tongue scrapers. *International Journal of Dental Hygiene*, (February 2011), doi: 10.1111/j.1601-5037.2011.00503.x. [Epub ahead of print], ISSN 1601-5029

Takahashi, Y.; Chai, J. & Kawaguchi, M. (1998). Effect of water sorption on the resistance to plastic deformation of a denture base material relined with four different denture reline materials. *The International Journal of Prosthodontics*, Vol.11, No.1, (January-February 1998), pp. 49-54, ISSN 0893-2174

Tarantino, L.; Tomassini, E.; Petti, S. & Simonetti D'Arca, A. (1997). Use of a microwave device for dental instrument sterilization: possibilities and limitations. *Minerva Stomatologica*, Vol.46, No.10, (October 1997), pp. 561-566, ISSN 0026-4970

Tate, W.H.; Goldschmidt, M.C.; Ward, M.T. & Grant, R.L. (1995). Disinfection and sterilization of composite polishing instruments. *American Journal of Dentistry*, Vol.8, No.5, (October 1995), pp. 270-272, ISSN 0894-8275

Tate, W.H.; Goldschmidt, M.C. & Powers, J.M. (1996). Performance of composite finishing and polishing instruments after sterilization. *American Journal of Dentistry*, Vol.9, No.2, (April 1996), pp. 61-64, ISSN 0894-8275

Thomas, C.J. & Webb, B.C. (1995). Microwaving of acrylic resin dentures. *The European Journal of Prosthodontics and Restorative Dentistry*, Vol.3, No.4, (June 1995), pp. 179-182, ISSN 0965-7452

Tuncer, N.; Tufekçioglu, H.B. & Calikkocaoglu, S. (1993). Investigation on the compressive strength of several gypsum products dried by microwave oven with different programs. *The Journal of Prosthetic Dentistry*, Vol.69, No.3, (March 1993), pp. 333-339, ISSN 0022-3913

Vergani, C.E.; Giampaolo, E.T.; Pavarina, A.C.; Silva, M.M.; Mima, E.G.O. & Machado, A.L. (2008). Efficacy of denture microwave disinfection in treating denture stomatitis. *Journal of Dental Research*, Vol.87, Special Issue Letter B, Abstract number 1289, ISSN 0022-0345.

Verran, J.; Kossar, S. & McCord, J.F. (1996). Microbiological study of selected risk areas in dental technology laboratories. *Journal of Dentistry*, Vol.24, No.1-2, (January-March 1996), pp. 77-80, ISSN 0300-5712

Verran, J. & Maryan, C.J. (1997). Retention of *Candida albicans* on acrylic resin and silicone of different surface topography. *The Journal of Prosthetic Dentistry*, Vol.77, No.5, (May 1997), pp. 535-539, ISSN 0022-3913

Wakefield, C.W. (1980). Laboratory contamination of dental prostheses. *The Journal of Prosthetic Dentistry*, Vol.44, No.2, (August 1980), pp. 143-146, ISSN 0022-3913

Watanabe, K.; Kakita, Y.; Kashige, N.; Miake, F. & Tsukiji, T. (2000). Effect of ionic strength on the inactivation of micro-organisms by microwave irradiation. *Letters in Applied Microbiology*, Vol.31, No.1, (July 2000), pp. 52-56, ISSN 0266-8254

Webb, B.C.; Thomas, C.J.; Harty, D.W. & Willcox, M.D. (1998). Effectiveness of two methods of denture sterilization. *Journal of Oral Rehabilitation*, Vol.25, No.6, (June 1998), pp. 416-423, ISSN 0305-182X

Webb, B.C.; Thomas, C.J. & Whittle, T. (2005). A 2-year study of *Candida*-associated denture stomatitis treatment in aged care subjects. *Gerodontology*, Vol.22, No.3, (September 2005), pp. 168-176, ISSN 0734-0664

Williams, H.N.; Falkler, W.A.Jr.; Hasler, J.F. & Libonati, J.P. (1985). The recovery and significance of nonoral opportunistic pathogenic bacteria in dental laboratory pumice. *The Journal of Prosthetic Dentistry*, Vol.54, No.5, (November 1985), pp. 725-730, ISSN 0022-3913

Woo, I.; Rhee, I. & Park, H. (2000). Differential damage in bacterial cells by microwave radiation on the basis of cell wall structure. *Applied and Environmental Microbiology*, Vol.66, No.5, (May 2000), pp. 2243-2237, ISSN 0099-2240

Wright, P.S.; Clark, P. & Hardie, J.M. (1985). The prevalence and significance of yeasts in persons wearing complete dentures with soft-lining materials. *Journal of Dental Research*, Vol.64, No.2, (February 1985), pp. 122-125, ISSN 0022-0345

Yannikakis, S.; Zissis, A.; Polyzois, G. & Andreopoulos, A. (2002). Evaluation of porosity in microwave-processed acrylic resin using a photographic method. *The Journal of Prosthetic Dentistry*, Vol.87, No.6, (June 2002), pp. 613-619, ISSN 0022-3913

Yeo, C.B.; Watson, I.A.; Stewart-Tull, D.E. & Koh, V.H. (1999). Heat transfer analysis of *Staphylococcus aureus* on stainless steel with microwave radiation. *Journal of Applied Microbiology*, Vol.87, No.3, (September 1999), pp. 396-401, ISSN 1364-5072

# Part 3

# Health and Environment

# 5

# Microwave Enhanced Advanced Oxidation in the Treatment of Dairy Manure

K.V. Lo and P.H. Liao
*Department of Civil Engineering, University of British Columbia, Vancouver, Canada*

## 1. Introduction

Microwave heating is a dielectric heating process, in which heat is generated via the interaction of dielectric materials with electromagnetic radiation. The dielectric constant is a measure of the material's capacity to retard microwave energy as it passes through, while the loss factor is a measure of the material's capacity to dissipate the energy. The materials with high loss factor are easily heated through microwave irradiation. Energy dissipation mechanism in the materials is via ionic conduction and dipolar rotation. Generally, the dielectric properties of a material are related to temperature, moisture content, density and material geometry. Microwave heating has the advantages of higher heating rates, no direct contact between the heating source and heating material, selective heating, reduced equipment size and better process control than those of conventional heating. Moreover, both DNA and bacterial cellular membranes can be damaged by microwave irradiation. Such an effect on the treatment process, besides heating, is referred to as the athermal effect (Hong et al. 2006).

Microwave heating has been applied in various processes and manufacturing industries, such as food process, wood drying and waste treatment process (Jones, et al., 2002). For the sewage wastewater treatment industry, there are many applications of microwave heating on sludge treatment for reducing volume, improving dewaterability, enhancing digestibility, enhancing nutrient release, pathogen destruction, and stabilizing heavy metal (Menendez, et al., 2002; Liao, et al., 2005; Wojciechowaska 2005; Hong et al., 2006; Eskicioglu et al., 2007; Hsieh et al., 2007; Yu, et al., 2010). It has also been reported that the overall treatment efficiency could be increased with a combination of microwave heating with chemicals (thermo-chemical) for the treatment of sludge (Chan, et al., 2007; Qiao, et al., 2008). The common chemicals used on thermo-chemical process are acid, base and oxidants.

The microwave enhanced advanced oxidation process (MW/$H_2O_2$-AOP) uses microwave irradiation in combination with hydrogen peroxide to generate hydroxyl radicals to react with organic compounds. As a result, the structure of the solid materials is altered, and nutrients can be released into solution in the process. During this treatment, the suspended solids (SS) content of the slurries is also reduced. Its mechanism is assumed to be similar to the wet-air oxidation process wherein the first process involves the breakdown of large particulate organic matters, such as carbohydrate and proteins into smaller and more soluble organic components, and the second process involves the further oxidation or gasification of some of the resulting organic products (Shanableh & Shimizu, 2000; Liao, et al., 2007). The factors affecting the performance of the MW/$H_2O_2$-AOP have been identified; they are microwave temperature, hydrogen peroxide dosage, microwave intensity, reaction time, and acid addition. The effects varied with the application, depending on the soluble materials of interest, and various waste organic slurries used in the process (Wong et al., 2007). Dairy manure contains fats, proteins, lignin, carbohydrates and inorganic residue, and is rich in a variety of nutrients including nitrogen, phosphorus, and minerals (Rico, et al., 2007) and it should be deemed as a valuable biological resource, rather than a waste product. Lignocellulosic components from dairy manure can be transformed into fermentable saccharides via enzyme, or acid hydrolysis, which can further be converted into ethanol and other valuable products (Wen, et al., 2004). Nutrients, such as phosphorus and ammonia-nitrogen can be recovered via a struvite crystallization process. Carbonaceous matters can also be subjected to anaerobic digestion for methane production, which is a valuable bioenegy source. However, most of the phosphorus in dairy manure (about 65%) is in a form that is not easily soluble, especially those in the form of structural makeup (Barnet, 1994). Besides, high suspended solids are typical of dairy manure; hence, a pre-treatment step is required to solublize the nutrients and to disintegrate solids first in order to achieve resource recovery.

The MW/$H_2O_2$-AOP, which can disintegrate solids and release nutrients into solution, can be a potential effective pre-treatment method for nutrient and energy recovery from dairy manure. The effectiveness of the MW/$H_2O_2$-AOP under both batch operation and a continuous mode of operation for solubilisation of nutrients and organic matters from dairy manure are therefore reported in this chapter.

## 2. Materials and methods

### 2.1 Apparatus

A lab-scale Milestone Ethos TC microwave oven digestion system (Milestone Inc., USA) was used for a batch mode operation. The system operates at a frequency of 2,450 MHz with a maximum power output of 1,000 W. The system has the capacity of handling up to 12 vessels in a single run: one reference and 11 sample vessels, each with a volume of 100 mL. A thermocouple is inserted into the reference vessel, providing real time temperature monitoring during the runs. The maximum operating temperature and pressure are 220 °C and 435 psig, respectively.

A household microwave oven (Panasonic Genius Prestige Countertop) was modified for use in a continuous mode operation. The system operates at a frequency of 2,450 MHz with a maximum power output of 1,000 W. A silicon tube with 6.35 mm diameter was wound into a continuous horizontal coil and held by a custom made perforated polypropylene shelf. The total length of the silicon tube inside the MW is 40 m and the volume is approximately 1.3 L. The dairy manure was pumped into the MW oven through the silicon tube and was collected from the outlet. An appropriate amount of concentrated sulfuric acid was added to the liquid manure to the desired acid concentration before the mixture was pumped into the MW oven. The desired $H_2O_2$ dosage in the process was controlled by the flow rate and the concentration of the prepared $H_2O_2$. Different exit temperatures after the MW treatment can be achieved by controlling the flow rate of both the liquid manure and $H_2O_2$; a higher flow rate yields a lower exit temperature. The overall heating rate of 5 °C/min was used in this study. The unit was not pressurized; the exit temperatures were therefore lower than the boiling point of water.

### 2.2 Substrate

Dairy manure was obtained from the UBC Dairy Education & Research Centre in Agassiz, British Columbia, Canada. The dairy manure used for the batch operation was the solid portion obtained after the solid-liquid separation. Once collected from the farm the dairy manure was stored in a closed container at 4°C. The dairy manure sample had total solids (TS) of 33%, which contained large amounts of sand, the bedding material, as well as undigested lignocellulosic materials. Distilled water was added to the dairy manure samples and subsequently decanted to remove the sand. The resulting dairy manure had a TS of 5.6% for the first part of the batch operation (Part 1), while 3.7 % TS for the second part of the study at pH 4 (Part 2).

For a continuous operation, the liquid manure after the solid-liquid separation was again passed through a U.S. Standard No. 18 sieve (1 mm openings) to remove large fibres and particles to avoid solids deposition in the silicon tube in the MW unit. It was then diluted four times with distilled water to have TS of 0.83%.

### 2.3 Experimental design

#### 2.3.1 Batch operation

For Part 1, few variables were held constant, including the dosage of hydrogen peroxide (1.5 mL of 30% concentration, or expressed as 0.28 g $H_2O_2$/g TS), treatment time (20 minutes), and treatment temperature (120 °C). In an effort to isolate the effects of pH and hydrogen peroxide on the microwave process, experiments under acidic, neutral, and basic conditions, as well as in the absence and presence of hydrogen peroxide were tested. The experimental conditions included microwave treatment only (MW), microwave treatment with hydrogen peroxide (MW/$H_2O_2$), microwave treatment in acidic condition (MW/$H^+$), microwave treatment with hydrogen peroxide in acidic condition (MW/$H_2O_2$/$H^+$), microwave treatment in basic condition (MW/$OH^-$), and microwave treatment with hydrogen peroxide in basic condition (MW/$H_2O_2$/$OH^-$). Four replicates were conducted for each experimental condition. In adjusting to acidic conditions, dilute $H_2SO_4$ was dripped into the dairy manure solution to reach a pH of 4; whereas for basic conditions, dilute NaOH solution was added to reach a pH of 10.

For Part 2, the acidified dairy manure was chosen. Variables including treatment temperature, treatment time, and hydrogen peroxide dosage at pH 4 were tested for their effects on solids disintegration and nutrient solubization. Based on a computer statistical program, the Box-Benken design for response surface plots was chosen, and a series of 15 trials was required (Sall, et al., 2005). The temperature range was between 80 and 160 °C, treatment time ranged from 10 to 20 min, while hydrogen peroxide dosage was from 0 to 0.14 g $H_2O_2$/g TS. The rate of temperature increase for all the experiments were set at 20 °C per minute up to the designated treatment temperature and subsequently held for the specified time period. The experimental design is listed in Table 1.

#### 2.3.2 Continuous operation

The MW/$H_2O_2$-AOP experiments were operated at temperatures of 60, 70, 80 and 90 °C. The $H_2O_2$ dosage was fixed at 0.1% (v/v) and the acid dosages were 0.2, 0.5 and 1% (v/v) for each temperature. A set of experiments without $H_2O_2$ was performed to serve as control. Overall, sixteen individual experiments each with three replicates were conducted.

| Set # | Temp °C | Heat time (min) | Dosage (mL) | g $H_2O_2$/g TS |
|---|---|---|---|---|
| 1 | 120 | 15 | 0.25 | 0.068 |
| 2 | 160 | 20 | 0.25 | 0.068 |
| 3 | 120 | 15 | 0.25 | 0.068 |
| 4 | 80 | 10 | 0.25 | 0.068 |
| 5 | 80 | 15 | 0 | 0.000 |
| 6 | 160 | 15 | 0.5 | 0.135 |
| 7 | 160 | 15 | 0 | 0.000 |
| 8 | 80 | 20 | 0.25 | 0.068 |
| 9 | 160 | 10 | 0.25 | 0.068 |
| 10 | 120 | 10 | 0 | 0.000 |
| 11 | 120 | 15 | 0.25 | 0.068 |
| 12 | 80 | 15 | 0.5 | 0.135 |
| 13 | 120 | 20 | 0 | 0.000 |
| 14 | 120 | 20 | 0.5 | 0.135 |
| 15 | 120 | 10 | 0.5 | 0.135 |

Table 1. JMP design for Part 2 of batch operation

### 2.4 Chemical analysis

The samples were centrifuged at 3500 rpm for 10 minutes first, and then their supernatants were extracted for analysis for soluble chemical oxygen demand (SCOD), orthophosphate, soluble ammonia, volatile fatty acids (VFA) and reducing sugar. For dairy manure, the orthophosphate analysis could be affected by dilution due to interference, resulting in erroneous values; in order to obtain correct results, the initial dairy manure sample was first diluted to 0.5% TS before proceeding with the analysis (Wolf et al., 2005). It was also proven in a previous study that after microwave treatment the results obtained are reliable (Kenge, et al., 2009). The initial dairy manure samples were also analyzed for TS, total chemical oxygen demand (TCOD), total phosphate (TP) and total Kjeldahl nitrogen (TKN). All of the chemical analyses, except that of reducing sugar, followed the procedures outlined in Standard Methods (APHA, 1998). The colorimetric method was employed to determine the total reducing sugar content using anthrone reagent (Raunkjer, et al., 1994).

All chemical analyses were determined by a flow injection system, except determinations of TS and COD (Lachat Quik-Chem 8000 Automatic Ion Analyzer, Lachat Instruments, USA). A Hewlett Packard 5890 Series II gas chromatograph, equipped with a flame ionization detector (FID), was used to measure VFA. Volatile separation was accomplished with an HP FFAP (free fatty acid phase) column (0.25 m × 0.31 mm with 0.52 μ film thickness). The injection temperature was set at 175 °C and the flame ionization detector was at 250 °C. Helium gas was used as the carrier at a head pressure of 69 kPa.

## 3. Results and discussion

### 3.1 Batch operation

#### 3.1.1 Part 1

The effects of pH and $H_2O_2$ on the nutrient release were evaluated in Part 1. The concentration of ortho-P and VFA increased with an addition of acid without the MW treatment, while concentration of both SCOD and reducing sugar decreased. It was suggested that the agglomeration of fine suspended particles of dairy manure occurred under the acidic condition, resulting in a decrease of the SCOD concentration. The reduction of reducing sugar concentration might be due to chemical reactions. There was no significant change of any initial concentration with the addition of NaOH (Table 3). Upon the addition of acid, the SCOD concentration decreased by about half. This may be attributed to chemical reactions between various soluble constituents in the samples. As a result, increased particle size caused the precipitation, and it was subsequently filtered out prior to this test. It appeared that acid addition aided the solubization of ortho-P, and the VFA production. However, it was not clear the role acid played in the reaction mechanism.

Reducing sugar was produced from all treatments. Microwave heating enhanced its yield, regardless of acidic or basic conditions. A significant amount of reducing sugar was produced in either the MW or MW/$OH^-$ treatment. The best yield was obtained with the MW/$H^+$/$H_2O_2$-AOP. The addition of hydrogen peroxide did not significantly help in producing more reducing sugar (Table 2). For the microwave pretreatment of dairy manure for anaerobic digestion, sugar (glucan/xylan) production was also reported (Jin et al., 2009). In their study, sugar yield was not affected by the different types of acid applied, sulfuric acid or hydrochloric acid, but was rather affected by the concentration of acid. The best yield in their study was produced either in 0.5% (v/v) of sulfuric acid or in 0.185% (v/v) of hydrochloric acid, but the yield did not increase further with a higher acid concentration. Sugar production was the worst in the given conditions with 2% $H_2SO_4$ addition. The comparison between the study by Jin et al. (2009) and this study should be made with the note that higher concentrations of acid, or base, and a lower hydrogen peroxide dosage were used in their study.

The low yield of reducing sugar was due to a diluted acid used in this study. Concentrated acid used for sugar production was more effective than diluted acid hydrolysis (Sun and Cheng, 2002). As high as 75% concentrated acid was used for treating dairy manure to produce more than 84% of glucose (g/100g cellulose) at 120 °C and 30 min of reaction time. The yield of glucose decreased with an increase of reaction time (Liao et al., 2006). Their results pointed out that the reaction period was also critical for the sugar yield.

When treated with MW alone, SCOD values increased due to the disintegration of organic particles. However, for the MW/$H^+$ and MW/$H^+$/$H_2O_2$ treatment, the SCOD value fell to below that of the MW treatment alone. On the other hand, it appeared that the disintegration of organic maters was enhanced in basic conditions. The high concentrations of SCOD were obtained for both MW/$OH^-$/$H_2O_2$ and MW/$OH^-$. These results were not quite similar to those of Jin et al. (2009). In their study, the high acid concentration (2% of acid) yielded the highest SCOD, about 35% of TCOD whereas the low acid concentration (0.5%) produced less SCOD than that of either NaOH or calcium oxide addition. In this study, high SCOD was obtained with the basic treatment. Despite the high SCOD yield with NaOH treatment, ortho-P, ammonia, and VFA release appeared to be lacking.

VFA was produced in a few of the trials. One common variable in the VFA production trials was the $H_2SO_4$ addition. It was clear that when acid was added, VFA was produced with or without MW. As hydrogen peroxide was added, VFA was reduced slightly. At the same time, TCOD was reduced for the acid trials with hydrogen peroxide, which indicated that instead of retaining carbon to form products such as VFA, carbon left the system as $CO_2$ as an end oxidation product. The presence of acid was also stabilizing the hydroxyl radicals to be more potent in oxidizing organics.

Phosphates have been known to be soluble in solution after treatment with the MW/$H_2O_2$-AOP. Orthophosphate was released in large amounts upon the addition of acid, at almost half the total phosphorus in solution. With the MW/$H_2O_2$-AOP treatment on top of acid addition, the release was even greater, at 62% of total phosphorus from the initial substrate. Soluble phosphate release was not significant for the basic treatment, with or without hydrogen peroxide addition. Dairy manure contained various ions, such as carbonate, ammonia, magnesium, calcium, sodium and sulfate, as well as phosphate. At a high pH (basic) condition, various phosphate precipitates can be formed in the sample, such as struvite, and many forms of calcium phosphate (Wang and Nancollas, 2008; Jin, et al., 2009); therefore, orthophosphate release after the MW treatment would be easily precipitated out of solution again. Basic condition would not be suitable for dairy manure treatment in terms of orthophosphate release. As mentioned earlier, basic conditions would be suitable for the purpose of solid disintegration, as a very high SCOD concentration was obtained in base treatment. The results indicated that in order to release considerable amounts of orthophosphate into solution, acid addition is required for the MW treatment. Hydrogen peroxide would also help an increase of phosphate release, but is not necessarily required in the process.

Ammonia was not produced in large quantities after the MW/$H_2O_2$-AOP treatment (Table 2). The initial ammonia concentration in solution barely increased after treatment with any chemical addition.

The results indicated that reducing sugar can be obtained from dairy manure with the MW treatment, regardless of treatment conditions. In order to solubilize phosphorus, dairy manure should be treated in acidic condition with an addition of hydrogen peroxide to maximize its yield. If the main goal is to disintegrate manure solids, the basic condition shall be chosen, which would yield the highest SCOD concentration.

| | o-$PO_4$ (mg P/L) | $NH_4^+$ (mg N/L) | SCOD (g/L) | Reducing sugar (mg/L) |
|---|---|---|---|---|
| initial concentration | 6.3 ± 0.2 | 600 ± 25 | 15 ± 0.6 | 672 ± 16 |
| initial acid addition | 130 ± 1.4 | 600 ± 28 | 6.0 ± 0.1 | 216 ± 31 |
| initial base addition | 4.6 ± 0.1 | 480 ± 0 | 17 ± 1.5 | 869 ± 69 |
| MW | 8.0 ± 0.8 | 830 ± 88 | 22 ± 1.9 | 1190 ± 95 |
| MW/$H_2O_2$ | 7.0 ± 0.3 | 830 ± 33 | 19 ± 2.0 | 1130 ± 91 |
| MW/$H^+$ | 240 ± 16 | 740 ± 220 | 14 ± 2.3 | 788 ± 133 |
| MW/$H^+$/$H_2O_2$ | 240 ± 4.5 | 790 ± 180 | 14 ± 2.3 | 1460 ± 62 |
| MW/$OH^-$ | 30 ± 1.9 | 670 ± 200 | 27 ± 1.5 | 1360 ± 126 |
| MW/$OH^-$/$H_2O_2$ | 13 ± 3.5 | 610 ± 110 | 25 ± 3.8 | 1180 ± 217 |

Table 2. Results from Part 1 of batch operation

### 3.1.2 Part 2

Based on the results from Part 1, an acidic condition was chosen for Part 2 of the study. The design was modeled after a study by Kenge et al. (2010), where a control without the addition of hydrogen peroxide was included. In that study,hydrogen peroxide (0.30-0.59 g $H_2O_2$/g TS) was in excess; its incremental effects on the solubilization of nutrients could not be fully evaluated. Large amounts of hydrogen peroxide were also not helpful in producing reducing sugar from dairy manure (Jin et al., 2009; Kenge et al., 2010). Therefore, lower dosages of hydrogen peroxide, ranging from 0 to 0.14 g $H_2O_2$/g TS were used in Part 2, instead of 0.28 g $H_2O_2$/g TS of hydrogen peroxide used in Part 1.

#### 3.1.2.1 Reducing sugar

The surface plot depicting the trend of reducing sugar production is showed in Figure 1. With the addition of hydrogen peroxide, reducing sugar release increased. It can be expected that with the addition of hydrogen peroxide, reducing sugar release increases due to its aid in the breakdown of lignocellulosic materials in dairy manure.

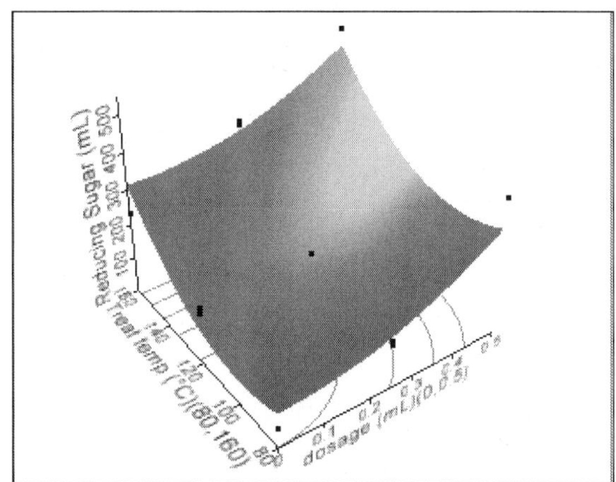


Fig. 1. Surface plot for Part 2 - reducing sugars

Reducing sugar concentration also increased with an increase of temperature. In the study by Kenge et al., (2010), temperature was the most significant factor affecting reducing sugar production; an increase of microwave temperature increased reducing sugar production. An increase in hydrogen peroxide dosage also increased reducing sugar yield at lower temperatures; however, its yield remained relatively constant at higher microwave temperatures. A much higher dosage of hydrogen peroxide was applied in the earlier study, causing solubilized reducing sugars to be further oxidized to form other oxidation products. Due to a balance between reducing sugars being solubilized and oxidized at the same time, reducing sugar concentration in solution remained relatively constant. The rates of reducing sugar production and sugar oxidation to form other compounds have not been measured, but it appeared that equilibrium between the two rates was reached in this case. However, results of this study showed that reducing sugar increased slightly with an increase of hydrogen peroxide at high operating temperatures. This might be due to the fact that there was no excess hydrogen peroxide in solution to engage and aid in further oxidation reactions.

The results obtained from this study and previously indicated that temperature and hydrogen peroxide dosage are factors affecting reducing sugar production (Kenge et al., 2010). The process could be operated with or without $H_2O_2$ for reducing sugar production; the substantial amounts of reducing sugar could also be obtained at a higher temperature and a longer heating period without $H_2O_2$.

**3.1.2.2 Solids disintegration and nutrient release**

SCOD release in Part 2 can be seen in Table 3. In general, higher SCOD concentrations were obtained at higher operating temperatures. Temperature was the most important factor for solids disintegration. At the same operating temperature, it could be operated either at a longer heating period without the addition of $H_2O_2$, or at a shorter heating period with an addition of $H_2O_2$ to yield the similar SCOD concentration. Hydrogen peroxide had some impact on the disintegration of solids, but it was not very significant in such low dosages. Microwave temperature was also a key factor for release of phosphates. Orthophosphate increased with an increase of microwave temperature (Figure 2).

| Set # | Reducing sugar (mg/L) | o-$PO_4$ (mg P/L) | $NH_4^+$ (mg N/L) | SCOD (g/L) | TCOD (g/L) | VFA (mg/L) |
|---|---|---|---|---|---|---|
| 1 | 106 ± 19 | 109 ± 4.8 | 43 ± 0.5 | 2.9 ± 0.2 | 20 ± 3.7 | 127 ± 6 |
| 2 | 344 ± 58 | 367 ± 4.0 | 71 ± 2.6 | 4.6 ± 0.2 | 35 ± 3.6 | 252 ± 7 |
| 3 | 126 ± 169 | 167 ± 3.9 | 57 ± 1.0 | 3.0 ± 0.4 | 33 ± 2.1 | 203 ± 5 |
| 4 | 65 ± 95 | 113 ± 4.0 | 40 ± 3.4 | 4.1 ± 0.3 | 19 ± 4.9 | 133 ± 14 |
| 5 | 60 ± 69 | 333 ± 5.0 | 50 ± 1.7 | 3.9 ± 0.1 | 37 ± 5.3 | 236 ± 26 |
| 6 | 517 ± 53 | 356 ± 4.3 | 73 ± 1.4 | 5.0 ± 0.1 | 36 ± 0.2 | 238 ± 9 |
| 7 | 230 ± 79 | 121 ± 0.5 | 51 ± 0.6 | 3.7 ± 0.4 | 19 ± 2.3 | 153 ± 20 |
| 8 | 104 ± 50 | 162 ± 8.4 | 51 ± 3.3 | 2.8 ± 0.6 | 38 ± 6.6 | 188 ± 14 |
| 9 | 296 ± 47 | 172 ± 3.4 | 66 ± 0.6 | 4.0 ± 0.2 | 37 ± 3.3 | 248 ± 12 |
| 10 | 110 ± 40 | 16 ± 6.7 | 54 ± 1.3 | 3.4 ± 0 | 31 ± 0.9 | 213 ± 19 |
| 11 | 172 ± 29 | 331 ± 6.0 | 56 ± 1.8 | 3.2 ± 0.1 | 34 ± 3.0 | 254 ± 10 |
| 12 | 387 ± 47 | 106 ± 3.8 | 42 ± 1.0 | 3.6 ± 0 | 20 ± 1.3 | 140 ± 5 |
| 13 | 111 ± 35 | 334 ± 10.8 | 55 ± 1.6 | 3.7 ± 0.5 | 37 ± 1.4 | 220 ± 22 |
| 14 | 183 ± 32 | 159 ± 6.5 | 56 ± 0.7 | 3.2 ± 0 | 32 ± 2.6 | 188 ± 16 |
| 15 | 149 ± 94 | 163 ± 3.6 | 55 ± 3.0 | 3.3 ± 0.3 | 28 ± 2.5 | 194 ± 17 |
| Initial | 322 ± 48 | 60 ± 3.2 | 16 ± 0.4 | 5.9 ± 0.1 | 36 ± 4.1 | 188 ± 19 |

Table 3. Results for Part 2 of batch operation

The polyphosphates contained in dairy manure could be broken down to form ortho-P at elevated temperature; the rate and extent of polyphosphates were dependent on the heating temperature (Kuroda, et al., 2002). Acid also help solublize organic phosphorus and polyphosphate (Chan, et al., 2007). A trend can be seen from Figure 2, where Treatment time highly favours the release of ortho-P, with a slight increase of hydrogen peroxide. Irrespective of hydrogen peroxide dosage and treatment temperature, ortho-P increased with treatment time (Figure 2). It can be seen from both parts of this study that acid addition is crucial to the release of ortho-P from dairy manure into solution. The orth-P results obtained in this study were consistent with results from previous studies (Kenge, et al., 2010; Yu, et al., 2010). It was reported by Jin, et al. (2010) that the optimal condition for orthophosphate release was at 135 °C and 26 min in microwave treatment of dairy manure. It seemed that the MW/$H_2O_2$-AOP was operated near the optimal condition for the release of orthophosphate and other soluble components in this study. In view of the results, the process should be operated at a high temperature region (120-160 °C), a long reaction time (15-20 min) and dosage of 0.07-0.14 g $H_2O_2$/g TS for maximizing soluble components from dairy manure.

Ammonia is one of the soluble products from the MW/$H_2O_2$-AOP, and it is a constituent of a potent fertilizer, struvite. No dramatic increases in ammonia production were found.

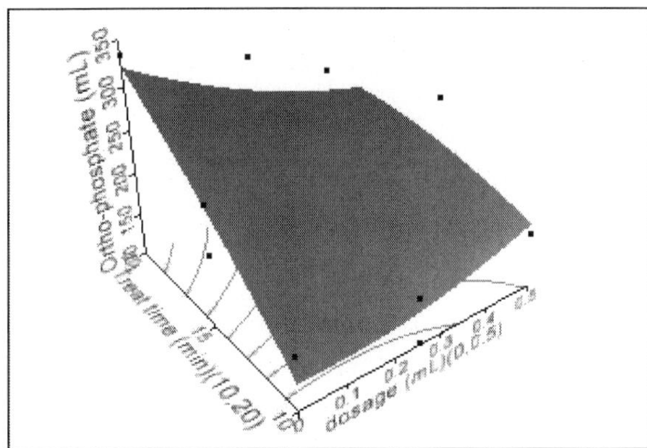


Fig. 2. Surface plot for Part 2 - orthophosphate

### 3.2 Continuous operation

The disadvantages of a batch operation are: (1) the materials must be transported twice from the reaction vessel, and (2) the amount of sample that can be treated at one time is limited by the batch reaction time and the sample vessel volume. Besides these shortcomings, hydrogen peroxide was also introduced prior to microwave irradiation, limiting the synergistic effects between $H_2O_2$ and microwave irradiation. The alternative to batch operation is a continuous process, which is more suitable for an industrial application.

The results are presented in Figures 3 to 7. The most pronounced effect on the solubilization of solids was due to microwave heating temperature (Fig. 3).

The SCOD solubilization remained relatively steady at a range of 60 to 70 °C, regardless of the acid concentration. Acid concentration aids the process to maintain or even increase the hydroxyl radical concentration by stabilizing them and thereby inhibiting their degradation. The narrow range of treatment temperature involved may mask the synergetic effects of adding acid to the process as the continuous mode of operation is not pressurized to exceed boiling point. The SCOD increased with an increase of microwave temperature to a higher range of 80 to 90 °C. This was due to the fact that heating would also increase the decomposition of $H_2O_2$ into hydroxyl radicals and therefore enhance the oxidation process when $H_2O_2$ was applied simultaneously with microwave heating (Eskicioglu, et al., 2008).

Heating alone can also add to the degradation of the substrate's structural integrity. The results showed that SCOD soliblization from dairy manure was most affected by MW temperature. To a lesser extent, it was also affected by acid concentration, especially at the higher temperature levels of 80 to 90 °C.

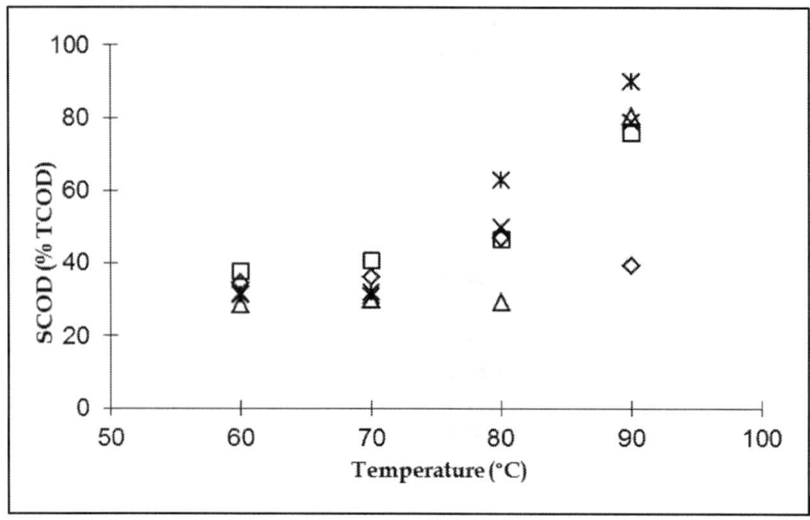


Fig. 3. Soluble COD as a per cent of TCOD (◇ influent; □ MW only; △ MW/$H_2O_2$-AOP with 0.2% acid; × MW/$H_2O_2$-AOP with 0.5% acid; ✳ MW/$H_2O_2$-AOP with 1.0% acid)

It was evident that an acidic condition was critical for phosphorus solubilization from dairy manure (Fig. 4). Without the acid addition, as microwave temperature increased from 60 to 90 °C, the ortho-P decreased from 34% to 11% of TP, which was lower than in the untreated substrate. Phosphorus solubilization was greatly improved with an addition of acid. However, acid dosages did not display significant effects on the solubilization of phosphorus; an acid addition of 0.5% appeared to be sufficient.

There was no significant release of ammonia over the range of 60 to 80 °C, regardless of the acid dosages (Fig. 5). However, up to 70% of TKN was solubilised at 90 °C. The effect of acid dosage on $NH_4$-N solubilization was similar to that of phosphorus solubilization; the MW/$H_2O_2$-AOP needed to be operated in an acidic condition; 0.5% of acid concentration was sufficient.

The VFA comprised of various organic acids, such as acetic, propionic, i-butyric, i-valeric, valeric, hexanoic, and heptanoic acids. However, acetic acid is the major component of the VFA, ranged from 46 to 53%. A considerable amount of VFA was released from the dairy manure after the MW/$H_2O_2$-AOP treatment. The component of acetic acid increased in the treated solutions, ranging from 53 to 84% of total VFA. Overall, the best VFA production of approximately 330 to 420 mg/L was solubilized with an acid dosage of 0.5%. Acid in combination with the MW/$H_2O_2$-AOP produced more VFA with increasing temperature, except for the highest acid dosage of 1%, where a slight decrease was found at the higher temperatures. This trend may not be conclusive to say higher acid dosage is detrimental to

VFA production, but it indicates that higher acid concentrations may have an effect in reducing VFA in dairy manure (Fig. 6).

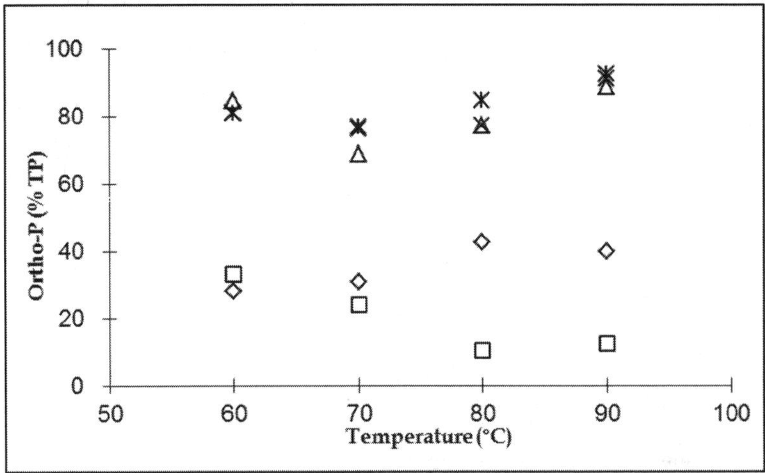


Fig. 4. Orthophosphate release as a per cent of TP (◇ influent; □ MW only; △ MW/$H_2O_2$-AOP with 0.2% acid; × MW/$H_2O_2$-AOP with 0.5% acid; ⋇ MW/$H_2O_2$-AOP with 1.0% acid)

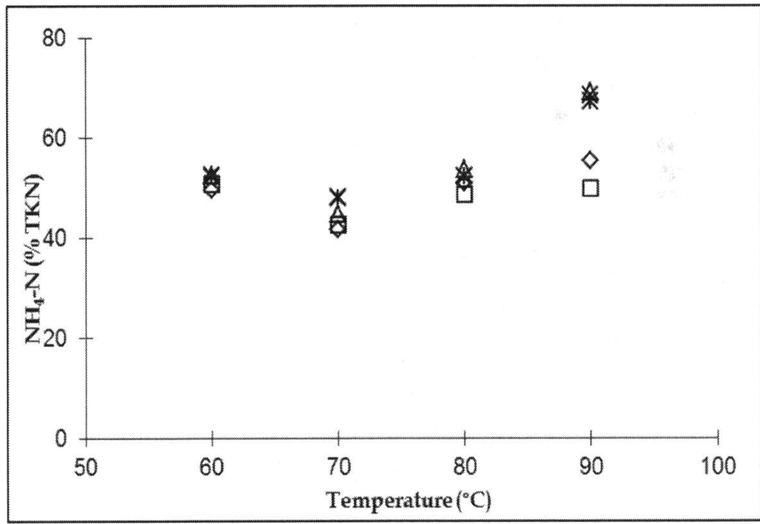


Fig. 5. Ammonia release as a per cent of TKN (◇ influent; □ MW only; △ MW/$H_2O_2$-AOP with 0.2% acid; × MW/$H_2O_2$-AOP with 0.5% acid; ⋇ MW/$H_2O_2$-AOP with 1.0% acid)

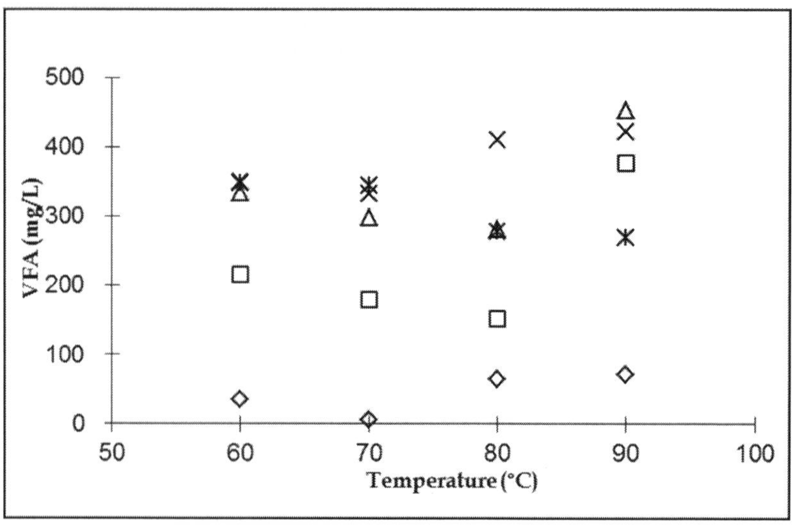


Fig. 6. Productions of volatile fatty acids (◇ influent; □ MW only; △ MW/$H_2O_2$-AOP with 0.2% acid; × MW/$H_2O_2$-AOP with 0.5% acid; ✳ MW/$H_2O_2$-AOP with 1.0%

For the MW only treatment, the reducing sugar concentrations remained similar to those in the influents; however, reducing sugar concentrations decreased after treatment with the MW/$H_2O_2$-AOP. For the MW/$H_2O_2$-AOP, the reducing sugar concentrations increased gradually with an increase of microwave temperature. With an increase in acid concentrations for every treatment temperature, reducing sugar concentrations were further decreased (Figure 7). The reducing sugars produced from dairy manure are from the hydrolysis of hemicellulose and cellulose, the resulting solution most likely contains a mixture of pentose and hexose. Indeed, reducing sugars of arabinose, xylose, galactose and glucose have been identified from the hydrolysis of dairy manure using an ion chromatograph (Wen, et al., 2004).

The decrease in reducing sugars could be explained as the further oxidation of soluble reducing sugars into VFA and $CO_2$ during the MW/$H_2O_2$-AOP. The oxidation process is more favourable at a lower pH, in other words, a higher acid dosage. Very low concentrations (less than 90 mg/L of reducing sugars) were obtained from different treatments at low microwave temperatures.

This was due to the manure being sieved and diluted; therefore, the lignocellulosic materials, a precursor to sugar production, were low in the manure used in this study. The other reason was that temperatures and acid concentration applied were much lower than that of optimal conditions. As indicated in conventional acid hydrolysis of lignocellulosic materials in dairy manure for the production of simple sugars, samples required pre-treatment with 75% acid concentration and 30 min of reaction time for the de-crystallization of fibre, and then were further hydrolyzed in 12.5% of acid at 135 °C for 10 minutes (Liao, et al., 2006). For the MW/$H_2O_2$-AOP treatment of dairy manure, a higher reducing sugar production was obtained at above 120 °C (Kenge, et al., 2010). The results show that the important factors affecting reducing sugar yield were temperature and acid concentration.

The presence of small amounts of hydrogen peroxide during the treatment process would decrease the degrees of polymerization and cellulose crystallinity, at the same time, igt would increase the accessible surface area and pore size of lignocellulosics. However, too much of hydrogen peroxide would decrease reducing sugar yield. It would favor the further oxidative reaction to form VFA or $CO_2$ (Kenge, et al., 2010).

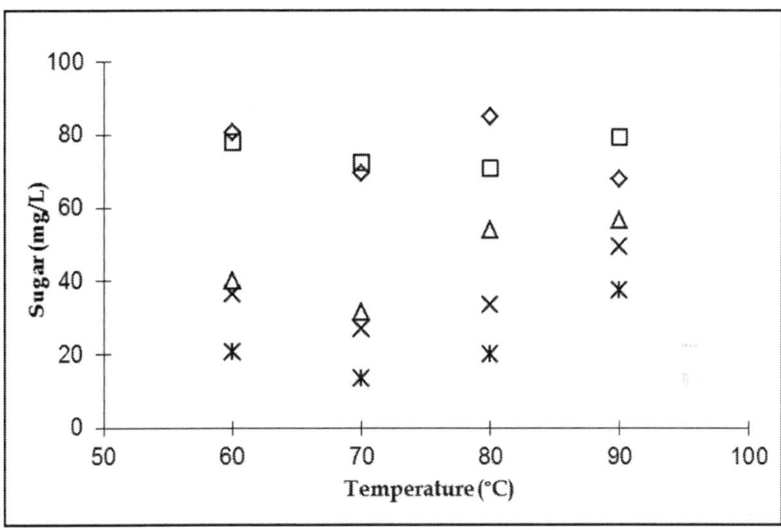


Fig. 7. Production of reducing sugars (◇ influent; □ MW only; △ MW/$H_2O_2$-AOP with 0.2% acid; × MW/$H_2O_2$-AOP with 0.5% acid; ✳ MW/$H_2O_2$-AOP with 1.0% acid)

### 3.2.1 Comparison of batch and continuous operations

Four experimental sets using a batch microwave unit were also conducted for the purpose of comparison to the continuous process. Three replicates were used at two testing temperatures of 60 and 90 °C. The rate of temperature increase was set at 5 °C/minute to attain the target temperature, at which time, the reaction was terminated. Two sets of experiments at each temperature were operated: one with both acid (0.2% v/v) and $H_2O_2$ (0.1% v/v), the other without.

The results of SCOD and ortho-P for both batch and continuous systems are presented in Table 4. Without the additions of $H_2O_2$ and acid, a simple decomposition process occurred under thermal treatment. As a result, there was little solids disintegration and little release for both nutrients. For the MW/$H_2O_2$-AOP, it was the thermal oxidation process that involved not only the breakdown of large particulate organic matters, but also, the further oxidation or gasification of some of the resulting organic products. Both SCOD and ortho-P concentrations for the MW/$H_2O_2$-AOP were higher than those of the MW treatment only (Table 4). A higher microwave temperature also increased phosphorus solubilization and solids disintegration. Higher ortho-P and SCOD concentrations were obtained from the continuous mode operation than from a batch system. The superior performance of a continuous system could be attributed to the synergistic effects of MW irradiation and $H_2O_2$. Since $H_2O_2$ was injected into the influent stream immediately prior to MW irradiation in a continuous mode unit, the synergistic effects between $H_2O_2$ and MW irradiation became manifested.

| Temperature (°C) | $H_2O_2$ and $H_2SO_4$ | Ortho-P Increase (%) | | SCOD Increase (%) | |
|---|---|---|---|---|---|
| | | Batch | Continuous | Batch | Continuous |
| 60 | No | -3.2 | 5.3 | -0.6 | 1.8 |
| | Yes | 26 | 57 | 12 | 37 |
| 90 | No | -7.1 | -14 | 0.1 | -6.6 |
| | Yes | 28 | 71 | 12 | 34 |

Table 4. Comparison of batch and continuous operation

This study also demonstrates the feasibility to produce reducing sugar from dairy manure for the subsequent processes, such as ethanol production, and useful fermentation products. To obtain a large quantity of reducing sugars from dairy manure, the MW process should be operated at a small amount of $H_2O_2$ or without $H_2O_2$. Nutrient is also released in the MW/$H_2O_2$-AOP, which can also be used for energy production, or struvite as a fertilizer.
However, the process has not been fully optimized yet, it was expected that under pressurized conditions, the continuous mode operation may be even more effective. More investigations should also be conducted to include a wider range of treatment temperatures. Based on the findings, a pilot-scale continuous mode MW/$H_2O_2$-AOP system is being conducted to further verify the feasibility of the technology for solids reduction and solubilisation of nutrients. The economic feasibility will also be conducted.

### 3.3 Economic implication

Applying the MW/$H_2O_2$-AOP to treat dairy manure, there will be significant increases of VFA and nutrients (N & P), and a decrease of solids content in the liquid filtrate. This will be very beneficial in that 1) with reduced solids content in the dairy manure, the more advanced anaerobic reactor, such as the fixe-film reactor, can be adopted for anaerobic digestion of the treated filtrate at much shorter HRTs, resulting in more methane production; and 2) with more VFA and other readily biodegradable substances in the treated dairy manure filtrate, the methane conversion efficiency will be much greater; the methane production rate from the manure filtrate can be as much as five to six times of that of the untreated manure filtrate (Lo, et al., 1985). The soluble phosphorus in the filtrate can be increased substantially by using the MW/$H_2O_2$-AOP. As a result, struvite production would also increase substantially. In addition, the MW/$H_2O_2$-AOP can destroy fecal coliforms in wastewater, making it possible for the final effluent available for barn cleaning and other purposes (Yu, et al., 2010). This not only reduces the water usage on farm, but also helps by reducing the pollution of fresh water resources.

## 4. Conclusion

The MW/$H_2O_2$-AOP could be an effective pre-treatment method to release nutrient, and disintegrate solids from dairy manure. It is recommended that the process shall be operated at high microwave temperatures (120-160 °C), a low hydrogen peroxide dosage (0.07-0.14 g $H_2O_2$/g TS) and a long treatment period (15-20 min).

Microwave temperature and acid concentration are factors affecting reducing sugars production. Hydrogen peroxide also plays a role in a less extent. Higher microwave temperatures favor the production of reducing sugars. The process could be operated with or without $H_2O_2$ for reducing sugar production; the substantial amounts of reducing sugars could be obtained at a higher temperature and a longer heating period without $H_2O_2$.

Temperature was the dominant factor for solids disintegration., At the same operating temperature, the process could be operated either at a longer heating period without addition of $H_2O_2$, or at a shorter heating period with an addition of $H_2O_2$ to yield the similar SCOD concentration. Heating time also affected orthophosphate release. Heating time and temperature were significant factors for ammonia release and VFA.

The MW/$H_2O_2$-AOP can be operated in a continuous operation, with equal or better results than a batch process. When the MW/$H_2O_2$-AOP is operated in a continuous mode, it maximizes the synergistic effects between $H_2O_2$ and MW irradiation, thereby promoting more nutrient solubilization and solid disintegration.

## 5. References

APHA. (1998). Standard Methods for the Examination of Water and Wastewater, 20th Ed. Am. Public Health Association, ISBN 0-87553-223-3, Washington, D.C., USA.

Barnett, G. (1994). "Phosphorus forms in animal manure". *Bioresource Technology*, 4, 139-147, ISSN 0960-8524

Chan, W; Wong, W.; Liao, P. & Lo, K. (2007). Sewage sludge nutrient solubilization using a single-stage microwave treatment. *Journal of Environ. Science & Health A*, 42, 59–63, ISSN 1093-4529

Eskicioglu, C.; Terzian, N.; Kennedy, K. & Droste, R. (2007). Athermal microwave effects for enhancing digestibility of waste activated sludge. *Water Research*, 41, 11, pp.2457-2466, ISSN 0043-1354

Eskicioglu, C.; Prorotb, A.; Marinc, J.; Drostec, R. & Kennedy, K. (2008). Synergetic pretreatment of sewage sludge by microwave irradiation in presence of $H_2O_2$ for enhanced anaerobic digestion. *Water Research*, *42*, 18, pp4674-4682.

Hong, S.; Park. J.; Teeradej, N.; Lee, Y.; Cho, T. & Park, C. (2006). Pretreatment of sludge with microwaves for pathogen destruction and improved anaerobic digestion performance. *Water Environmental Research*, 78, pp. 76-81, ISSN 1061-4303

Hsieh, C.; Lo, S.; Chiuch, P.; Kuan, W. & Chen, C. (2007). Microwave-enhanced stabilization of heavy metal sludge. *Journal of Hazardous Materials*, B139, pp. 160-166, ISSN 0043-1354

Jin, Y.; Hu, Z. & Wen, Z. (2009). Enhancing anaerobic digestibility and phosphorus recovery of dairy manure through microwave-based thermochemical pretreatment. *Water Research*, 43, 3493-3502, ISSN 0043-1354

Jones D.; Lelyveld, T.; Mavrofidis, S.; Kingman, S. & Miles, N. (2002). *Resource, Conservation and recycling*, 34, pp. 75-90, ISSN 0921-3449

Kenge, A.; Liao P. & Lo K. (2009). Solubilization of municipal anaerobic sludge using microwave-enhanced advanced oxidation process. *Journal of Environ Science & Health A*, 44, pp. 502-506, ISSN 1093-4529

Kenge A; Liao P. & Lo K. (2010). Nutrient release from solid dairy manure using the microwave advanced oxidation process. *Biomass & Bioenergy*, ISSN 0961-9534 (Submitted)

Kuroda, A.; Takiguchi, N.; Gotanda, T.; Normura, K.; Kato, J.; Ikeda, T. & Ohtake, H. (2002). A simple method to release to release polyphosphate from activated sludge for phosphorus reuse and recycling. *Biotechnology and Bioengineering,* 78, pp. 333-338, ISSN 0006-3592

Liao, P.; Wong, W. & Lo, K. (2005). Release of phosphorus from sewage sludge using microwave technology, *Journal of Environmental Engineering and Science,* 4, pp. 77-81, ISSN 1496-256X

Liao, P.; Lo, K..; Chan, W & Wong, W. (2007). Sludge Reduction and Volatile Fatty Acid Recovery Using Microwave Advanced Oxidation Process. *Journal of Environ. Science & Health A,* 42, pp. 633–639, ISSN 1093-4529

Liao, W.; Liu Y.; Liu, C.; Wen, Z. & Chen, S. (2006). Acid hydrolysis of fibers from dairy manure. *Bioresource Technology,* 97, pp. 1687-1695, ISSN 0960-8524

Lo, K. & Liao, P. (1985). High-rate anaerobic digestion of screened dairy manure. *Journal of Agricultural Engineering Research, 32,* pp. 349-358, ISSN 0021-8634

Menendez, J.; Inguanzo, M. & Pis, J. (2002). Microwave-induced pyrolysis of sewage sludge. *Water research,* 36, pp. 3261-3264, ISSN 0043-1354

Qiao, W.; Wang, W.; Xun, R.; Lu, W. & Yin, K.. (2008). Sewage sludge hydrothermal treatment by microwave irradiation combined with alkali addition 43, *Journal Mater Science,* 43, pp. 2431-2436, ISSN 0022-2461

Raunkjer, K..; Hvitved-Jacobsen, T. & Nielsen, P. (1994). Measurement of pools of protein, carbohydrate and lipid in domestic wastewater. *Water Research,* 28, pp. 251-262, ISSN 0043-1354

Rico J.; Garcia H,; Rico C, & Tejero I. (2007). Characterisation of solid and liquid fractions of dairy manure with regard to their component distribution and methane production. *Bioresource Technology,* 98, pp. 971–979, ISSN 0960-8524

Sall J, Creighton, L., Lehman, A. 2005. JMP Start Statistics: A guide to statistics and data analysis using JMP and JMP IN software. 3rd edition. Brooks/Cole-Thomason Learning. ISBN 0-534-99749-X, Belmont, California.

Shanableh, A. & Shimizu, Y. (2000). Treatment of sewage using hydrothermal oxidation technology application challenge. *Water Science and Technology, 41,* pp. 85-92, ISSN 0273-1223

Sun, Y. & Cheng, J. (2002). Hydrolysis of lignocellulosic materials for ethanol production: a review. *Bioresource Technolog,* 83, pp. 1-11, ISSN 0960-8524

Wang, L. & Nancollas, G. (2008). Calcium orthophosphates: crystallization and dissolution. *Chemical Reviews,* 108, pp. 4628-4669, ISSN 0009-2665

Wojiechowaska, E. (2005). Application of microwaves for ssewage sludge conditioning. *Water Research,* 39, pp. 4749-4574, ISSN 0043-1354

Wolf A.; Kleinman P.; Sharpley A. & Beegle D. (2005). Development of a water-extractable phosphorus test for manure: An interlaboratory study. *Soil Sci. Soc. Am. J.,* 69, pp. 695-700, ISSN 0361-5995

Wen, Z.; Liao, W. & Chen, S. (2004). Hydrolysis of animal manure lignocellulosics for reducing sugar production. *Bioresource Technology, 91,* 31-39. ISSN 0960-8524

Wong, W.; Lo, K. & Liao, P. (2007). Factors affecting nutrient solubilization from sewage sludge in microwave advanced oxidation process. *Journal of Environ. Science & Health A,* 42, 825-829, ISSN 1093-4529

Yu, Y.; Chan, W.; Liao, P. & Lo, K. (2010). Disinfection and solubilization of sewage sludge using the microwave enhanced advanced oxidation process. *Journal of Hazardous Materials* 181, pp. 1143-1147, ISSN 0304-3894

Yu. Y.; Lo, I..; Liao, P. & Lo, K. (2010). Treatment of dairy manure using the microwave enhanced advanced oxidation process under a continuous mode operation. *Journal of Environmental Science and Health B* 45, pp. 838-843, ISSN 0360-1234

# 6

# Microwave Processing of Meat

M.S. Yarmand[1] and A. Homayouni[2]
[1]*Department of Food Science and Technology, Faculty of Biosystems Engineering, Agriculture and Natural Resources Campus, University of Tehran, Tehran,*
[2]*Department of Food Science and Technology, Faculty of Health and Nutrition, Tabriz University of Medical Sciences, Tabriz,*
*I.R. Iran*

## 1. Introduction

In recent years, the rapid growth of chilled and frozen food market is due to rapid reheating instruments such as domestic or commercial microwave oven as well as conventional ovens. The goal of these products manufacturers is to cook the products to a temperature that would ensure the reduction of pathogenic bacteria to a safe level. It is a common way to heat the chilled foods in a microwave oven to a minimum temperature of 70°C for at least 2 minutes or an equivalent temperature/time condition, e.g. 75 °C for 30 seconds, 65 °C for 10 minutes, etc. A small change in heating time in a microwave oven has a large effect on the survival of Salmonella spp. In inoculated meat loaves, Salmonella survived after 3.5 minutes heating but it was not found when the heating time was increased to 4 minutes (European Chilled Food Federation, 1996; Farber et al., 1998; Levre & Valentini, 1998).

The ever-increasing range of processed foods is produced with microwave reheating instructions surround the world as well as increased microwave applications. In industrial applications, the microwave systems have been developed for drying, pre-cooking of bacon/meat, pasteurization of ready meals and the tempering of meat and fish. There is, however, a growing new application such as blanching, baking and microwave phyto-extraction. Microwave heating was primly applied in agriculture, in grain drying and insect control. In addition, it was used for food drying, blanching, pasteurization and cooking (Chen et al., 1971; Lin & Li, 1971; Maurer et al., 1971; Bhartia et al., 1973; Nelson, 1973; Jaynes, 1975; Avisse & Varaquaux, 1977; Nykvist & Decareau, 1976; Sobiech, 1980; Shivhare et al., 1994; Tulasidas et al., 1995; Begum & Brewer, 2001; Hong et al., 2001; Sumnu, 2001; Ramesh et al., 2002; Beaudry et al., 2003; Severini et al., 2004; Williams et al., 2004; Zhang et al., 2004).

Frozen materials are usually thawed or tempered before further processing in food industry. Thawing is usually regarded as complete when all the material has reached 0 °C and no free ice is present. This is the minimum temperature at which the meat can be boned or other products cut or separated by hand. Lower temperatures (e.g. -5 to -2 °C) are acceptable for mechanical chopping of product, but such material is `tempered' rather than thawed. The two mentioned processes are not same with together because tempering only constitutes the initial phase of a complete thawing process while thawing is often considered as simply the reversal of the freezing process. However, inherent in thawing is a major problem that it does not occur in freezing operation. Thawing operation has some steps, at first the surface areas of food to rise in temperature and bacterial multiplication can restart. On large objects surface spoilage can occur before the centre regions have fully thawed. It is interesting to know that, there is basic difference between conventional thawing and tempering systems with microwave. Conventional thawing and tempering systems supply heat to the surface and then rely on conduction to transfer that heat into the centre of product but microwave systems use electromagnetic radiation to generate heat within the food. The main detrimental effect of freezing and thawing on meat is the large increasing in amount of proteinaceous fluid (drip) released on final cutting. Several studies showed that fast thawing rates would increase amount of drip. Microwave thawing has some advantages: in this way, the ice formation during freezing breaks up cell structure and fluids are reduced during thawing but microwave thawing has a better quality product than thawing at 4 or 21 °C. It is necessary to more investigate the microwave processing because of large applications of microwave in meat processing (tempering, thawing and cooking) as well as quality control of meat and meat products (Riihonen & Linko, 1990).

## 2. Definition

High-frequency heating is considered a thermal process which causes oscillation of water molecules, friction, and resultant heat generation. In several studies it was permitted the radiofrequency heating include 13.56, 27.12, and for microwave applications is 40.68 MHz and 433, 915, 2450, and 5800 MHz. Radiation is a mechanism for heating meat by electromagnetic energy. When the electrons in atom move from a higher to a lower energy state, it sends energy as waves. These waves are not in the same energy level and frequencies. Lower energy electro-magnetic radiation (microwave, radio, TV) occurs as very long waves with frequencies ranging from 300MHz to 300 GHz. Unlike gamma and X-rays, non-ionizing MW energy is sufficient to move the atoms of a molecule, but can not change its chemical bounds. Also MW move at the same speed as light waves essentially. Metallic objects reflect them; some dielectric materials absorb them while others transmit them. Water, carbon, and foods which are high in water are good MW absorbers, while thermoplastics, glass and ceramics allow them to pass through and they can not absorb it completely. When the polar molecules such as $H_2O$ place in a magnetic field, they line up with the field. If the field alternates, the polar molecules alternate at the MW frequency to maintain this alignment. As they rotate, they disrupt H-bonds between adjacent water molecules, generating heat. Freedom of the polar molecules plays an important role in the rate of heat generation. Because the movement of water molecules in ice is restricted, it is a poor MW energy absorber. In an electric field kinetic energy accelerates the ions in solution ($Na^+$, $Cl^-$ and $Ca^{++}$). They crash with other ions and give up heat. The more ions in solution, the higher the kinetic energy release (Decareau and Peterson, 1986; Hugas et al. 2002; Aymerich et al. 2008).

All microwave ovens have a similar design that includes a magnetron device as a power source and a waveguide to bring radiation to a heating chamber. A radiofrequency oven which is known as RF applicator is equipped with a generator coupled with a pair of electrodes. Microwaves can penetrate to a depth of 5 to 7 cm, which results in faster cooking and fewer nutrient changes compared with conventional ovens. It was showed 2 $\log_{10}$ CFU/g reductions in total microbial flora after microwaving meat balls at 2450 MHz frequency, in an 800 W oven, for 300 seconds. Conveyor zed systems have been applied to thawing of meat, in some cases using water surrounding the material to aid temperature

uniformity. Electromagnetic (900-3000 MHz) waves are directed at the product through waveguides without electrodes. Potentially very rapid, the application is limited by thermal instability and penetration depth. Instability results from preferential absorption of energy by warmer sections and by different ingredients, such as fat. Warmer sections may be present at the start of the process; for example, the surface temperature may be warmer than the middle, or they may be produced during the process. In the extreme, such warming can lead to some parts of the food being cooked while others remain frozen. In addition to these problems, the high capital costs of equipments have greatly limited commercial applications. Attempts to avoid runaway heating have involved low-power (and hence longer duration) microwaving, cycling of power on and off to allow equalization periods, and cooling of surfaces with air or liquid nitrogen. Penetration depth depends upon temperature and frequency, being generally much greater at frozen temperatures and greater at lower frequencies (Murano 2003; Yilmaz et al. 2005; James & James, 2010).
As mentioned before, microwaves include waves with wavelengths from 3mm to 3m. In the electromagnetic spectrum, microwaves are located between radio waves at low frequencies and infrared at higher frequencies (Fig. 1). In addition to industrial applications of microwaves in meat tempering, thawing and cooking, microwave was used to detect the fraudulent addition of water in meat products. By using microwave, a nondestructive method was developed for measuring fat in fish and meat, measuring water activity in protein gels and also sensory controlling meat and fish freshness based on the change in dielectric properties (Kent et al. 1993; 2000, 2001 & 2002; Kent 1990; Boggaard et al. 2003; Clerjon et al. 2003; Tejada et al. 2007; Clerjon & Damez 2007).

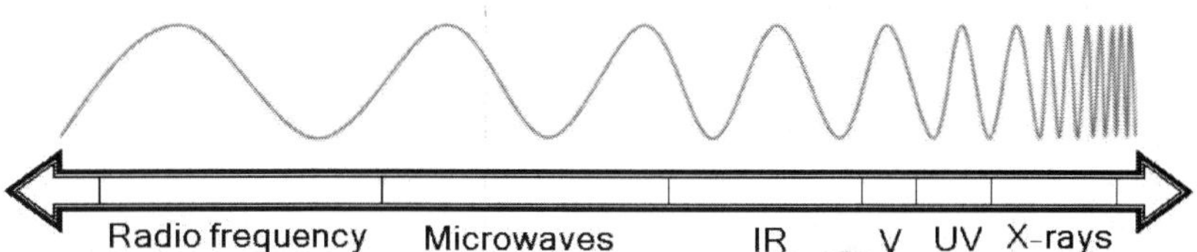


Fig. 1. The electromagnetic spectrum.

### 2.1 Dielectric properties of foods

Microwave with 915MHz frequencies, is used for industrial heating, and 2450 MHz, in domestic microwave ovens worldwide. Ice, having a very small loss factor, is almost transparent to microwaves. Oils are esters of long-chain fatty acids which have much less mobility compared to water molecules in response to oscillating electromagnetic fields. The dielectric constant and loss factor of oil are therefore very small compared to free water. The air voids in some foods reduce the loss factor and increase the penetration depth of microwaves at 915MHz and 2450 MHz. Variation in dielectric properties among high protein products can be as large as among other groups of food at 915 MHz and 2450 MHz. Temperature and salt are important factors on dielectric properties of meat products such as cooked ham and beef. The loss factor of cooked ham is much larger than cooked beef. The penetration depth of microwaves at 915MHz and 2450 MHz in ham is less than 0.5 cm. The dielectric constants and penetration depth of microwaves in these foods decrease, while the loss factors increase, along with increase in temperatures at 915 MHz and 1800 MHz. In general, predicting the dielectric properties of complex food products is difficult and its direct measurement needs to be made over specific composition, temperature and frequency ranges (Wang et al., 2003; Guan et al., 2004).

### 2.2 Water migration adjustment as temperature controller

Adjusting the meat composition in a microwave burger/bun combination can offer flexibility to temperature during microwave heating. It is important to adjust the food with emulsifiers, gums, proteins, and sugars to manage the water content of the meat and the bun separately so that, when processed together in a microwave environment, the bun does not burn while the meat is cooking. With the addition of vegetable fiber, there is swelling of the fiber and fat absorption which interacts with the beef protein to form a complex matrix preventing fat and moisture release (Anderson & Berry, 2001; Lyng et al., 2002; Parker & Vollmer, 2004).

### 2.3 Microwave interactions with food components

Variations in electric fields, food constituents and the location of the food in a MW oven can lead to no uniform heating, allowing for less-than-ideal interaction of food components and survival of microorganisms. A number of techniques to improve uniformity of MW heating, such as rotating and oscillating foods, providing an absorbing medium (water) around the product, cycling the power (pulsed power), and varying the frequency and phase, can improve the situation; however, dielectric properties of the food must be known in order to develop effective processes (Yang & Gunasekaran, 2001; Guan et al., 2004).

## 3. Commercial microwave systems

A wide range of batch and continuous commercial microwave systems for thawing and tempering have been produced since the 1970s and now microwave drying and cooking are used for food processing. The systems were applied to temper meat, fish, butter and berries. A number of the batch systems were used either as one stage of a hybrid microwave/conduction system or to augment large conduction systems by fast tempering of small batches of urgently required material. Industrial systems can be batch or continuous and range in size from small systems that can process one block (25 kg) of food to large continuous tunnels processing up to 8 tones per hour. A single 25 kg block of frozen meat can be tempered in 65 s. the throughput is very dependent on product composition and the final temperature achieved after tempering. The systems are fabricated out of stainless steel for food applications with a uniform load distribution within the oven; a multi-mode cavity applicator develops a uniform heat distribution in the entire oven. The belt material and configuration are selected based on the nature of the product being heated. Each end of the conveyor is provided with a special vestibule to suppress any microwave energy leakage into the environment. To assure uniform heat distribution in a large variety of load configurations, each oven section is provided with a waveguide splitter with dual microwave feed points and mode stirrers (Swain & James, 2005).

## 4. Application of microwave heating for meat processing

### 4.1 Microwave tempering

Tempering of frozen food, can make food easier to slice and reduce drip loss. Frozen meat is heated to just below the freezing point (-2 to -4 °C) can be tempered by MW energy in tempering process then allowed to thaw fully at refrigeration temperature. At temperatures slightly below 0 °C, the outer layer of the meat can absorb significant amounts of energy, resulting in overheating near the surface. Frozen foods thawed in the MW may experience `runaway heating' due to selective heating of the liquid phase (opposed to the crystalline phase of the ice). MW power penetration is greater at 915MHz than at 2450MHz so it is more suitable for heating thick masses of materials. Tempering frozen food has been more successful at 915MHz than at 2450MHz, partly because of surface overheating at the latter frequency. This can be balanced, to some degree, by circulating cold air around the product; however, this may increase drip losses by up to 10%. MW tempering requires less time and space, produces little or no weight loss, increases juice retention, and reduces bacterial growth (Decareau & Peterson, 1986; Rosenberg & Boe, 1987).

There are widespread applications of industrial microwave systems for tempering meat. The blocks processed directly from frozen storage can be acceptably tempered in a batch microwave unit to a mean temperature of approximately -3 °C (range -5 to 0 °C) with no hot spots. In general, microwave tempering of blocks for 8 h in ambient temperatures is unsatisfactory. If it is not possible to produce a uniform power/time combination, surface temperatures, especially at the corners of the meat blocks, rose to unacceptable levels and there was substantial drip loss from thawed surfaces. Trials have to be carried out to determine the correct power and time setting for each type of block in optimal tempering. Blocks sorted into batches of similar type should be processed directly from frozen storage under the predetermined conditions. Successful tempering can be made in some minutes using a microwave system, compared with the 1 to 14 days required in industrial air tempering systems. Continuous conveyorized microwave tempering systems using either a single 60 kW magnetron or two 40 kW magnetrons can temper 2 to 2.5 tones per hour depending upon fat content. In large-throughput operations the continuous microwave tempering plant provides considerable flexibility, in that changes in raw material requirements, for the post-tempering processes, can be accommodated in minutes. Using air tempering systems, at least one day and up to eight days are required to accommodate equivalent changes. Microwave systems bring higher product yield due to reductions in evaporative and drip loss during tempering. Since the majority of conventional plants temper material in a wrapped form, evaporative losses are insignificant, while substantial periods at air temperatures above 0 °C would be required before thawing of surface tissues occurred and drip became apparent (Swain & James, 2005).

### 4.2 Microwave thawing

Microwave thawing utilizes electromagnetic waves directed at the product through waveguides without using electrodes. Whilst the thawing of frozen food by microwave energy is a very fast method, its application is limited by thermal instability. In microwave heating, parts of the food may be cooked whilst the rest remains frozen. The absorption of electromagnetic radiation by frozen food can increase the temperature, especially at -5 °C. During irradiation, if a region of the material is slightly hotter than its surroundings, proportionately more energy will be absorbed. Enthalpy increasing cause the absorption to increases and the unevenness of heating worsens at an ever-increasing rate. If irradiation is continued after reaching the hot spot to its initial freezing point, the temperature rises at appalling rate. Lowering the power density to allow thermal conduction to even out the enthalpy distribution through the food can reduce such runaway heating. Since the main instability tends to occur at the surface, attempts have been made to cool the surface during thawing using air or liquid nitrogen. A hybrid microwave/vacuum thawing system was developed in the 1980s in which boiling of surface water at low temperature was used to cool the surface. The system consisted of a cylindrical vacuum chamber approximately 1 meter in diameter and 1 meter long. The chamber could be evacuated to absolute pressures as low as 10 mbar by a water ring pump in series with a rotary pump. Microwaves at a frequency of 915 MHz were introduced into the chamber via two waveguides positioned near the top at the front and rear of the plant. The microwaves were produced by a 25kW generator, though power had to be limited to 2.5kW to avoid arcing problems in the chamber. A large circular twisted disc was rotated within the chamber during thawing to produce a more even microwave field. After thawing, the average meat temperature was 9.4 °C (ranging from -1 to 23.3 °C between blocks) and the maximum surface temperature measured on any block was 28.4 °C. Weight losses averaged 7.6%, which appeared large, but weight loss in commercial systems was stated to range from 2 to 10%. The overall energy efficiency of the plant was 49% with 24% of the energy being absorbed by the structure of the chamber. The combination of microwave energy and cold air with different ambient temperatures can reduce thawing time and avoid runaway heating during microwave-assisted thawing. The microwave power was cycled on and off using hot and cold points control schemes to a predetermined temperature gradient. When appropriate conditions were used, thawing time was reduced by a factor of seven compared to convective thawing at ambient temperature. The rate of sublimation with low microwave power at the surface resulted low weight losses during the process. The temperatures claimed at the end of thawing processes lasting between 24.5 and 34.25 minutes ranged from -1.1 to -2.0 °C. The shielded region was in fluid communication with the cavity so that thawed material may flow from the cavity into the shielded region. Even if microwave-thawing systems have not been commercially successful but microwave-tempering systems have found successful applications in the meat and dairy industry (Bialod et al., 1978; Virtanen et al. 1997; James, 1984; Yagi & Shibata, 2002; Swain & James, 2005).

### 4.2.1 Vacuum-heat thawing (VHT)

A vacuum-heat thawing system operates by transferring the heat of condensing steam under vacuum to the frozen product. Theoretically, a condensing vapor in the presence of a minimum amount of a non-condensable gas can achieve a surface film heat transfer coefficient far higher than that achieved in water thawing. The principle of operation is that when steam is generated under vacuum, the vapor temperature will correspond to its equivalent vapor pressure. For example, if the vapor pressure is maintained at 1106 $Nm^{-2}$, steam will be generated at 15 °C. The steam will condense onto any cooler surface such as a frozen product. The benefits of latent heat transfer cooking without any problems which would occur at atmospheric pressure. Thawing cycles are very rapid with thin materials, enabling high daily throughputs to be achieved. The advantage of a high h-value is increasing less marked as material thickness, and beef quarters or 25 kg meat blocks require thawing times permitting no more than one cycle per day. The cost of largest capacity units (10-12 tones) can restrict its application. The frozen product is continuously tumbled with vacuum tumble thawing systems while steam under vacuum condenses on the exposed surfaces of the food. Very fast thawing is claimed to be obtained with small bulk-frozen individual products. However, delicate products such as fish fillets cannot be thawed in such systems (Swain & James, 2005).

### 4.2.2 Radio frequency thawing

Heat is produced in the frozen foodstuff during radio frequency thawing because of dielectric losses when a product is subjected to an alternating electric field. A homogeneous regular slab is placed between parallel electrodes in an idealized case of radio frequency heating without any heat exchanging. When an alternating electromotive force (emf) is applied the electrodes thawing and tempering using microwave processing the resulting field in the slab is uniform, so the energy and the resultant temperature rise is identical in all parts of the food. Foodstuffs are not generally in the shape of perfect parallelepipeds; for example, frozen meat consists of at least two components fat and lean. During loading, frozen meats pick up heat from the surroundings, the surface temperature rises and the dielectric system is not presented with the uniform temperature distribution required for even heating. By using a conveyorized system to keep the product moving past the electrodes and/or surrounding the material by water, commercial systems have been produced for blocks of oily fish and white fish. Successful thawing of 13 cm thick meat, and 14 cm thick offal, blocks has also been reported but the temperature range at the end of thawing (44 minutes) was stated to be -2 °C to 19 °C and -2 °C to 4 °C respectively, and the product may not have been fully thawed. To overcome runaway heating with slabs of frozen pork bellies, workers have tried coating the electrodes with lard, placing the bellies in oil, water and saline baths and wrapping the meat in cheesecloth soaked in saline solution. Only the last treatment was successful but even that was not deemed practical. An industrial system was installed to thaw frozen blocks of boned-out poultry in the late 1980s. The continuous plant had a throughput of 450 kg per hour with four 12.5kW generators operating at a frequency of 27.12 MHz providing the heating. It was claimed that the process reduced thawing time from 3 to 4 days to less than 2 hours and cut weight loss from 7 to 0.5% (Satchell and Doty, 1951; Jason & Sanders, 1962; Sanders, 1966; Swain & James, 2005).

### 4.3 Microwave cooking

Microwave or radio frequency cooking are newer methods that have been introduced to the meat industry. Roasts cooked by microwave took less time to reach endpoint temperatures in comparison with conventional methods. Meat cooked with microwaves does not have typical browned surface associated with other methods of cooking because of short cooking time. Radio frequency as a volumetric form of heating is another rapid cooking alternative in which heat is generated within the product, which reduces cooking times and could potentially lead to a more uniform heating. Microwave oven cooking tends to retain higher amounts of vitamins such as retinol, thiamin, and riboflavin compared with earth-oven-cooked meat. The difference in vitamin retention could be due to a higher cooking temperature of earth-oven cooking compared with microwave cooking (Welke et al., 1986; Kumar & Aalbersberg, 2006; Zhang et al. 2006).

In cooking process, the contractile proteins of meat (myosin and actin) will denature and the connective tissue (collagen) under goes to solubilize. MW heating can solubilize more collagen (percentage of hydroxyproline) than does boiling. In general, MW-cooked meat and poultry have higher cooking losses than those cooked by conventional methods (Moody et al., 1978; Riffero & Holmes, 1983; Yarmand & Homayouni, 2009); however, species (beef, pork, and poultry) and cut have important role in losses and palatability. Cooking losses, evaporation and drip losses were greater for steaks cooked in MW convection ovens compared with those cooked in forced air convection or conventional ovens, though they found no difference in juiciness, tenderness, beef flavor, or external color of the steaks.

Evaporative losses were higher for oven dry-roasted beef than for MW- convection roasts, which were higher than MW roasted samples. Drip losses were highest for MW, followed by oven roasted, then by MW-convection roasted samples. Although Instron shear values did not differ, oven dry-roasted samples were judged as tender by taste panelists. Lean meat cook yield was lower for MW-cooked rib eye roasts while it was the same for round and chuck roasts, whether cooked conventionally or in the MW. Tenderness, softness, natural flavor and tenderness (shear force) were unaffected. MW-cooked rib eye roasts were browner and less juicy. Visual color and tenderness improved for meat products cooked in the MW oven. There were no differences in drip loss of MW-cooked chops, though evaporative and total losses were lower than for broiling. Flavor scores were highest for broiled chops and did not differ due to MW power level. Chops cooked at low MW power were juiciest and tenderness varied inversely with cooking rate. Overall acceptability was lowest for chops cooked at high MW power level. Textural analysis showed greater peak forces for hot air oven-cooked products compared to MW-cooked products. MW cooking time was shorter and produced more uniform heating, though flavor was better in hot air oven-cooked products. Cooking time of chicken breasts increase with decreasing power level, but cooking losses are not affected. Both sensory and instrumental tenderness (Instron compression) were best at 60% power level, while juiciness, mealiness and flavor were unaffected by power level. It was reported that convection-MW-cooked chicken was more tender, juicy and acceptable than MW-cooked chicken even if its flavor intensity was similar (Hines et al., 1980; Fulton & Davis, 1983; Payton & Baldwin, 1985; Hammernick-Oltrogge & Prusa, 1987; Howat et al., 1987; Barbeau & Schnepf, 1989; Yarmand & Homayouni, 2009).

Thiamin retention ranged from 77% in conventionally cooked chicken breasts to 98% in MW-cooked chicken legs. Exposure of pre-rigor broiler muscle to MW energy can decrease glycogen metabolism but it has no effect on ATP retention and is not improve tenderness. When meat is cooked in the MW, addition of fluids and other ingredients, and cooking from the frozen state, appear to improve the overall outcome. Curing of buffalo meat using a polyphosphate solution may increase pH from 5.70 to 6.12 and reduce cooking loss from 17.5 to 10.3%. When cooking started from the frozen state, roasts cooked by MW on low power were comparable to those conventionally cooked in sensory quality. Roasts cooked from the frozen state by MW on high power had lower palatability scores (except flavor) and higher shear values. While MW cooking resulted in lower fat contents, and higher water-holding capacity, TBA value and cook yield than conventional oven cooking. Visually, MW-heated beef may appear to be unevenly cooked. MW-thawed and cooked roasts and steaks may be redder in the interior, and lighter and more yellow on the exterior, than their broiled counterparts. The irregular shape of many whole meat cuts results in significantly non-uniform heating. In addition, bone reflects MW, causing overheating in the regions adjacent to the bone (Moody et al., 1978; Dunn & Heath, 1979; Drew et al., 1980; Riffero & Holmes, 1983; Mendiratta et al., 1998; Hoda et al., 2002).

The temperatures experienced by a MW-heated beef slab can vary by as much as 50 °C. This variation is a function of slab thickness (between 1 and 3 cm) and time, or a function of time alone for slabs >4 cm thick. Thermal insulation at the faces of the slab increases this variation. MW is used primarily for reheating precooked meat products, making cook loss even more problematic. Because fluid losses appear to be the primary roadblock in producing precooked, reheatable meat products, adding fluid (enhancement) has been evaluated in an effort to produce a juicy product. Pumping fresh meat to 10% over original weight to produce 1% salt and 0.3% phosphate in the final product can produce an acceptably palatable, value-added, precooked product (boneless pork roast) designed for MW reheating. However, flavor may suffer during MW reheating. Reheating of precooked frozen turkey in MW and conventional ovens tends to create stale, aldehyde-like aroma. Meaty-brothy flavor and aroma were more intense in MW-reheated meat. MW-reheated light meat was flat or bland flavored, less juicy and higher in moisture content than conventionally reheated meat. MW cooking had no effect on chicken flavor or aroma, though MW reheating resulted in less warmed-over flavor than conventional reheating. TBA analyses showed that multiple reheating of dark turkey meat using MW energy retarded lipid oxidation, including that caused by NaCl (2.0%) which was a pro-oxidant compared to KCl (2.0%). MW reheating did not influence warmed-over aroma or flavor or TBA values of cooked stored roast beef. Development of off-flavors may be somewhat species-specific (Cipra & Bowers, 1971; Hsieh & Baldwin, 1984; Steiner et al., 1985; Boles & Parrish, 1990; King & Bosch, 1990).

Cooking fresh meat in a microwave can represent a challenge since the cool air surrounding the meat results in lack of browning. While the meat heats up inside, there is mass transfer from the interior to the outside, resulting in a tough, dry and flavourless product. Ingredient selection can target these problems, such as a salt-based coating to attract microwave energy to the surface of the meat along with a colouring agent, a water binding ingredient (such as starch) to reduce moisture loss and an enzyme to retain tenderness. Flavours can be added for overall improvements. In general it is recommended to keep the salt concentration as low as possible to improve the penetration depth and slow the heating at the surface while providing more uniform heating and avoiding thermal runaway. On the other hand, the addition of salt may be recommended in some cases as it can increase the heating rate at the surface for particular applications (Schiffmann, 1986; Taki, 1991; Schiffmann, 1997).

## 5. Microstructure of microwave processed meats

Quality control methods and microstructural evaluation were employed to study the qualitative parameters of meat and meat products. Among these methods, the environmental scanning electron microscopy (ESEM) has its own advantages for evaluation of microstructural changes in meat especially for comparing the effect of various heat treatments (Arvanitoyannis & van Houwelingen-Koukaliaroglou, 2003; Papadima, et al., 1999; Tzouros & Arvanitoyannis, 2001; Yarmand & Homayouni, 2009; 2010).

### 5.1 Optical microscopy

Microscopical techniques are used to characterize meat structure. Optical microscopy achieved the observation of fat globules distribution and protein gel in emulsion-type buffalo meat sausages. It revealed that caseinate and modified whey form distinct dairy protein gel regions within meat batters, and this could explain their ability to enhance the textural properties of the meat batters compared with the other dairy proteins (Krishnan and Sharma 1990; Barbut, 2006).

### 5.2 SEM and ESEM

Scanning electron microscopy (SEM) and environmental scanning electron microscopy (ESEM) have been used for the examination of living and fresh botanical samples including fungal mycelium and cross sections of stems from different plant sources. ESEM has been known as one of the most interesting new developments in the field of electron microscopy. A number of studies have demonstrated the use of ESEM for hydrated biological samples.

Some research compared unprocessed ESEM specimens and samples prepared by conventional methods. Much research has been done on the study of wool fiber. Processes for industrial wool commence with sheep breeding and conclude with the study of finished fabric. Investigations in many of these processes can be largely enhanced using ESEM microscopes and techniques. Techniques of ESEM have been demonstrated by Baumgarten (1990) and Peters (1990). In this study ESEM has been applied to study the microstructure of goat semimembranosus muscle, using raw muscle as a control and comparing it to different heat treatments including conventional and microwave heating. Scanning electron microscopy was useful to show structure differences in low-fat sausages. The environmental scanning electron microscopy (ESEM) is capable of examine specimens in a gaseous environmental saturated with water vapor and higher resolution micrographs have been illustrated at the presence of gas. The environmental scanning microscope has been described as a scanning electron microscopy technique to retain a minimum water vapor pressure of at least 609 pa in the chamber specimen. ESEM creates the possibility of testing practically any sample which is wet (Fig. 2). The difference of ESEM from conventional SEM is its capacity to examine materials consisting of liquids and oils in their natural situation without any initial preparation for the samples (Danilatos, 1981; Danilatos & Postle, 1982; Danilatos, 1989; Danilatos, 1991; Klose, et al., 1992; O'Brien, et al., 1992; Wallace, et al., 1992; Uwins, 1994; Morin et al. 2004; Cáceres et al. 2008).

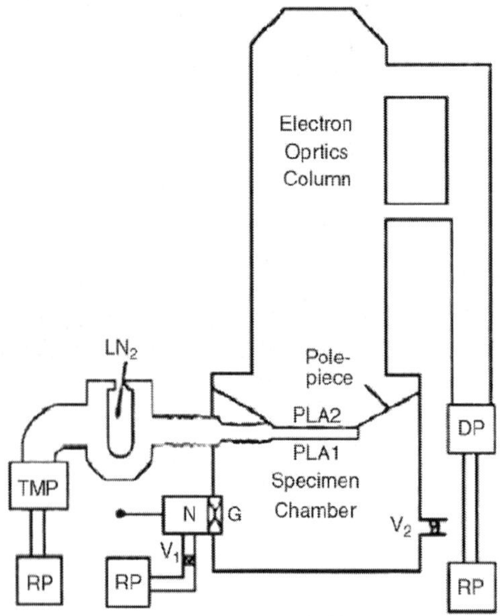


Fig. 2. Diagrammatic representation of a two-stage differential pumping system for an ESEM. Source: (Danilatos, 1991).

The two main parts of the instrument include an electron gun chamber or electron opticsolumn and specimen chamber. The gun chamber is located at the top part of instrument and provides a flow of electrons by heating a tungsten filament, lanthanum hexaboride filament or field emission source. The specimen chamber is capable of working at a very poor vacuum unlike any other types of ESEM which require high vacuum. Therefore this specimen will not dry and ESEM is used to observe specimens in the fresh state. Some important advantages have been shown for the gas around the samples in the specimen chamber. Accumulation of charge on insulating samples can be recognized as a basic advantage for this technique. This phenomenon occurs by ionization of the gases inside the specimen chamber and nowadays much work has been done on it. The gas itself can be employed as a detector in the microscope system which is another advantage for ESEM (Parsons et al., 1974; Moncrieff et al., 1978; Crawford, 1979; DaniJatos, 1983, 1986, 1990; Uwins, 1994).

Study of the microstructure of goat SM muscle was carried out by Environmental Scanning Electron Microscopy (Yarmand & Homayouni, 2009; 2010). The result shows bundles of muscle fiber in raw SM muscle that is parallel to each other (Fig. 3 and Fig. 4).

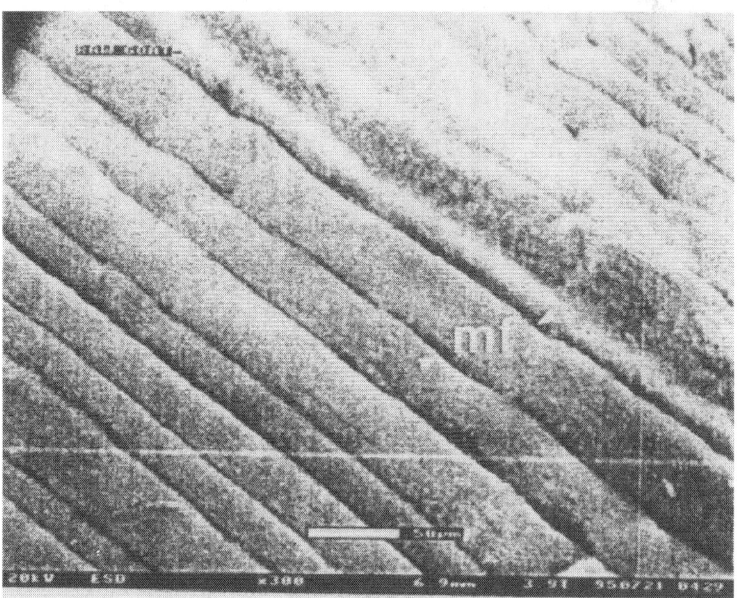


Fig. 3. Illustration of structure of raw goat SM muscle by ESEM (mf: muscle fiber).

The collagen fibers surrounding the muscle fiber are not clear in some Figures, but in Fig. 5 the collagen fibers appear clearer and the myofibril surfaces seem normal. In order to observe collagen more clearly than before, Miller stain was used. This technique increased contrast by increasing the conductivity of the material (Miller, 1971).

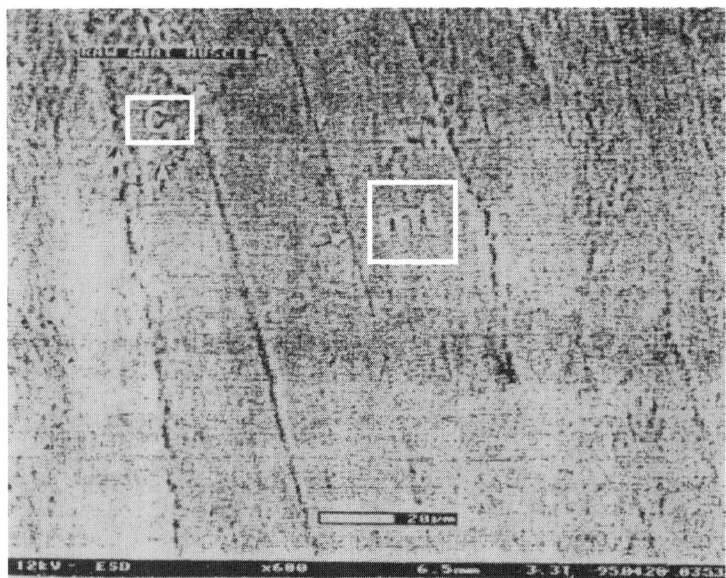


Fig. 4. Another illustration of structure of raw goat SM muscle by ESEM (c: collagen, mf: muscle fiber).

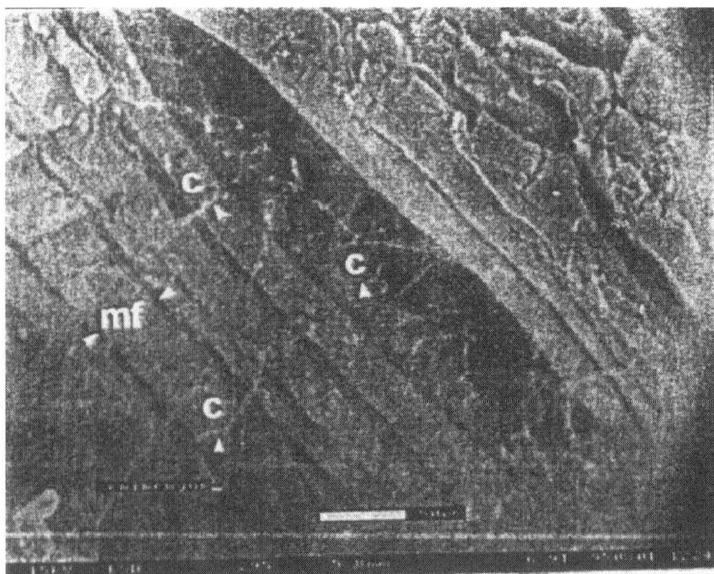


Fig. 5. Structure of SM muscle using Miller stain. Collagen network appears bright (c: collagen, mf: muscle fiber).

As shown in Fig. 6 collagenous fibers are clear and appeared as a white network with covering bundles of SM muscle fiber. Cross section of SM muscle has been shown in Fig. 7 In application of ESEM, it is important to consider possibility of dehydration of the fresh tissues which might occur in the sample chamber. There is also the possibility which might occur from electron beam on the sample if the operating system is not clearly set and adjusted prior to imaging. To prevent dehydration and localized beam damage in correctly specimen, it can be maintained in semi-wet condition by monitoring the system at a lower voltage (2-5 kv) or at lower temperature by applying a cooling step (-20 to 30 °C) and at higher gas pressure up to 20 torr. (Uwins, 1994).

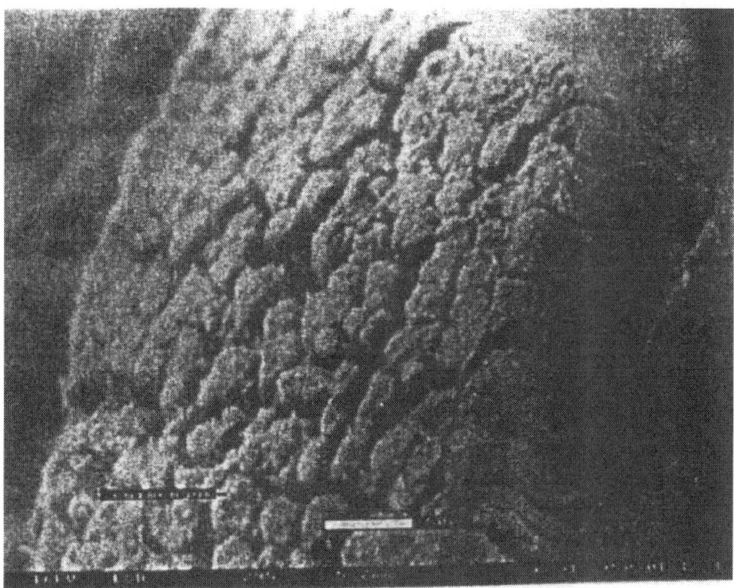

Fig. 6. Cross section of SM muscle.

In the study of the microstructure of SM muscle, more damage was observed when domestic microwave heating was used. This resulted in more shrinkage and breakdown in the SM muscle. The myofibril surface did not seem to be normal but showed little evidence of tissue damage. The cross section of SM (Fig. 8) illustrated the damage at the surface of specimen, further application of the ESEM illustrated that domestic microwave heating causes more physical damage to connective tissue and myofibril elements compared to roasting technique (Fig. 9). Results showed that domestic microwave heating caused little hydrolysis in connective tissue.

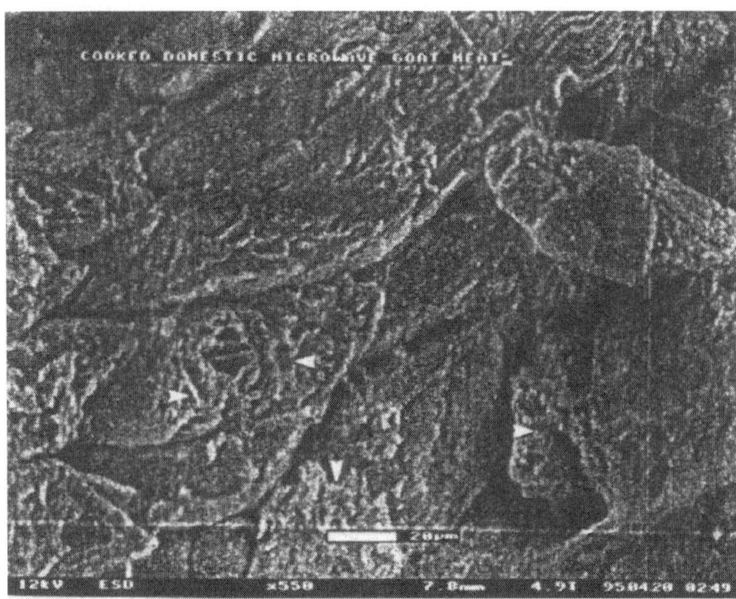


Fig. 7. Environmental Scanning Electron Microscopy of domestic microwave heated SM muscle. Surface damages were shown by arrows in cross section view.

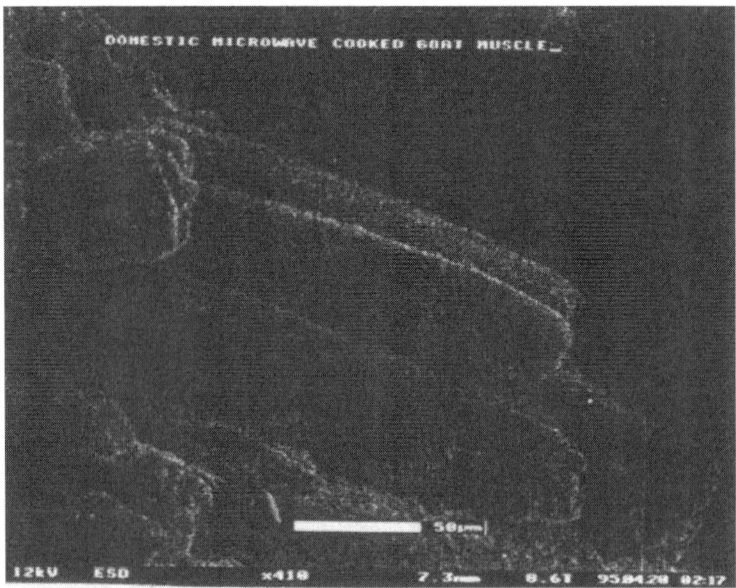


Fig. 8. Another view of domestic microwave cooked (700 W) SM muscle by ESEM.

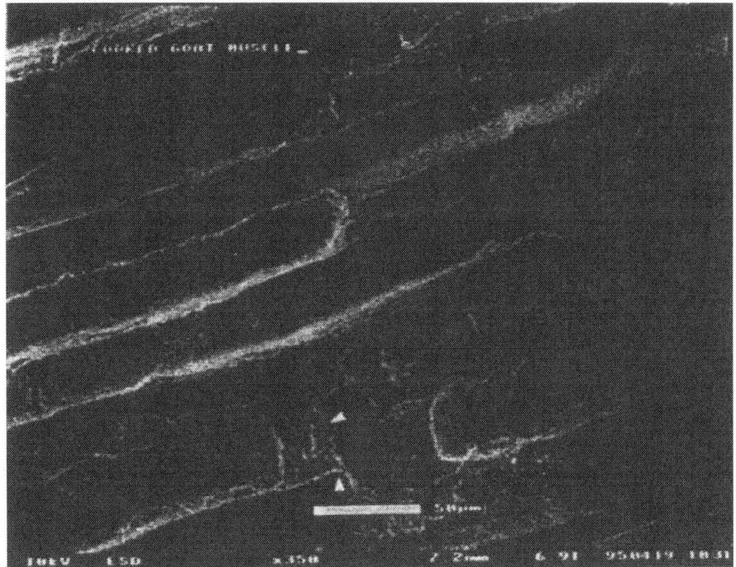

Fig. 9. Effect of conventional heating on SM muscle by ESEM. Transverse breakage was shown by arrows.

As a result of these investigations, it is possible to identify and characterize the fine structure of lamb, veal and goat semimembranosus muscle. Moreover to study and compare the effect of various heat treatments such as conventional and microwaves at two wattages levels (700 and 12000 W) on structure of this muscle. In lamb and goat SM muscles results show that microwave heating cause more structural damage at both levels comparing to conventional heating and heat distribution in microwave heating is responsible for the surface damage to muscle fiber and separation of some parts and denaturation of collagen. While in veal SM muscle because of original texture of the veal, results are different. Domestic microwave heating caused less physical damage to the connective tissue network and myofibril elements comparing to the conventional cooking results. Further research is still required to study the fine structure of semimembranosus or other muscle from various animals (Yarmand & Homayouni, 2009; 2010).

### 5.3 CSLM

Confocal microscopy was described at 1957 (Fig. 10). It has some advantages in composition with conventional light microscopy. Since the laser as a light source was not invented on that time, it was not possible to improve Confocal Microscopy. In addition lack of a suitable computer due to collecting data of image presentation, also causes a problem in the progress of this instrument. Reflection Scanning Light Microscopy (TRSLM) was made as the second confocal system. For the first time the sectioning effect in confocal microscopy (CM) was presented. Generally the major construction of the second microscope was entirely different from the first microscope suggested by Minsky (1957). This instrument was not satisfactory because of poor image quality and resolution.

In spite of the sectioning in TRSLM, the original could not be called as three-dimensional image (Sheppard and Wilson, 1978).
Early presentation of the imaging abilities and the initial improved confocal was displayed in publication by Brakenhoff (1979). Sheppard et al (1977) and Brakenhoff (1979) both employed laser as a light source. They also used a mechanical scanning that moves the specimens through the stationary confocal point. The nucleoid structure of the Echerichia Coli was studied by Brakenhoff microscope. Many experiments described the advantage of the combination of fluorescent techniques and high aperture confocal microscopy for the 3D imaging of biological structure. Fluorescence confocal microscopy was used for food emulsions and nerve cells. Fluorescence confocal microscopy also applied in the investigation of living cells. Calcium probes for the study of calcium spread in cells and searching SH groups (Brakenhoff, 1979; Grynkiewicz et al., 1985; Wijnaeadts Van Resandt et al., 1985).

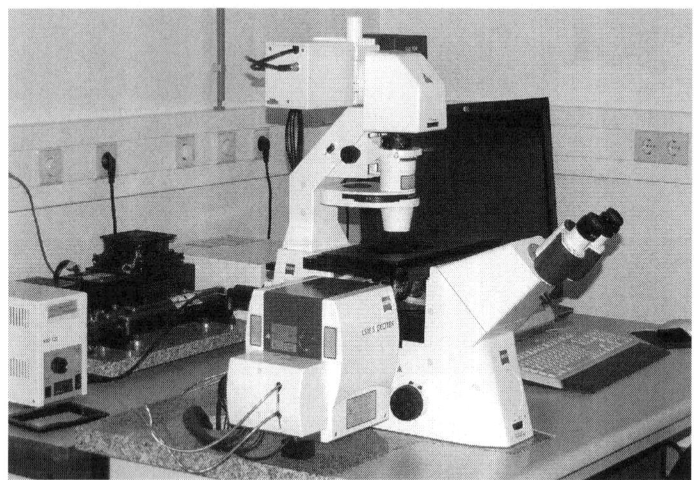

Fig. 10. Confocal microscopy

Confocal scanning laser microscopy has been applied as a new technique in the study of microstructure of food during recent years. sheppard and Gu (1993) applied confocal microscopy for presentation of three dimensional fluorescence images of muscle fibers. According to them, a simple model of striated muscle fiber is quite useful for such image modeling. Fluorescence confocal scanning laser microscopy (FCSLM) model DMRBE was used for study of structure of goat SM muscle after various heat treatments. All specimens were stained with Picro Sirius Red technique. Confocal scanning laser microscopy has several advantages over conventional microscopes. According to Brakenhoff et al., (1988) Confocal scanning light microscopy forms a bridge between conventional light microscopy with its limited resolution (but capable of imaging hydrated, possibly live, specimens) on the one side, and electron microscopy with its higher resolution on the other. The principal of the basic confocal microscope has been explained by Sheppard and Choudhury (1977). This microscope might be operated in two techniques: reflection or in fluorescence. Optical sectioning and 3D reconstruction might be considered as a very important superior of confocal microscopy. Optical sectioning is non-invasive and allows for the imaging of thick objects and a three dimensional reconstruction of the sample. The Performance of confocal microscope suggests new possibilities in microstructure studies of food by the disturbance free visualizing of the three-dimensional internal structure. Such a set of observations from different sections of a muscle specimen will present complete information about the three-dimensional microstructure of semimembranosus muscle. Confocal microscopy is able to provide X-Z images which are images perpendicular to normal image and finally provide ability to observe the depth of materials. Despite that, using CSLM has another benefit that of improved resolution, superior fluorescence and improved quality of sample. Improved sample quality caused by optical sectioning which diminishes limitations of physical sectioning. The system is equipped with fluorescence techniques which exposes the specimen to the fluorescence. Fluorescence techniques provide the ability for the CSLM to choose CSLM the wavelength band of fluorescence light which contributes to better images. Confocal microscopy can show that the size of fat droplets varies with gum type in minced ostrich meat batter. In chicken meat gels, it showed that low-fat protein gels obtained by pressure and containing microbial transglutaminase had a more compact and homogeneous microstructure compared with controls that were pressurized but contained no MTGase (Chattong et al. 2007; Trespalacios and Pla, 2007).

CSLM used in fluorescence gives an excellent perspective of relationship between muscle fiber and collagen network including endomysial collagen fibers. An image of raw SM visualized by three dimensional construction of this muscle is shown in Fig. 11.

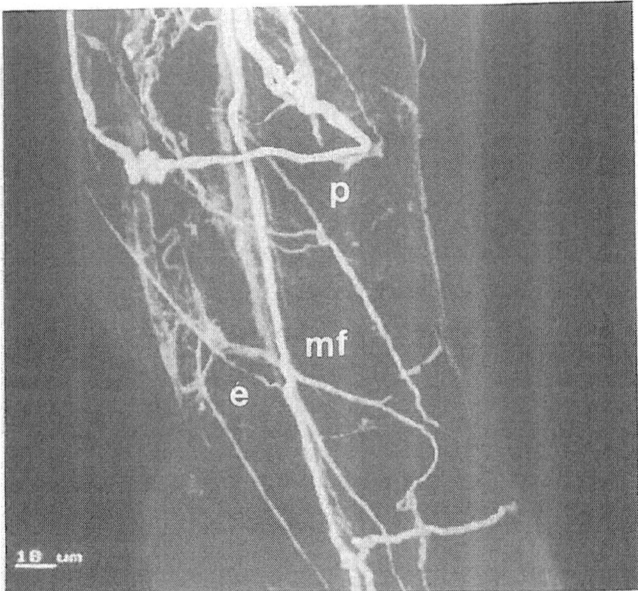


Fig. 11. Illustration of raw SM muscle of goat by FCSLM. Perimysial (P) and endomysial (e) collagenous network were appeared. Mf: muscle fiber.

As mentioned previously optical sectioning through specimen is responsible for the 3D construction of images. Fluorescence confocal scanning laser microscopy (FCSLM) has been used in the study of semimembranosus muscle in goat (Yarmand & Homayouni, 2009; 2010). Four treatments raw (control), conventional and microwave heating at two power levels 700 and 12000 W were examined. As mentioned previously the main advantage of CSLM is the capability of preparing three dimensional pictures which enables the relative distribution of connective tissue to be visualized. Because of the ability to 3D image muscle when heat treatment reduces the length or width of a muscle structure, it will be apparent in the images. Fig. 12 represents how FCSLM build this construction. Overlap of sequence images in Fig.12; give us better view of 3D images which illustrate the depth of material by providing X-Z images in final observation. As described earlier in the principle of confocal scanning electron microscopy, X-Z images are perpendicular to normal images which provide the ability for this microscope technique to observe the depth of material.

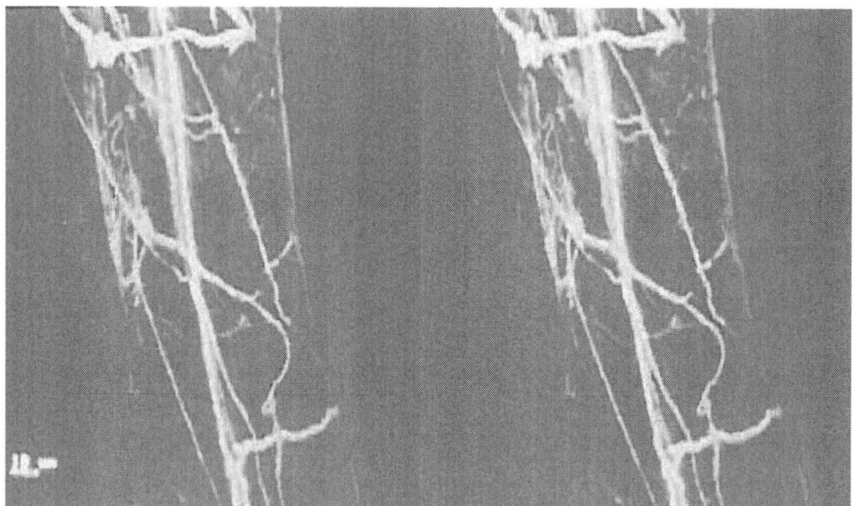

Fig. 12. Sequence of images from raw SM muscle of goat. Overlap of these images visualized better view of 3D images in FCSLM.

Conventional heating micrograph shows less damage to the structure of SM muscle (Fig. 13 and Fig. 14). As it shows in Fig. 13, muscle fiber appeared red which surrounded by a collagenous tissue network. Connective tissue is seen as fluorescent yellow color. Sequences of images indicate the manner of 3D construction of conventional micrographs by FCSLM. In the confocal pictures the signal from collagen fluorescence stained with Picro Sirius Red give a yellow golden hue whereas the muscle fluorescence is illustrated as red. Using the fluorescence property with CSLM allowed us to select the wavelength band of fluorescence light contributing to the image. In this way, we are able to distinguish various structures inside the SM muscle and study their 3-D relationships. Conventional heating micrograph shows less damage to the structure of SM muscle (Fig. 13). As it shows in Fig. 13, muscle fiber appeared red which was surrounded by connective tissue network. Connective tissue seems fluorescent with yellow color.

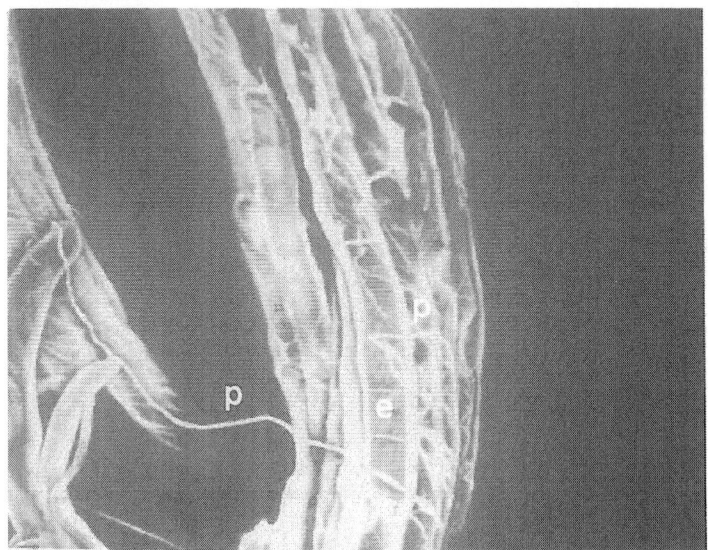


Fig. 13. Conventional heating image of goat SM muscle by FCSLM Perimysial (p) and endomysial (e) connective tissue.

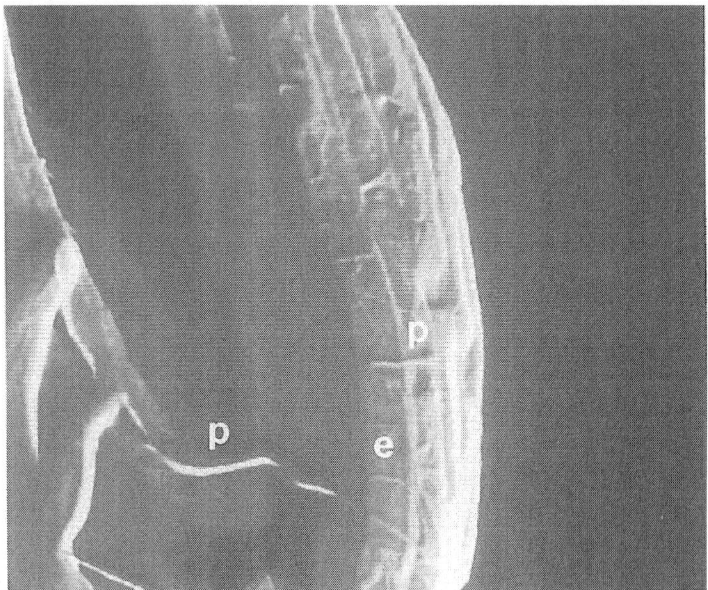


Fig. 14. Another view of conventional heating of goat SM muscle. Perimysial (p) and endomysial (e) connective tissue have been shown by FCSLM.

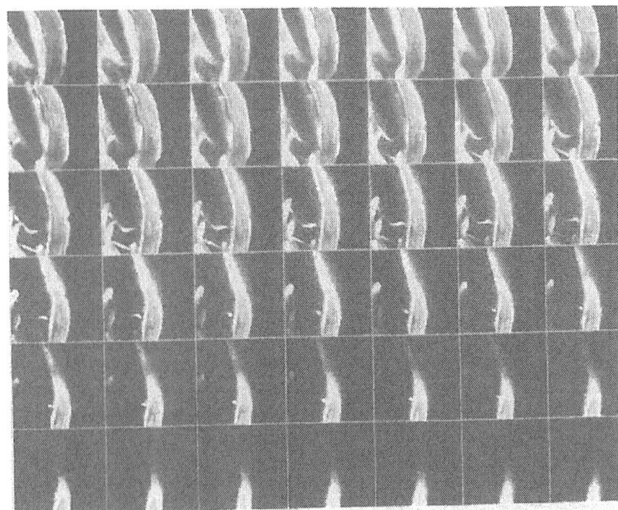

Fig. 15. Demonstration of domestic microwave heated of goat SM muscle by FCSLM. Separation of connective tissue network has been shown in micrograph. Mf: muscle fiber, c: connective tissue.

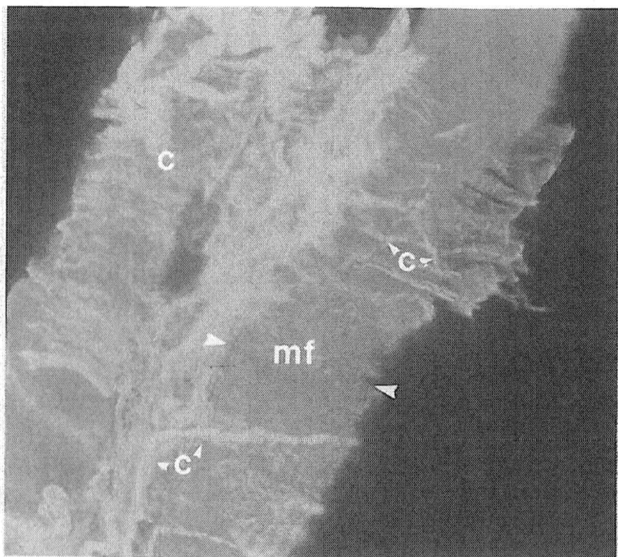


Fig. 16. Presentation of industrial microwave heated of goat SM muscle by FCSLM. Fluorescence Fluorescence property of perimysial (p) and endomysial (e) connective tissue have been illustrated. Arrows indicate degradation and breakdown area of muscle fiber. mf: muscle fiber, scale bar: 10

More damage was observed in the structure of goat SM muscle after using industrial microwave heat treatment. Degradation and breakdown also appeared in microwave heating (Fig. 15 and 16). This result is in agreement with environmental scanning electron microscopy of goat SM muscle. Fig. 17 illustrated another view of industrial microwave cooked. The same result is provided by FCSLM for domestic microwave muscle. The sequence of images provided as a set of observations by fluorescence confocal scanning laser microscope, is shown in Fig. 12. It seems that in raw (control) specimen, fluorescence property of connective tissue is more apparent and it seems that it slightly decreases after heating by various techniques. This reduction was visualized in Fig. 13 and appeared more in microwave heating Fig. 15 and 16. This suggests that particular heat transfer occurred. However, still further researches are needed in order to understand the effect of heat on fluorescence property of collagen.

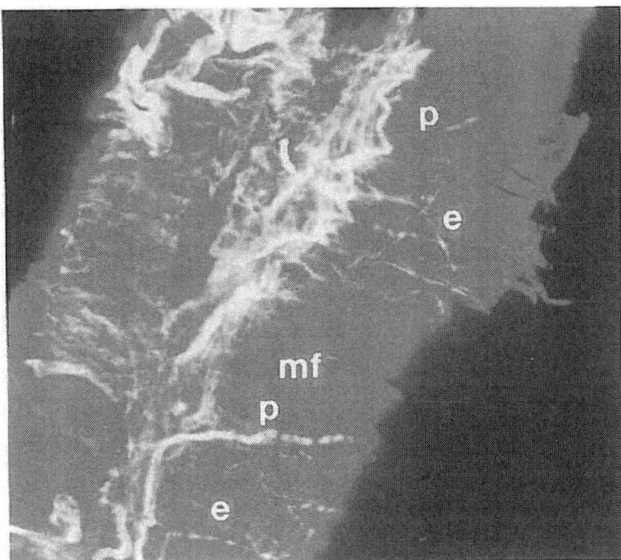


Fig. 17. Another microscopic view of industrial microwave heating of goat SM muscle. mf: muscle fiber, c: connective tissue, scale bar: 10 microns.

## 6. Conclusion

Microstructural evaluation was employed to study the qualitative parameters of meat and meat products. Microscopical techniques are used to characterize meat structure. Optical microscopy can be used for observation of fat globules distribution and protein gel in sausages. Among microscopical methods, the environmental scanning electron microscopy (ESEM) has its own advantages for evaluation of microstructural changes in meat especially for comparing the effect of various heat treatments. Because of large applications of microwave in meat processing (tempering, thawing and cooking) as well as quality control of meat and meat products, more studies are needed to investigate the effects of microwave processing on these products.

## 7. References


Anderson, E. T. & Berry, B. W. (2001). Effects of inner pea fiber on fat retention and cooking yield in high fat ground beef, Food Research International 34(8): 689-694.

Arvanitoyannis, I. S., & van Houwelingen-Koukaliaroglou, M. (2003). Implementation of chemometrics for quality control and authentication of meat and meat products. Critical Reviews in Food Science and Nutrition 43: 173-218.

Avisse, C. & Varaquaux, P. (1977). Microwave blanching of peaches, Journal of Microwave Power 12(1): 73-77.

Aymerich, T., Picouet, P. A. & J. M. Monfort. (2008). Decontamination technologies for meat products. Meat Science 78: 114-129.

Barbeau, W. E. & Schnepf, M. (1989). Sensory attributes and thiamine content of roasting chickens cooked in a microwave, convection microwave and conventional electric oven, Journal of Food Quality 12(3): 203-213.

Barbut, S. (2006). Effects of caseinate, whey and milk powders on the texture and microstructure of emulsified chicken meat batters. LWT-Food Science and Technology 39(6): 660-664.

Baumgarten, N. (1990). Introduction to the environmental scanning electron microscope. Scanning, 12: 36-37.

Beaudry, C., Raghavan, G. S. V. & Rennie, T. J. (2003). Microwave finish drying of osmotically dehydrated cranberries, Drying Technology 21(9): 1797-1810.

Begum, S. & Brewer, M. S. (2001). Chemical, nutritive and sensory characteristics of tomatoes before and after conventional and microwave blanching and during frozen storage, Journal of Food Quality, 24(1): 1-15.

Bhartia, P., Stuchly, S. S. & Hamid, M. A. K. (1973). Experimental results for combinatorial microwave and hot air drying, Journal of Microwave Power 8: 245-252.

Bialod, D., Jolion, M. & Legoff, R. (1978). Microwave thawing of food products using associated surface cooling. Journal of Microwave Power 13(3): 269.

Boggaard, C., Christensen, L. B. & Jespersen, B. L. (2003). Reflection mode microwave spectroscopy for on-line measurement of fat in trimmings. Presented at 49th International Congress of Meat Science and Technology. 31 August-5 September in Campinas, Brazil.

Boles, J. A. & Parrish, F. C. (1990). Sensory and chemical characteristics of precooked microwave-reheatable pork roasts, Journal of Food Science 55(3): 618-620.

Brakenhoff, G. J. (1979). Imaging modes in confocal scanning light microscopy. Journal of microscopy 117: 233-236.

Brakenhoff, G. J., van der Voort, G. T. M., Spronsen, E. A. & Nanniga, N. (1988). Three-dimensional imaging of biological structures by high resolution confocal scanning laser microscopy. Scanning Microscopy 2: 33-40.

Cáceres, E., García, M. L. & Selgas, M. D. (2008). Effect of pre-emulsified fish oil-As source of PUFA n-3 -on microstructure and sensory properties of mortadella, a Spanish bologna-type sausage. Meat Science 80 (2): 183-193.

Chattong, U., Apichartsrangkoon, A. & Bell, A. E. (2007). Effects of hydrocolloid addition and high pressure processing on the rheological properties and microstructure of a commercial ostrich meat product "Yor" (Thai sausage). Meat Science 76 (3): 548-554.

Chen, S. C., Collins, J. L., Mccarty, I. E. & Johnston, M. R. (1971). Blanching of white potatoes by microwave energy followed by boiling water, Journal of Food Science, 36: 742-743.

Cipra, J. S. & Bowers, J. A. (1971). Flavour of microwave- and conventionally-reheated turkey, Poultry Science 50(3): 703-706.

Clerjon, S., & Damez, J. L. (2007). Microwave sensing for meat and fish structure evaluation. Meat Science and Technology 18 (4): 1038-1045.

Clerjon, S., Daudina, J. D. & Damez, J. L. (2003). Water activity and dielectric properties of gels in the frequency range 200 MHz-6 GHz. Food Chemistry 82 (1): 87-97.

Crawford, C. K. (1979). Charge neutralization using very low energy ions. Scanning Electron Microscopy 2: 31-46.

Danilatos, G. D. (1981). The examination of fresh or living plant material in an environmental scanning electron microscope. Journal of Microscopy 121: 235-238.

Danilatos, G. D. (1983). A gaseous detector device for an environmental SEM. Micron and Microscopica Acta 14: 307-318.

Danilatos, G. D. (1989). Environmental SEM; A new instrument, a new dimension. Proceedings of the Institute of Physics Electron Microscopy and Royal Microscopy Society Conference 98(1): 455-458.

Danilatos, G. D. (1990). Theory of the gaseous detector device in the ESEM. Advances in Electronics and Electron Physics 78: 1-102.

Danilatos, G. D. (1991). Review and outline of environmental SEM at present. Journal of Microscopy 162: 391-402.

Danilatos, G. D., & Postle, R. (1982). The environmental scanning electron microscope and its application. Scanning Electron Microscopy 1: 1-16.

Decareau, R. V. & Peterson, R. A. (1986). Current state of microwave processing. In *Microwave-Processing and Engineering*, Ellis Horwood, Chichester, UK, pp. 22-23.

Drew, F., Rhee, K. S. & Carpenter, Z. L. (1980). Cooking at variable microwave power levels, Journal of American Dietetic Association 77(4): 455-459.

Dunn, N. A. & Heath, J. L. (1979). Effect of microwave energy on poultry tenderness, Journal of Food Science 44(2): 339-342.

European Chilled Food Federation (1996). Guidelines for Good Hygienic Practice in the Manufacture of Chilled Foods. London, European Chilled Food Federation, pp. 59.

Farber, J. M., Daoust, J. Y., Diotte, M., Sewell, A. & Daley, E. (1998). Survival of Listeria spp. on raw whole chickens cooked in microwave ovens. Journal of Food Protection 5: 1465-1469.

Fulton, L. & Davis, C. (1983). Roasting and braising beef roasts in microwave ovens, Journal of American Dietetic Association 83(5): 560-563.

Grynkiewicz, G., Poenie, M. & Tsien, R. Y. (1985). A new generation of $Ca^{2+}$ indicators with greatly improved fluorescence properties. Journal of Biologival Chemistry 260: 3440-3450.

Guan, D., Cheng, M., Wang, Y. & Tang, J. (2004). Dielectric properties of mashed potatoes relevant to microwave and radio-frequency pasteurization and sterilization processes, Journal of Food Science 69(1): 30-37.

Hammernick-Oltrogge, M. & Prusa, K. J. (1987). Sensory analysis and Instron measurements of variable-power microwave-heated baking hen breasts, Poultry Science 66(9): 1548-1551.

Hines, R. C., Ramsey, C. B. & Hoes, T. L. (1980). Effects of microwave cooking rate on palatability of pork loin chops, Journal of Animal Science 50(3): 446-451.

Hoda, I, Ahmad, S, & Srivastava, P. K. (2002). Effect of microwave oven processing, hot air oven cooking, curing and polyphosphate treatment on physico-chemical, sensory and textural characteristics of buffalo meat products, Journal of Food Science and Technology 39(3): 240-245.

Hong, N., Yaylayan, V., Raghavan, G. S. V., Parea, J. & Bealanger, J. (2001). Microwave-assisted extraction of phenolic compounds from grape seeds, Natural Product Letters 5(3): 197-204.

Howat, P. M., Gros, J. N., Mcmillin, K. W., Saxton, A. M. & Hoskins, F. (1987). Comparison of beef blade roasts cooked by microwave, microwave convection and conventional oven, Journal of Microwave Power 22(2): 95-98.

Hsieh, J. H. & Baldwin, R. E. (1984). Storage and microwave reheating effects on lipid oxidation of roast beef, Journal of Microwave Power 19(3): 187-194.

Hugas, M., Garriga, M. & Monfort, J. M. (2002). New mild technologies in meat processing: High pressure as a model technology. Meat Science 62: 359-371.

James, C. & James, S. J. (2010).Freezing/Thawing in: F. Toldrá (ed.). *Handbook of Meat Processing*, A John Wiley & Sons, Inc., Publication, pp.105-124.

James, C. (2000) Optimising the microwave cooking of bacon, Meat and Poultry 2000 Seminar, CCFRA (Campden & Chorleywood Food Research Association), 17 July 2000.

James, S. J. & Crow, N. (1986). Thawing and tempering: Industrial practice. Meat Thawing/Tempering and Product Quality, IFR-BL: Subject Day.

James, S. J. & James, C. (2002). Thawing and tempering. In Meat Refrigeration, Woodhead Publishing, Cambridge, 159-190.

James, S. J. (1984). Thawing meat blocks using microwaves under vacuum. Proceedings of the 30th European Meeting of Meat Research Workers, Bristol, Paper 2: 61-62.

James, S. J. (1986). Microwave thawing and tempering of frozen meat. BNCE Heating and Processing 1-3000 MHz, Cambridge, 4-7.

Jason A. C. & Sanders, H. R. (1962). Dielectric thawing of fish. Food Technology 16: 101-107.

Jason A. C. (1974). Microwave thawing-some basic considerations. In Meat Freezing; Why and How? Meat Research Institute Symposium 3: 31-43

Jaynes, H. O. (1975). Microwave pasteurization of milk, Journal of Milk and Food Technology 38: 386-387.

Kent, M. (1990). Hand-held instrument for fat/water determination in whole fish. Food Control 1: 47-53.

Kent, M., Knökel, R. Daschner, F. & Berger, U. K. (2001). Composition of foods including added water using microwave dielectric spectra. Food Control 12: 467-482.

Kent, M., Knökel, R., Daschner, F. & Berger, U. K. (2000). Composition of foods using microwave dielectric spectra. European Food Research and Technology 210: 359-366.

Kent, M., Lees, A. & Roger, A. (1993). Estimation of the fat content of minced meat using a portable microwave fat meter. Food Control 4: 222-227.

Kent, M., Peymann, A., Gabriel, C. & Knight, A. (2002). Determination of added water in pork products using microwave dielectric spectroscopy. Food Control 13: 143-149.

King, A. J. & Bosch, N. (1990). Effect of NaCl and KCl on rancidity of dark turkey meat heated by microwave, Journal of Food Science 55(6): 1549-1551.

Klose, M. J., Webb, R. I. & Teakle, D. S. (1992). Studies on the virus pollen association of tobacco streak virus using ESEM and MDD techniques. Journal of Computer Assisted Microscopy 4: 213-220.

Krishnan, K. R. & Sharma, N. (1990). Studies on emulsion-type buffalo meat sausages incorporating skeletal and offal meat with different levels of pork fat. Meat Science 28 (1): 51-60.

Kumar, S. & Aalbersberg, B. (2006). Nutrient retention in foods after earth-oven cooking compared to other forms of domestic cooking 2. Vitamins. Journal of Food Composition and Analysis 19: 311-320

Levre, E. & Valentini, P. (1998). Inactivation of Salmonella during microwave cooking. Zentralblatt fuÈr Hygiene und Umweltmedizin 6: 431-436.

Lin, C. C. & Li, C. F. (1971). Microwave sterilization of oranges in glass pack, Journal of Microwave Power 6: 45-48.

Lyng, J. G., Scully, M., Mckenna, B. M., Hunter, A. & Molloy, G. (2002). The influence of compositional changes in beef burgers on their temperatures and their thermal and dielectric properties during microwave heating, Journal of Muscle Foods 13(2): 123-142.

Maurer, R. L., Trembley, M. R. & Chadwick, E. A. (1971). Use of microwave energy in drying alimentary pastes, Proceedings of Microwave Power Symposium IMPI, Monterey, CA, USA.

Meisel, N. (1972). Tempering of meat by microwaves, Microwave Energy Application Newsletter 5(3): 3-7.

Mendiratta, S. K., Kumar, S., Keshri, R. C. & Sharma, B. D. (1998). Comparative efficacy of microwave oven for cooking of chicken meat, Fleischwirtschaft 78(7): 827-829.

Miller, P. G. (1971). An elastin stain. Medical Laboratory Technology 28: 148-149.

Moncrieff, D. A., Robinson, V. N. E., & Harris, L. B. (1978). Charge neutralization of insulating surfaces in the SEM by gas ionization. Journal of Physics D: Applied Physics 11(17): 2315-2325.

Moody, W. G., Bedeau, C. & Langlois, B. E. (1978). Beef thawing and cookery methods: Effect of thawing and cookery methods, time in storage and breed on the microbiology and palatability of beef cuts, Journal of Food Science 43(3): 834-838.

Morin, L. A., Temelli, F. & McMullen. L. (2004). Interactions between meat proteins and barley (Hordeum spp.): B-glucan within a reduced-fat breakfast sausage system. Meat Science 68 (3): 419-430.

Murano, P. (2003). Understanding Food Science and Technology. Belmont, Calif.: Thompson Wadsworth.

Nelson, S. O. (1973). Insect control studies with microwaves and other radio frequency energy, Bulletin of the Entomological Society of America 19(3): 157-163.

Nykvist, W. E. & Decareau, R. V. (1976). Microwave meat roasting, Journal of Microwave Power, 11(1): 3-24.

O'Brien, G. P., Webb, R. K., Uwins, P. J. R., Desmarchelier, P. M., & Imrie, B. (1992). Suitability of the environmental scanning electron microscope for studies of bacteria on Mung bean seeds. Journal of Computer Assisted Microscopy 4: 225-229.

Papadima, S. N., Arvanitoyannis, I. S., & Bloukas, J. G. (1999). Chemometric model for describing Greek traditional sausages. Meat Science 51: 271-277.

Parker, K. & Vollmer, M. (2004). Bad food and good physics: the development of domestic microwave cookery, Physics Education 39(1): 82-90.

Parsons, D. F., Matriccardi, V. R., Moretz, R. C., & Turner, J. N. (1974). Electron microscopy and diffraction of wet unstained and unfixed biological objects. Advances in Biological and Medical Physics 15: 161-271.

Payton, J. & Baldwin, R. E. (1985). Comparison of top round steaks cooked by microwave-convection, forced-air convection and conventional ovens, Journal of Microwave Power 20(4): 255-259.

Peters, K. R. (1990). Introduction to the technique of environmental scanning electron microscopy. Scanning 90 abstracts (pp. 71-72). FACMS Inc.

Ramesh, M. N., Wolf W., Tevini, D. & Bognar, A. (2002). Microwave blanching of vegetables, Journal of Food Science 67(1): 390-398.

Riffero, L. M. & Holmes, Z. A. (1983). Characteristics of pre-rigor pressurized versus conventionally processed beef cooked by microwaves and by broiling, Journal of Food Science 48(2): 346-350.

Riihonen, L. & Linko, P. (1990). Effect of thawing on the quality of frozen mechanically deboned meat. Journal of Agricultural Science in Finland 62(5): 407-415.

Rosenberg, U. & Boe, G. L. W. (1987). Microwave thawing, drying and baking in the food industry, Food Technology 41(6): 85.

Sanders, H. R. (1966). Dielectric thawing of meat and meat products. Journal of Food Technology 1: 183.

Satchell, F. E. & Doty, D. M. (1951). High-frequency dielectric heating for defrosting frozen pork bellies. Bulletin of the American Meat Institute Foundation 12: 95.

Schiffman, R. F. (1986), Food product development for microwave processing, Food Technology 40(6): 94.

Schiffman, R. F. (1997). Microwave technology, a half-century of progress, Food Product Design pp. 1-15.

Severini, C., Baiano, A., De Pilli, T., Romaniello, R. & Derossi, A. (2004). Microwave blanching of sliced potatoes dipped in saline solutions to prevent enzymatic browning, Journal of Food Biochemistry 28(1): 75-88.

Sheppard C. J. R. & Gu, M. (1993). Modeling of three-dimensional fluorescence images of muscle fibers: an application of the three-dimensional optical transfer function. Journal of microscopy 168: 339-345.

Sheppard, C. J. R. & Choudhury, A. (1977). Image formation in the scanning microscope. Optica Acta 24: 1051-1073.

Sheppard, C. J. R. & Wilson, T. (1978). Depth of field in the scanning microscope. Optics Letter 3: 115-117.

Shivhare, U. S., Raghavan, G. S. V. & Bosisio, R. G. (1994). Modeling the drying kinetics of maize in a microwave environment, Journal of Agricultural Engineering Research 57: 199-205.

Sipahioglu, O., Barringer, S. B., TAUB, T. & Yang, A. P. P. (2003). Characterization and modeling of dielectric properties of turkey meat, Journal of Food Science 68(2): 521-527.

Sobiech, W. (1980). Microwave-vacuum drying of sliced parsley root, Journal of Microwave Power 15(3): 143-154.

Steiner, J. B., Johnson, R. M. & Tobin, B. W. (1985). Effect of microwave and conventional reheating on flavor development in chicken breasts, Microwave World 6(2): 10-12.

Sumnu, G. (2001). A review on microwave baking of foods, International Journal of Food Science and Technology 36: 117-127.

Swain, M. & James, S. (2005). Thawing and tempering using microwave processing in: Schubert, H. & Regier, M. (ed) The microwave processing of foods, Woodhead Publishing Limited, pp. 189-206.

Taher, B. J. & Farid, M. M. (2001). Cyclic microwave thawing of frozen meat: experimental and theoretical investigation. Chemical Engineering and Processing 40(4): 379-389.

Taki, G. H. (1991). Functional ingredient blend produces low-fat meat products to meet consumer expectations, Food Technology 71-74.

Tejada, M., De las Heras, C. & Kent, M. (2007). Changes in the quality indices during ice storage of farmed Senegaleses ole (Solea senegalensis). European Food Research Technology 225: 225-232.

Trespalacios, P. & Pla, R. (2007). Simultaneous application of transglutaminase and high pressure to improve functional properties of chicken meat gels. Food Chemistry 100(1): 264-272.

Tulasidas, T. N., Raghavan, G. S. V. & Mujumdar, A. S. (1995). Microwave drying of grapes in a single mode cavity at 2450 MHz I: drying kinetics, Drying Technology, 13: 1949-1972.

Tzouros, N. E., & Arvanitoyannis, I. S. (2001). Agricultural produces: Synopsis of employed quality control methods for the authentication of foods and application of chemometrics for the classification of foods according to their variety or geographical origin. Critical Reviews in Food Science and Nutrition 41: 287-319.

Uwins, P. J. R. (1994). Environmental scanning electron microscopy. Materials Forum, 18: 51-75.

Virtanen, A. J., Goedeken, D. L. & Tong, C. H. (1997). Microwave assisted thawing of model frozen foods using feed-back temperature control and surface cooling. Journal of Food Science 62(1): 150-154.

Wallace, H. M., Uwins, P. J. R. & McChonchie, C. A. (1992). Investigation of stigma interaction in Macadamia and Grevillea using ESEM. Journal of Computer Assisted Microscopy 4: 231-234.

Wang, Y., Wig, T., Tang, J. & Hallberg, L. M. (2003). Dielectric properties of food relevant to RF and microwave pasteurization and sterilization, Journal of Food Engineering 57(3): 257-268.

Welke, R. A., Williams, J. C., Miller, G. J. & Field, R. A. (1986). Effect of cooking method on the texture of epimysial tissue and rancidity in beef roasts. Journal of Food Science 51(4): 1057-1060.

Wijnaendts-van-Resandt, R. W., Marsman, H. J. B. Kaplan, R. Davoust, J. Stelzer, E. H. K. & Stricker, R. (1985). Optical fluorescence microscopy in three dimensions: microtomoscopy. Journal of microscopy 138: 29-34.

Williams, O. J., Raghavan, G. S. V., Orsat, V. & Dai, J. (2004). Microwave assisted extraction of capsaicinoids from capsicum fruit, Journal of Food Biochemistry 28(2): 113-122.

Yagi, S. & Shibata, K. (2002). Microwave defrosting under reduced pressure. US Patent Application 20020195447.

Yang, H. W. & Gunasekaran, S. (2001). Temperature profiles in a cylindrical model food during pulsed microwave heating, Journal of Food Science 66(7): 998-1004.

Yarmand, M. S. & Homayouni, A. (2009). Effect of microwave cooking on the microstructure and quality of meat in goat and lamb. Food Chemistry, 112, 782-785.

Yarmand, M. S. & Homayouni, A. (2010). Quality and microstructural changes in goat meat during heat treatment. Meat Science, 86, 451-455.

Yarmand, M. S. & Sarafis, V. (1997). An improved maceration technique with polarized light microscopy in the study of muscle. Egyptian Journal of Food Science, 25: 425-431.

Yilmaz, I., Arici, M. & Gümüs, T. (2005). Changes of microbiological quality in meatballs after heat treatment. European Food Research and Technology 221: 281-283.

Zagrodzki, S., Niedzielski, Z. & Kulogawska, A. (1977). Meat thawing under controlled conditions. Acta Alimentaria Polonica 3(21): 1-11.

Zhang, H., FU, C., Zheng, X., XI, Y., Jiang, W. & Wang, Y. (2004). Control of post harvest rhizopus rot of peach by microwave treatment and yeast antagonist, European Food Research and Technology 218(6): 568-572.

Zhang, L., Lyng, J. G. & Brunton, N. P. (2006). Quality of radio frequency heated pork leg and shoulder ham. Journal of Food Engineering 75: 275-287.

# Part 4

## Technology

# 7

# Reproducibility and Scalability of Microwave-Assisted Reactions

Ángel Díaz-Ortiz et. al.*

*Facultad de Ciencias Químicas, Universidad de Castilla-La Mancha, Ciudad Real*
*Johnson & Johnson Pharm. Res. and Develop., Division of Janssen-Cilag S.A., Toledo*
*Spain*

## 1. Introduction

High-speed microwave synthesis has attracted a considerable amount of attention in the last two decades.

Since 2000 the number of publications related to MAOS has increased dramatically. One of the reasons for the increased interest in the use of microwave heating was the introduction, at the dawn of the 21st Century, of dedicated monomode and multimode instruments with appropriate temperature and pressure controls, a development that allows reproducibility of results. However, when the microwave methodology was introduced twenty years ago, most reactions were performed in domestic ovens without appropriate temperature control.

In recent years, a number of reports have disclosed the reproducibility of results between monomode and multimode microwave instruments for solution chemistry, the application in parallel and combinatorial chemistry, and some comparisons between homogeneous and heterogeneous systems and reactions performed in closed or open vessels. In the same way, it has been shown that solvent-free reactions can be updated and applied in microwave reactors and the influence of the polarity on the reproducibility of these processes has been highlighted. Furthermore, computational calculations can assist in explaining and/or predicting whether a reaction will be reproducible or not.

Possible approaches to scale-up microwave-assisted reactions include continuous-flow reactors, small-scale batch stop-flow protocols or large-scale single-batch reactors. However, the scalability of microwave-induced processes represents an important obstacle due to numerous factors, but principally owing to increased heat loss, changes in absorption, limited penetration depth of the radiation into the reaction medium (only a few centimetres at 2.45 GHz) and the additional reflection of the microwaves. Such intrinsic complications have prevented the use of microwave reactors for volumes larger than a few litres, thus inhibiting the use of this approach for the production of larger quantities of material.

Alternatively, some researchers and microwave manufacturers have explored the potential of continuous flow microwave systems. Such an approach offers many advantages in terms

* Antonio de la Hoz, Jesús Alcázar, José R. Carrillo, M. Antonia Herrero, Juan de M. Muñoz, Pilar Prieto and Abel de Cózar.
*Facultad de Ciencias Químicas, Universidad de Castilla-La Mancha, Ciudad Real; Johnson & Johnson Pharm. Res. and Develop., Division of Janssen-Cilag S.A., Toledo; Spain.*

of processing versatility, safety, reaction monitoring and optimization. Importantly, this processing method also avoids the limitations associated with the design of scaled microwave cavities, including the associated costs.

In this chapter we intend to highlight the most important reports studying and solving problems concerning the reproducibility and scalability of microwave-assisted processes, from the early reactions performed in unmodified domestic ovens to the recent syntheses utilizing dedicated apparatus or suitable flow instrumentation.

## 2. Reproducibility under microwaves

The microwave applicator is the component of a processing system in which energy is applied to the product to be heated. In a domestic oven the radiation power is generally controlled by on-off cycles of the magnetron. Large switching periods are undesirable in chemistry because of the absence of irradiation between switching steps. In this situation some areas may receive large amounts of energy whereas others may receive very little energy. Typically it is not possible to monitor the reaction temperature in a reliable way. Heating organic solvents in open vessels can lead to violent explosion induced by electric arcs inside the cavity or sparking as a result of switching of the magnetron. On the other hand, working with sealed vessels without monitoring of the pressure can also lead to serious accidents. For these reasons, the use of such equipment in chemistry cannot be recommended. In contrast, dedicated microwave reactors for synthesis feature built-in magnetic stirrers, direct temperature measurement of the reaction mixture with the aid of fibre-optic probes or IR sensors, and software that allows on-line temperature/pressure control by regulation of the microwave power (Kappe & Stadler, 2005).

Multimode instruments allow the use of bigger reaction vessels or an increase in reaction throughput by the use of multivessel rotors for parallel synthesis or scale-up. A general disadvantage associated with multimode instruments is the poor level of performance in small-scale experiments (<3 mL). While the generated microwave power is high (1000–1400 W), the power density of the field is generally rather low.

Monomode instruments generate a single, highly homogeneous energy field of high power density. These systems couple efficiently with small samples and the maximum power output is in most cases limited to 300 W (Ondruschka et al., 2006).

The question of reproducibility and scale-up will always involve the question of reaction conditions. In addition, the reaction medium (phase) plays a much more important role for this kind of power input compared with classical reactions. Besides the molecular mass, the polarity of the reaction mixture is a critical factor for the absorption of microwave power.

### 2.1 Reproducibility in scaled-up microwave-assisted reactions

One of the limitations of microwave scale-up technology is the restricted penetration depth of microwave irradiation into absorbing materials. This means that solvent or reagents in the centre of large reaction vessel are heated by convection and not by direct 'in core' microwave dielectric heating. For this reason, many researchers have studied the scale-up and reproducibility in batch of the reactions performed on a small scale.

In 1998 Hamelin studied the large-scale synthesis of dioxolanes, dithiolanes and oxathiolanes from 2,2-dimethoxypropane and 3,3-dimethoxypentane in the absence of solvent. These reactions had previously been performed in a Prolabo Synthewave 402 apparatus on a 10 mmol scale and were reproduced in a Synthewave 1000 reactor on a 2 mol scale (Perio et al., 1998) (Scheme 1). The authors found that the 2-mol scale was easier than the 10-mmol scale owing the possibility of continuous distillation of methanol under irradiation in the Synthewave 1000 with a packed column, a set-up that is not possible in the smaller Synthewave 402.

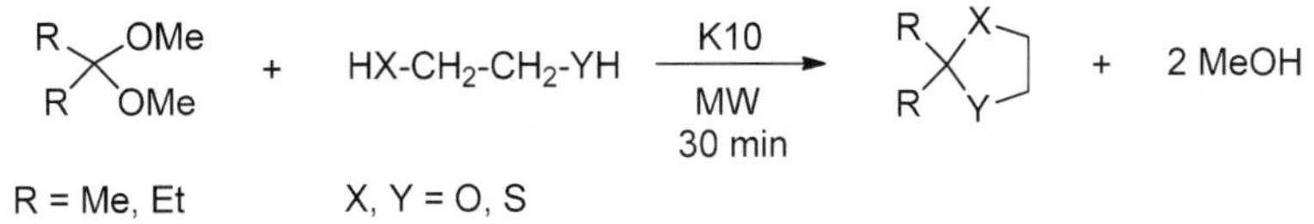

$R_2C(OMe)_2$ + HX-$CH_2$-$CH_2$-YH —(K10, MW, 30 min)→ cyclic product + 2 MeOH

R = Me, Et    X, Y = O, S

Scheme 1. Synthesis and scale-up of dioxolanes, dithiolanes and oxathiolanes

Similar results were reported by Loupy for the reproducibility and scale-up of several reactions (potassium acetate alkylation, regioselective phenacylation of 1,2,4-triazole, deethylation of 2-ethoxyanisole, and typical examples in carbohydrate chemistry) to several hundred grams in a larger batch reactor (Synthewave 1000), with yields equivalent to those obtained under similar conditions (temperature, reaction time) in laboratory-scale experiments (Synthewave 402) (Cléophax et al., 2000).

Luthman studied the appropriate reaction conditions in terms of choice of solvent, reaction temperature and reaction time to allow the fast and reproducible production of Mannich bases from small (2 mmol) to large (40 mmol) scale reactions in moderate to high yields (18–60%) and high purity (F. Lehmann et al., 2003) (Scheme 2).

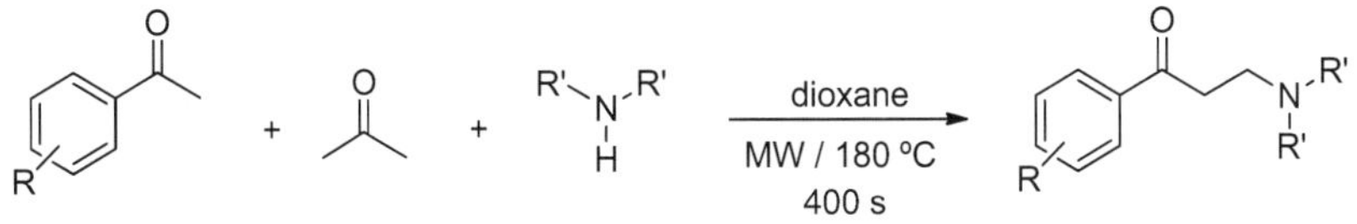

dioxane, MW / 180 °C, 400 s

Scheme 2. Microwave-assisted Mannich reactions using substituted acetophenones

Two different microwave systems designed for large-scale operation, such as the Anton Paar 3000 multimode batch reactor (8 × 100 mL PTFE vessels in a ceramic vessel jacket, filling volume of 60–70 mL for each vessel) and the CEM Voyager SF stop-flow reactor (80 mL glass vessel, filling volume 50 mL), were evaluated by Lehmann (H. Lehmann & LaVecchia, 2005) for special use in a kilolab. As a model reaction, the aromatic substitution of aryl halides with nucleophiles like phenolates or amines was chosen (Scheme 3).

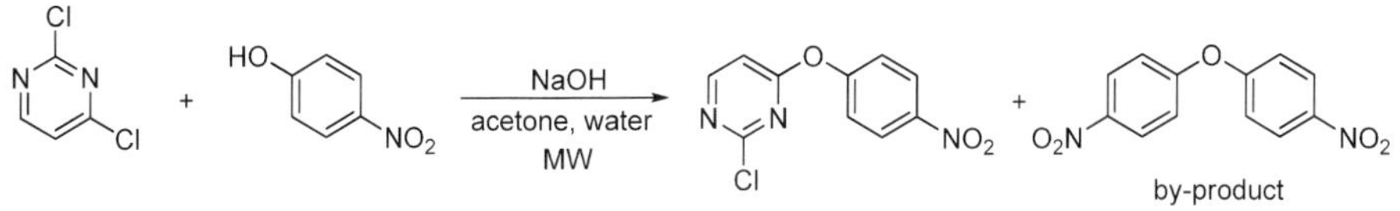

NaOH, acetone, water, MW; by-product

Scheme 3. Aromatic substitution of 2,4-dichloropyrimidine with *p*-nitrophenol

These reactions were previously performed in a small-scale Emrys Optimizer microwave reactor (20 mL glass vials). The study was focused on temperature/pressure limits, handling of suspensions, ability for rapid heating and cooling, robustness, and overall processing time.

A green approach to *N*-heterocyclization reactions ranging in scale from 20 mmol (CEM Discover) to 1 mol (CEM MARS) performed under microwave irradiation in open vessels was investigated by Barnard (Barnard et al., 2006) (Scheme 4). On using water as the solvent and in the absence of transition metal catalysts, *N*-heterocycles were formed in a fraction of the time needed on using classical heating. These reactions had previously been performed on a small scale by Varma in sealed vessels (Ju & Varma, 2006). Analogously, representative Suzuki couplings in water using low catalyst concentrations in conjunction with microwave heating have been transferred from sealed-vessel to open-vessel reaction conditions, and scale-up to the mole scale using the dedicated multimode apparatus CEM MARS (Arvela et al., 2005; Leadbeater et al., 2006) (Scheme 5). In both processes, single-mode and multimode cavities were used for open-vessel synthesis without changing the reaction times and these produced similar yields.

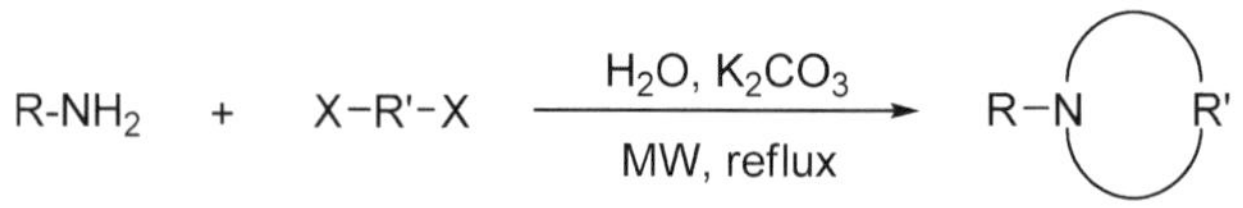

$$R\text{-}NH_2 + X\text{-}R'\text{-}X \xrightarrow[\text{MW, reflux}]{H_2O,\ K_2CO_3} R\text{-}N\ \ R'$$

Scheme 4. *N*-Heterocyclization reactions in water

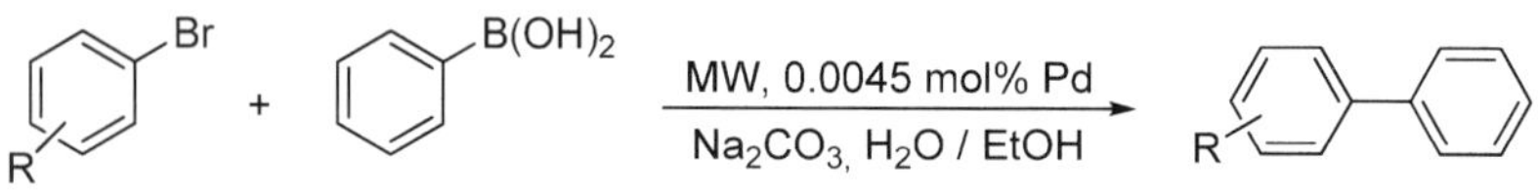

$$\text{R-C}_6\text{H}_4\text{-Br} + \text{C}_6\text{H}_5\text{-B(OH)}_2 \xrightarrow[Na_2CO_3,\ H_2O\ /\ EtOH]{\text{MW, 0.0045 mol\% Pd}} \text{R-C}_6\text{H}_4\text{-C}_6\text{H}_5$$

Scheme 5. Open-vessel microwave-promoted Suzuki reactions in water

### 2.2 Reproducibility in parallel microwave-assisted reactions

In general, two approaches have been used for microwave-assisted parallel and combinatorial chemistry: On the one hand, series' of compounds can be prepared sequentially in an automated single-mode instrument, which allow for control of temperature and pressure in each reaction independently. On the other hand, compounds can be prepared in parallel arrays using a multimode instrument. The use of a single-mode instrument offers the advantage of full control of each reaction. However, in this case, all reactions must be processed sequentially and this could lead to a bottleneck in productivity, especially for large series' of compounds. Multimode instruments offer the possibility of performing multiple reactions in one irradiation experiment, but reactions are usually performed without appropriate control of the temperature, which in turn limits the reproducibility of the experiments - especially when unmodified domestic ovens are used. In an effort to overcome these problems, several dedicated microwave reactors have been developed since 2003.

Nüchter demonstrated that a multiPREP reactor system (36 pressure reactors or 15 reflux reactors), combiCHEM (plates with 24, 48 and 96 reactions vessels sealed with Teflon-laminated silicon mats), and an HPR system (6–10 small autoclaves for pressurized reactions at up to 50 bar and 250 °C) allow parallel reactions to be carried out with up to 96 reactor chambers in a microwave field with total reproducibility in yield and/or selectivity (Nüchter & Ondruschka, 2003a). The authors studied different processes such as nucleophilic substitutions, condensation reactions, oxidations, and multi-centre reactions.

Also in 2003, Kappe scaled different organic reactions from the 1 mmol to 100 mmol scale using a prototype multimode microwave reactor from Anton Paar that allowed parallel processing in either quartz or PTFE-TFM vessels with maximum operating limits of 300 °C and 80 bar of pressure (eight-vessel rotor employing either 100 mL PTFE-TFM or 80 mL quartz vessels) (Stadler et al., 2003). The transformations included multicomponent chemistries, transition metal-catalyzed carbon-carbon cross-coupling protocols, solid-phase organic synthesis, and Diels–Alder cycloaddition reactions using gaseous reagents in prepressurized reaction vessels.

Alcázar investigated the possibility of scaling up the synthesis of several compounds simultaneously in one irradiation experiment (Alcázar et al., 2004). This parallel approach would allow reduced reaction times in comparison with the sequential irradiation of single samples in single-mode instruments. The authors studied an *N*-alkylation reaction and used optimized reaction conditions to probe the total reproducibility of five parallel reactions performed in a multimode system (3 mmol scale) and the same reaction carried out in a single-mode reactor on one-tenth of the scale (0.3 mmol) (Scheme 6, Figure 1).

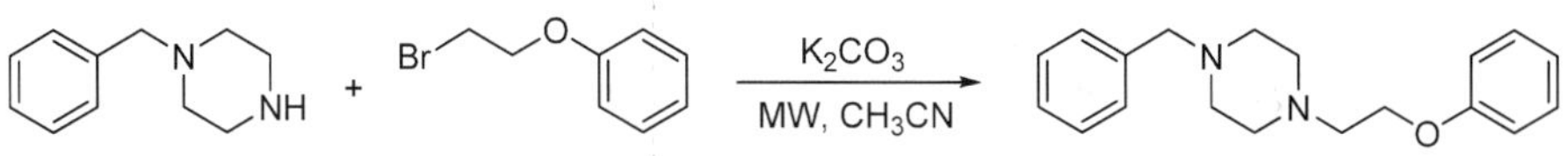

Scheme 6. *N*-Alkylation of *N*-benzylpiperazine

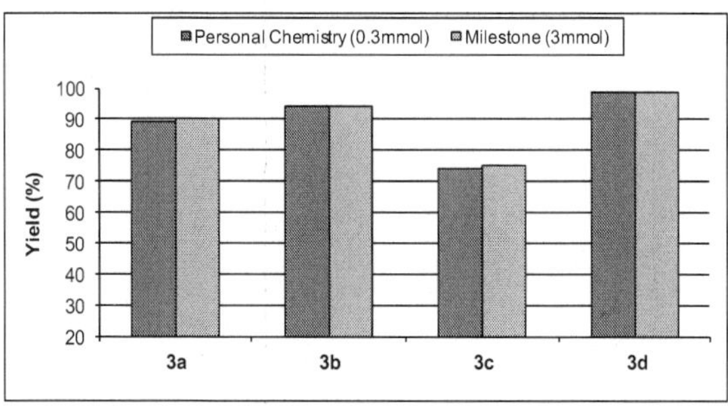


Fig. 1. Yield comparison between systems for *N*-alkylations of *N*-benzylpiperazine

On the basis of these results, a multimode combiCHEM reactor was used to assess the heating homogeneity of a multiwell plate with 24 equal reactions (Alcázar, 2005). Interestingly, reaction yields were slightly higher in the multimode instrument than in a single-mode set-up, and the author concluded that the former system offered enough reproducibility to develop further parallel chemistry. Once the performance of the multiwell plate had been evaluated, a set of 24 different compounds was prepared in parallel using the fully optimized conditions by combining four amines with six alkylating agents. In order to make an appropriate correlation across instruments, the same reactions were performed in a single-mode reactor. The results showed that good reproducibility was achieved between the two instruments: on average, the difference in yield was only 2.3%. However, to complete the whole sequence of reactions the single-mode instrument required 2.5 h, whereas the multimode instrument required only 40 min to achieve comparable results. This system combined the advantages of the parallel approach and microwave heating.

As mentioned above, for the scale-up from gram to kilogram scales two different approaches have been used historically - batch and continuous-flow. The major limitation of the continuous microwave reactor is that these systems are unsuitable for heterogeneous mixtures and viscous liquids. For these reasons, Maes performed a comparative study between a stop-flow system (CEM Voyager system based on the Discover single-mode platform equipped with an 80 mL glass vessel, a peristaltic pump and two valves) and two multi-mode platforms that allow parallel processing of several vessels per batch: Milestone MicroSYNTH equipped with a rotor with 10 vessel positions (vessel volume 100 mL) and CEM MARS equipped with a fourteen rotor positions (80 mL) (Loones et al., 2005). The reaction model used was the Buchwald–Hartwig amination (Scheme 7). They found that rapid Pd-catalyzed amination of aryl chlorides under microwave irradiation can be easily scaled-up on both single-mode and multi-mode platforms with similar yields if BTF (trifluoromethylbenzene) is used as the solvent. However, the stop-flow Voyager system was preferred since it is a completely automated unit that allows the continuous production of reaction product without the need to manually load and unload reaction vessels. Moreover, the Voyager system allows pumping of heterogeneous mixtures, which is problematic in continuous-flow units.

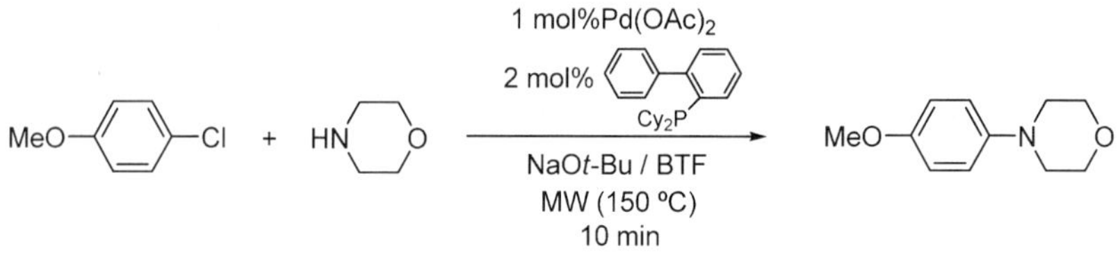

MeO–C6H4–Cl + HN(morpholine) → MeO–C6H4–N(morpholine)

1 mol%Pd(OAc)$_2$
2 mol% (biphenyl)Cy$_2$P
NaO*t*-Bu / BTF
MW (150 °C)
10 min

Scheme 7. Scale-up of the Pd-catalyzed amination of 4-chloroanisole with morpholine

### 2.3 Reproducibility across instruments

Reworking some of the reactions described in the literature that were originally performed in household microwave devices showed that it is absolutely necessary to analyze any result obtained in these specific devices. Due to the incomplete description of many reactions reported in microwave chemistry, reproducibility is difficult to assess in most cases. Only in rare cases are identical microwave devices available. Additionally, devices of the same series can have different field homogeneity. Only in a few cases are all reaction conditions known precisely and reproducibly. Even the use of different amounts of starting materials can produce different results, a factor that can usually be neglected for conventionally heated reactions. Moreover, the applied settings of the microwave devices are often insufficiently reported or the importance of these parameters is completely ignored (Nüchter et al., 2003b). In 1994 Login reported that commercial domestic ovens have limitations that can result in non reproducible fixation results in pathology and he developed a calibration protocol to identify the best locations for fixation within household microwave ovens (Login & Dvorak, 1994).

Microwave-assisted extraction (MAE) is a process in which microwave energy is used to heat solvents in contact with a sample in order to partition analytes from the sample matrix into the solvent. In most cases, recoveries of analytes and reproducibility are improved compared to conventional techniques. The basic principles and main studies, including reproducibility, on the use of microwave energy for extraction have been reviewed (Eskilsson & Björklund, 2000).

Mastragostino reported the microwave-assisted synthesis of $Ag_2V_4O_{11}$ starting from $V_2O_5$ and $Ag_2CO_3$ (Arbizzani et al., 2007). Although acceptable reproducibility may be attained in the exploratory phase by using low-cost domestic ovens modified to read the temperature of the reaction mixture (the irradiation was manually stopped when the temperature of the sample reached a set value), properly controlled synthesis conditions can be achieved only with scientific microwave systems with control of irradiation power and reaction temperature (such as the CEM Discover single-mode reactor).

The influence of microwave energy on the rate of cleavage of -C-S- and -S-S- bonds by azide ions has been investigated using a domestic oven. It was found that microwaves do not accelerate the cleavage of the -C-S- bond whereas -S-S- bond cleavage is faster. However, the reproducibility of this reaction is very poor and the microwave technique is not recommended for quantitative determination (Kurzawa & Stachowiak, 2001).

When the microwave methodology was introduced twenty-five years ago, most reactions were performed in domestic ovens without appropriate temperature and pressure controls. As a consequence, there is a plethora of interesting organic transformations that have been reported to take place in domestic ovens. However, reactions performed in this kind of instrument, i.e. without appropriate temperature and pressure controls, are generally considered as not reproducible, thus limiting the application of such reactions.

The polarity of the solvent is the most important parameter to consider when microwave reactions are performed in solution. Polar solvents absorb the microwave radiation directly and the polarity of the substrate is relatively unimportant. In the case of non-polar solvents, however, the radiation is absorbed by the substrates, but the differences in absorption of the substrates are moderated by the solvent, especially in dilute solution. In both cases, the reaction temperature is limited by the solvent boiling point and microwave-assisted reactions in solution are easily controlled. In neat reactions, radiation is again absorbed by the substrates but there is no solvent to stabilize the temperature. In this situation, the nature of the substituents and the polarity of the substrates influence the absorption of microwave energy and, hence, the yield. Moreover, in the absence of solvent the temperature is not limited by the boiling point of the solvent. A process involving highly polar reagents, intermediates or transition states can hardly be controlled under microwave irradiation. Hence, it does not follow that a solvent-free reaction previously performed in a domestic oven will be controllable and reproducible in dedicated microwave reactors in a similar way to reactions in solution, as the field density is very different from one instrument to another (Nüchter et al., 2004).

In 2007 Díaz-Ortiz & Alcázar showed for the first time that solvent-free reactions performed in domestic ovens could be reproduced, scaled and parallelized in controlled microwave monomode and multimode reactors (Díaz-Ortiz et al., 2007). The 1,3-dipolar cycloaddition of nitrile *N*-oxides with nitriles (Scheme 8) was selected as a model reaction, and this was previously performed by the same authors using a domestic oven (Díaz-Ortiz et al., 1996). The study was structured in four steps. The first step involved the translation of the 1,3-dipolar cycloaddition reaction conditions from the domestic oven to the single-mode CEM Discover microwave reactor by increasing the temperature gradually in successive experiments to prove that a linear dependence between the yield and temperature is observed for this solvent-free reaction (Figure 2). The second step was to study the reproducibility of the reaction under non-optimized conditions because under such conditions reactions are more sensitive to temperature differences. In this case, four identical reactions were performed in parallel in the MicroSYNTH multimode reactor at 100 °C on a scale five times larger than the one used in the monomode instrument. It can be concluded that there is good reproducibility between the two systems (Figure 2).

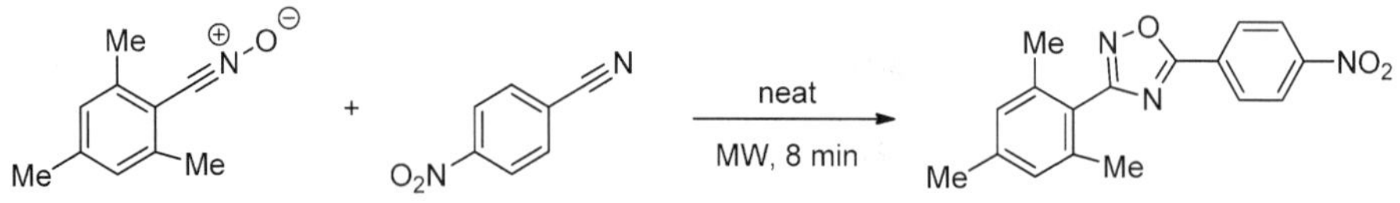

Scheme 8. 1,3-Dipolar cycloaddition of nitrile with nitrile *N*-oxide

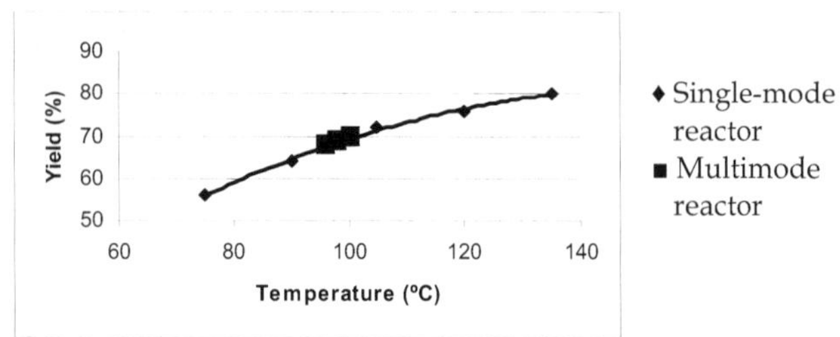


Fig. 2. Plot of yield versus temperature for 1,3-dipolar cycloaddition reactions

The third step in the study concerned the scalability of the reaction. For this purpose, eight different nitriles were reacted with a nitrile *N*-oxide using optimized reaction conditions. Reactions were performed in both single-mode and multimode microwave ovens. On average, the difference in yield between the two instruments was 2.2%. The fourth step was to study the preparation of 24 different compounds in a multiwell plate under microwave irradiation to provide an effective combination of productivity and speed. To overcome the problem of different absorption levels in each reaction, due to the different polarity induced by the substituents, the authors took advantage of a Weflon well plate (Nüchter, 2003a). Finally, the results obtained for the 8 substrates used in all experiments (single-mode, multimode, Weflon plate and isolated yield) were compared and good reproducibility was found for each compound under each set of conditions: the average yield difference was 3.75% (Figure 3). Similar conclusions were reported by Leadbeater in a study into the rapid optimization of reaction conditions (Leadbeater & Schmink, 2007).

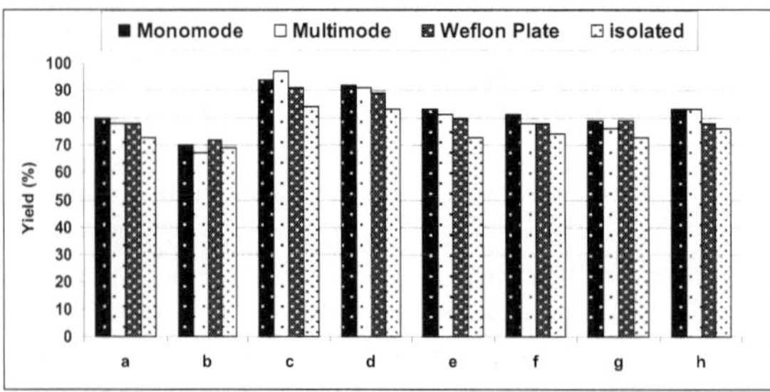


Fig. 3. Comparative results for eight nitriles in different applications and isolated yields obtained

In order to extend the preliminary results to see whether any solvent-free reaction previously performed in a domestic oven can be reproduced in dedicated apparatus, the same authors studied four new reactions that cover a wide range of chemical transformations (Díaz-Ortiz et al., 2011). It was found that *N*-alkylation of (1*H*)-benzotriazoles, condensation of anilines with urea (or thiourea) and Beckmann rearrangement of ketoximes are reproducible in both single-mode and multimode microwave instruments using temperature-controlled conditions. However, all attempts to reproduce and control the oxidation of benzylic bromides with pyridine *N*-oxide in single-mode or multimode reactors failed (Scheme 9).

$$\text{Pyridine } N\text{-oxide} + ArCH_2Br \xrightarrow[\text{2 min}]{\text{MW}} \left[ \text{Py}^{+}\text{-}OCH_2Ar \; Br^{-} \right] \longrightarrow \text{Pyridine}\cdot HBr + ArCH{=}O$$

Scheme 9. Oxidation of benzylic bromides

With the aim of explaining why the oxidations of benzylic bromides are not reproducible processes under microwave irradiation, a computational study on this reaction and the aforementioned 1,3-dipolar cycloaddition - which is easily reproduced - was performed (Table 1). The results show that reactions involving a moderate or medium increase in polarity in the pathway from reactants to products (Table 1, Entry 1) are relatively easy temperature-controlled processes under microwaves. In contrast, large increases in polarity during the reaction path (Table 1, Entries 2 and 3) give rise to extreme absorptions of microwave energy and make these processes more difficult to control and reproduce.

| Entry | Process | Dipole Moments (Debyes) | |
|---|---|---|---|
| | | Reactants | Intermediate |
| 1 | 1,3-Dipolar cycloaddition solution (acetonitrile) | 5.51 | 11.27 |
| 2 | Oxidation of benzylic bromides $S_N2$ solution (bromobenzene) | 5.05 | 25.02 |
| 3 | Oxidation of benzylic bromides $S_N1$ solution (bromobenzene) | 5.55 | 52.04 |

Table 1. Dipole moments (Debyes) of selected stationary points

A combined (multimode and single-mode) microwave device (MLS ETHOS 1200 Combi ) was used by Ondruschka to perform a comparative study of both single-mode and multimode microwave irradiation methods (Nüchter et al., 2003b). The authors studied a lipase-catalyzed transesterification process, and Biginelli and Hantzsch reactions. The results were similar for both single-mode and multimode microwave devices in all reactions. Differences in the yields of the pure products were due to different workup procedures and small systematic errors. The authors completed their work with a useful study of the heating behaviour of polar systems in the multimode microwave field: energy input in pure substances, and mixtures of polar and non-polar substances.

## 3. Large-scale production under microwaves

Most examples of microwave-assisted chemistry published to date have been performed on a scale of less than 1 g, with a typical reaction volume of 1–5 mL. The main applications have been to accelerate and optimize well-known and established synthetic procedures. In microwave-assisted synthesis there is a need to develop techniques that can ultimately provide products on a multikilogram scale before this approach can become a fully accepted industrial technology. Thus, the further development of large reactors is required, at least on the pilot plant scale to enable multikilogram production of lead compounds. It was not until the mid-1990s that the issue of scale-up was first investigated (Raner et al., 1995; Cablewski et al., 1994; Roberts & Strauss, 2005). Since that time, a significant number of prototypical and commercial microwave scale-up reactors have been reported, for both batch and continuous operation, and most of these were described by chemical engineering groups.

The scale-up of microwave chemistry clearly has several benefits to offer over conventional heating. However, there are some problems that make the scale-up of microwave chemistry difficult to achieve. The big challenge for process scale-up involving microwave technology is to establish a reliable and safe process setup where issues like physical properties (i.e., penetration depth), temperature control, and reactor design have to be carefully considered. On using batch reactors, the user is confronted with problems in heating large volumes due to the limited penetration depth of microwave irradiation (Nüchter et al., 2004).

The scale-up of microwave-mediated transformations can be defined in different ranges, leading to the use of different concepts and varying instrumentation. Depending on the user, the term "scale-up" will have different meanings. In the case of method development, scale-up starts with processing a 50 mL reaction mixture, corresponding to a 10- to 100-fold scale-up of reactions performed in standard single-mode microwave vials with a processing volume of 0.5–5.0 mL. A possibility for further scale-up would be the use of the "numbering up" approach, i.e. running the same reaction several times in sequence. This approach is aided by existing robotic equipment, which allows unattended vessel transfer in and out of the microwave cavity. Alternatively, such reactions can also be performed by parallel synthesis in corresponding multivessel rotor systems on switching to multimode instruments. From an industrial viewpoint, scale-up means process development and the highest possible throughput, a situation that virtually excludes the use of batch reactors. In fact, it is the productivity and not the size of the vessel that is important, which clearly indicates that flow systems, regardless of whether applied in stop-flow (SF) or continuous flow (CF) mode, have distinct benefits over batch process reactors.

The approaches can be categorized as: scale-up in parallel, scale-up in sequence, and scale-up by continuous flow (Figure 4). Hybrid approaches have also been devised.

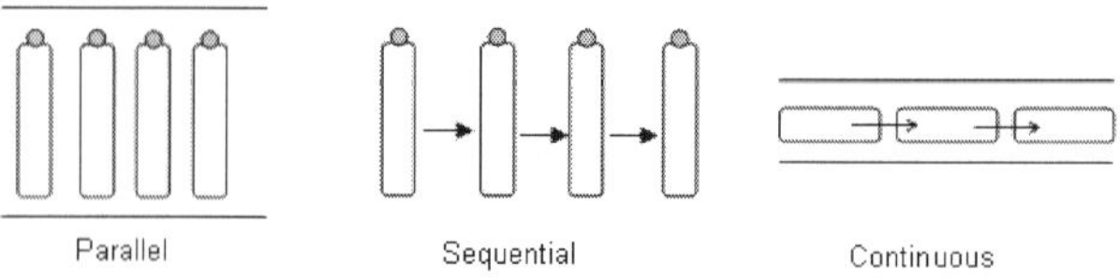


Fig. 4. Strategies for microwave scale-up

### 3.1 Scale-up in batch

The advantages of microwave heating in chemistry results in the need for scale-up possibilities. Moreover, the direct homogenous heating of reaction mixtures under microwave irradiation is believed to facilitate direct scaling without heat and mass transport issues (Kappe & Stadler, 2005). In fact, direct scaling of microwave-assisted synthesis in batch mode, without the need for process optimization, has already been demonstrated in organic synthesis on using open or closed reaction vessels (Leadbeater, 2010).

#### 3.1.1 Scale-up in a single large open vessel

When considering batch scale-up, one approach is to use a single, larger, reaction vessel. This avoids the problem of tedious charging and discharging of multiple reaction vessels. However, penetration depth and/or power density could become issues as the vessel size increases.

If the reaction can be carried out in the absence of solvents, or superheated solvents are not required, or if the process can be performed successfully below the boiling point of the solvent, the large reaction can be run in an open-vessel model. At present there are several commercially available microwave ovens to carry out this scale-up mode, including CEM MARS, MILESTONE Micro-SYNTH and MINILABOTRON SAIREN.

This methodology has some disadvantages. For example, the penetration depth or power density may be an issue and of course reactions that need elevated temperatures and pressures cannot be carried out. However, considerable scale-up can be - and has been - achieved in such reactors.

Open-vessel conditions prove advantageous for driving equilibrium reactions to completion when the product or by-product is a more volatile component than the starting materials. A Milestone MicroSYNTH reactor has been used to perform a fractional distillation in the synthesis of a tricyclic heterocycle (Razzaq & Kappe, 2007). *N*-Methylaniline was reacted with two moles of diethyl malonate in a cyclocondensation reaction with elimination of four moles of ethanol to give the product pyranoquinoline on a 0.2 molar scale. On this scale, the distillation took only 82 min as compared to 3-5 h conventionally. When the reaction was performed in a sealed-vessel microwave system only <5% product formation was achieved (Scheme 10).

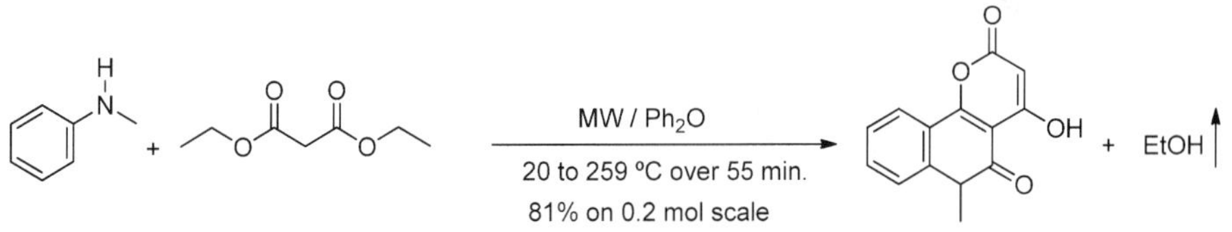

MW / $Ph_2O$
20 to 259 °C over 55 min.
81% on 0.2 mol scale
+ EtOH

Scheme 10. Reaction between *N*-methylaniline and diethyl malonate

The Hantzsch 1,4-dihydropyridine synthesis (Bowman et al., 2008a) was also scaled up successfully to the 0.5–1 mol level in the MARS unit and this involved heating up to 1 L of reaction mixture. One batch provided 250 g of dihydropyridine product in less than 1 h.

Leadbeater et al. evaluated a new batch reactor, designed by AccelBeam Synthesis, that allows reactions to be performed on scales from 2–12 L (Schmink et al., 2010). A number of reactions have been investigated, including palladium-mediated transformations, condensation reactions, nucleophilic aromatic substitution reactions, and alkylations. One important aspect of this work was that a linear scaling approach was taken and changes were not made to the protocol on moving from the small, developmental scale to larger scales. In some cases reactions were scaled over 18,000-fold on moving from small (0.1–1 mmol) to large (1–18 mol) runs. It is noteworthy that in order to simulate a situation where multiple sequential microwave steps were employed to reach a desired target compound, these authors developed a sequence of reactions as a medicinal chemist might do at the 10 mmol scale in order to synthesize a drug-like molecule (Scheme 11). Overall, the sequence employed three microwave steps and afforded 473 g of the target molecule in 38% yield, almost identical to the 39% obtained in the small scale process.

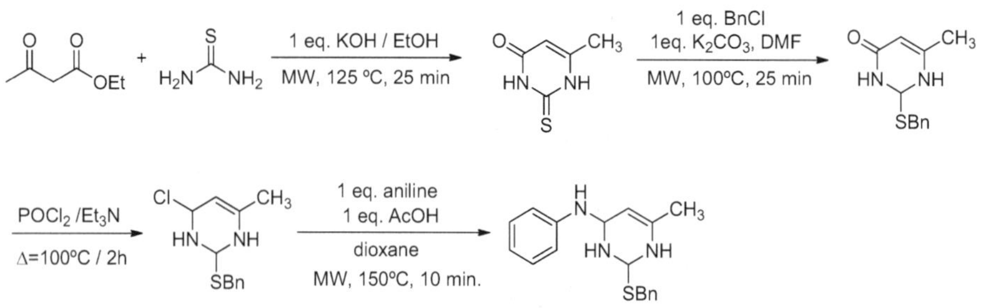

1 eq. KOH / EtOH
MW, 125 °C, 25 min
1 eq. BnCl
1eq. $K_2CO_3$, DMF
MW, 100°C, 25 min
$POCl_2$ /$Et_3N$
Δ=100°C / 2h
1 eq. aniline
1 eq. AcOH
dioxane
MW, 150°C, 10 min.

Scheme 11. Preparation of 2-(benzylthio)-6-methyl-4-(phenylamino)pyrimidine

Horikoshi et al. employed a batch reactor to synthesize silver nanoparticles in aqueous media by reduction of a diaminosilver(I) complex with carboxymethylcellulose (Horikoshi et al., 2010). The microwave process yielded smaller silver particles with a narrower distribution than the conventional heating process.

### 3.1.2 Scale-up in a single large sealed vessel

Another approach to perform the scale-up involves the use of a single large sealed vessel. In this case, it is possible to carry out reactions in heterogeneous conditions, in the presence of solvents, and at pressure. Despite this versatility, the limiting factor is the vapour pressure generated by a superheated solvent due to safety concerns. While moderately high pressures (~20 bar) are easily contained in small scale microwave reactors, the use of large vessel sizes requires more safety features and greater engineering expertise in the system design. This results in more complex reactors that are less easy to use and more expensive. In this sense, to contain the pressures likely to be generated, strong materials must be used to construct the reaction vessels, but they must also be microwave transparent. This rules out the use of metals on this scale and leaves thick-walled (quartz) glass, ceramics, and polytetrafluoroethene (PTFE or Teflon), or combinations thereof. However, these materials are all thermal insulators. In either case, cooling is further compromised by the necessarily thick-walled vessels such that, for larger microwave reactors, the cooling time is often notably longer than the combined heat-up and reaction periods. This situation increases the cycle time per batch and reduces the overall output, thus lessening the benefit of rapid microwave heating. Some of these problems have been solved in certain systems. For example, flash evaporation using the mechanical pressure of superheated solvent in the vessel can discharge the reaction mixture while it is still hot into a collection vessel, thus allowing the cooling to occur off-line from the reactor.

Commercially available microwave reactors in which sealed-vessel scale-up can be performed are the BIOTAGE ADVANCER 350, Batch SYNTH, Ultraclave, MILESTONE ETHOS 1600, Anton Paar monowave 300, and Anton Paar Masterwave BTR.

Scale-up in a single sealed vessel represents perhaps the least utilized scale-up option. This may be because the reactors are large and expensive and that more expertise is required for their use.

One particular example of pharmaceutical interest was the Grandberg synthesis of 2-methyltryptamine (Scheme 12) (Bowman et al., 2008b). A conventional approach to this reaction had already been scaled up to 20 kg at Novartis (Slade et al., 2007). However, the Advancer reactor was used to give results on a 0.2 mol scale that are comparable to those achieved by the Novartis group.

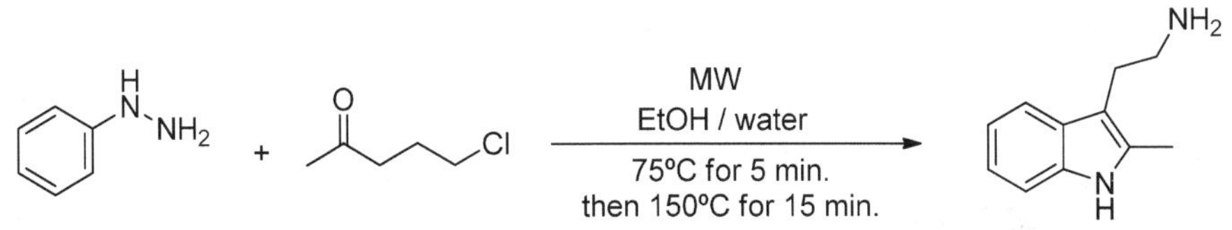

MW
EtOH / water
75°C for 5 min.
then 150°C for 15 min.

Scheme 12. Grandberg synthesis of 2-methyltryptamine

Schubert reported ring-opening polymerizations in batch mode of 2-ethyl-2-oxazoline on scales up to 100 g without the need for process optimization (Paulus et al., 2006). Nevertheless, it should be noted that this pressurized process cannot be scaled indefinitely in batch mode for safety reasons.

Deetlefs reported the use of a CEM MARS system for the synthesis of a wide range of ionic liquids based on nitrogen-containing heterocycles (Deetlefs & Seddon, 2003). The process can be performed on a range of reaction scales (50 mmol to 2 mol) in either sealed or open vessels.

### 3.2 Scale-up in parallel synthesis

Laboratory-scale batch microwave reactors generally offer a maximum batch size of 1 L reaction volume and in most cases this is divided into several smaller reaction vessels (multivessel rotor systems). This approach does not exactly match the "one vessel" philosophy, but the parallel setup allows for the processing of several batches of either the same reaction mixture or of closely related mixtures. In fact, this combination of batch synthesis and parallel processing can furnish either compound libraries on the gram scale or a larger amount of one single compound in a short time. Although the Milestone UltraCLAVE system provides a 3.5 L single vessel cavity, it has proven to be more effective to heat smaller volumes in parallel rather than one big batch, given that identical microwave output power is applied (Loones et al., 2005; Cléophax et al., 2000).

The demands made on industry, especially the pharmaceutical industry, are changing at an unprecedented pace, making speed critical in modern chemistry. For this reason, high-speed microwave synthesis and combinatorial chemistry have attracted a considerable amount of attention in recent years. These approaches offer significant advantages to the synthetic chemist: reduced reaction times and improved yields as a result of microwave heating and increased productivity due to the implementation of combinatorial chemistry. However, the evolution of microwave instrumentation is delivering new systems that allow reactions to be

carried out in specialized multiwell plates under temperature and pressure controlled conditions (Nüchter & Ondruschka, 2003; Alcázar, 2005; Kremsner et al., 2007).
Caliendo et al. disclosed the synthesis of series of piperazine derivatives as 5-HT1A agonists using microwave irradiation in all the reactions (Caliendo et al., 2001; Caliendo et al., 2002) (Scheme 13). They compared the results obtained with microwave heating with those obtained using traditional heating. In all cases, compounds were obtained in higher overall yields and reaction times were reduced from hours to minutes on using microwave irradiation. Reactions were performed in a multimode instrument using parallel racks that allow the preparation of sets of compounds in a single irradiation experiment.

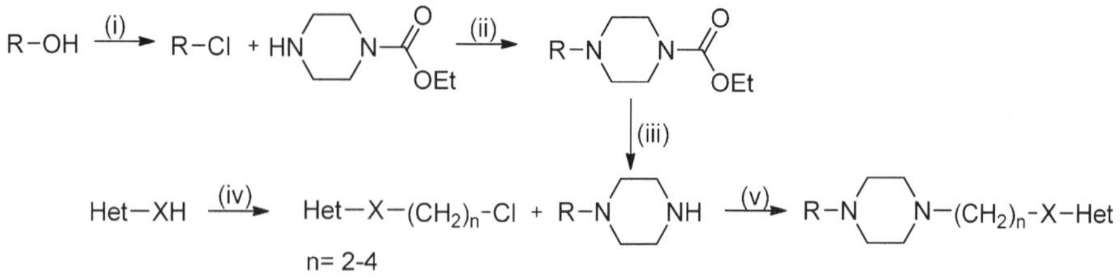

Scheme 13. Synthesis of *N*-substituted piperazine derivatives. (i) $SOCl_2$, toluene, 55 °C, 20 min, MW; (ii) $K_2CO_3$, NaI, DMF, reflux, 1 h, MW; (iii) KOH, water, methanol, reflux, 1 h, MW; (iv) $Br(CH_2)_nCl$, $K_2CO_3$, DMF, MW; (v) $K_2CO_3$, DMF, 70 min, MW

Researchers at Johnson & Johnson Pharmaceuticals have described tropanylidene derivatives as combined mu/delta opioid receptor agonists (Costas et al., 2004). The preparation of these analogues was achieved by developing a parallel solution phase strategy under microwave irradiation, which allows multiple sequential reactions to occur in a single reaction vessel (Scheme 14).

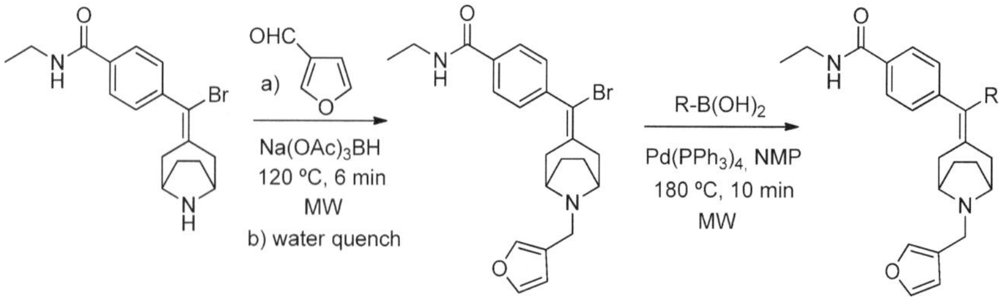

Scheme 14. Preparation of tropanylidene derivatives

This approach allowed the synthesis of 192 compounds in a quicker and more efficient manner than using the corresponding solid phase approach. Additionally, the parallel solution phase approach allowed rapid scale-up of selected compounds for further *in vivo* studies.
Caliendo et al. described an efficient, facile, and practical parallel combinatorial synthesis of substituted-benzoxazines under microwave irradiation (Caliendo et al., 2004). The procedure involved the use of a specially designed microwave oven for organic synthesis that was suitable for the parallel synthesis of solution libraries. As a demonstration, a library

of 19 substituted *N,N*-dimethyl- and *N*-methyl-benzoxazineamide derivatives, structurally related to the potassium channel opener cromakalim, was generated by both conventional and microwave procedures, with a reduction in the library generation time for the microwave approach from 7 h to 30–36 min (Figure 5).

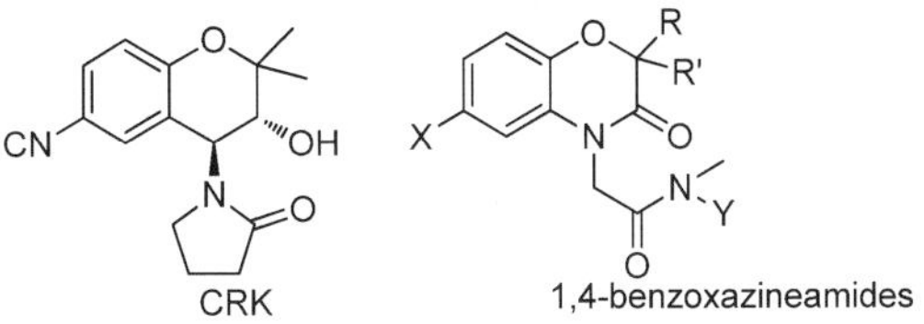

Fig. 5. Structure of cromakalin (CRK) and 1,4-benzoxazineamides

To facilitate the preparation of β-peptide libraries in parallel, Murray and Gellman adapted the reaction conditions for the solid-phase synthesis of 14-helical β-peptides for use in a multimode microwave reactor (Murray & Gellam, 2006). The low temperature and pressure requirements of microwave-assisted β-peptide synthesis were found to be compatible with multiwall filter plates composed of polypropylene. Microwave heating of the 96-well plate was sufficiently homogeneous to allow the rapid preparation of a β-peptide library in acceptable purity.
An environmentally friendly method based on microwave radiation has been developed for the synthesis of a library of *N*-acylhydrazones (Andrade & Barros, 2010). With this protocol, the use of solvents and catalysts can be avoided, and the products are obtained in short reaction times and in very good yields. A variety of *N*-acylhydrazones were synthesized under microwave irradiation within 2.5–10 min, starting from benzo, salicyloyl, and isonicotinic hydrazides. The results are reproducible on a 500 mg to 5 g scale and this approach could be applied to a large number of ketones and aldehydes. The reported method could be a useful synthetic path to obtain *N*-acylhydrazones on an industrial scale (Scheme 15).

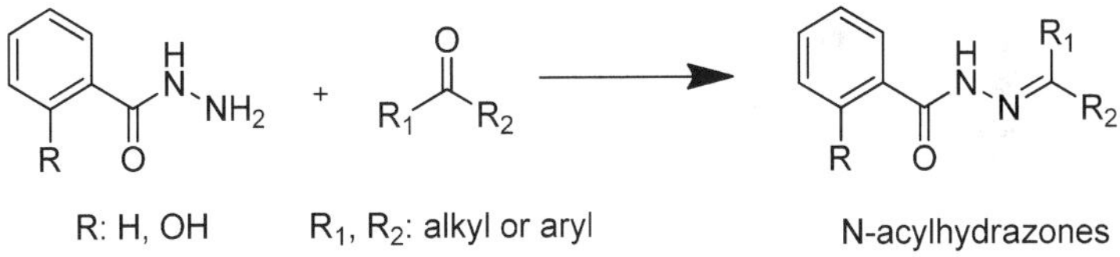

Scheme 15. *N*-acylhydrazones synthesized under microwave irradiation

The combination of parallel synthesis and microwave chemistry has clearly improved the efficiency in the preparation of derivatives. The synthesis of compound libraries in single-mode instruments fully integrated into automatic platforms in a sequential way is well established. Additionally, integration of reaction performance in multiwell plates under microwave irradiation with automatic platforms for sample preparation and work up is expected to further optimize library synthesis.

### 3.3 Scale-up in continuous flow systems

Although dedicated modern microwave instruments for MAOS are very successful in small-scale operations, microwave scale-up in batch mode beyond a reaction volume of approx. 1

L is not feasible (Glasnov & Kappe, 2007). The penetration depth of only a few centimetres and the limited dimensions of the standing wave cavity are the main reasons for the development of continuous or stop-flow microwave reactors. In these systems the reaction mixture is passed through a relatively small microwave-heated flow cell, which avoids the aforementioned drawbacks.

It was recognised early on in the development of microwave reactors that flow-based applications offered tremendous advantages in terms of processing capabilities. Two of the early pioneers of microwave chemistry extolled the virtues of continuous microwave reactors as early as the 1990s (Chen et al., 1990; Cablewski et al., 1994). However, such pioneering developments received little recognition or further development due to a combination of inferior technology, poor quality manufacture and the fact that the adoption of continuous flow protocols was in direct contradiction to the emerging, at that time, Combinatorial Chemistry (small scale multiparallel batch processes).

The key requirement of a continuous flow microwave reactor is the ability to continuously monitor and adjust the reaction parameters whilst in operation. This facilitates easy reaction optimization, introduction of automated safety controls and ensures a reliable and prolonged processing capability. Additionally, the combination of such processing techniques with the enabling technology of immobilized reagents, scavengers and catalysts to effect multi-step synthetic transformations has allowed the generation of novel pharmaceutical architectures and more complex natural products.

One of the first successful examples of scale-up was developed by Strauss (Cablewski et al., 1994). These authors prepared hundred gram quantities of a cyclopentenone intermediate using the continuous microwave reactor that they developed themselves (Scheme 16). The standard purification protocol for the product (distillation to remove small residual quantities of the starting material) was ineffective for large scale syntheses and, as a result, a scavenging procedure using a bisulfate-functionalized ion exchange resin to remove the reaction base (NaOH) was employed. The enhanced processing capability provided by this solid phase purification approach demonstrates the superior throughput that can be achieved by an innovative combination of technologies.

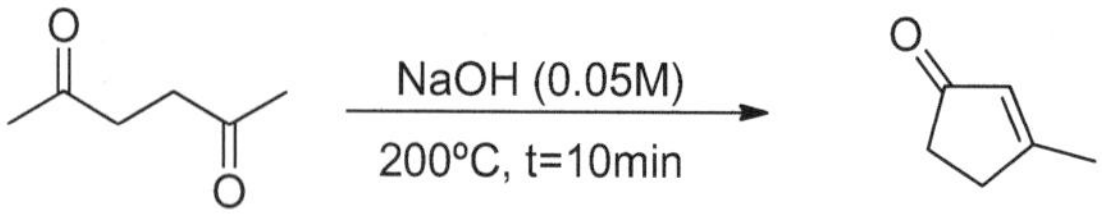

NaOH (0.05M)

200°C, t=10min

Scheme 16. Preparation of 3-methylcyclopent-2-enone using a flow microwave reactor

Recently published examples of continuous flow organic microwave synthesis involve, for example, the synthesis of 5-amino-4-cyanopyrazoles (Smith et al., 2007). The design consists of fluorinated polymer tubing wound around a Teflon core that is fitted with a dummy pressure cap. This flow device has the basic shape of a 20 mL vial suitable for the Biotage EXP single-mode instrument. The input tube is connected to an HPLC pump and a 7 bar back-pressure regulator is placed at the exit. Both connections are physically located at the bottom of the microwave unit. In addition, with this system, purification can be facilitated by passing the exiting flow stream through columns packed with polymer-supported reagents or scavengers. The synthesis of various 5-amino-4-cyanopyrazoles by reaction of a set of hydrazines with ethoxymethylene malononitrile in methanol was performed at temperatures between 100–120 °C with a residence time of 0.8–4 min (Scheme 17).

Subsequent passage of the reaction mixture through a column with supported benzylamine to scavenge any unreacted malononitrile, followed by a column packed with activated carbon to remove coloured impurities, furnished the desired product in high purities, good to excellent yields and in quantities up to 250 g. A benefit of this flow device is its versatility, since different tubing lengths can be wrapped in the system to provide reactors with different internal volumes. The application of multiple bundle tubings allows different reactions to be performed in parallel in one single reactor.

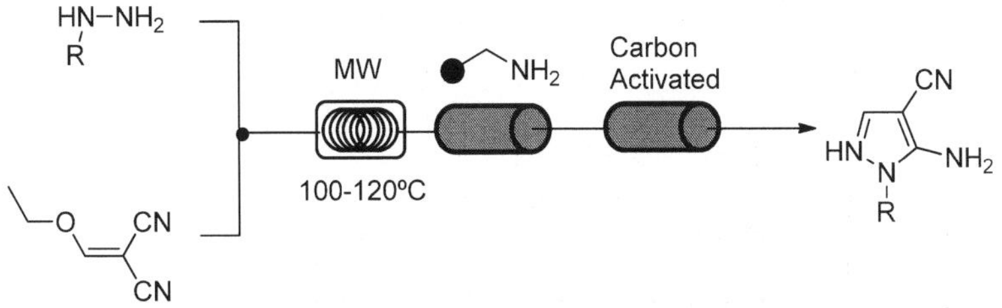

Scheme 17. Flow microwave synthesis of 5-amino-4-cyanopyrazoles

Microwave heating seems to be particularly competitive with transition metal-catalyzed processes, not only bringing long reaction times down to minutes but also minimizing the levels of undesired side products and preventing collapse of the catalytic system (Nikbin et al., 2007).

The combination of reactor design with immobilization techniques is very important for the flow process as it facilitates maximal interaction between reagents and catalysts without causing clogging problems. Moreover, the immobilized catalysts could be used over several cycles without a significant drop in activity. Kappe described a Heck cross-coupling reaction using a continuous-flow reactor based on a megaporous glass carrier material with suitable polymer functionalization introduced into the pore volume of this support (Kunz et al., 2005). For this purpose, a basic ion-exchange resin-loaded monolith was used in order to create the close neighbourhood of ionic sites and Pd(0) sites (Scheme 18). Pure products were collected without the need for extensive purification steps. The combination of composite-based flow-through reactors with microwave irradiation may lead to new and effective methods to scale up organic reactions.

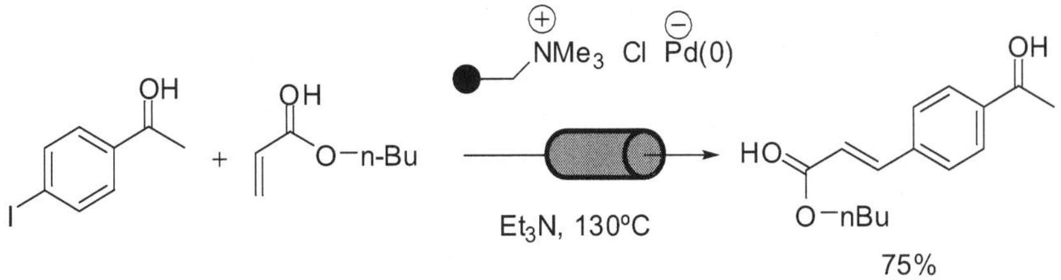

Scheme 18. Heck transformation applying a continuous-flow reactor

Recently, Organ has prepared highly porous Pd films composed of nanometer-sized particles (Comer & Organ, 2005). These Pd films served as an excellent catalyst for Heck reactions under continuous flow microwave conditions. The authors showed that 10 mg of product can be obtained from this reactor system within 1 min. Gram quantities of the products were obtained by flowing reaction mixtures through a single capillary for about 90 min.

Another interesting study was described by Wilson (Wilson et al., 2004), who developed a continuous microwave reactor that eliminates the potential reaction parameter re-optimization (time and temperature) typically required when methods are transferred from small-volume, single-mode systems to larger (but limited)-volume multimode systems (Figure 6). The representative chemistries explored include $S_NAr$, esterification, and the Suzuki cross-coupling reaction, all of which were successfully and safely scaled up to multigram quantities using a home-made continuous flow microwave cell (Scheme 19).

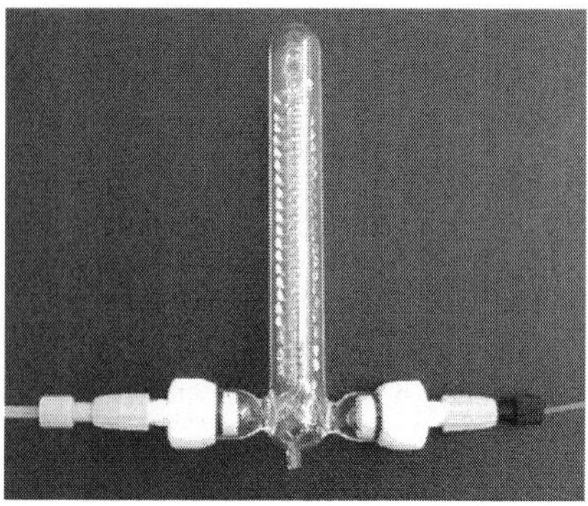

Fig. 6. Glass coiled flow cell

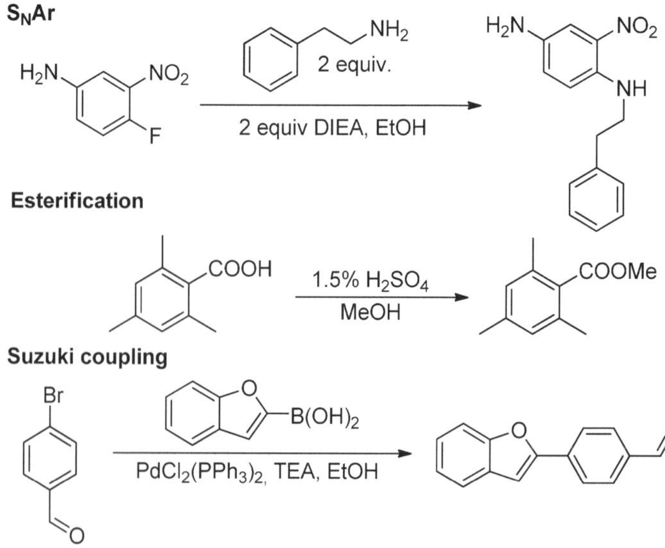

Scheme 19. $S_NAr$, esterification, and a Suzuki cross-coupling reaction in a microwave-flow reactor

When reactions need to be performed on several hundred gram or kilogram scales, typical for pilot or manufacturing plants, microwave synthesis is not easily scalable (Singh et al., 2008). As a result, the development of stop-flow or continuous flow microwave reactors, capable of working with slurries, are key for the successful translation of the reaction conditions. In this way Moseley (Marafield & Moseley, 2010) used $S_NAr$ reactions to evaluate the stop-flow approach, where reagents were loaded into the microwave vial, the reaction proceeded and finally the vial was emptied in an automatic process without manual intervention. This approach gave production rates of >0.5 Kg per day, which would meet the manufacturing requirements of early clinical pharmaceutical development.

The same transformation shown in Scheme 20 was used in a recent article to validate the Milestone FlowSYNTH, a commercially available continuous flow microwave reactor designed for working at manufacturing scale (Bergamelli et al., 2010). This instrument gave productivities up to 0.65 mol/h (~170 g/h). Even higher productivities were achieved for the ortho Claisen rearrangement, in which nearly 30 kg of material per day could be produced in one unit. The rearrangement was included in a three-step microwave synthesis of naphthofuran, which was successfully scaled up, although the first reaction, naphthol *O*-alkylation, was performed in a 2L batch reactor as pumps became blocked by the heavy slurry (Scheme 21).

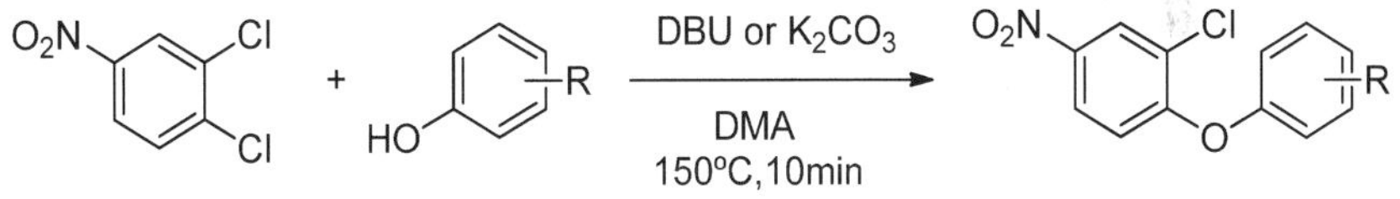

Scheme 20. $S_NAr$ evaluated in a stop-flow reactor

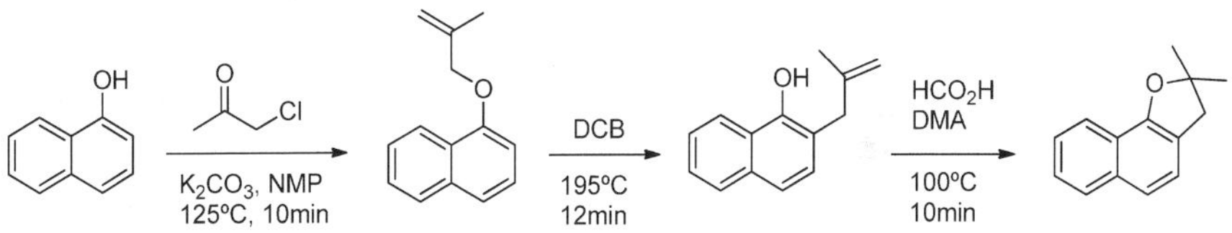

Scheme 21. Synthesis of naphthofuran

In order to improve the performance of the system, Dressen modified five key points to reduce the potential blockage by solids (Dressen et al., 2010). The modified system was assessed in the enantiomerically selective esterification of (*R*,*S*)-1-phenylethanol with vinyl acetate using Novozym 435 and the esterification of (*S*)-pyroglutamic acid with *n*-decanol (Scheme 22). Despite these improvements, plugging problems still occurred as a result of solid reactant /product deposition. Despite these issues the authors concluded that scaling up heterogeneous reactions by microwave irradiation is still feasible as the results obtained are comparable to those from the batch procedure.

To sum up this section, the scale up of processes using microwave irradiation in continuous flow systems has proved to be a valuable tool when homogeneous reactions are involved. In the case of heterogeneous reactions, solid reagents in cartridges have proven efficacy on the multigram scale. In the case of kilogram scale heterogeneous reactions, improvements in the handling of heavy slurries have been achieved, even though further development of this technology is still needed to overcome clogging and plugging problems. Other factors that must be taken into account, as introduced by different authors (Bergamelli et al., 2010; Dressen et al., 2010), concern the concept of energy efficiency. Energy cost is a key factor when deciding between different technologies.

**Enantiomerically selective esterification of (R,S)-1-phenylethanol.**

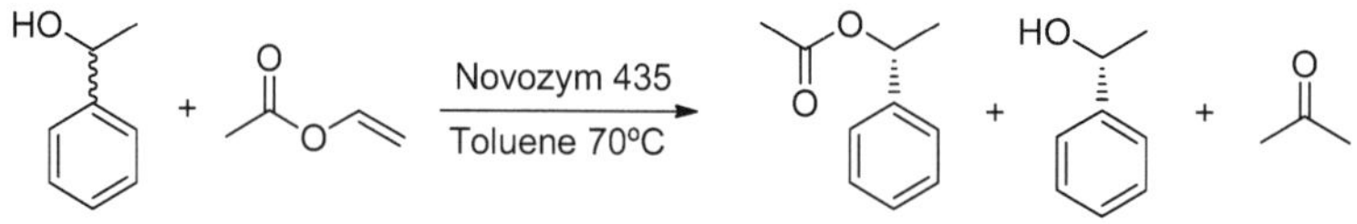

Novozym 435
Toluene 70°C

**Esterification of (S)-pyroglutamic acid with n-decanol.**

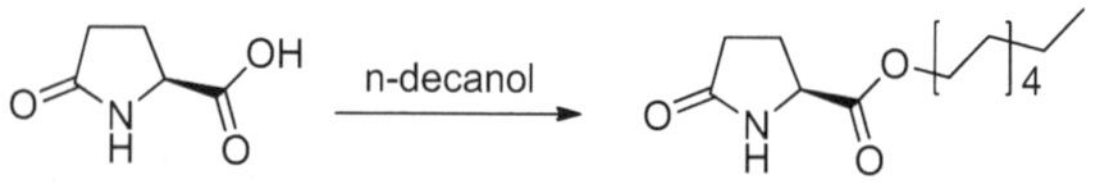

n-decanol

Scheme 22. Esterification reactions of (*R*,*S*)-1-phenylethanol and (*S*)-pyroglutamic acid

## 4. Conclusions

The application Microwave Assisted Organic Synthesis on the laboratory scale has been very successful - especially since the introduction of dedicated microwave instruments. However, the expansion of this technique as the first choice to perform a chemical synthesis has been hindered by the following limitations:

- The use of various instruments from different companies and with different characteristics (power density, volume, waveguides etc.).
- The indistinct use of monomode and multimode instruments with and without control of the incident power, respectively.
- The general use, until 2000, of domestic kitchen-type microwave instruments, the results of which are considered to be non-reproducible.
- The low penetration depth of microwave irradiation, which hinders the scale up of reactions assisted by microwaves.

In this first part of the review we have described the efforts made to solve the problems of reproducibility. Reproducibility between microwave instruments can be achieved through accurate control of the reaction conditions, especially when the temperature of the reaction is the parameter to be controlled. This reproducibility can be extended from kitchen-type microwave instruments to monomode and multimode reactors and even when the reaction is scaled-up. Reproducibility can be attained regardless of the polarity of the solvent and also in solvent-free conditions, where the polarities of the substrates have a dramatic influence on the absorption of microwave irradiation. The only exceptions are the reactions in which the polarity increases dramatically along the reaction coordinate.

In the second part of the review, the strategies for scaling-up microwave reactions have been covered. One approach involves the use of large microwave cavities (volumes up to 12 L) in batch and, in some cases, with specially designed multimode microwave instruments.

A second approach concerns parallel reactions in multimode systems and probably the most promising is the use of continuous flow systems that combine the advantages of microwave irradiation, volumetric heating with a high heating rate, and flow reactors, especially the efficient heat transfer and the possible application of the same reaction conditions from laboratory scale to pilot plant.

## 5. Acknowledgments

Financial support from DGCYT of Spain through project CTQ2010-14975/BQU and Consejería de Educación JCCM through projects PII2I-0100 and PEII11-0049 is gratefully acknowledged.

## 6. References

Alcázar, J.; Diels, G. & Schoentjes, B. (2004). Reproducibility Across Microwave Instruments: First Example of Genuine Parallel Scale-Up of Compounds under Microwave Irradiation. *QSAR & Combinatorial Chemistry*, Vol.23, No.10, (December 2004), pp. 906-910, ISSN 1611-020X.

Alcázar, J. (2005). Reproducibility Across Microwave Instruments: Preparation of a Set of 24 Compounds on a Multiwell Plate under Temperature-Controlled Conditions. *Journal of Combinatorial Chemistry*, Vol.7, No.3, (May/June 2005), pp. 353-355, ISSN 1520-4766.

Andrade, M. M. & Barros, M. T. (2010). Fast Synthesis of N-Acylhydrazones Employing a Microwave Assisted Neat Protocol. *Journal of Combinatorial Chemistry*, Vol.12, No.2, (March/April 2010), pp. 245-247. ISSN 1520-4766.

Arbizzani, C.; Beninati, S.; Damen, L. & Mastragostino, M. (2007). Power and Temperature Controlled Microwave Synthesis of SVO. *Solid State Ionics*, Vol.178, No.5-6, (March 2007), pp. 393-398, ISSN 0167-2738.

Arvela, R. K.; Leadbeater, N. E. & Collins, M. J. Jr. (2005). Automated Batch Scale-Up of Microwave-Promoted Suzuki and Heck Coupling Reactions in Water Using Ultra-Low Metal Catalyst Concentrations. *Tetrahedron*, Vol.61, No.36, (September 2005), pp. 9349-9355, ISSN 0040-4020.

Barnard, T. M. ; Vanier, G. S. & Collins, M. J. Jr. (2006). Scale-Up of the Green Synthesis of Azacycloalkanes and Isoindolines under Microwave Irradiation. *Organic Process Research & Development*, Vol.10, No.6, (November 2006), pp. 1233-1237, ISSN 1083-6160.

Bergamelli, F.; Iannelli, M.; Marafie, J. A. & Moseley, J. D. (2010). A Commercial Continuous Flow Microwave Reactor Evaluated for Scale-up. *Organic Process Research & Development*, Vol.14, No.4, (July-August 2010) pp. 926-930, ISSN 1083-6160.

Bowman, M. D. ; Holcomb, J. L. ; Kormos, C. M. ; Leadbeater, N. E. & Williams, V. A. (2008). Approaches for Scale-Up of Microwave-Promoted Reactions. *Organic Process Research & Development*, Vol.12, No.1, (January/February 2008), pp. 41-57, ISSN 1083-6160.

Bowman, M. D. ; Scmink, J. R. ; McGowan, C. M. ; Kormos, C. M. & Leadbeater, N. E. (2008). Scale-Up of Microwave-Promoted Reactions to the Multigram Level Using a Sealed-Vessel Microwave Apparatus. *Organic Process Research & Development*, Vol.12, No.6, (November-December 2008) pp. 1078-1088. ISSN 1083-6160.

Cablewski, T.; Faux, A. F. & Strauss, C. R. (1994). Development and Application of a Continuous Microwave Reactor for Organic Synthesis. *Journal of the Organic Chemistry*, Vol.59, No.12, (June 1994), pp. 3408–3412, ISSN 0022-3263.

Caliendo, G.; Fiorino, F.; Perissutti, E.; Severino, B.; Gessi, S.; Cattabriga, E.; Borea, P.A. & Santagada, V. (2001). Synthesis by Microwave Irradiation and Binding Properties of Novel 5-HT1A Receptor Ligands. *European Journal of Medicinal Chemistry*, Vol.36, No.11-12, (November/December 2001), pp. 873-866, ISSN 0223-5234.

Caliendo, G.; Fiorino, F.; Perissutti, E.; Severino, B.; Scolaro, D.; Gessi, S.; Cattabriga, E.; Borea, P. A. & Santagada, V. (2002). A Convenient Synthesis by Microwave Heating and Pharmacological Evaluation of Novel Benzoyltriazole and Saccharine Derivatives as 5-HT1A Receptor Ligands. *European Journal of Pharmaceutical Science*, Vol.16, No.1-2, (July 2002), pp. 15-28, ISSN 0928-0987.

Caliendo, G.; Perissutti, E.; Santagada, V.; Fiorino, F.; Severino, B.; Cirillo, D.; D'Emmanuele de Villa Bianca, R.; Loppolis, L.; Pinto, A. & Sorrentino, R. (2004). Synthesis by Microwave Irradiation of a Substituted Benzoxazine Parallel Library with Preferential Relaxant Activity for Guinea Pig Trachealis. *European Journal of Medicinal Chemistry*, Vol.39, No.10, (October 2004), pp. 815-826, ISSN 0223-5234.

Chen, S.-T.; Chiou, S.-H. & Wang, K.-T. (1990). Preparative Scale Organic Synthesis Using a Kitchen Microwave Oven. *Journal of the Chemical Society, Chemical Communications*, No.11, (June 1900), pp. 807-809, ISSN 0022-4936.

Cléophax, J.; Liagre, M.; Loupy, A. & Petit, A. (2000). Application of Focused Microwaves to the Scale-Up of Solvent-Free Organic Reactions. *Organic Process Research & Development*, Vol.4, No.6, (November/December 2000), pp. 498-504, ISSN 1083-6160.

Comer, E. & Organ, M. G. (2005). A Microreactor for Microwave-Assisted Capillary (Continuous Flow) Organic Synthesis. *Journal of the American Chemical Society*, Vol.127, No.22, (June 2005), pp. 8160-8167, ISSN 0002-7863.

Cotas, S. J.; Schulz, M. J.; Carson, J. R.; Codd, E. E.; Hlasta, D. J.; Pitis, P. M.; Stone, D. J.; Zhang, S. P.; Colburn, R. W. & Dax, S. L. (2004). Parallel Methods for the Preparation and SAR Exploration of N-ethyl-4-[(8-alkyl-8-aza-bicyclo[3.2.1]oct-3-ylidene)-aryl-methyl]-benzamides, Powerful mu and delta Opioid Agonists. *Bioorganic & Medicinal Chemistry Letters*, Vol.14, No.22, (November 2004), pp. 5493-5498, ISSN 0960-894X.

Deetlefs, M. & Seddon, K. (2003). Improved Preparations of Ionic Liquids Using Microwave Irradiation. *Green Chemistry*, Vol.5, No.2, (March 2003), pp. 181-186, ISSN 1463-9262.

Díaz-Ortiz, A.; Díez-Barra, E.; de la Hoz, A.; Moreno, A.; Gómez-Escalonilla, M. J. & Loupy, A. (1996). 1,3-Dipolar Cycloaddition of Nitriles under Microwave Irradiation in Solvent-Free Conditions. *Heterocycles*, Vol.43, No.5, (May 1996), pp. 1021-1030, ISSN 0385-5414.

Díaz-Ortiz, A.; de la Hoz, A.; Alcázar, J.; Carrillo, J. R.; Herrero, M. A.; Fontana, A. & Muñoz, J. de M. (2007). Reproducibility and Scalability of Solvent-Free Microwave-Assisted Reactions: From Domestic Ovens to Controllable Parallel Applications. *Combinatorial Chemistry & High Throughput Screening*, Vol.10, No.3, (March 2007), pp. 163-169, ISSN 1386-2073.

Díaz-Ortiz, A.; de la Hoz, A.; Alcázar, J.; Carrillo, J. R.; Herrero, M. A.; Fontana, A.; Muñoz, J. de M.; Prieto, P. & de Cózar, A. (2011). Influence of Polarity on the Scalability and Reproducibility of Solvent-Free Microwave-Assisted Reactions. *Combinatorial Chemistry & High Throughput Screening*, Vol.14, No.2, (February 2011), pp. 109-116, ISSN 1386-2073.

Dressen, M. H. C. L.; Kruijs, B. H. P. van de; Meuldijk, J.; Vekemans, J. A. J. M. & Hulshof, L. A. (2010). Flow Processing of Microwave-Assisted (Heterogeneous) Organic Reactions. *Organic Process Research & Development*, Vol.14, No.3, pp. 351-361, ISSN 1083-6160.

Eskilsson, C. S. & Björklund, E. (2000). Analytical-Scale Microwave-Assisted Extraction. *Journal of Chromatography A*, Vol.902, No.1, (December 2000) pp. 227-250, ISSN 0021-9673.

Glasnov T. N. & Kappe, C. O. (2007). Microwave-Assisted Synthesis under Continuous-Flow Conditions. *Macromolecular Rapid Communications*, Vol.28, No.4, (February 2007), pp. 395-410, ISSN 1022-1336.

Horikoshi, S. ; Abe, H. ; Torigoe, K. ; Abe, M. & Serpone, N. (2010). Access to Small Size Distributions of Nanoparticles by Microwave-Assisted Synthesis. Formation of Ag Nanoparticles in Aqueous Carboxymethylcellulose Solutions in Batch and Continous-Flow Reactors. *Nanoscale*, Vol.2, No.8, (May 2010), pp. 1441-1447. ISSN 2040-3364.

Ju, Y. & Varma, R. S. (2006). Aqueous N-Heterocyclization of Primary Amines and Hydrazines with Dihalides: Microwave-Assisted Syntheses of N-Azacycloalkanes, Isoindole, Pyrazole, Pyrazolidine, and Phthalazine Derivatives. *Journal of the Organic Chemistry*, Vol.71, No.1, (January 2006), pp. 135-141, ISSN 0022-3263.

Kappe, C. O. & Stadler, A. (2005). *Microwaves in Organic and Medicinal Chemistry*, Wiley-VCH, ISBN 3-527-31210-2, Weinheim.

Kremsner, J. M.; Stadler, A. & Kappe, C. O. (2007). High-Throughput Microwave-Assisted Organic Synthesis: Moving from Automated Sequential to Parallel Library-Generation Formats in Silicon Carbide Microtiter Plates. *Journal of Combinatorial Chemistry*, Vol.8, No.2, pp. 298-291, ISSN 1520-4766.

Kunz, U.; Kirschning, A.; Wen, H.-L.; Solodenko, W.; Cecilia, R.; Kappe, O. & Turek, T. (2005). Monolithic Polymer/Carrier Materials: Versatile Composites for Fine Chemical Synthesis. *Catalysis Today*, Vol.105, No.3-4, (August 2005), pp. 318-324, ISSN 0920-5861.

Kurzawa, J. & Stachowiak, S. (2001). Investigation of Influence of Microwave Energy on Cleavage of S-S and C-S Bonds by Azide Ions. *Chemia Analityczna*, Vol.46, No.6, (March 2001), pp. 907-910, ISSN 0009-2223.

Leadbeater, N. E.; Williams, V. A.; Barnard, T. M. & Collins, M. J. Jr. (2006). Open-Vessel Microwave-Promoted Suzuki Reactions Using Low Levels of Palladium Catalyst: Optimization and Scale-Up. *Organic Process Research & Development*, Vol.10, No.4, (July 2006), pp. 833-837, ISSN 1083-6160.

Leadbeater, N. E. & Schmink, J. R. (2007). Use of a Scientific Microwave Apparatus for Rapid Optimization of Reaction Conditions in a Monomode Function and then Substrate Screening in a Multimode Function. *Tetrahedron*, Vol.63, No.29, (July 2007), pp. 6764-6773, ISSN 0040-4020.

Leadbeater, N. E. (CRC Press) (2010). Microwave Heating as a Tool for Process Chemistry: From Microwave Heating as a Tool for Sustainable Chemistry, Taylor & Francis Group, ISBN 978-1-4398-1269-3, Boca Raton.

Lehmann, F.; Pilotti, A. & Luthman, K. (2003). Efficient Large Scale Microwave Assisted Mannich Reactions Using Substituted Acetophenones. *Molecular Diversity*, Vol.7, No.2-4, (June 2003), pp. 145-152, ISSN 1381-1991.

Lehmann, H. & LaVecchia, L. (2005). Evaluation of Microwave Reactors for Prep-Scale Synthesis in a Kilolab. *Journal of the Association for Laboratory Automation*, Vol.10, No.6 , (December 2005), pp. 412-417, ISSN 1540-2452.

Login, G. R. & Dvorak, A. M. (1994). Application of Microwave Fixation Techniques in Pathology. *Journal of Neuroscience Methods*, Vol.55, No.2, (December 1994), pp. 173-182, ISSN 0165-0270.

Loones, K. T. J.; Maes, B. U. W.; Rombouts, G.; Hostyn, S. & Diels, G. (2005). Microwave Assisted Organic Synthesis: Scale-Up of Palladium-Catalyzed Aminations Using Single-Mode and Multi-Mode Microwave Equipment. *Tetrahedron*, Vol.61, No.43, (October 2005), pp. 10338-10348, ISSN 0040-4020.

Marafield, J. A. & Moseley, J. D. (2010). The Application of Stop-Flow Microwave Technology to Scaling-Out $S_NAr$ Reactions Using a Soluble Organic Base. *Organic & Biomolecular Chemistry*, Vol.8, No.9, (February 2010), pp. 2219-2227, ISSN 1477-0520.

Murray, J. K. & Gellam, S. H. (2006). Microwave-Assisted Parallel Synthesis of a 14-Helical β-Peptide Library. *Journal of Combinatorial Chemistry*, Vol.8, No.1, (January-February 2006), pp. 58-65, ISSN 1520-4766.

Nikbin, N.; Ladlow, M. & Ley, S. V. (2007). Continuous Flow Ligand-Free Heck Reactions Using Monolithic Pd [0] Nanoparticles. *Organic Process Research & Development*, Vol.11, No.3, (May-June 2007), pp. 458–462, ISSN 1083-6160.

Nüchter, M. & Ondruschka, B. (2003). Tools for Microwave-Assisted Parallel Syntheses and Combinatorial Chemistry. *Molecular Diversity*, Vol.7, No.2-4, (June 2003), pp. 253-264, ISSN 1381-1991.

Nüchter, M.; Müller, U.; Ondruschka, B.; Tied, A. & Lautenschläger, W. (2003). Microwave-Assisted Chemical Reactions. *Chemical & Engeneering Technology*, Vol.26, No.12, (December 2003), pp. 1207-1216, ISSN 0930-7516.

Nüchter, M.; Ondruschka, B.; Bonrath, W. & Gum, A. (2004). Microwave-Assisted Synthesis - A Critical Technology Overview. *Green Chemistry*, Vol.6, No.3, (February 2004), pp. 128-141, ISSN 1463-9262.

Ondruschka, B.; Bonrath, W. & Stuerga, D. (2006). Chapter 2, In: *Microwaves in Organic Synthesis, 2nd. Edition*, Loupy, A., Ed., pp. 62-107, Wiley-VCH, ISBN 3-527-31452-0, Weinheim.

Paulus, R. M. ; Erdmenger, T.; Hoogenboom, R. ; Pilotti, A. & Schubert, U. S. (2006). Scale-Up of Microwave-Assisted Polymerizations in Batch Mode: The Cationic Ring-Opening Polymerization of 2-Ethyl-2-oxazoline. *Macromolecular Rapid Communications*, Vol.27, No.18, (September 2006), pp. 1556-1560, ISSN 1022-1336.

Perio, B.; Dozias, M. J. & Hamelin, J. (1998). Ecofriendly Fast Batch Synthesis of Dioxolanes, Dithiolanes, and Oxathiolanes without Solvent under Microwave Irradiation. *Organic Process Research & Development*, Vol.2, No.6, (November-December 1998), pp. 428-430, ISSN 1083-6160.

Raner, K. D.; Strauss, C. R.; Trainor R. W.; Thorn, J. S. (1995). A New Microwave Reactor for Batchwise Organic Synthesis. *Journal of the Organic Chemistry*, Vol.60, No.8, (April 2005), pp. 2456-2460, ISSN 0022-3263.

Razzaq, T. & Kappe, C. O. (2007). Rapid Preparation of Pyranoquinolines Using Microwave Dielectric Heating in Combination with Fractional Product Distillation. *Tetrahedron Lett.*, Vol.48, No.14, (April 2007), pp. 2513-2517, ISSN 0040-4039.

Roberts, B. A. & Strauss, C. R. (2005). Toward Rapid, "Green", Predictable Microwave-Assisted Synthesis. *Accounts of Chemical Research*, Vol.38, No.8, (August 2005), pp. 653–661, ISSN 0001-4842.

Schmink, J. R. ; Kornos, C. M. ; Devine, W. G. & Leadbeater, N. E. (2010). Exploring the Scope for Scale-Up of Organic Chemistry Using a Large Batch Microwave Reactor. *Organic Process Research & Development*, Vol.14, No.1, (January-February 2010), pp. 205-214, ISSN 1083-6160.

Singh, B. K.; Kaval, N.; Tomar, S.; Van der Eycken, E. & Parmar, V. S. (2008).Transition Metal-Catalyzed Carbon-Carbon Bond Formation Suzuki, Heck, and Sonogashira Reactions Using Microwave and Microtechnology. *Organic Process Research & Development*, Vol.12, No.3, (May-June 2008), pp. 468-474, ISSN 1083-6160.

Slade, J. ; Parker, D. ; Girgis, M. ; Wu, R. ; Joseph, S. & Repic, O. (2007). Optimization and Scale-Up of the Grandberg Synthesis of 2-Methyltryptamine. *Organic Process Research & Development*, Vol.11, No.4, (July-August 2007), pp. 721-725, ISSN 1083-6160.

Smith, C. J.; Iglesias-Sigüenza, F. J.; Baxendale, I. R. & Ley, S. V. (2007). Flow and Batch Mode Focused Microwave Synthesis of 5-Amino-4-cyanopyrazoles and Their Further Conversion to 4-Aminopyrazolopyrimidines. *Organic & Biomolecular Chemistry*, Vol.5, No.17, (July 2007), pp. 2758-2761, ISSN 1477-0520.

Stadler, A.; Yousefi, B. H.; Dallinger, D.; Walla, P.; Van der Eycken, E.; Kaval, N. & Kappe, C. O. (2003). Scalability of Microwave Assisted Organic Synthesis. From Single-Mode to Multimode Parallel Batch Reactors. *Organic Process Research & Development*, Vol.7, No.5, (September-October 2003), pp. 707-716, ISSN 1083-6160.

Wilson, N. S.; Sarko, C. R. & Roth, G. P. (2004). Development and Applications of a Practical Continuous Flow Microwave Cell. *Organic Process Research & Development*, Vol.8, No.3, (May-June 2004), pp. 535-538, ISSN 1083-6160.

# 8

# Use of Microwave Heating in Coal Research and in Materials Synthesis

Mohindar S. Seehra and Vivek Singh
*Department of Physics, West Virginia University*
*United States*

## 1. Introduction

In this chapter, the use of microwave heating in addressing some problems in coal research and in the synthesis of inorganic materials are reviewed. The review covers the underlying theory of microwave heating and how it has been used in specific cases. This review is divided into three sections: Section 2 dealing with the fundamental theory behind microwave heating and section 3 and 4 dealing with its use in coal research and in the synthesis of materials respectively. References to selected papers are also given. However this list of references is not exhaustive and readers are asked to consult these references for earlier studies in specific areas.

## 2. Fundamentals of microwave heating

When a time varying electric field $E = E_o\exp(i\omega t)$ is applied to a liquid or solid, polarization develops which in general can come from three sources viz. dipolar, ionic and electronic (Kittel, 1996). The dipolar contribution is present in polar materials viz. materials with permanent dipoles such as water as these dipoles re-orient in the direction of the applied field E. The relaxation time $\tau$ associated with this reorientation is in picoseconds (ps) and therefore this phenomena is observable when frequency $f = (\omega/2\pi)$ is in the microwave range (f = 0.1 GHz to 100 GHz). The ionic contribution to the polarization results from the various normal modes of vibration of the atoms relative to each other. These frequencies in the IR range are from 300 GHz to 100 THz. Finally, the electronic contribution resulting from the relative displacements of the nucleus with positive charge and electron shell of the atoms occurs in the ultraviolet (UV) region of the frequencies. Thus microwave heating as shown below is most significant in materials having permanent dipoles such as water.

A number of experimental studies (Kaatze, 1989; Merabat & Bose, 1988) have shown that dielectric relaxation in water is adequately described by the Debye relation (Debye, 1929)

$$\varepsilon(\omega) = \varepsilon(\infty) + [\varepsilon(0) - \varepsilon(\infty)]/(1 + i\omega\tau) \tag{1}$$

where $\varepsilon(0)$ and $\varepsilon(\infty)$ are the dielectric constant in the limit $\omega\rightarrow 0$ and $\omega\rightarrow\infty$ respectively. For water, $\varepsilon(\infty)$ includes the polarization effects at the infrared and optical-uv frequencies. Separating the in phase and the out of phase components of $\varepsilon = \varepsilon' - i\varepsilon''$ in Eq.(1) leads to

$$\varepsilon' = \varepsilon(\infty) + [\varepsilon(0) - \varepsilon(\infty)]/(1 + \omega^2\tau^2) \quad (2)$$

and

$$\varepsilon'' = \omega\tau[\varepsilon(0) - \varepsilon(\infty)]/(1 + \omega^2\tau^2) \quad (3)$$

Eliminating $\varepsilon(\infty)$ through substitutions in Eq. (2) and (3) leads to the following relation between $\varepsilon'$ and $\varepsilon''$:

$$\varepsilon' = \varepsilon(0) - \varepsilon''\cdot\omega\cdot\tau \quad (4)$$

Once experimental data for $\varepsilon'$ and $\varepsilon''$ are available for different $\omega$, a plot of $\varepsilon'$ vs. $\varepsilon''\cdot\omega$ allows one to determine the relaxation time $\tau$. Such an analysis for water in the frequency range of 1 GHz to 60 GHz and in the temperature range of 0°C and 60°C have shown (Kaatze, 1989) that at T = 25°C, $\tau$ = 8.27 ps, $\varepsilon(0)$ = 78.36 and $\varepsilon(\infty)$ = 5.2 and that with increase of temperature, $\varepsilon(0)$, $\varepsilon(\infty)$ and $\tau$ decrease rapidly so that at 60°C, $\varepsilon(0)$ = 66.7, $\varepsilon(\infty)$ = 4.2 and $\tau$ = 4.0 ps.
Microwave heating involves absorption of electrical power by water from the microwaves. The power absorbed P can be shown to be given by (Agmon, 1996)

$$P = 2\pi|E_o|^2\varepsilon_o f\cdot\varepsilon'', \quad w/m^3 \quad (5)$$

where $\varepsilon_0 = 8.854 \times 10^{-12}$ A s/V m, $E_o$ is the amplitude of the electric field and $\omega = 2\pi f$. The important point to note is that P is proportional to $f\cdot\varepsilon''$ so that P depends upon the frequency f of the microwaves and $\varepsilon''$ of the medium (water). From Eq. (3) it can be easily shown that $\varepsilon''$ peaks at $\omega\tau = 1$.
In order to provide visualization of the frequency variation of the quantities $\varepsilon'$, $\varepsilon''$ and $f\cdot\varepsilon''$, the latter representing the power absorbed P, we have used the data for water given in a tabular form (Kaatze, 1989). The simulated plots of the frequency dependence of $\varepsilon'$, $\varepsilon''$ and $f\cdot\varepsilon''$ for the data valid at 25°C are shown as lines in Fig. 1 where the circles are experimental values (Kaatze, 1989). The data up to 60 GHz fit quite well with the Debye relaxation with a single $\tau$. Later studies have suggested that two relaxation times are necessary to explain the data at frequencies higher than 100 GHz (Barthel et al., 1990; Agmon, 1996).
An interesting feature of Fig. 1 is the frequency dependence of $f\cdot\varepsilon''$ representing power absorbed. This plot, usually not avaliable in literature, shows that power absorbed saturates only near 100 GHz. At f = 2.45 GHz, the conventional frequency used in microwave ovens, domestic and industrial, the absorbed power is only about 1.6% of the power absorbed at 100 GHz. However, the wavelengths of microwaves $\lambda$ = 0.3 cm at 100 GHz compared to $\lambda$ = 12.2 cm at 2.45 GHz, thus reducing the size of the microwave ovens to impractical size of only about 0.3 cm at the higher frequency. At 17 GHz where $\varepsilon''$ has the maximum value of 36, $\lambda$ = 1.76 cm and power absorbed is nearly 50% of the maximum value at 100 GHz.
As water heats up, $\tau$, $\varepsilon(0)$ and $\varepsilon(\infty)$ decrease (Kaatze, 1989) affecting the peak positions for $\varepsilon''$ and $f\cdot\varepsilon''$. In Fig. 2 simulated frequency dependence of $\varepsilon'$, $\varepsilon''$ and $f\cdot\varepsilon''$ is compared at two temperatures viz. 25°C and 60°C using the magnitudes of $\varepsilon(0)$, $\varepsilon(\infty)$ and $\tau$ (Kaatze, 1989). The important point to note is that the power absorbed at a fixed frequency below about 30 GHz decreases as the temperature of water increases.
The frequency derivative of power absorbed by water given by $\partial(f\cdot\varepsilon'')/\partial f$, is plotted in Fig. 3. The maximum slope occurs at 11 GHz for 25°C and the position of this maximum changes to 23 GHz at 60°C. Analytically, using relation (3), it can be shown that $\partial(f\cdot\varepsilon'')/\partial f$ is maximum at $f = 1/\pi\tau\sqrt{12}$.

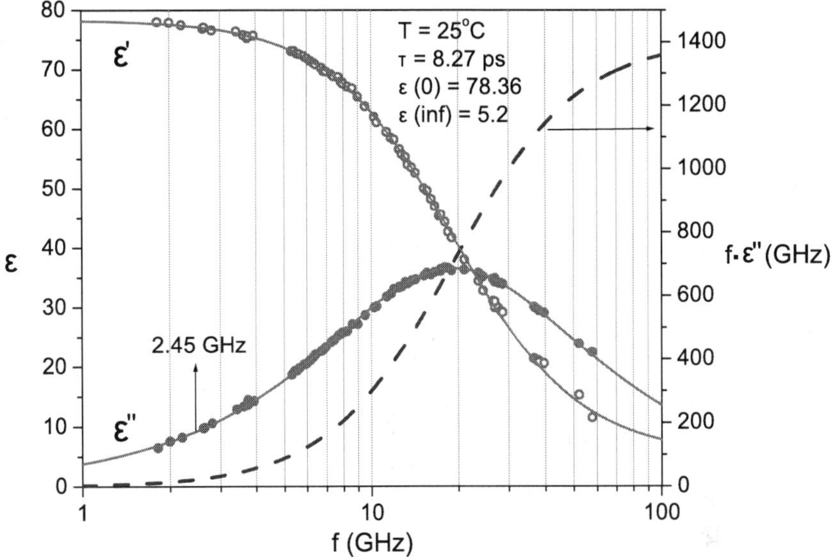


Fig. 1. The simulated plots of the frequency dependence of ε′, ε″ and f·ε″ for water using the data (circles) at 25°C from Kaatze (1989).

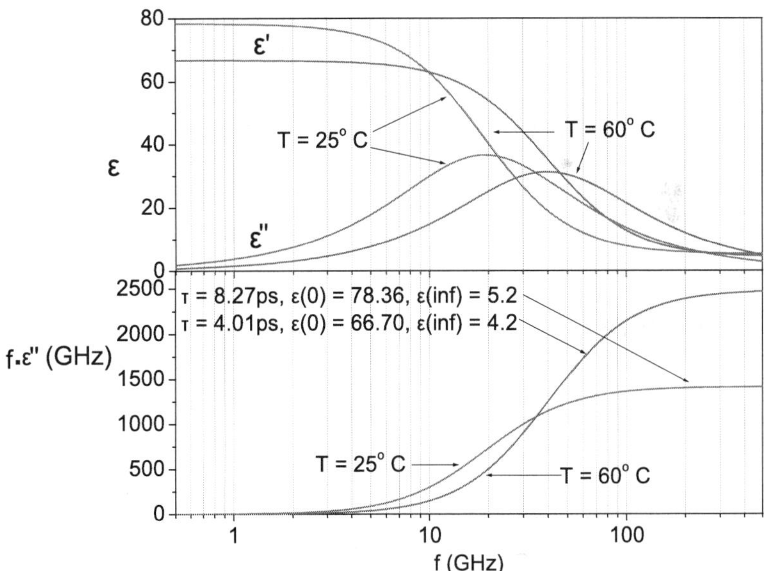


Fig. 2. Simulated frequency dependence of ε′, ε″ and f·ε″ is compared for water at two temperatures viz. 25°C and 60°C using the magnitudes of ε(0), ε(∞) and τ

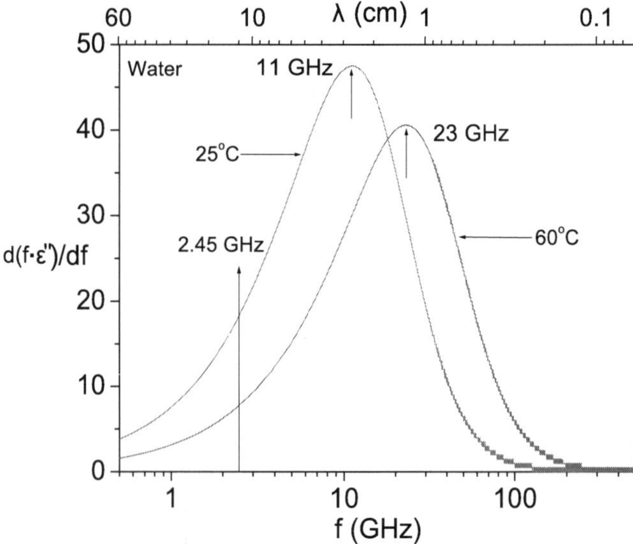


Fig. 3. Derivatives of power absorbed from Fig. 2 vs. frequency and wavelength at 25°C and 60°C.

## 3. Microwave heating in coal research

### 3.1 Microwave properties of coals

In the recent paper, (Marland et al., 2001) have reported the measurements of ε′and ε″ at 60°C for over a dozen coals and for the ash components normally found in most coals. In these studies, data were reported at f = 0.615 GHz, 1.413 GHz and 2.216 GHz for the coals and the minerals pyrite, quartz, dolomite, kaolin, mica and calcite. Although there is some variations in the magnitudes of ε′ and ε″ from coal to coal, the range of values were between 1.6 to 3.5 for ε′ and 0.07 to 0.31 for ε″at f = 2.216 GHz. The frequency dependence among the three frequencies was found to be relatively weak. Among the minerals the highest magnitude was found to be for pyrite with ε′= 7.07 and ε″= 1.06 at 2.216 GHz and at 60°C. For the other minerals, ε′ ~ 2 to 3 and ε″ ~ 0.05-0.06. Moisture levels in these coals varied from a low of 1% to a high of 13%. From Fig. 1 ε″ $\simeq$ 10 for water at 2.45 GHz. Since power P absorbed is proportional to f•ε″ (Eq. 1), the highest power absorbed will be by moisture, followed by pyrite. Thus the largest effect of microwave heating of coal is the removal of moisture and heating of pyrite $FeS_2$.

Pyrite $FeS_2$ when heated above 250°C is known to convert to pyrrhotites $Fe_{1-x}S$ ($x$ = 0 - 0.125) whose mangetization at room temperature is ferrimagnetic like, with the magnitude depending on the $x$ value (Jagadeesh & Seehra, 1981). Whereas $FeS_2$ is only paramagnetic with feeble magnetization (Burgardt & Seehra, 1977) pyrrhotites $Fe_{1-x}S$ are strongly magnetic (Jagadeesh & Seehra, 1981). Therefore conversion of pyrites to pyrrhotites after microwave heating make them amenable to separation by magnetic methods. Thus microwave heating of coals at an appropriate frequency followed by magnetic separation of the impurities can lead to both dewatering and desulphurization, since a major source of sulphur in coals is pyrites. A combination of these techniques is likely being used in the recently developed and marketed technology by CoalTek (http://www.coaltek.com/).

### 3.2 Measuring moisture content by microwaves

Another area in which microwaves have been used in coal research is the on-line measurements of moisture in coals employing the frequencies of 2.45 GHz (Cutmore et al., 2000) and 1 GHz (Ponte et al., 1996). In these techniques, microwave horns are used to transmit and receive microwaves through the coal sample. The use of the lower microwave frequencies and hence corresponding larger wavelengths of $\lambda$ = 12.2 cm (30 cm) for 2.45 GHz (1 GHz) allows the use of larger samples encountered at coal power plants. As microwaves pass through a coal sample, large $\varepsilon'$ of water produces a frequency and phase shift of the microwaves and $\varepsilon''$ results in the attenuation of the microwave signal. Both phase shift and attenuation are measured by microwave network analyzer. However, experiments show that phase shift provides a better linear correlation with the moisture content (Ponte et al., 1996; Cutmore et al., 2000). Since attenuation is due to absorption of microwaves primarily by moisture present, this technique can also result in sufficient dewatering of the coals. This topic of dewatering or drying of coals by using microwave is discussed next.

### 3.3 Microwave dewatering of coals

Dewatering or drying of coals is an important consideration in coal industry since removal of moisture reduces the transportation costs and improves the BTU value of the coals. The use of microwaves for dewatering of coals has been discussed in a number of earlier studies (Lindroth, 1985; Chatterji & Mishra, 1991; Perkin, 1980; Standish el al., 1988; Marland el al., 2000; Lester & Kingman, 2004) and in a more recent study from author's laboratory (Seehra et al., 2007). In microwave dewatering of coals, the basic mechanism for power absorption by water discussed in Section 2 is operative.

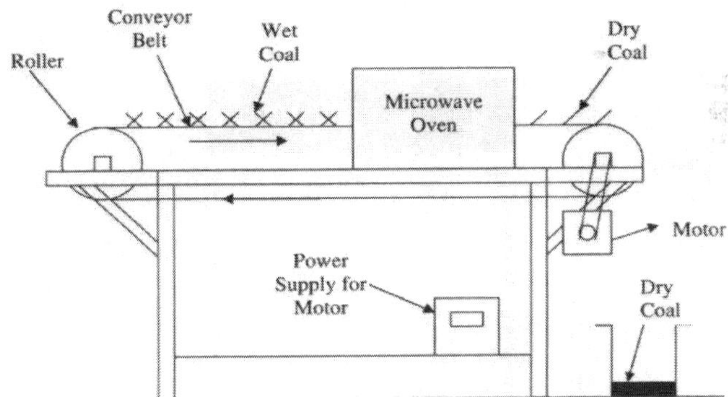


Fig. 4. Block diagram of the bench-scale microwave dewatering unit (reproduced from the paper by Seehra et al., Fuel 2007).

In our studies of microwave dewatering (Seehra et al., 2007), the focus was on dewatering fine coal slurries containing nearly 50% moisture. A domestic microwave oven operating at 2.45 GHz was modified, Fig. 4 and Fig. 5, so that a conveyer belt could be used to continously feed wet coal into the oven and dried coal collected at the other end. The whole apparatus was shielded with a microwave shield. This bench scale unit was designed to simulate a practical situation at a coal plant. The website of CoalTek shows such a conveyer belt approach and perhaps the only way to achieve large scale continous operation.

Fig. 5. (Left) Picture of the bench-scale microwave dewatering units with microwave shield in place. (Right) Close-up of the unit without the microwave shield showing the conveyor belt for feeding wet coal into the microwave oven (reproduced from the paper by Seehra et al., Fuel 2007).

In our experiment using a 0.8 kW domestic oven, the amount of moisture removed depended strongly on the belt speed; The slower belt speed removed more moisture but consumed more energy. This data on the relation between belt speed, electrical energy consumed and more moisture lost is shown in Fig. 6 (Seehra et al., 2007). Belt speed of

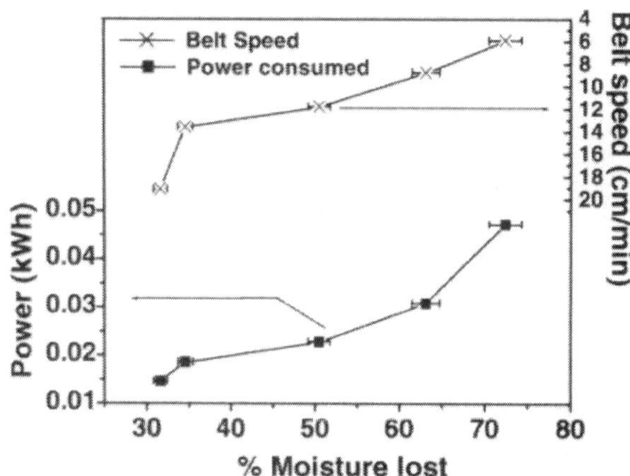


Fig. 6. Power consumed and belt speed (cm/min) vs. % moisture lost. The region of the least slope (13 cm/min) yields the optimum speed (reproduced from the paper by Seehra et al., Fuel 2007).

around 13 cm/min was found to be most energy efficient (~ 83%) for removal of water, the remaining power perhaps lost to surroundings and small heating of the minerals in coal and coal itself. In the experiments of Lindroth using higher power magnetrons (9.7 kW) operating at 2.45 GHz, the reported efficiency varied from 52% to 97% for drying of fine coals. A cost estimate of $ 3/ton to remove 10% moisture from coal was made based only on the amount of electrical power consumed (Seehra et al., 2007).

Dewatering efficiencies of microwave heating vis-à-vis thermal heating for the coal slurry samples are compared in Fig. 7 (Seehra et al, 2007). In this comparison, 20 gms of coal slurry sample was heated in a thermal oven set at 110°C followed by taking the sample out after every 30 s for quick weighing and reinserting the sample in the oven in less than 5 s. Similar experiments were done using a domestic microwave oven. The results of weight loss vs. time in Fig. 7 show that complete dewatering of the coal slurry takes about 5 min with microwave heating compared to about an hour with thermal heating. Thus this experiment shows an order of magnitude reduction in time in microwave heating vis-à-vis thermal heating.

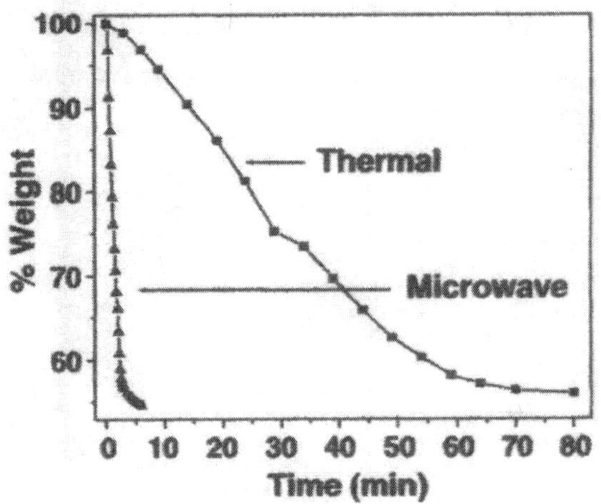


Fig. 7. Comparison of the efficiencies of thermal heating vis-a`-vis microwave heating using 20 g sample in each case (reproduced from the paper by Seehra et al., Fuel, 2007).

## 4. Use of microwaves in material synthesis

In section 2, the mechanism of microwave heating was described in terms of dielectric constants ε′and ε″ with specific examples given for microwave heating of water. Same argument can be used for all polar materials with known ε′and ε″. As an example, Barthel et al. (1990) have discussed the microwave heating using lower alcohols (methanol, ethanol, 1-propanol and 2-propanol).

In the chemical synthesis of a variety of materials, use of a conventional heating which has been a norm for a long time, is being replaced by microwave heating. In conventional heating, energy is first delivered to the surface of a material, followed by heat transfer to the material through conduction, convection and radiation. This results in non-uniform heating producing poor crystallinity and often wider distribution of particle sizes. By contrast, in microwave heating, energy is transferred throughout the material via molecular interaction thus producing uniform heating without thermal gradients and stress. This often results in materials with finer microstructures and particles with narrow size distribution. The reader is referred to several recent papers (Barison et al., 2010; Wu et al., 2008; Prasad et al. 2002; Utchariyajit et al. 2010; Chang et al. 2009; Gopinath et al. 2002; Vincente et al 2010) where examples of these advantages are described in detail. Thus, compared to conventional thermal heating, microwave heating has the advantages of speed, energy efficiency, finer microstructures and more uniform particle size distribution in the synthesis of materials in line with the data shown in Fig. 7.

Microwave heating in combination with hydrothermal processing has shown positive effects on crystallinity of samples. Samples prepared by conventional hydrothermal treatment showed poor crystallinity in an amorphous phase, and they required higher temperatures and longer times than that of microwave hydrothermal treatment (Kim et al. 2001). Crystalline samples have been obtained without increase in particle size (Addamo et al. 2008). Other observations have shown the particles to have higher crystallinity indicating larger particle size (Vincente et al. 2009), and it avoids the post treatment such as annealing required after sol-gel synthesis (Hu et al. 2007). The treatment time under microwave heating has shown increase in the crystallite size due to Oswald mechanism of crystallite growth. Thus for applications which involve surface and phase characteristics, and the presence of porosity used in catalytic applications, microwave synthesis has been very successful (Wilson et al. 2006).

From a review done by Yanshou et al. in 2008, synthesis using microwave heating has been found to have the Si/Al ratio increased from 1.4 to 1.8 in zeolite membranes. It is clearly pointed out that the difference between microwave heating and conventional heating is due to different heating mechanisms (Li &Yang, 2008). Shape effects have also been noticed by Chen et al. 2009, where the concentration of solvents of diols have significantly affected the length of the Si-MFI crystals compared to the widths and thickness having a lesser effect. This effect could not be reproduced using a conventional synthesis method (Chen et al., 2009). In general, microwave hydrothermal treatment has proved to be more efficient than conventional hydrothermal treatment because of energy saving and time. For preparing microwave-assisted organic materials, readers are referred to the review by Lidström et al. (2001).

Experimentally, all materials placed inside a microwave oven during operation must be "microwave safe" , i.e.. they cannot be good electrical conductors, as is well known in the domestic use of microwave cooking. This is because the concentrated electric field associated with the microwaves directly flows through the good conductor resulting in electrical sparks and even fires. Ceramic and polymer containers which are good insulators and so "microwave safe" are now avaliable for use in material synthesis.

Historically, the earlier use of microwave heating in material synthesis involved the use of domestic microwave oven appropriately modified by each investigator. In recent years, microwave ovens specially designed for synthesis of materials are now avaliable in which temperatures and even pressures in an autoclave can be monitored in situ. The use of autoclaves inside the microwave ovens producing materials under hydrothermal conditions has greatly expanded the use of microwave heating. (e.g. see Chang et al. 2009 & Kim et al. 2001). Suppliers of such commercial units can be easily located by the readers in each country through internet search.

In the rest of this chapter on the use of microwave heating in the synthesis of materials, we have organized the available examples of such cases in a tabular form. The following table lists the name of the materials synthesized, conditions used in the synthesis and the corresponding reference covering the period from 2010 to 1998 with the most recent listed first. From this list, it is evident that the use of microwave heating in material synthesis is now becoming ubiquitous and it is sure to expand even more in the future.

| SYSTEM | CONDITIONS | REFERENCES |
|---|---|---|
| $BaCe_{0.65}Zr_{0.20}Y_{0.15}O_{3-\delta}$ | P = 350, 500 W | Barison S., Fabrizio M., Fasolin S., Montagner F. &Mortalò C., (2010). A microwave-assisted sol-gel Pechini method for the synthesis of $BaCe_{0.65}Zr_{0.20}Y_{0.15}O_{3-\delta}$powders. *Mater. Res. Bull.*, Vol. 45, pp. 1171-1176. |
| $Zn_4O(BDC_3)$ | t = 30 min, P = 1 kW | Lu C., Liu J., Xiao K.& Harris A., (2010). Microwave enhanced synthesis of MOF-5 and its $CO_2$ capture ability at moderate temperatures across multiple capture and release cycles. *Chem. Eng. J.*, Vol. 156, pp. 465-470. |
| Capillary MFI-type Zeolite-Ceramic Membrane | T = 70 - 180 °C, P = 250 or 400 W, t = 10 - 150 min | Sebastian V., Mallada R., Coronasa J., Julbe A., Terpstra R.&Dirrix R., (2010). Microwave-assisted hydrothermal rapid synthesis of capillary MFI-type zeolite-ceramic membranes for pervaporation application.*J. Membrane Sci.*, Vol. 355, pp. 28-35. |
| Gold Nanoparticles | t = 30 sec, P = 800 W t = 9 min, P = 1 kW | Vargas-Hernandez C., Mariscal M., Esparza R. &Yacaman M., (2010). A synthesis route of gold particles with using a reducing agent.*Appl. Phys. Lett.* Vol. 96, pp. 213115-(1-3). |
| Silicalite (Si-MFI) Crystals | T = 180 °C, P = 400 W, t = 10 min | Chen X., Yan W., Cao X. &Xu R., (2010). Quantitative correlation between morphology of silicalite-1 crystals and dielectric constants of solvents. *Microporous andMesoporous Materials*, Vol. 131, pp. 45-50. |
| Saponite $M_x[Mg_6Al_xSi_{8-x}O_{20}(OH)_4]$ (M = Na, Li, $NH_4$) | T = 453 K, t = 6 hrs | Vicente I., Salagre P., Cesteros Y., Medina F.&Sueiras J., (2010). Microwave-assisted synthesis of saponite. *Appl. Clay Sci.*, Vol. 48, pp. 26-31. |
| $K_8Ln_3Si_{12}O_{32}NO_3 \cdot H_2O$ (Ln = Eu, Tb, Gd, Sm). | T = 363 K, t = 20 min | Wang X., Li J., Wang G., Han Y., Su T., Li Y., Yu J.&Xu R., (2010). Synthesis, characterization and properties of microporous lanthanide silicates:$K_8Ln_3Si_{12}O_{32}NO_3 \cdot H_2O$ (Ln = Eu, Tb, Gd, Sm). *Solid State Sciences*, Vol. 12, pp. 422-427. |

| | | |
|---|---|---|
| Calcia Stabilized Zirconia Nanoparticles | T = 200 °C, t = 90 min | Rizzuti A., Corradi A., Leonelli C., Rosa R., Pielaszek R. &Lojkowski W., (2010). Microwave technique applied to the hydrothermal synthesis and sintering of calcia stabilized zirconia nanoparticles. *J. of Nanopart. Res.*, Vol. 12, pp. 327-335. |
| Double-walled Carbon Nanotubes | T = 200 °C, t = 20 min | Dunens O. M., MacKenzie K. J. & Harris A. T., (2010). Large-Scale Synthesis of Double-Walled Carbon Nanotubes in Fluidized Beds. *Ind. Eng. Chem.Res.*, Vol. 49, No. 9, pp. 4031-4035). |
| One-Dimensional Nano-Structured Silver Titanates | T = 200 °C, P = 300 W, t = 1 hrs | Li Q., Kako T. & Ye J., (2010). Strong adsorption and effective photocatalytic activities of one-dimensional nanostructured silver titanates.*Appl. Catalysis A: General*, Vol. 375, pp. 85-91. |
| Mesoporous Zeotype SAPO-5 (Silicoaluminophosphate) | T = 180 - 200 °C, t = 0.5 - 2 hrs | Utchariyajit K. &Wongkasemji S., (2010) Effect of synthesis parameters on mesoporous SAPO-5 with AFI-type formation via microwave radiation using alumatrane and silatrane precursors. *Microporous and Mesoporous Materials*, Vol. 135, pp. 116-123. |
| $Ru/La_{0.75}Sr_{0.25}Cr_{0.5}Mn_{0.5}O_{3-\delta}$ | P = 750 W, t = 1 min | Barison S., Fabrizio M., Mortalò C., Antonucci P., Modafferi V.&Gerbasi R., (2010). Novel $Ru/La_{0.75}Sr_{0.25}Cr_{0.5}Mn_{0.5}O_{3-\delta}$ catalysts for propane reforming in IT-SOFCs. *SolidState Ionics*, Vol. 181, pp. 285-291. |
| Tungsten Oxide Nanorods | T = 180 °C, t = 20 min | Li Y., Su X., Jian J. & Wang J., (2010). Ethanol sensing properties of tungsten oxide nanorods prepared by microwave hydrothermal method. *Ceramics International*, Vol. 36, pp. 1917-1920. |
| Saponites with $Mg^{2+}$ | T = 180 °C, P = 600 W, t = 1, 2, 3, 4, 8 or 16 hrs | Trujillano R., Rico E., Vicente M., Herrero M. & Rives V., (2010). Microwave radiation and mechanical grinding as new ways for preparation of saponite-like materials. *Appl. Clay Sci.*, Vol. 48, pp. 32-38. |
| Titanate Nanotubes | T = 403 K, P = 70, 400 or 700 W, t = 3 hrs | Chen Y., Lo S. &Kuo J., (2010). Pb(II) adsorption capacity and behavior of titanate nanotubes made by microwave hydrothermal method.*Colloids and Surfaces A:Physicochemical and Engineering Aspects*, Vol. 361, pp. 126-131. |

| | | |
|---|---|---|
| Mg, Al-Layered Double Hydroxides | T = 125 °C, t = 10 - 180 min | Benito P., Herrero M., Labajos F. & Rives V., (2010). Effect of post-synthesis microwave hydrothermal treatment on the properties of layered double hydroxides and related materials. *Appl. Clay Sci.*, Vol. 48, pp. 218-227. |
| $Y_2O_3$ and Yb-$Y_2O_3$ | T = 140, 170 °C, P = 150 - 300 W, t = 5 - 22 min, 2 hrs | Serantoni M., Mercadelli E., Costa A., Blosi M., Esposito L.&Sanson A., (2010). Microwave-assisted polyol synthesis of sub-micrometer $Y_2O_3$ and Yb-$Y_2O_3$ particles for laser source application. *Ceramics International*, Vol. 36, pp. 103-106. |
| Co-Zn-Al catalysts | T = 125 °C, t = 10 - 300 min | Benito P., Herrero M., Labajos F. M., Rives V., Royo C., Latorre N. &Monzon A., (2009). Production of carbon nanotubes from methane: Use of Co-Zn-Al catalystsprepared by microwave-assisted synthesis. *Chemical Engineering Journal*, Vol. 149, pp. 455-462. |
| Hectorite $M_x[Li_xMg_{6-x}Si_8O_{20}(OH)_4]$ (M = Na, Li, $NH_4$) | T = 393 K, t = 16 hrs | Vicente I., Salagre P., Cesteros Y., Guirado F., Medina F.&Sueiras J., (2009). Fast microwave synthesis of hectorite. *Appl. Clay Sci.*, Vol. 43, pp. 103-107. |
| Titania Nanowires | T = 210 °C, P = 350 W, t = 5 hrs | Chang M., Lin C., Deka J., Chang F. & Chung C., (2009). Nanomechanical characterization of microwave hydrothermally synthesized titania nanowires. *J. Phys. D:Appl. Phys.*, Vol. 42, pp. 145105. |
| Graphitic Carbon Particle Chains | T = 160 - 220 °C, t = 40 min | Harris A., Deshpande S. &Kefeng X., (2009). Synthesis of graphitic carbon particle chains at low temperatures undermicrowave irradiation. *Materials Lett.*,Vol. 63, pp. 1390-1392. |
| Silicalite (Si-MFI) | T = 80 °C, P = 250 W, t = 90 min T = 180 °C, P = 400 W, t = 60 min | Chen X., Yan W., Cao X., Yu J. &Xu R., (2009). Fabrication of silicalite-1 crystals with tunable aspect ratios by microwave assisted solvothermal synthesis. *Microporous and Mesoporous Materials*, Vol. 119, pp. 217-222. |
| $ZnWO_4$ Nanoparticles | T = 160 °C, t = 1, 3 hrs | Bi J., Wu L., Li Z., Ding Z., Wang X. & Fu X., (2009). A facile microwave solvothermal process to synthesize ZnWO4 nanoparticles. *J. Alloys and Compd.*, Vol. 480, pp. 684-688. |
| Multiferroic $BiFeO_3$ | t = 30 min, P = 800 W | Prado-Gonjal J., Villafuerte-Castrejón M., Fuentes L. &Morán E., (2009). Microwave-hydrothermal synthesis of the multiferroic $BiFeO_3$. *Mater. Res. Bull.*, Vol. 44,pp. 1734-1737. |

| | | |
|---|---|---|
| $NaYF_4:Yb^{3+},Tm^{3+}$ | T = 180 ˚C, t = 4 hrs | Chen X., Wang W., Chen X., Bi J., Wu L., Li Z.&Fu X., (2009). Microwave hydrothermal synthesis and upconversion properties of $NaYF_4:Yb^{3+},Tm^{3+}$ with microtube morphology. *Mater. Lett.*,Vol. 63, pp. 1023-1026. |
| Carboxylate-Intercalated Layered Double Hydroxides | T = 85, 100, 125 ˚C, t = 10, 30, 180 min | Benito P., Labajos F., Mafra L., Rocha J. &Rives V., (2009). Carboxylate-intercalated layered double hydroxides aged under microwave-hydrothermal treatment. *J. Solid State Chem.*, Vol. 182, pp. 18-26. |
| Rapid Synthesis of Titania Nanowires | T = 210 ˚C, P = 350 W, t = 2 hrs | Chung C., Chung T. & Yang T., (2008). Rapid Synthesis of Titania Nanowires by Microwave-Assisted Hydrothermal Treatments. *Ind. Eng. Chem. Res.*, Vol. 47, No. 7, pp. 2301-2307. |
| $TiO_2$ Photocatalysts | T = 393, 423 K, t = 15, 30, 60 min | Addamo M., Bellardita M., Carriazo C., Paola A., Milioto S., Palmisano L. & Rives V., (2008). Inorganic gels as precursors of $TiO_2$photocatalysts prepared by low temperature microwave or thermal treatment. *Appl. Catalysis B: Environmental*, Vol. 84,pp. 742-748. |
| Titanate Nanowires | T = 210 ˚C, P = 350 W, t = 5 hrs | Chang M., Chung C., Deka J., Lin C. & Chung T., (2008). Mechanical properties of microwave hydrothermally synthesized titanate nanowires. *Nanotechnology*, Vol. 19, pp. 025710. |
| Mesoporous$AlPO_4$-5 (AFI) Zeotype | T = 180 - 200 ˚C, t = 0.5 - 2 hrs | Utchariyajit K. &Wongkasemjit S., (2008). Structural aspects of mesoporous $AlPO_4$-5 (AFI) zeotype using microwave radiation and alumatrane precursor. *Microporous and Mesoporous Materials*, Vol. 114, pp. 175-184. |
| Zn,Al-$CO_3$ Layered Double Hydroxides | T = 100, 125 ˚C, t = 30, 60, 180, 300 min | Benito P., Guinea I., Labajos F., Rocha J. & Rives V., (2008). Microwave-hydrothermally aged Zn,Alhydrotalcite-like compounds: Influence of the composition and the irradiation conditions. *Microporous and Mesoporous Materials*, Vol. 110, pp. 292-302. |
| Ni-Al, Zn-Al Hydrotalcite | T = 100, 125, 150, 175 ˚C, t = 5 - 300 min | Benito P., Herrero M., Barriga C., Labajos F. & Rives V., (2008). Microwave-Assisted Homogeneous Precipitation of Hydrotalcites by Urea Hydrolysis. *Inorg. Chem.*, Vol. 47, No. 12, pp. 5453-5463. |

| | | |
|---|---|---|
| Synthesis, Passivation and Stabilization of Nano (particles, rods & wires) | P = 650 W (60%), t = 10 sec to mins | Abdelsayed V., Panda A. B., Glaspell G. P. & El-Shall M. S., (2008). Synthesis, Passivation and Stabilization of Nanoparticles, Nanorods and Nanowires by Microwave Irradiation. *ACS Symposium Series*, Vol. 996, pp. 225-247. |
| $LaPO_4$:RE (RE=Ce,Tb,Eu) and $In_2O_3$:Sn (ITO) | T = 300 °C, t = 0 - 30 sec | Buhler G., Stay M. &Feldmann C., (2007). Ionic liquid based approach to nanoscale functional materials. *Green Chemistry*, Vol. 9, pp. 924-926. |
| $ZnIn_2S_4$ Submicrospheres | T = 200 °C, t = 10 min | Hu X., Yu J., Gong J. &Quan Li., (2007). Rapid Mass Production of Hierarchically Porous $ZnIn_2S_4$Submicrospheres via a Microwave-SolvothermalProcess.;*Cryst. GrowthDes.*,Vol. 7, No. 12, pp. 2444-2448. |
| $Co^{2+}$ in Layered Double Hydroxides | T = 100, 125 °C, t = 10, 30, 60, 180 min | Herrero M., Benito P., Labajos F. &Rives V., (2007).Stabilization of $Co^{2+}$ in layered double hydroxides (LDHs) by microwave-assisted ageing. *J. Solid State Chem.*, Vol. 180, pp. 873-884. |
| MFI Zeolite Membranes | T = 80, 120 °C, P = 250, 400 W, t = 60, 90 min | Motuzas J., Heng S., Lau P., Yeung K.,Beresnevicius Z. &Julbe A., (2007). Ultra-rapid production of MFI membranes by coupling microwave-assisted synthesis with either ozone or calcination treatment. *Microporous and Mesoporous Materials*, Vol. 99, pp. 197-205. |
| Titanate Nanotubes | T = 110, 130, 150, 175 °C, P = 70, 200, 400, 700 W, t = 1.5, 3, 6, 12 hrs | Ou H., Lo S. &Liou Y., (2007).Microwave-induced titanate nanotubes and the corresponding behavior after thermal treatment. *Nanotechnology*, Vol. 18, pp. 175702. |
| $\alpha$-$Fe_2O_3$ Nanoarchitectures | T = 120 - 220 °C, t = 10 min | Hu X., Yu J. &Gong J., (2007). Fast Production of Self-Assembled Hierarchical $\alpha$-$Fe_2O_3$Nanoarchitectures. *J. Phys. Chem. C*, Vol.111 , No. 30, pp. 11180-11185. |
| Nanosized Cobalt Oxides | T = 100 °C, t = 10, 30, 60, 180, 300 min | Herrero M., Benito P., Labajos F. & Rives V., (2007). Nanosize cobalt oxide-containing catalysts obtained through microwave-assisted methods. *Catalysis Today*, Vol. 128, pp. 129-137. |
| Nanorods and Nanoplates of Rare Earth Oxides | P = 650 W (70%) | Panda A. B., Glaspell G. & El-Shall M. S., (2007). Microwave Synthesis and Optical Properties of Uniform Nanorods and Nanoplates of Rare Earth Oxides. *The Journal of Physical Chemistry B Letters*. Vol. 111, pp. 1861-1864. |

| | | |
|---|---|---|
| $AlPO_4$-18 Films | T = 150 - 200 °C, P = 1 kW | Vilaseca M., Yagüe C., Coronas J. &Santamaria J., (2006).Development of QCM sensors modified by $AlPO_4$-18 films. *Sensors and Actuators B: Chemical,* Vol. 117, pp. 143-150. |
| Modification of $TiO_2$ | T = 145 °C, P = 1.2 kW, t = 5 -360 min | Wilson G., Matijasevich A., Mitchell D., Schulz J. &Will G., (2006). Modification of $TiO_2$ for Enhanced Surface Properties: Finite Ostwald Ripening by a Microwave Hydrothermal Process. *Langmuir,* Vol. 22, No. 5, pp. 2016-2027. |
| Nano-sized Hydroxyapatite ($Ca_{10}(PO_4)_6(OH)_2$) | T = 85 - 300 °C, P = 250 - 650 W, t = 4 min | Han J., Song H., Saito F. &Lee B., (2006). Synthesis of high purity nano-sized hydroxyapatite powder by microwave hydrothermal method. *Mater. Chem. Phys.,* Vol. 99, No. 2-3, pp. 235-239. |
| Ba-exchanged zeolite A | T = 150, 200, 230 °C, P = 1kW, t = 40, 60, 120 min | Ferone C., Esposito S. &Pansini M., (2006). Microwave assisted hydrothermal conversion of Ba-exchanged zeolite A into metastable paracelsian. *Microporous andMesoporous Materials,* Vol. 96, pp. 9-13. |
| Nanocrystalline Anatase $TiO_2$ | P = 950 W, t = 2 - 6 min | Murugan A., Samuel V. &Ravi V., (2006).Synthesis of nanocrystallineanatase $TiO_2$ by microwave hydrothermal method. *Materials Lett.,*Vol. 60, pp. 479-480. |
| Semiconductor Rods and Wires | P = 650 W (60%), t = 30 sec to 3 mins | Panda A. B., Glaspell G. & El-Shall M. S., (2006). Microwave Synthesis of Highly Aligned Ultra Narrow Semiconductor Rods and Wires. *J. Am. Chem. Soc.,* Vol. 128, No. 9, pp. 2790-2791. |
| $TiO_2$ Nanocrystals | P = 650 W (33%), t = 30 mins | Glaspell G., Panda A. B. & El-Shall M. S., (2006). Reversible paramagnetism to ferromagnetism in transition metal-doped $TiO_2$ nanocrystals prepared by microwave irradiation. *Journal of Applied Physics,* Vol. 100, pp. 124307-12412. |
| Silicalite-1 Seeds | T = 70 - 180 °C, P = 250, 400 W, t = 10 - 150 min | Motuzas J., Julbe A., Noble R. D., Guizard C., Beresnevicius Z. J. &Cot D., (2005).Rapid synthesis of silicalite-1 seeds by microwave assisted hydrothermal treatment. *Microporous and MesoporousMaterials,*Vol. 80, pp. 73-83. |
| Coaxial Ag/C Nanocables | T = 200 °C, t = 20 min | Yu J., Hu X., Lib Q. &Zhang L., (2005).Microwave-assisted synthesis and in-situ self-assembly of coaxial Ag/C nanocables. *Chem. Commun.,* pp. 2704-2706. |

| | | |
|---|---|---|
| M-Doped ZnO (M = co, Cr, Fe, Mn, Ni) | P = 650 W (33%), t = 10 min | Glaspell G., Dutta P. & Manivannan A., (2005). A Room-Temperature Synthesis of M-Doped ZnO (M = Co, Cr, Fe, Mn & Ni), *Journal of Cluster Science*, Vol. 16, No. 4, pp. 523-536. |
| Au and Pd Nanoparticle Catalysts | P = 650 W (33%), t = 10 mins | Glaspell G., Fuoco L. & El-Shall M. S., (2005). Microwave Synthesis of Supported Au and Pd Nanoparticle Catalysts for CO Oxidation. *The Journal of Physical Chemistry B Letters*, Vol. 109, pp. 17350-17355. |
| $AlPO_4$-5 | T = 180 °C, P = 500 W, t = 17 min | Lin C., Dipre J. &Yates M., (2004).Novel Aluminum Phosphate-5 Crystal Morphologies Synthesized by Microwave Heating of a Water-in-Oil Microemulsion. *Langmuir*, Vol. 20,No. 4, pp. 1039-1042. |
| Crystallization of Zeolite A | P = 180, 360, 900, 1260 W, t = 10, 30, 40, 50 min P = 360, 900 W, t = 5,10, 20,30, 50 min P = 360 W, t = 15, 30, 40 min | Bonaccorsi L. &Proverbio E., (2003).Microwave assisted crystallization of zeolite A from dense gels. *J. Cryst. Growth*, Vol. 247, No. 3-4, pp. 555-562. |
| Sodalite/$\alpha Al_2O_3$ Membranes | T = 160 - 200°C, P = 250 W, t = 15 - 180 min | Julbe A., Motuzas J., Cazevielle F., Volle G. &Guizard C., (2003). Synthesis of sodalite/$\alpha Al_2O_3$ composite membranes by microwave heating. *Separation andPurification Technology*, Vol. 32, pp. 139-149. |
| $BaTiO_3$ Film | T = 100 °C, t = 5, 10, 15, 30, 60 min | Lee J., Kumagai N., Watanabe T. &Yoshimura M., (2002). Direct fabrication of oxide films by a microwave-hydrothermal method at low temperature. *Solid State Ionics*, Vol. 151, pp. 41-45. |
| Titanium Silicates | T = 175 °C, P = 800 W | Prasad M., Kamalakar G., Kulkarni S., Raghavan K., Rao K., Prasad P. &Madhavendra S., (2002). An improved process for the synthesis of titanium-rich titanium silicates (TS-1) under microwave irradiation. *Catalysis Commun.*,Vol. 3, pp. 399-404. |

| | | |
|---|---|---|
| Hydrodechlorination of Chlorobenzene on Pd/$Nb_2O_5$ Catalysts | T = 200 °C, P = 650 W, t = 5 min | Gopinath R., Rao K., Prasad P., Madhavendra S., Narayanan S. &Vivekanandan G., (2002).Hydrodechlorination of chlorobenzene on $Nb_2O_5$-supported Pd catalysts: Influence of microwave irradiation during preparation on the stability of the catalysts. *J. of Mol. Catalysis A: Chemical*, Vol. 181, pp. 215-220. |
| Co-, Co-Zn and Ni-Zn Ferrite Powders | T = 120, 150, 180, 200 °C, t = 30 min | Kim C. K., Lee J. H., Katoh S., Murakami R. &Yoshimura M., (2001). Synthesis of Co-, Co-Zn and Ni-Zn ferrite powders by the microwave hydrothermal method. *Mater. Res. Bull.*, pp. 2241-2250. |
| Uniform Pure and Dye-Loaded $AlPO_4$-5 Crystals | T = 145 - 170 °C, P = 1 kW, t = 5 - 60 min | Ganschow M., Schulz-Ekloff G., Wark M., Wendschuh-Josties M. &Wohrle D., (2001). Microwave-assisted preparation of uniform pure and dye-loaded $AlPO_4$-5 crystals with different morphologies for use as microlaser systems. *J.Mater. Chem.*, Vol. 11, pp. 1823-1827. |
| Synthesis of $AlPO_4$-5 | T = 180 - 190 °C, P = 900 W, t = 8 min | Braun I., Schulz-Ekloff G. &Lautenschläger W., (1998).Synthesis of AlPO4-5 in a microwave-heated, continuous-flow, high-pressure tube reactor. *Microporous andMesoporous Materials*, Vol. 23, pp. 79-81. |

Table 1. Summary of selected materials prepared using microwave heating with conditions and apporiate references listed.

## 5. Acknowledgement

The authors are grateful to the U.S. Department of Energy for financial support of this work provided through the Center for Advanced Separation Technologies.

## 6. References

Addamo M., Bellardita M., Carriazo C., Paola A., Milioto S., Palmisano L. & Rives V., (2008). Inorganic gels as precursors of $TiO_2$ photocatalysts prepared by low temperature microwave or thermal treatment. *Appl. Catalysis B: Environmental*, Vol. 84, pp. 742-748.

Agmon N., (1996). Tetrahedral Displacement: The Molecular Mechanism behind the Debye relaxation in Water. *J. Phys. Chem.*, Vol. 100, pp. 1072-1080.

Barison S., Fabrizio M., Fasolin S., Montagner F. & Mortalò C., (2010). A microwave-assisted sol-gel Pechini method for the synthesis of $BaCe_{0.65}Zr_{0.20}Y_{0.15}O_{3-\delta}$ powders. *Mater. Res. Bull.*, Vol. 45, pp. 1171-1176.

Barthel J., Bachhaber K., Buchner R. & Hetzenauer H., (1990). Dielectric Spectra of some common solvents in the Microwave Region: Water and Lower Alcohols. *Chem. Phys. Lett.*, Vol. 165, pp. 369-373.

Burgardt P. & Seehra M. S., (1977). Magnetic susceptibility of iron pyrite ($FeS_2$) between 4.2 and 620 K. Solid State Commun., Vol. 22, No. 2, pp. 153-156.

Chang M., Lin C., Deka J., Chang F. & Chung C., (2009). Nanomechanical characterization of microwave hydrothermally synthesized titania nanowires. *J. Phys. D:Appl. Phys.*, Vol. 42, pp. 145105.

Chatterjee I & Misra M., (1991). Electromagnetic and thermal modeling of microwave drying of fine coal. *Miner Metall Process*, Vol. 8, pp. 110-114.

Chen X., Yan W., Cao X., Yu J. & Xu, R, (2009) Fabrication of silicate-1 crystals with tunable aspect ratios by microwave-assisted solvothermal synthesis. *Microporous and Mesoporous Materials*, Vol. 119, pp. 217-222.

Cutmore N. G., Evans T. G., McEwan A. J., Roger C. A. & Stoddard S. L., (2000). Low Frequency Microwave Technique for On-Line Measurement of Moisture. *Mineral Engineering*, Vol. 13, pp. 1615-1622.

Debye P., (1929). *Polar Molecules*, The Chemical Catalog Company, NY.

Gopinath R., Rao K., Prasad P., Madhavendra S., Narayanan S. &Vivekanandan G., (2002). Hydro dechlorination of chlorobenzene on $Nb_2O_5$-supported Pd catalysts: Influence of microwave irradiation during preparation on the stability of the catalysts. *J. of Mol. Catalysis A: Chemical*, Vol. 181, pp. 215-220.

Hu X., Yu J. & Gong J., (2007). Fast Production of Self-Assembled Hierarchical $\alpha$-$Fe_2O_3$ Nanoarchitectures. *J. Phys. Chem. C*, Vol. 111 , No. 30, pp. 11180-11185.

Jagadeesh M. S. & Seehra M. S., (1981). Thermomagnetic studies of conversion of pyrite and marcasite in different atmospheres (vacuum, $H_2$, He and CO). *J. Phys. D: Appl. Phys.*, Vol. 14, pp. 2153-2167.

Kaatze U., (1989). Complex permittivity of water as a function of frequency and temperature. *J. Chem. Eng. Data*, Vol. 34, pp. 371-374.

Kim C. K., Lee J. H., Katoh S., Murakami R. &Yoshimura M., (2001). Synthesis of Co-, Co-Zn and Ni-Zn ferrite powders by the microwave hydrothermal method. *Mater. Res. Bull.*, pp. 2241-2250.

Kittel C., (1996). *Introduction to Solid State Physics*, (Seventh Edition), John Wiley & Sons, Inc.), NY.

Lester E. & Kingman S., (2004). The effect of microwave preheating on five different coals. *Fuel*, Vol. 83, pp. 1941-1947.

Li Y., &Yang W., (2008). Microwave synthesis of zeolite membranes: A review. *Journal of Membrane Science*, Vol. 316 , pp. 3-17.

Lidström P., Tierney J., Wathey B., & Westman J., (2001). Microwave assisted organic synthesis - a review. *Tetrahedron*, Vol. 57, pp. 9225-9283.

Lindroth D. P., (1985). Microwave drying of fine coal. *US Bureau of Mines Report # 9005*, US Department of Interior.

Marland S., Han B., Merchant A. & Rowson N., (2000). The effect of microwave radiation on coal grindability. *Fuel*, Vol. 79, pp. 1283-1288.

Marland S., Merchant A. & Rowson N., (2001). Dielectric properties of coal. *Fuel*, Vol. 80, pp. 1839-1849.

Merabet M. & Bose T. K., (1988). Dielectric Measurements of Water in the Radio and Microwave Frequencies by Time Domain Reflectometry. *J. Phys. Chem.*, Vol. 92, pp. 6149-6150.

Perkin R. M., (1980). The heat and mass transfer characteristics of boiling point drying using radio frequency and microwave electromagnetic fields. *Int J Heat Mass Transfer*, Vol. 23, pp. 687-95.

Ponte D. G., Prieto I. F., Vair P. F. & Luengo J. C. G., (1996). Determination of moisture content in power station coal using microwaves. *Fuel*, Vol. 75, No. 2, pp. 133-138.

Prasad M., Kamalakar G., Kulkarni S., Raghavan K., Rao K., Prasad P. &Madhavendra S., (2002). An improved process for the synthesis of titanium-rich titanium silicates (TS-1) under microwave irradiation. *Catalysis Commun.*,Vol. 3, pp. 399-404.

Seehra M. S., Kalra A. & Manivannan A., (2007). Dewatering of fine coal slurries by selective heating with microwaves. *Fuel*, Vol. 86, pp. 829-834.

Standish N, Worner H. & Kaul H., (1988). Microwave drying of brown coal agglomerates. *J. Microwave Power EM Energy*, Vol. 23, pp. 171-175.

Utchariyajit K. &Wongkasemji S., (2010) Effect of synthesis parameters on mesoporous SAPO-5 with AFI-type formation via microwave radiation using alumatrane and silatrane precursors. *Microporous and Mesoporous Materials*, Vol. 135, pp. 116-123.

Vicente I., Salagre P., Cesteros Y., Guirado F., Medina F.&Sueiras J., (2009). Fast microwave synthesis of hectorite. *Appl. Clay Sci.*, Vol. 43, pp. 103-107.

Vicente I., Salagre P., Cesteros Y., Medina F.&Sueiras J., (2010). Microwave-assisted synthesis of saponite. *Appl. Clay Sci.*, Vol. 48, pp. 26-31.

Wilson G., Matijasevich A., Mitchell D., Schulz J. &Will G., (2006). Modification of $TiO_2$ for Enhanced Surface Properties: Finite Ostwald Ripening by a Microwave Hydrothermal Process. *Langmuir*, Vol. 22, No. 5, pp. 2016-2027.

Wu L., Bi J., Li Z., Wang X. &Fu X., (2008). Rapid preparation of $Bi_2WO_6$ photocatalyst with nanosheet morphology via microwave-assisted solvothermal synthesis. *Catalysis Today*, Vol. 131, No. 1-4, pp. 15-20.

Yanshou L. & Weishen Y., (2008). Microwave synthesis of zeolite membranes: A review. *Journal of Membrane Science*, Vol. 316, pp. 3-17.

# 9

# Synthesis and Processing of Biodegradable and Bio-Based Polymers by Microwave Irradiation

Guido Giachi, Marco Frediani, Luca Rosi and Piero Frediani
*University of Florence*
*Italy*

## 1. Introduction

Biodegradable polymers, as well as polymers produced from renewable feedstocks, are attracting increasing interests as possible substitutes for conventional plastics: a higher energy efficiency in synthesis and processing steps must be continuously pursued in order to maximize the intrinsic environmental benefits brought by this class of materials. Microwave assisted organic synthesis (MAOS) is nowadays a major topic in green chemistry and a great (yet rising) number of papers can already be found that report striking advantages over conventional thermal heating. Nonetheless microwave (MW) energy sources are recently being chosen also for several polymerization reactions. Indeed, reduced heating times and superior homogeneity provided by MW reactors may play a central role in optimizing production processes, with a dramatic improvement in the environmental performance. In the introduction of this chapter we're briefly recalling some theoretical principles of microwave-matter interaction; several experimental setups are then examined and, eventually, thermal and non-thermal specific microwave effects are described and commented. The following paragraph is dedicated to a comprehensive survey of synthesis examples found in scientific literature and categorized by polymerization technique, in which particular relevance is given to products of increasing commercial importance like poly(ε-caprolactone) (PCL) and poly(lactic acid) (PLA). A third part of the chapter deals with the employment of microwave heating for chemical modification and processing of polymers; the last paragraph summarizes advantages and drawbacks of microwave assisted polymer chemistry, stressing the energy efficiency topic and drawing conclusions.

### 1.1 Theoretical principles of microwave irradiation

Microwaves are electromagnetic radiations characterized by frequencies that extend from 0.3 GHz (corresponding to an energy of $1.2 \times 10^{-6}$ eV) to 30 GHz ($1.2 \times 10^{-3}$ eV), covering the portion of the spectrum between the less energetic Radio waves and the more energetic Infrared waves. Due to an intensive employment by the information and telecommunication technologies, only few frequencies among the whole microwave band are actually available for scientific, industrial and household heating applications: in order to avoid any interference, the vast majority of devices operates at 2.45 GHz ($1.0 \times 10^{-5}$ eV). Comparing the latter value to the energy associated with some common covalent bonds (C-C single bond:

3.61 eV; C=C double bond: 6.35 eV) or even with a weaker interaction like hydrogen bond (0.04 - 0.44 eV) it is clear that microwaves do not posses enough energy to induce a chemical reaction by breaking any of these. Very often, indeed, the impressive shortening of reaction times, reported by the authors cited in the following paragraphs, is ascribed to peculiarities in the heating process like homogeneity, inverted temperature gradient (i.e. vial walls are colder than reactants in the bulk) and extremely fast temperature increase. The interaction between microwaves and the electric dipoles of irradiated reactants or solvent molecules is what underlies "dielectric heating". Since the order of magnitude of molecular rotational frequencies is lower, but quite close to microwave frequencies, the tendency to align with the oscillating electromagnetic field produces an immediate and simultaneous motion that can't indefinitely follow the oscillating field and eventually results in thermalization. Kinetic energy is transformed into heat via phenomena (i.e. phase shifts and dielectric losses deriving from molecular friction) whose mechanism depends on both molecular mass (more correctly on rotational moment of inertia) and dielectric coefficient "ε" of each component of the irradiated reaction mixture. Dielectric coefficient (or complex permittivity) consists of a real component and of an imaginary component ($\varepsilon = \varepsilon' + i\varepsilon''$): widely adopted parameters to describe the efficiency of radiation-to-heat conversion are the "loss factor"(ε") and the "dissipation factor", indicated with "tan(δ)" or "D" and defined as:

$$\tan(\delta) = D = \varepsilon''/\varepsilon' \tag{1}$$

Polar solvents and reactants normally possess high values of tan(δ), indicating strong microwave absorption; on the other hand low values shown by most non-polar substances indicate transparency and inefficient energy transfer (Fig. 1).

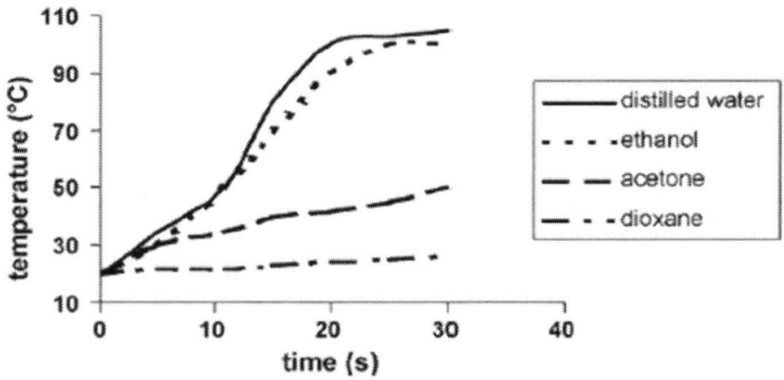


Fig. 1. Microwave heating curves of different solvents in a multimodal cavity (power = 150 W) (from C. Leonelli, T.J. Mason, Microwave and Ultrasonic Processing: Now a Realistic Option for Industry, Chemical Engineering and Processing, 2010, 49, 885–90. Copyright Elsevier Ltd. Reproduced with permission).

Even if, commonly, homogeneous heating represents a topic feature of microwave devices it should be noted that highly dielectric materials rapidly diminish the incident radiation intensity, hence recourse to large volumes may cause shielding effects and colder matter at the core. This is the reason why "penetration depth", defined as the point where only 37% (1/e) of the incident MW radiation is still present and calculated as the reciprocal of loss factor, turns out to be another useful parameter in the characterization of an experiment. Besides absorption and transmission, a third phenomenon can take place: radiation can be reflected by the surface of a body (highly conducting metals are important examples). This eventuality should generally be avoided in setting up a reaction since, while only negligible heating is generated in the reflecting material, microwave constructive interferences may arise elsewhere amid the irradiated matter and cause uneven temperature profiles and "hot spots" on account of local super-heating effects.

### 1.2 Microwave heating setups

Despite papers display sometimes poor technical detail, in microwave assisted synthesis (much more than in conventional chemistry) the choice of the experimental apparatus with its engineering specifications is a matter of chief importance. Reproducibility strongly depends on how accurately actual reaction parameters correspond to the user's setup. Being able to properly tune and monitor the most important variables gives reproducible results that can be fruitfully compared with literature. Unquestionably huge improvements have been achieved in this regard during the last decades: with microwave assisted synthesis gaining more and more attention through the years, several dedicated reactors were commercialized. Nowadays these devices have replaced former domestic MW ovens in many laboratories, delivering new standards in terms of safety, efficacy and parameter regulation. Reactors are frequently subdivided in multi- and mono-modal depending on the way in which microwaves are directed from the magnetron to the sample. Multimodal devices generate a fairly homogeneous electromagnetic field distributing radiations towards different directions through a large cavity in which interference with reflected waves averages field intensity; monomodal devices, on the other hand, feature precisely calibrated waveguides in which radiation is reflected in phase to generate standing waves with a maximum placed at the sample compartment position. Domestic MW ovens are examples of multimodal instruments as well as some high-capacity scientific device, while monomodal applicators are found exclusively among dedicated scientific equipment. Some authors (Nuchter et al., 2004) point out that the aforementioned distinction holds true only until the reaction mixture is placed into the cavity, perturbing the emitted field and turning monomodal reactors into (partly) multimodal ones: a more useful classification should be based, in their view, on radiation intensity or power density. Nonetheless it is generally agreed that monomodal reactors are best suited for small loads (note that, for this class, viable volumes are limited by the dimension of the wavelength itself) since allow to minimize the occurrence of hot spots, a detrimental phenomenon that disturbs reactions in multimodal devices when uneven temperature profiles are not averaged out by a large amount of absorbing matter. This is the reason why it is often recommended (Bogdal et al., 2003) to fill multimodal cavities with samples occupying more than 50% of the available volume. When large volumes and reliable field homogeneities are required, applicators with radiant geometries may also be profitably employed: the possibility to place antennas in the cavity makes the field much less subject to perturbation by the load (Leonelli & Mason, 2010). It is worth noting that a central technical feature for synthetic purpose, generally offered by a professional apparatus, is the capability of emitting continuous (i.e. unpulsed) radiation, so that the reaction mixture is heated at a user-assigned power level that is constant with time. Inexpensive domestic ovens operate irradiating at their maximum power in a pulsed profile, with the on/off time ratio depending on the chosen power level: this means being able to tune average MW intensity in the whole irradiation period, but obviously exposes to the risk of momentary overheating. Temperature control represents a critical engineering point: some authors (F. Liu et al., 2010; Sharma & Mishra, 2010) still choose to perform the measure immediately after microwave exposure (even a conventional thermometer may be used in this case), however a real-time monitoring is usually favored, nowadays, in order to operate a more accurate control. Three configurations are adopted in the vast majority of the reported examples: at a laboratory scale (industrial application is dampened by their cost and fragility) the preferred and most affordable probes seem to be fiber-optic thermometers inserted in the reaction mixture using transparent tubes. The operative range is 0-300 °C and some authors report that after few hours at 250 °C permanent deterioration is already observable (Nuchter et al., 2004); nonetheless high temperature measurement (up to approx. 2000 °C) can be performed using sapphire based fiber-optic thermometers. Optical (IR) pyrometers represent a universally viable solution (widespread in commercial technical devices as well as utilizable for large scale industrial reactors) and offer a much wider temperature range (-40 °C to 1000 °C), but necessarily measure the temperature at the vial wall, which means they invariably underestimate the value. Shielded thermocouples are the cheapest option, can withstand temperatures up to 300 °C but are quite bulky and require lossy (polar) solvents to avoid excessive MW absorption by their metallic shields which would result in overheating effects and biased measurement. The choice of the proper solvent for a MW assisted reaction must take into account its dielectric coefficient (ε) and it is worth reminding that in most cases dielectric constants (ε') significantly decrease with temperature (Fig. 2): an example is pressurized water at 300 °C showing a value of 20 which is very close to that of acetone at room temperature (Strauss & Trainor, 1995).

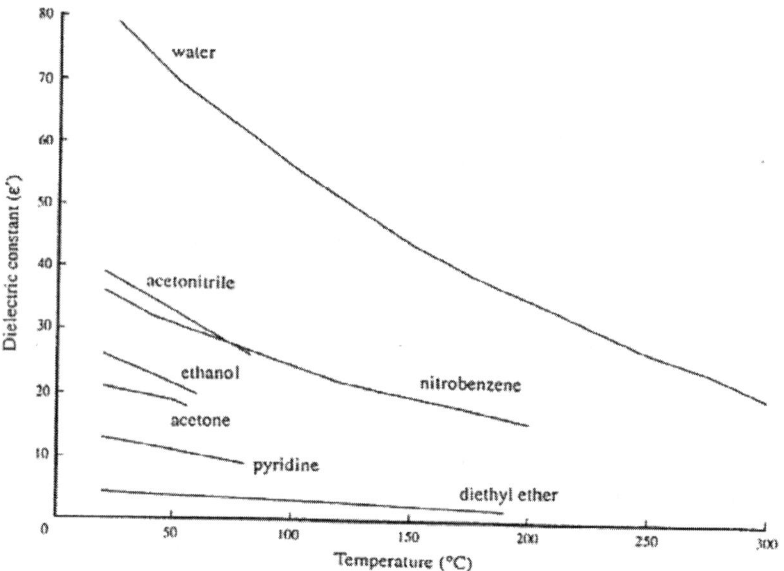


Fig. 2. Dielectric constants vs. temperature for different solvents
(from Dean, J. A. Ed.; *Lange's Handbook of Chemistry,* 13th ed.; McGraw-Hill, New York, 1985; p.99, ISBN: 0070161925. Copyright McGrawHill Education. Reproduced with permission).

Non polar solvents may be employed improving their heating efficiency by small additions (few percents in weight) of miscible liquids with high loss factors or even salts, preferably with a sufficient solubility in order to assure homogeneous conditions (Lidström et al., 2001). A similar approach underlies the diffuse exploitation as heating elements of highly MW absorbing solids, like silicon carbide (Kremsner & Kappe, 2006) (Fig. 3), carbon black (F. Liu et al., 2010) or silica particles (P. Liu & Su, 2005). Microwave assisted reactions may also be conducted in environmentally friendly water-miscible ionic liquids which are characterized by high dielectric constants and offer the possibility of isolating hydrophobic products by the addition of water to the homogeneous solution (Guerrero-Sanchez et al., 2007).

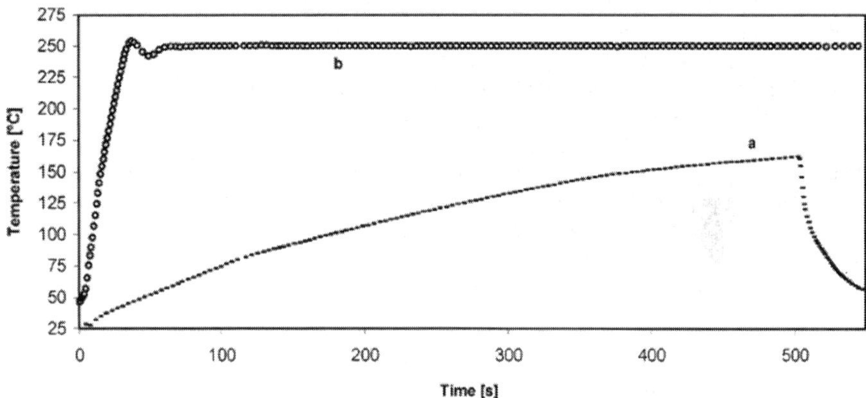


Fig. 3. Microwave heating curves for a 0.2 M solution of allyl phenyl ether in toluene: a) no heating element; b) with SiC as heating element (from J. M. Kremsner, C. O. Kappe, Silicon Carbide Passive Heating Elements in Microwave-Assisted Organic Synthesis, J. Org. Chem., 2006, 71, 4651-4658. Copyright American Chemical Society (ACS). Reproduced with permission).

When possible, the solvent-free option should always be favored on the basis of the green chemistry principles: the bulk ring opening polymerization of lactide to yield poly(lactic acid) is a notable case (Frediani et al., 2010). Early experiments were often performed in open vessels, at a temperature below the boiling point of high-boiling solvents (e.g. DMSO or DMF). This simplest configuration may as well comply with solvent-free reactions not requesting modified atmosphere, in which at least one of the reactants is a high-boiling liquid or a low-melting solid capable of efficient microwave absorption. Even if open vessels were tentatively adopted with more volatile solvents like toluene, increasing fire hazards brought by this choice should be seriously considered. Pressurized vials allow to use solvents at a temperature above their boiling point, considerably increasing reaction rates, though it should be noted that gas-tight sealed caps might cause explosions: technical devices usually feature dedicated vials equipped with proper septa capable of withstanding pressures up to a determinate value. Examples of customized reflux systems are frequently found in literature (Waugh & Parkin, 2010; Zhu et al., 2003): this particular setup proved to be safe and easy to scale-up, on the other hand it precludes overheating limiting the advantages of microwave irradiation over conventional heating. It is eventually worth mentioning that industrial-scale MW generators can be engineered to work efficiently under conditions from full vacuum to 1.03 MPa and from 30 °C to 200 °C (Leonelli et al., 2007). Microwave assisted synthesis is gaining increasing appeal for industrial applications, therefore the possibility of scaling-up reactions becomes a pivoting issue. One major limitation that prevents the load of large volumes at one time is the penetration depth of microwaves through lossy solvents or reactants: as already stated, the shielding of radiation by the peripheral matter produces inefficient heating at the core. Besides, heating large quantities of liquids in sealed vessels augments explosion hazard and obligate to employ thick, robust walls that must be MW transparent at the same time. It is eventually worth noting that massive quantities contained in thick vials are very difficult to cool rapidly and evenly with water or pressurized air. This critical points may be circumvented by loading high volumes in smaller, separated vials in parallel (a tedious configuration for large scale) or more efficiently by switching to a continuous flow reactor equipped with a relatively thin reaction tube through which reactants and solvents move in a necessarily homogeneous solution (Moseley & Woodman, 2008; Glasnow & Kappe, 2007). Open vials were successfully chosen for the irradiation of volumes up to 1000 mL (Deetlefs & Seddon, 2003) though it should be pointed out that MW penetration depth and the boiling point of the mixture limit the feasibility of such a layout. When only high-boiling substances are employed, solvent free reactions may also be performed with vials sealed under vacuum. As an example, the bulk ring opening polymerization of Caprolactone was recently performed loading five 1.8 L customized glass tubes with about 800 g of monomer each and irradiating simultaneously into a large multimodal microwave heater (Xu et al., 2010).

### 1.3 "Thermal" and "non-thermal" specific microwave effects

Since the early works, the astonishing increase in microwave-assisted reaction rates observed by many authors posed a number of questions about the rational origins of this phenomenon. A large variety of differences have been described between experiments conducted under conventional or microwave conditions: normally these findings are explained on the basis of rational hypotheses, but the lack of a rigorous and generally adopted terminology sometimes generates ambiguity when it comes to surveying the related literature. One frequent distinction is made between "thermal" and "non thermal" effects, sometimes called "specific microwave" effects. Precisely, also the "thermal" effects are microwave-specific: despite being based on heating phenomena they cannot be obtained by conventional oil baths. Dealing with variations in reaction kinetics, former works (Perreux & Loupy, 2001) sensibly suggested to refer to the Arrhenius equation:

$$k = A \exp(-\Delta G^{\neq}/RT) \tag{2}$$

In the equation three variables may be influenced by the use of microwaves as energy source, hence the rate (k) can be enhanced by three contributions: higher pre-exponential factor (A) due to increased collision efficiency; lower activation energy ($\Delta G^{\neq} = \Delta H^{\neq} - T\Delta S^{\neq}$) due to stabilized polar transition states and eventually a higher actual temperature. The latter can obviously be listed as a "thermal" effect, and comprises the phenomena of local superheating and hotspots, inverted temperature gradient (i.e. wall effect), selective heating and similar. The variations of A and $\Delta G^{\neq}$, on the other hand, may stem from the peculiar mechanism of interaction between matter and electromagnetic field, and therefore could be catalogued as "non thermal" effects. It should be pointed out that this approach holds also for the quite frequent cases of altered or enhanced reaction selectivities: since thermodynamic parameters do not depend upon the energy source, selectivity could be identified, here, as a kinetic competition between two or more reaction pathways, keeping in mind that different molecules may absorb microwaves with a different efficiency. This fairly rigorous paradigm is sometimes not applied in the interpretation of real life experiments since relating the observed phenomena to the terms of the equation is not always straightforward. Some authors, indeed, simply compare the real temperatures monitored during conventional and microwave heating procedures and try to estimate whether thermal differences are important enough to justify the discrepancies in their results (Jing et al., 2006). While the occurrence of "non thermal" effects is sometimes claimed and sometimes ruled out, remaining a controversial topic, the factual relevance of "thermal" microwave effects is nowadays widely recognized (though an exhaustive rationalization of their occurrence is still to come). Descriptions of these phenomena are occasionally found and highlighted among the papers surveyed in the following paragraphs.

## 2. Microwave assisted synthesis of biodegradable and bio-based polymers

In this paragraph many works are mentioned and catalogued on the basis of the polymerization process: the next two sub paragraphs are dedicated, respectively, to step growth polymerizations and ring opening polymerizations. All the examples cited here deal with biodegradable or bio-based polymers (i.e. polymers totally or partly derived from renewable feedstocks) to emphasize the continuous diffusion and development of products with a more environmentally friendly perspective.

### 2.1 Step growth polymerizations

Step growth polymerizations generally involve bifunctional monomers yielding linear chains with functionalized terminations, capable of reacting with more monomer units. On account of the same principle, multifunctional monomers may be utilized to generate branched structures and thermosetting materials. The most common polymerizations in this class belong to the family of polycondensations, in which every coupling reaction implies the release of a small molecule (i.e. water).

#### 2.1.1 Aliphatic polyesters

Among the conspicuous category of aliphatic polyesters many important examples of biodegradable and bio-based polymers can be found. An eco-friendly, solvent-free synthesis of poly(butylene succinate) (PBS) with a $M_w$ of $1.03 \times 10^4$ g/mol and $2.35 \times 10^4$ g/mol was achieved by the use of microwave heating at a temperature of, respectively, 200 °C and 220 °C (Scheme 1). An irradiation time of 20 min was needed, while poly(buthylene succinate) with a $M_w$ of $1.02 \times 10^4$ g/mol was obtained by conventional heating in 5 h. The polycondensation was performed under atmospheric pressure of $N_2$, in the presence of 1,3-dichloro-1,1,3,3-tetrabutyldistannoxane as catalyst (Velmathi et al., 2005).

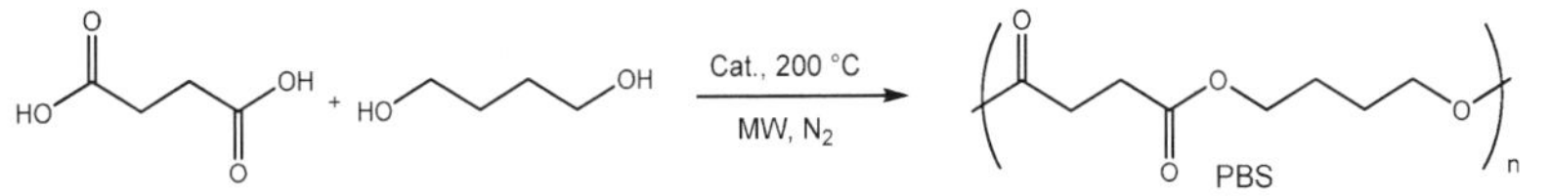

HO OH + HO OH —Cat., 200 °C / MW, $N_2$→ PBS

Scheme 1. Synthesis of poly(butylene succinate) (PBS).

Several authors accomplished the direct polycondensation of lactic acid to yield poly(lactic acid) (PLA) under microwave conditions. Ring opening polymerization of Lactide (the cyclic dilactone of lactic acid) is usually favored for the synthesis of relatively high molecular weight PLAs (see paragraph 2.2.3), nonetheless this single-step route is often attempted on account of its simplicity and low cost. In a former work D,L-Lactic acid was polymerized in bulk in a domestic microwave oven for 30 min. at a power level of 650 W, yielding oligomers with molecular masses ranging from 500 to 2000 g/mol. Reaction time was significantly shortened comparing to 24 h of conventional heating at 100°C and molecular masses were found to increase with irradiation time, while yields followed an opposite trend; after 20 min. cyclic oligomers were also formed, affecting the physical properties of the product (Keki et al., 2001). More recently, P(L)LAs with markedly higher molecular masses ($M_w \approx 16,000$ g/mol) were obtained within 30 min. of microwave irradiation employing a single-mode CEM *Discover* reactor set at a temperature of 200 °C. A binary catalyst of $SnCl_2$/p-TsOH was used and the polycondensation was conducted at a reduced pressure (30 mmHg) in order to remove water that progressively formed, avoiding the hydrolysis of the growing chain. Cyclic oligomers were still present as byproducts (Nagahata et al., 2007). Further enhancement was proposed in a later work by the use of solid super-acid (SSA) as a green heterogeneous catalyst: PLAs with molecular masses as high as 30,000 g/mol were synthesized in 2-3 h irradiation times, after a water removal step under vacuum, at a temperature of 260 °C reached gradually. Interestingly, the authors claimed to save about 90% energy with respect to the time consuming (24-48 h) conventional heating procedure (Lei et al., 2009). A phenyl substituted derivative of lactic acid, precisely L-2-hydroxy-3-phenylpropanoic acid, was also polymerized in a domestic microwave oven with a maximum power of 510 W: molecular masses, quite low in any case, were found to increase with irradiation time, ranging from 1,800 g/mol within 30 min. to 5,400 g/mol within 2.5 h (L. Liu et al. 2001).

### 2.1.2 Other step growth polymerizations

The polycondensation of sebacic acid and ω-amino alcohols was performed in a variable frequency microwave reactor to yield poly(amide-ester)s with impressive mechanical properties yet susceptible to degradability by proteolytic enzymes. Temperature was set at 180 °C, 200 °C and 220 °C depending on the chosen amino alcohol and regulated by an on - off power switching; power level was kept relatively low (50-70 W) due to the strong absorption of the starting materials. Reaction time was reduced from 3h of conventional heating to 1h of microwave irradiation, while product yields, molecular masses, and physico-chemical properties were found to be equal or higher (Borriello et al. 2007). Poly(anhydride)s are studied since long time as drug delivery systems with a tunable surface erosion: the synthesis of several poly(anhydride)s from aliphatic and aromatic diacids were conducted under microwave conditions to reduce reaction times from 3 h to 6-20 min at atmospheric pressure. Molecular weights were similar to the product of conventional heating; working under vacuum, as well as the isolation of an intermediate acetylated prepolymer, was not necessary, drastically simplifying the synthetic procedure (Vogel et al., 2003). With the aim of valorizing renewable feedstocks, interesting poly(ether)s were synthesized from isosorbide (1,4:3,6-dianhydrosorbitol) and 1,8-dibromo- or 1,8-dimesyl-octane using phase transfer catalysis under microwave irradiation. A monomodal *Synthwave 402* reactor, manufactured by Societè Prolabo, allowed to regulate temperature by modulating the emitted field and to monitor the reaction temperature on the surface of the vials by an infrared detector; toluene (1 mL for 5 mg of reagents) was used as a solvent to enhance temperature control and lower the mixture viscosity. Several advantages using microwave heating were reported: similar yields were obtained in 30 min instead of 1 day of reaction in an oil bath; products showed higher molecular weights and better homogeneities; a different termination mechanism generates ethylenic instead of hydroxylated chain ends (Chatti et al., 2002). Few years later the same group described the synthesis of novel poly(ether-ester)s from adipoyl chloride and aliphatic diols derived from isosorbide: polycondensations under microwave conditions proceeded five times faster (95% yield after 5 min at 150 °C) and the polymer showed higher molecular weights ($M_w$ = 4,200 g/mol vs 4,050 g/mol obtained by thermal heating). It is worth noting that, considering the identical temperature profiles generated by the two heating procedures, the better results shown by microwave irradiation are ascribed by the authors to "specific microwave features" such as the extension of polarity in the transition state comparing to the ground state (Chatti et al., 2006). Biodegradable, low molecular weight poly(alkylene hydrogen phosphonate)s were synthesized by a transesterification reaction of dimethyl hydrogen phosphonate with PEG 400 under microwave irradiation. Slightly higher temperatures were reached during the polymerization and markedly reduced reaction times (55 min vs 9h of conventional heating) were observed, which avoided the undesirable thermal degradation of dimethyl hydrogen phosphonate (Bezdushna et al., 2005).

### 2.2 Ring opening polymerizations

Ring opening polymerizations (ROPs) are nowadays widespread and very popular synthetic routes. The use of cyclic monomers exhibits some unquestionable advantages for the preparation of high molecular weight polymers since no byproduct (e.g. water) is released during the propagation step, avoiding chain scission as well as competition with monomers and consequential early chain termination. In some cases ring strain may also play a role in favoring the polymerization thermodynamically.

#### 2.2.1 Poly(ε-caprolactone)

Poly(ε-caprolactone) (PCL) is not produced from renewable resources, nonetheless, in the continuously developing class of biodegradable polymers, it proved to be one of the most useful and attractive: the number of works in scientific literature concerning this polymer is indeed impressive, and several successful microwave assisted syntheses are reported. An early example of microwave assisted synthesis employed titanium tetrabutylate as catalyst and an impressively refined custom monomodal reactor: conversion, measured by means of viscosity build-up, and number average molecular weights were close to those observed for conventionally heated procedure and did not indicate any advantage in using microwave irradiation (Albert et al., 1996). Poly(caprolactone) with weight-average molar mass ($M_w$) over $4 \times 10^4$ was prepared by metal free, microwave assisted ring opening polymerization initiated by benzoic acid: the authors indicate that the ratio of polymer chain propagation was directly proportional to temperature between 160 °C and 230 °C, while above this interval the synthesized PCL degraded. A multimodal domestic MW oven was employed: power was modulated by power on-off cycles, temperature was monitored by a tin-grounded thermocouple and the reactions were performed in vacuum sealed ampoules. As stated by the authors, the advantage of microwave irradiation was the significantly enhanced propagation ratio of PCL chain while the formation of growing centers was greatly inhibited (Yu & Liu, 2004). Earlier, the same group compared the synthesis of PCL initiated by benzoic acid and chlorinated acetic acids (Fig. 4): in both cases higher $M_w$ values and lower polydispersity indexes were observed with respect to conventional heating for an equal reaction time, nonetheless the molar masses of polymers synthesized employing chlorinated acids were lowered by degradation processes (Yu et al., 2003).

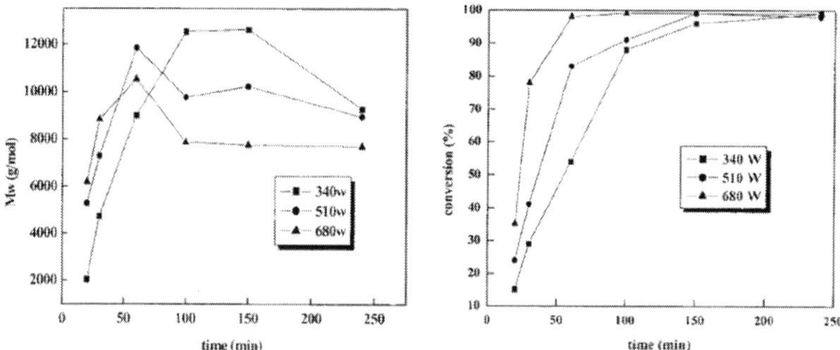


Fig. 4. Mw (left) and conversion (right) vs irradiation time for the ROP of ε-caprolactone initiated by chloroacetic acid (M/I = 15) (from Z. J. Yu, L. J. Liu, R. X. Zhuo, Microwave-Improved Polymerization of ε-Caprolactone Initiated by Carboxilic Acids, Journal of Polymer Science: Part A: Polymer Chemistry, 2003, 41, 13-21. Copyright Wiley-VCH Verlag GmbH & Co. KGaA. Reproduced with permission).

A lipase-catalyzed microwave assisted synthesis of PCL was attempted in several solvents at their boiling point and remarkably different behaviors were observed: while the reactions in toluene and benzene proceeded slower compared to oil bath heating, the reaction in diethyl ether showed an increased rate. It is known that the activity of the catalyst Novozyme 435, depends on temperature and on the polarity of the medium: no polymerization was indeed observed in polar solvents like THF or dioxane. Microwave irradiation was performed with a monomodal CEM-*Discover* device, equipped with a CEM fiber optic temperature sensor, for 90 min under reflux conditions (Kerep & Ritter, 2006). A "green" synthesis was described to obtain poly(ε-caprolactone)s with molar masses ranging from 3,000 g/mol to 16,000 g/mol: no solvent was used and non toxic, biologically acceptable lanthanide halides were chosen as initiators. Two different microwave equipments were employed: a monomodal Prolabo *Synthwave S402* equipped with an IR temperature sensor (15 to 45 min of irradiation at 180 - 200 °C) and a Sairem *GMP 12T* provided with a Marconi *F1192-12* water-cooled circulator (3 to 20 min of irradiation at 230 °C). Interestingly, the degradation rate of PCL was also assessed: it was found that the slow hydrolytic process is accelerated by the catalyst, while the faster enzymatic process is inhibited (Barbier-Baudry et al., 2003). The $Sn(Oct)_2$ catalyzed microwave assisted polymerization of ε-caprolactone (ε-CL) can be initiated by methacrylic acid or acrylic acid yielding radical polymerizable PCL macromonomers. Reactions were carried out in a monomodal CEM *Discover* reactor at 180 °C for 90 min in bulk, under air; it is worth noting that the high temperature reached under microwave irradiation did not imply uncontrolled free radical polymerization processes, though in this case a comparison with classical thermal activation showed no significant acceleration effect (Sinwell et al., 2006). The kinetics of pure and zinc catalyzed ring opening polymerization of ε-caprolactone was also investigated in comparison to conventional heating syntheses and a model based on continuous distribution kinetics with time/temperature-dependent rate coefficients was proposed. From these coefficients the activation energies were calculated: while for pure thermal heating and catalyzed thermal heating the values were respectively 24.3 and 23.4 kcal/mol, the value found for catalyzed microwave heating was 5.7 kcal/mol, reflecting the observed increment in the polymerization rate (Sivalingam et al., 2004). Kinetic investigations were also performed comparing the rate constant for the stannous octoate catalyzed ROP of ε-CL under microwave conditions and conventional heating. Several equilibrium temperatures were reached by means of different power levels: while conventional polymerizations followed Arrhenius' law, the rate of microwave assisted reactions (Fig. 5) showed an abnormal peak value for 180 °C (at 500 W), inducing the authors to suggest the hypothesis of non-thermal microwave effects (Li et al., 2007).

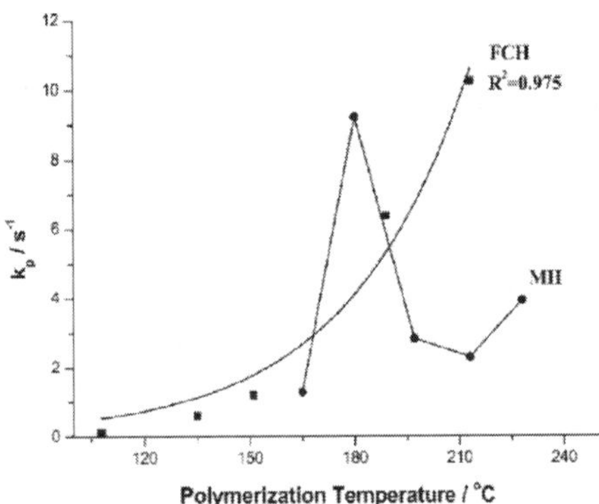


Fig. 5. Chain propagation rate constant ($k_p$) for the ROP of ε-caprolactone by microwave heating (circles) and conventional heating (squares) (from H. Li, L. Liao, L. Liu, Kinetic Investigation into the Non-Thermal Microwave Effect on the Ring Opening Polymerization of ε-Caprolactone, Macromol. Rapid Commun., 2007, 28, 411-416. Copyright Wiley-VCH Verlag GmbH & Co. KGaA. Reproduced with permission).

Combining ring opening polymerization and microwave assisted "click" chemistry, acetylene-functionalized PCL was synthesized and reacted with heptakis-azido-β-cyclodextrin to obtain star shaped polymers. A tetra-arm metallopolymer was also achieved by the "supramolecular click" reaction between di(pyridinyl)piridazine-functionalized macroligands and Copper (I) ions (Hoogenboom et al., 2006). The 1-butyl-3-methylimidazolium tetrafluoroborate ionic liquid has been successfully used as solvent for the ROP of ε-caprolactone: on account of its efficient microwave energy absorption. PCL with $M_w$ of 28,500 g/mol was obtained by means of a multimodal domestic oven at 85 W for 30 min, using Zinc oxide as catalyst (Liao et al., 2006). A large scale synthesis was recently performed using up to 2.45 Kg of monomer to assess the industrialization possibilities of microwave assisted ε-CL polymerization. A self-designed industrial microwave oven, capable of a maximum power of 6,000 W, was employed at different power levels and loaded with different amount of reactants: in any case the stannous octoate catalyzed ring opening polymerizations proceeded smoothly, yielding PCL with $M_w$ ranging from 66,000 g/mol to 122,000 g/mol (Xu et al., 2010). The effects of the presence of diisopropyl hydrogen phosphonate (2.9%) on the microwave assisted synthesis of PCL was examined, concluding that the ROP was initiated by traces of water but could be catalyzed by the oligo(ε-caprolactone)-substituted hydrogen phosphonate resulting from transesterification reactions. A polymer with molecular weight of 8,100 g/mol was obtained in 100 min at 510 W of irradiation power (Tan et al., 2009).

### 2.2.2 Poly(ε-caprolactone) copolymers

One of the first examples of microwave assisted synthesis of PCL copolymers was performed by means of a variable frequency device operated in the range from 0.4 to 3 GHz: poly(ε-caprolactam-*co*-ε-caprolactone) was obtained by the anionic catalyzed ring opening of the two cyclic monomers (both showed effective radiation absorption). Compared to conventional procedures, higher yields and higher amide content were observed, the latter implying a higher value for the glass transition temperature ($T_g$) of the copolymer (Fang et al., 2000). More recently, a series of completely biodegradable poly(vinyl alcohol)-*graft*-poly(ε-caprolactone) comb-like copolymers were prepared employing poly(vinyl alcohol) as an initiator for the microwave assisted ROP of ε-CL. The authors report a degree of polymerization (DP) for PCL in the range of 3-24 and a degree of substitution (DS) of PVA from 0.35 to 0.89 and specify that both values were enhanced by microwave heating: irradiation was carried out at three power levels (340, 510 and 680 W), finding that the higher power provided higher DPs and DSs (Yu & Liu, 2007). Amphiphilic di-block PEG-*b*-PCL copolymers were synthesized in a one pot procedure under microwave irradiation using PEG monomethyl ester as oxydrilic initiator and boron trufluoride, Sodium hydride or stannous octoate as alternative catalysts, the latter yielding polymers with the highest purity (Ahmed et al., 2009). By means of a similar procedure, different authors obtained PEG-*b*-PCL copolymers with $M_w$ values ranging from 5,800 g/mol to 14,500 g/mol and very low polydispersity indexes (≤ 1.19). A comparison with oil-bath heating synthesis showed a dramatic reduction of reaction time under microwave conditions: analogous conversions were obtained after 45 min instead of 1 day (Karagöz & Dinçer, 2010). Biodegradable poly(ε-caprolactone)-*co*-poly(*p*-dioxanone) random copolymers with $M_w$ in the 120,000-180,000 g/mol range and polydispersity indexes around 2 were synthesized by ring opening polymerization of the two cyclic co-monomers under microwave irradiation. The reaction was completed within 20 min at 140 °C: after 30 min the conversion of ε-CL remained constant, while the conversion of *p*-DO slightly decreased (Chen et al., 2010).

### 2.2.3 Poly(lactic acid)

Even though poly(lactic acid) can be synthesized by direct polycondensation of lactic acid (see paragraph 2.1.1), polymers with higher molecular weights and lower polydispersity indexes are commonly obtained by ring opening polymerization of lactide (i.e. the cyclic dilactone of lactic acid) (Scheme 2). LL- and DD-lactide yield semi crystalline polymers named P(L)LA and P(D)LA, while *rac*-lactide is used to synthesize amorphous polymers named P(LD)LA. On account of its attractive physico-chemical properties and its biodegradability, PLA is widely used as bio-based, large scale packaging material and in a number of biomedical applications, hence it represents a subject of extensive research. A stannous octoate catalyzed microwave assisted synthesis of P(LD)LA was reported without the need of an inert atmosphere during irradiation. Using a domestic oven set at a power level of 450 W for 30 min a polymer with viscosity-average molecular weight over 200,000 g/mol was obtained in 85% yield; the reaction was carried out in bulk, using open vessels and adding carborundum

(SiC) to the mixture as microwave absorber. It is worth noting that, comparing the temperature in microwave and conventional conditions, the occurrence of both thermal and non-thermal microwave effects is claimed by the authors (Jing et al., 2006). In a different study, by means of an analogous procedure (except in this case the polymerization took place under vacuum) P(LD)LA with weight-average molecular weight ranging from 97,000 g/mol to 185,000 g/mol and polydispersity indexes around 2 were obtained in 90% yield within 10 min.

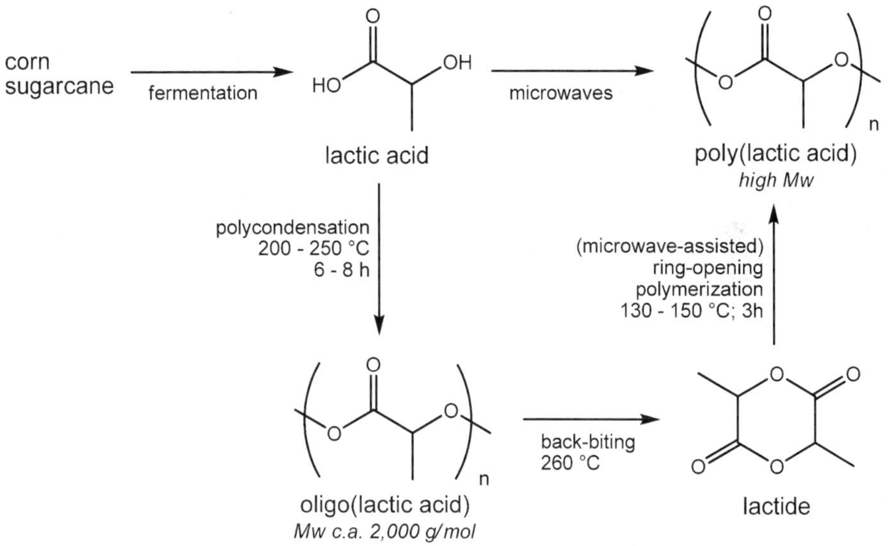

corn
sugarcane
fermentation
lactic acid
microwaves
poly(lactic acid)
*high Mw*
polycondensation
200 - 250 °C
6 - 8 h
(microwave-assisted)
ring-opening
polymerization
130 - 150 °C; 3h
oligo(lactic acid)
*Mw c.a. 2,000 g/mol*
back-biting
260 °C
lactide

Scheme 2. Synthetic routes to high Mw poly(lactic acid).

Optimal results were achieved setting the power level at 255 W, while reactions carried out at 340 W or more showed a significant degradation of the product (C. Zhang et al., 2004). According to other authors, the reaction rate in the ROP of L-lactide increases linearly with power output (tested up to 1,300 W), though molecular weights (and consequently degradation phenomena) were not monitored for P(L)LA (Koroskenyi & McCarthy, 2002). The synthesis of relatively high molecular weight P(LD)LA (ranging from 26,700 g/mol to 112,500 g/mol) was recently carried out in a CEM *Discover* monomodal MW reactor set at 100 °C for 30 min. Stannous octoate in low quantities with respect to monomer (0.1 mol%; 0.02 mol% and 0.01 mol%) was used as catalyst; the reaction was carried out in bulk, under reduced pressure (Nikolic et al., 2010). P(LD)LA with a high viscosity-average molecular weight (159,200 g/mol) could also be obtained, in similar conditions, by means of a modified domestic oven capable of a continuous microwave irradiation: the authors report a good yield (81.1%) after only 10 min at a low power level of 90 W (Y. Zhang et al., 2009). In a recent work an innovative calix[4]arene-Ti(IV) catalyst (Fig. 6) was employed in the bulk ring opening polymerization of L-lactide under microwave irradiation. A CEM *Discover* S-class monomodal reactor was operated at a maximum power level of 200 W for variable reaction times: a conversion of 85 % was registered after 40 min of irradiation reaching 95 % after 80 min; longer times led to decomposition phenomena. Compared to conventional synthesis, microwave heating markedly reduced reaction times but slighltly lowered the control over molecular weights and polydispersity indexes (Frediani et al., 2010).

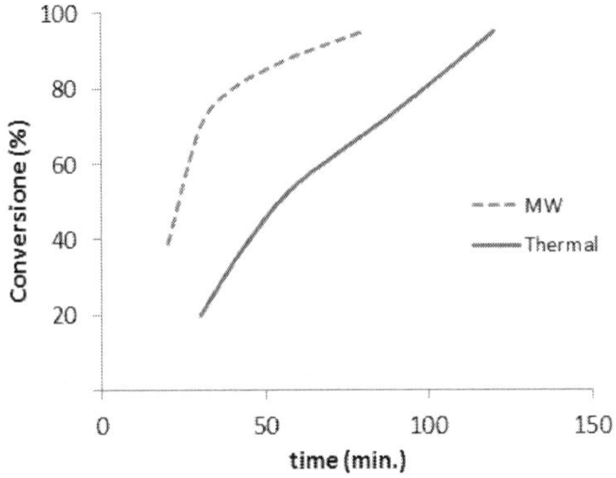


Fig. 6. Conversion vs reaction time for the ROP of L-lactide (left) and the employed calix[4]arene-Ti catalyst (see Frediani et al., 2010).

### 2.2.4 Poly(lactic acid) copolymers

Fully biocompatible and biodegradable poly(glycolic acid-*co*-lactic acid) copolymers with potential applications in the biomedical field were synthesized by ring opening polymerization: stannous octoate was used as catalyst (0.03 mol%) and octadecanol as the oxydrylic initiator (0.01 mol%). Microwave irradiation was performed in a *MAS-II* scientific reactor at a temperature of 120 °C for a variable time (3 min to 9 min), under nitrogen atmosphere: different glycolic acid (GA) : lactic acid (LA) feed ratios were assessed and yields ranged from 69% to 86%. It is worth noting how the products showed higher GA content compared to the feed ratio, and that copolymers with higher GA content were characterized by decreased glass transition temperature and increased melting temperature (G. Li et al., 2010). Amphiphilic, biodegradable PLLA-*b*-PEG-*b*-PLLA triblock copolymers were synthesized by microwave assisted ring opening polymerization of L-lactide employing poly(ethylene glycol) (PEG2000) as a difunctional oxydrylic initiator (Scheme 3).

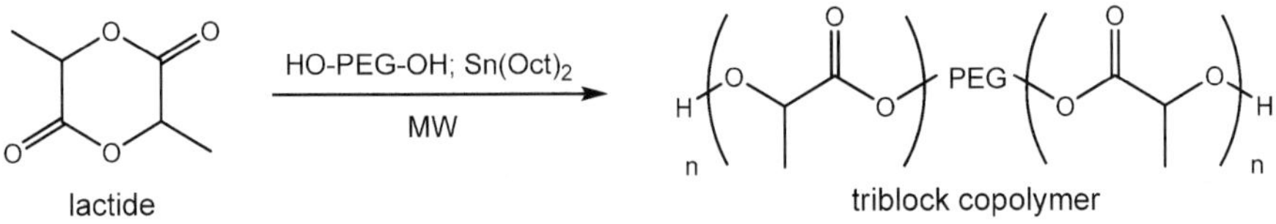

lactide — HO-PEG-OH; $Sn(Oct)_2$, MW → triblock copolymer

Scheme 3. Synthesis of PLA-*b*-PEG-*b*-PLA triblock copolymer.

A surprisingly short irradiation time of 3 min at 100 °C in a CEM *Discover* reactor was enough to achieve 92% lactide conversion, yielding a copolymer with a number average molecular weight of 28,230 g/mol. It was observed that longer PLLA segments (i.e. higher PLLA content) induced higher glass transition and melting temperatures (C. Zhang et al., 2007).

### 2.2.5 Other ring opening polymerizations

Stannous octoate catalyzed ring opening polymerization of *p*-dioxanone was conducted under microwave irradiation to obtain (63% yield) biodegradable PDDO with a viscosity average molecular weight of 156,000 g/mol. A domestic oven was employed for this study and several power level tested: 270 W for 25 min proved to be the most favorable choice, allowing a dramatic increase in the reaction rate compared to the 14 h of heating required by conventional procedures in analogous conditions (Y. Li et al., 2006). Poly(trimethylen carbonate) (PTMC) is a promising, biocompatible poly(alkyl carbonate) with good mechanical properties: a PTMC-*b*-PEG-*b*-PTMC triblock copolymers with tunable degradation properties were synthesized by catalyst-free microwave assisted ring opening polymerization of TMC using PEG600, PEG1000 and PEG2000 as a difunctional oxydrylic initiators. It was observed that the molar mass of the copolymer and the TMC conversion reached a maximum after 30 min at 120 °C (Fig. 7): further irradiation or higher temperatures induced product degradation phenomena (Liao et al., 2008).

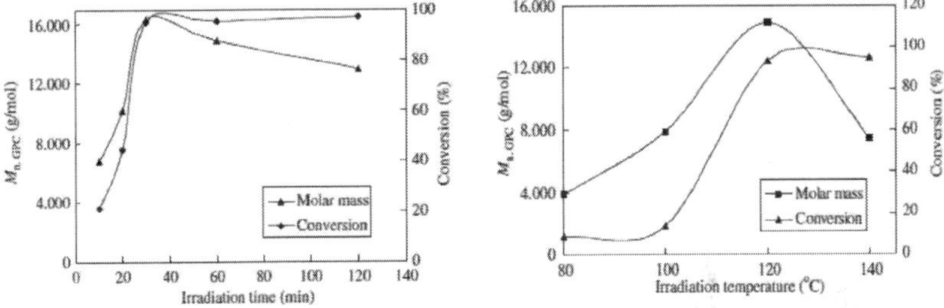


Fig. 7. Synthesis of a PTMC-*b*-PEG-*b*-PTMC copolymer: effect of irradiation time (left) and temperature (right) on Mn and conversion (from Liqiong Liao, Chao Zhang, Shaoqin Gong, Preparation of Poly(trimethylene carbonate)-*block*-poly(ethylene glycol)-*block*-poly(trimethylene carbonate) Triblock Copolymers under Microwave Irradiation, Reactive & Functional Polymers, 2008, 68, 751-758. Copyright Elsevier Ltd. Reproduced with permission).

The ring opening of cyclohexene-oxide in the presence of $CO_2$ for the synthesis of biodegradable aliphatic poly(ether-carbonate)s was investigated in order to employ a troublesome waste product as feedstock for the production of useful materials. Double metal cyanide complex catalysts based on $Zn_3[Co(CN)_6]$ were profitably utilized and microwave irradiation was chosen as heating method, with strong beneficial effects on polymerization times and $CO_2$ incorporation grade (Dharman et al., 2008). Poly(oxazoline)s are relatively non toxic polymers with attractive properties for biomedical applications, nonetheless they cannot be classified as biodegradable or bio-based polymers: comprehensive reviews about microwave assisted synthesis in this class of products can be found elsewhere (Adams & Schubert, 2007; Makino & Kobayashi, 2010; Wiesbrock et al., 2004).

## 3. Microwave assisted modification and processing

In this paragraph we review a number of papers exemplifying how microwave irradiation can also be employed for the chemical modification and processing of natural abounding or synthetic polymers. Some studies deal with non biodegradable oil-based polymers: these works are reported anyway, since our attention here mostly focuses on the processing technique rather than the polymer itself.

### 3.1 Grafting and chemical modification of natural polymers

Chemical modification of natural polymers, most commonly polysaccharides, is a clever way to obtain new materials from abundant renewable feedstocks. A very effective approach to this task is the design of graft copolymers in which the former natural polymer constitutes the backbone and is covalently bound to numerous branches of synthetic polymer. For example, in order to obtain a biodegradable amphoteric material, chitosan was first reacted with phtalic anhydride to protect the amino groups along the chain and then used as an oxydrylic initiator for the stannous octoate catalyzed ROP of ε-caprolactone: after deprotection a chitosan-*g*-PCL copolymer with grafting percentage above 100% by weight was obtained and characterized (Scheme 4). The polymerization step was carried out smoothly under microwave irradiation using a domestic oven and best results in terms of increased reaction rate were obtained with high power levels of 450 W and 600 W (L. Liu et al., 2005). The β-cyclodextrin functionalization of chitosan was also performed under microwave irradiation yielding graft copolymers with a superior lead adsorption capacity if compared to analogous products obtained by conventional heating. The authors report that a superior yield was achieved despite the fact that no redox initiator was required (Sharma & Mishra, 2010). It is worth noting that chitosan itself can be obtained from chitin via a microwave mediated deacetylation reaction (Sahu et al., 2009). With a similar approach, PCL side chains were grafted onto pea starch (35% amylose, 65% amylopectin) in a microwave assisted (255 W for 3 min) one pot synthesis yielding a thermoformable hydrophobic copolymer with promising properties as eco-friendly or biomedical material. Regrettably, not much about the utilized microwave heating setup is reported by the authors (Chang et al., 2009). An efficient superabsorbent material capable of a high swelling ratio in distilled water was synthesized grafting sodium acrylate onto cornstarch under microwave irradiation and using potassium persulfate as initiator. Microwave power is reported to be the most influent factor to obtain best performing copolymers (i.e. with higher swelling ratio and lower solubility): 10 min of irradiation at 85-90 W proved to be the most suitable heating conditions (Tong & al., 2004). The microwave assisted preparation of a poly(acrylamide)-*g*-starch copolymer with promising properties as a flocculant for water treatment was recently described using ceric ammonium nitrate as initiator. The authors report high grafting rates and superior yields compared to the conventional procedure. Despite the use of a domestic 800 W microwave oven, and the manual switching off of irradiation whenever the reaction mixture was about to boil, reproducibility is claimed to be one of the advantages over ordinary methods (S. Mishra et al., 2011). Cellulose, being the most abundant natural polysaccharide, is also attracting a huge attention and indeed several examples of its chemical modification can be found in literature. Microwave heating was employed for the synthesis of common cellulose derivatives like esters (Satgé et al., 2002) and carbammates (Guo et al., 2009) as well as for the graft copolymerization of methyl methacrylate initiated with ceric amonium nitrate (Wan et al., 2011). Guar Gum is an edible galactomannan isolated from a legume native of the northwestern India and finds wide application as water thickener due to the ability of forming highly viscous colloidal dispersions. In order to tune its physicochemical and degradation properties a number of microwave assisted derivatizations were successfully carried out by the same group within short irradiation times. We report, among these, a poly(ethyl acrylate) graft copolymer for efficient removal of cadmium(II) from aqueous phases (Singh et al., 2009); a poly(acrylonitrile) graft copolymer with high water absorbing capacity (Singh et al., 2004) and the first example of poly(acrylamide) graft copolymer (Singh et al., 2004). It is worth noting how the authors achieve these results in the absence of redox initiators or catalysts: this is due, in their view, to non thermal effects and to radical formation as a consequence of molecular friction under microwave heating. Salient papers regarding microwave assisted synthesis deal also with grafting of methyl methacrylate onto carrageenan in which backbone decomposition is prevented by short reaction times (Prasad et al., 2006); poly(acrylonitrile)-*g*-xyloglucan copolymers by an initiator-free preparation method (A. Mishra et al., 2008) and poly(acrylic acid) grafting of *Cassia Javanica* seed gum to obtain a sorbent material for Hg(II) removal (Singh et al., 2010).

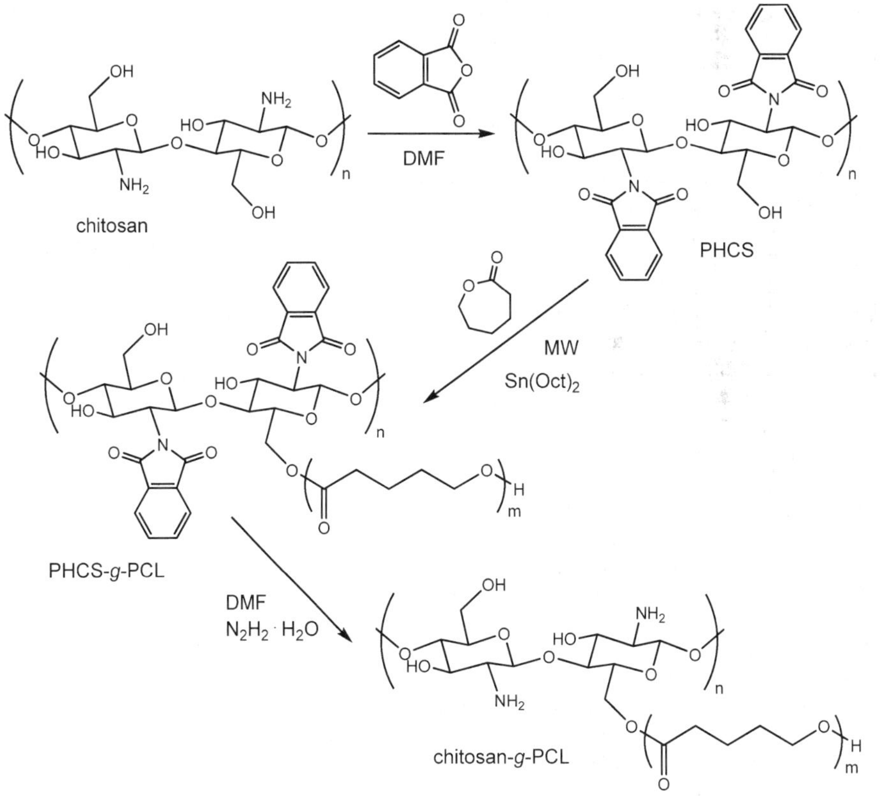

Scheme 4. Synthesis of a chitosan-*g*-PCL copolymer.

### 3.2 Polymer processing

With the development of industrial scale dedicated devices, microwave polymer processing techniques are nowadays sensible alternatives to conventional heating methods, setting new benchmarks in terms of time and energy efficiency. Comprehensive reviews about this chief research field can be found in literature (Clark & Sutton, 1996; Das et al., 2008; Leonelli & Mason, 2010; Thostenson & Chou, 1999); our aim here is to introduce some representative examples dealing with a variety of applications ranging from polymer crosslinking to welding of thermoplastics. Short irradiation times and a relatively low temperature (100 °C) turned out to be sufficient to carry out the one-pot crosslinking reaction of poly(vinyl alcohol) in the presence of metal salts, yielding polymer-inorganic nanocomposites with several different morphologies. On account of the good control over nanostructure's shape and size distribution and considering that similar products were previously prepared via hydrothermal treatment at 200 °C in 72 h, the efficiency of microwave heating applied to these syntheses results clear (Polshettiwar et al., 2009; Varma & Nadagouda, 2007). Polymer surface modification has been performed by a microwave induced plasma treatment (Leonelli & Mason, 2010): to cite an example, poly(dimethylsiloxane) (PDMS) slabs were irreversibly sealed within 30-60s by means of a 30 W plasma generated in glass bottles containing 2-3 Torr of oxygen (Hui et al., 2005). Similar procedures are reported in papers dealing with modification of PDMS surface hydrophilicity (Ginn & Steinbock, 2003) and polyaniline-aided welding of high density poly(ethylene) (Wu & Benatar, 1997). Welding of thermoplastics such as ultra high molecular weight polyethylene (UHMW PE), polycarbonate and ABS could be achieved by focused microwave irradiation of the specimen joint interface (Yarlagadda & Chai, 1998). An interesting comparison between thermal curing and microwave curing of the E44 epoxy resin showed a significant reduction of processing time and temperature; a lower quantity of curing agent (i.e. maleic anhydride) was needed under microwave irradiation and products with improved compressive strength and bending strength are obtained by the authors (Zhou et al., 2003). In a recent work epoxies of linseed oil and dehydrated castor oil were blended with PVA within 2 min of microwave irradiation (75 °C; 350 W) with the aim to develop new bio-based materials with desirable properties for packaging and mulching: as usual, the chief advantage over conventional blending procedures is reported to be the extremely short processing time (Riaz et al., 2011). According to several authors, microwave foaming under vacuum (Jaja & Durance, 2008) or at atmospheric pressure (Torres et al., 2007) proved to be an effective route to obtain porous materials for tissue engineering applications from polymeric gels. It is worth reminding that microwave transparent polymers, such as HDPE, may easily be processed employing microwaves as energy source in the presence of highly absorbing fillers and additives like carbon black (F. Liu et al., 2010). Theoretical models have also been developed to investigate the optimal conditions for an efficient heating: in a recent study the application of microwaves on natural rubber and Nylon 66 slabs was analyzed highlighting the role of irradiation pulsing parameters and ceramic support plates (Durairaj & Basak, 2009).

## 4. Conclusion

During the last decade, the diffusion of reliable research devices turned microwave assisted polymer chemistry from pioneering curiosity to realistic alternative heating technique, while increased knowledge and engineering contributed to optimize well established polymer processing methods. Increased reaction rates under microwave conditions are reported by virtually every paper in the related literature and time saving is widely recognized as the most obvious advantage over conventional heating. Even though the explanations for this beneficial effect are sometimes controversial, many authors nowadays tend to exclude the occurrence of non-thermal phenomena and rather focus on faster energy transfer, higher local temperatures or solvents overheating. Higher products yields are often, yet not univocally, mentioned as one of the relevant features provided by microwave reactors, as well as altered selectivities, probably related to differential radiation absorption, are described in some cases. Some experiments show how products purity may be improved under microwave conditions, minimizing detrimental side reactions by a more homogeneous heating and reduced reaction times. Despite these remarkable points, assessing how "green" microwaves are in terms of energy efficiency seems not to be straightforward: indeed several papers report efficiencies which closely compare with conventional oil baths and mantles (Devine & Leadbeater, 2011; Gronnow et al., 2005; Moseley & Kappe, 2011; Razzaq & Kappe, 2008). The choice of a specific experimental setup notably influences the overall balance and particularly important appear to be the reaction scale, the use of monomodal or multimodal devices and open or pressurized vials, nonetheless the relatively low efficiency of magnetrons (40% to 65%) generally counterweight the much shorter reaction times. All things considered, energy efficiency of microwave assisted chemistry should not be taken for granted but carefully evaluated case by case.

## 5. Acknowledgment

Authors would like to thank MIUR (PRIN 2008 project 200898KCKY – 'Inorganic nanohybrids based on bio-polyesters from renewable resources') and Regione Toscana ('Competitività regionale e occupazione' 2007–2013 – project TeCon@BC, POR-FESR 2007–2013) for financial support.

## 6. References

Adams, N. & Schubert, U. S. (2007). Poly(2-oxazolines) in Biological and Biomedical Applications. *Advanced Drug Delivery Review,* Vol. 59, No. 15 (December 2007), pp. 1504-1520, ISSN: 0169-409X

Ahmed, H.; Trathnigg, B.; Kappe, O. & Saf, R. (2009). Characterization of Poly(ethylene glycol)-*b*-poly(ε-caprolactone) by Liquid Chromatography under Critical Conditions: Influence of Catalysts and Reaction Conditions on Product Composition. *European Polymer Journal,* Vol. 45, No. 8 (August 2009), pp. 2338-2347, ISSN: 0014-3057

Albert, P.; Warth, H.; Mülhaupt, R. & Janda, R. (1996). Comparison of Thermal and Microwave-Activated Polymerization of ε-Caprolactone with Titanium Tetrabutylate as Catalyst. *Macromolecular Chemistry and Physics,* Vol. 197, No. 5 (May 1996), pp. 1633-1641, ISSN: 1022-1352

Barbier-Baudry, D.; Brachais, L.; Cretu, A.; Gattin, R.; Loupy, A. & Stuerga, D. (2003). Synthesis of Polycaprolactone by Microwave Irradiation - an Interesting Route to Synthesize this Polymer via Green Chemistry. *Environmental Chemical Letters,*Vol. 1, No. 1, (January 2003), pp. 19-23, ISSN: 1610-3661

Bezdushna, E.; Ritter, H. & Troev, K. (2005). Poly(alkylene hydrogen phosphonate)s by Transesterification of Dimethyl Hydrogen Phosphonate with Poly(ethylene glycol). *Macromolecular Rapid Communications,* Vol. 26, No. 6 (March 2005), pp. 471-476, ISSN: 1521-3927

Bogdal, D.; Penczek P.; Pielichowsky, J. & Prociack, A. (2003). Microwave Assisted Synthesis, crosslinking and Processing of Polymeric Materials. *Advances in Polymer Science,* Vol. 163 (2003), pp. 193-263, ISSN: 0065-3195

Borriello, A.; Nicolais, L.; Fang, X.; Huang, S. J. & Scola, D. A. (2007). Synthesis of Poly(amide-ester)s by Microwave Methods. *Journal of Applied Polymer Science,* Vol. 103, No. 3 (February 2007), pp. 1952-1958, ISSN: 0021-8995

Chang, P. R.; Zhou, Z.; Xu, P.; Chen, Y.; Zhou, S. & Huang, J. (2009). Thermoforming Starch-*graft*-polycaprolactone biocomposites via One-pot Microwave Assisted Ring Opening Polymerization. *Journal of Applied Polymer Science,* Vol. 113, No. 5 (September 2009), pp. 2973-2979, ISSN: 0021-8995

Chatti, S.; Bortolussi, M.; Loupy, A.; Blais, J. C.; Bogdal, D. & Majdoub, M. (2002). Efficient Synthesis of Polyethers from Isosorbide by Microwave-Assisted Phase Transfer Catalysis. *European Polymer Journal,* Vol. 38, No. 9 (September 2002), pp. 1851-1861, ISSN: 0014-3057

Chatti, S.; Bortolussi, M.; Bogdal, D.; Blais, J. C. & Loupy, A. (2006). Synthesis and Properties of New Poly(ether-ester)s Containing Aliphatic Diols Based on Isosorbide. Effects of the Microwave-Assisted Polycondensation. *European Polymer Journal,* Vol. 42, No. 2 (February 2006), pp. 410-424, ISSN: 0014-3057

Chen, R.; Hao, J. Y.; Xiong, C. D. & Deng, X. M. (2010). Rapid Synthesis of Biodegradable Poly(epsilon-caprolactone-*co-p*-dioxanone) Random Copolymers under Microwave Irradiation. *Chinese Chemical Letters,* Vol. 21, No. 2 (February 2010), pp. 249-252, ISSN: 1001-8417

Clark, D. E. & Sutton, W. H. (1996). Microwave Processing of Materials. *Annual Review of Material Science,* Vol. 26, pp. 299-331, ISSN: 0084-6600

Das, S.; Mukhopadhyay, A. K.; Datta, S. & Basu, P. (2008). Prospects of Microwave Processing: an Overview. *Bulletin of Material Science,* Vol. 31, No. 7 (December 2008), pp. 943-956, ISSN: 0973-7669

Devine, W. G. & Leadbeater, N. E. (2011). Probing the Efficiency of Microwave Heating and Continuous-flow Conventional Heating as Tools for Organic Chemistry. *Arkivoc,* Vol. 2011, Part (v), pp. 127-143, ISSN:1424-6376

Dharman, M. M.; Ahn, J. Y.; Lee, M. K.; Shim, H. L.; Kim, K. H.; Kim, I & Park, D. W. (2008). Moderate Route for the Utilization of $CO_2$ – Microwave Induced Coploymerization with Cyclohexene Oxide Using Higly Efficient Double Metal Cyanide Complex Catalysts Based on $Zn_3[Co(CO)_6]$. *Green Chemistry,* Vol. 10, No. 6 (2008), pp. 678-684, ISSN: 1463-9270

Deetlef, M. & Seddon, K. R. (2003). Improved Preparation of Ionic Liquids Using Microwave Irradiation. *Green Chemistry,* Vol. 5, No. 2 (2003) pp. 181-186, ISSN: 1463-9270

Durairaj, S. & Basak, T. (2009). Analysis of Pulsed Microwave Processing of Polymer Slabs Supported With Ceramic Plates. *Chemical Engineering Science,* Vol. 64, No. 7 (April 2009), pp. 1488-1502, ISSN: 0009-2509

Fang, X.; Hutcheon, R. & Scola, D. A. (2000). Microwave Syntheses of Poly(ε-caprolactam-*co*-ε-caprolactone). *Journal of Polymer Science - Part A: Polymer Chemistry,* Vol. 38, No. 8 (April 2000), pp. 1379-1390, ISSN: 1099-0518

Frediani, M.; Sémeril, D.; Matt, D.; Rizzolo, F.; Papini, A. M.; Frediani, P.; Rosi, L.; Santella, M. & Giachi, G. (2010). L-Lactide Polymerization by Calix[4]arene-titanium (IV) Complex Using Conventional Heating and Microwave Irradiation. *e-Polymers,* No. 019 (February 2010) pp. 1-8, ISSN: 1618-7229

Ginn, B. T. & Steinbock, O. (2003). Polymer Surface Modification Using Microwave-Oven-Generated Plasma. *Langmuir,* Vol. 19, No. 19 (September 2003), pp. 8117-8118, ISSN: 0743-7463

Glasnow, T. N. & Kappe, C. O. (2007). Microwave-Assisted Synthesis Under Continuous Flow Conditions. *Macromolecular Rapid Communications,* Vol. 28, No. 4 (February 2007), pp. 395-410, ISSN: 1521-3927

Gronnow, M. J.; White, R. J.; Clark, J. H. & Macquarrie, D. J. (2005). Energy Efficiency in Chemical Reactions: a Comparative Study of Different Reaction Techniques. *Organic Process Research & Engineering,* Vol. 9, No. 4 (July 2005), pp. 516-518, ISSN: 1083-6160

Guerrero-Sanchez, C.; Lobert, M.; Hoogenboom, R. & Schubert, U. S. (2007). Microwave Assisted Homogeneous Polymerizations in Water-Soluble Ionic Liquids: an Alternative and Green Approach for Polymer Synthesis. *Macromolecular Rapid Communications,* Vol. 28, No. 4 (February 2007), pp. 456-464, ISSN: 1521-3927

Guo, Y.; Zhou, J.; Song, Y. & Zhang, L. (2009). An Efficient and Environmentally Friendly Method for the Synthesis of Cellulose Carbammate by Microwave Heating. *Macromolecular Rapid Communications,* Vol. 30, No. 17 (September 2009), pp. 1504-1508, ISSN: 1521-3927

Hoogenboom, R.; Moore, B. C. & Schubert, U. S. (2006). Synthesis of Star-Shaped Poly(ε-caprolactone) via 'Click' Chemistry and 'Supramolecular Click' Chemistry. *Chemical Communications,* No. 38 (2006), pp. 4010-4012, ISSN: 1364-548X

Hui, A. Y.; Wang, G.; Lin, B. & Chan, W. T. (2005). Microwave Plasma Treatment of Polymer Surface for Irreversible Sealing of Microfluidic Devices. *Lab on a Chip,* Vol. 5, No. 10 (October 2005), pp. 1173-1177, ISSN: 1473-0189

Jaja, S. & Durance, T. D. (2008). Tailor-made Biopolymers Porous Scaffold Fabrication for Tissue Engineering: Application of Radiant Energy in the Form of Microwave under Vacuum. *Bio-medical Materials and Engineering,* Vol. 18, No. 6, pp. 357-366, ISSN: 0959-2989

Jing, S.; Peng, W.; Tong, Z. & Baoxiu, Z. (2006). Microwave-Irradiated Ring-Opening Polymerization of D,L-Lactide Under Atmosphere. *Journal of Applied Polymer Science,* Vol. 100, No. 3 (may 2006), pp. 2244-2247, ISSN: 0021-8995

Karagöz, A. & Dinçer, S. (2010). Microwave Assisted Synthesis of Poly(ε-caprolactone)-*block*-poly(ethylene gycol) and Poly(lactide)-*block*-poly(ethylene glycol). *Macromolecular Symposia,* Vol. 295, No. 1 (September 2010), pp. 131-137, ISSN: 1521-3900

Keki, S.; Bodnar, I.; Borda, J.; Deak, G. & Zsuga, M. (2001). Fast Microwave-Mediated Bulk Polycondensation of D,L-Lactic Acid. *Macromolecular Rapid Communications,* Vol. 22, No. 13 (September 2001), pp. 1063-1065, ISSN: 1521-3927

Kerep, P. & Ritter, H. (2006). Influence of Microwave Irradiation on the Lipase-Catalyzed Ring Opening Polymerization of ε-Caprolactone. *Macromolecular Rapid Communications,* Vol. 27, No. 9 (May 2006), pp. 707-710, ISSN: 1521-3927

Koroskenyi, B. & McCarthy S. (2002). Microwave-Assisted Solvent-Free or Aqueous-Based Synthesis of Biodegradable Polymers. *Journal of Polymers and the Environment,* Vol. 10, No. 3 (July 2002), pp. 93-104, ISSN: 1566-2543

Kremsner, J. M. & Kappe, C. O. (2006). Silicon Carbide Passive Heating Elements in Microwave-Assisted Organic Synthesis. *Journal of Organic Chemistry,* Vol. 71, No. 12 (June 2006), pp. 4651-4658, ISSN: 0022-3263

Lei, H. C.; Wang, P. & Yuan, W. B. (2009). Microwave-Assisted Synthesis of Poly(L-Lactic Acid) via Direct Melt Polycondensation using Solid Super Acids. *Macromolecular Chemistry and Physics,* Vol. 210, No. 23 (December 2009), pp. 2058-2062, ISSN: 1521-3935

Leonelli, C.; Veronesi, P. & Grisoni, F. (2007). Numerical Simulation of an Industrial Microwave Assisted Filter Dryer: Criticality Assessment and Optimization. *Journal of Microwave Power and Electromagnetic Energy,* Vol. 41, No. 3 (October 2007), pp. 5-13, ISSN: 0832-7823

Leonelli, C. & Mason, T. J. (2010). Microwave and Ultrasonic Processing: Now a Realistic Option for Industry. *Chemical Engineering and Processing,* Vol. 49, No. 9 (September 2010), pp. 885-900, ISSN: 0255-2701

Li, G.; Zhao, N.; Bai, W.; Chen, D. L. & Xiong, C. D. (2010). Microwave-Assisted Ring-Opening Polymerization of Poly(glycolic acid-co-lactic acid) Copolymers. *E-Polymers,* No. 051 (May 2010), pp. 1-6, ISSN: 1618-7229

Li, H.; Liao, L. & Liu, L. (2007). Kinetic Investigation into the Non-Thermal Microwave Effect on the Ring Opening Polymerization of ε-Caprolactone. *Macromolecular Rapid Communications,* Vol. 28, No. 4 (February 2007), pp. 411-416, ISSN: 1521-3927

Li, Y.; Wang, X. L.; Yang, K. K. & Wang, Y. Z. (2006). A Rapid Synthesis of Poly(*p*-dioxanone) by Ring Opening Polymerization under Microwave Irradiation. *Polymer Bulletin,* Vol. 57, No. 6 (October 2006), pp. 873-880, ISSN: 0170-0839

Liao, L.; Liu, L. & Zhang, C. (2006). Microwave-Assisted Ring Opening Polymerization of ε-Caprolactone in the Presence of Ionic Liquid. *Macromolecular Rapid Communications,* Vol. 27, No. 24 (December 2006), pp. 2060-2064, ISSN: 1521-3927

Liao, L.; Zhang, C. & Gong, S. (2008). Preparation of Poly(trimethylene carbonate)-*block*-poly(ethylene glycol)-*block*-poly(trimethylen carbonate) Triblock Copolymers under Microwave Irradiation. *Reactive and Functional Polymers,* Vol. 68, No. 3 (March 2008), pp. 751-758, ISSN: 1381-5148

Lidström, P.; Tierney, J.; Wathey, B. & Westman, J. (2001). Microwave Organic Synthesis - a Review. *Tetrahedron,* Vol. 57, No. 45 (November 2001), pp. 9225-9283, ISSN: 0040-4020

Liu, F.; Qian, X.; Wu, X.; Guo, C.; Ley, Y. & Zhang, J. (2010). The Response of Carbon Black Filled High-density Polyethylene to Microwave Processing. *Journal of Materials Processing Technology,* Vol. 210, No. 14 (November 2010), pp. 1991-1996, ISSN: 0924-0136

Liu, L.; Liao, L. Q.; Zhang, C.; He, F. & Zhuo, R. X. (2001). Microwave-Assisted Polycondensation of L-2-hydroxy-3-phenylpropanoic Acid. *Chinese Chemical Letters,* Vol. 12, No. 9 (September 2001), pp. 761-762, ISSN: 1001-8417

Liu, L.; Li, Y.; Fang, Y. & Chen, L. (2005). Microwave-Assisted Graft Copolymerization of ε-Caprolactone onto Chitosan via the Phtaloyl Protection Method. *Carbohydrate Polymers,* Vol. 60, No. 3 (May 2005), pp. 351-356, ISSN: 0144-8617

Liu, P.; Su, Z. (2005). Thermal Stabilities of Polystyrene/Silica Hybrid Nanocomposites via Microwave-assisted In-situ Polymerization. *Materials Chemistry and Physics,* Vol. 94, No. 3 (December 2005), pp. 412-415, ISSN: 0254-0584

Makino, A. & Kobayashi, S. (2010). Chemistry of 2-Oxazolines: a Crossing of Cationic Ring-Opening Polymerization and Enzymatic Ring-Opening Polyaddition. *Journal of Polymer Science - Part A: Polymer Chemistry,* Vol. 48, No. 6 (March 2010), pp. 1251-1270, ISSN: 1099-0518

Mishra, S.; Mukul, A.; Sen, G. & Jha, U. (2011). Microwave Assisted Synthesis of Polyacrylamide Grafted Starch (St-*g*-PAM) and its Applicability as Flocculant for Water Treatment. *International Journal of Biological Macromolecules,* Vol. 48, No. 1 (January 2011), pp. 106-111, ISSN: 0141-8130

Mishra, A.; Clark, J. H.; Vij, A. & Daswal, S. (2008). Synthesis of Graft Copolymers of Xyloglucan and Acrylonitrile. *Polymers for Advanced Technologies,* Vol. 19, No. 2 (February 2008), pp. 99-104, ISSN: 1042-7147

Moseley, J. D. & Woodman E. K. (2008). Scaling-Out Pharmaceutical Reactions in an Automated Stop-Flow Microwave Reactor. *Organic Processes Research & Development,* Vol. 12, No. 5 (September 2008), pp. 967-981, ISSN: 1083-6160

Moseley, J. D. & Kappe, C. O. (2011). A Critical Assessment of the Greenness and Energy Efficiency of Microwave-assisted Organic Synthesis. *Green Chemistry,* Advance Article, ISSN: 1463-9270

Nikolic, L.; Ristic, I,; Adnadjevich, B.; Nikolic, V.; Jovanovich, J. & Stankovich, M. (2010). Novel Microwave-Assisted Synthesis of Poly(D,L-Lactide): the Influence of Monomer/Initiatior Ratio on the Product Properties. *Sensors,* Vol. 10, No. 5 (May 2010), pp. 5063-5073, ISSN: 1424-8220

Nadagouda, M. N. & Varma, R. S. (2007). Preparation of Novel Metallic and Bimetallic Cross-Linked Poly(vinyl alcohol) Nanocomposites under Microwave Irradiation. *Macromolecular Rapid Communications,* Vol. 28, No. 4 (February 2007), pp. 465-472, ISSN: 1521-3927

Nagahata, R.; Sano, D.; Suzuki, H. & Takeuchi, K. (2007). Microwave-Assisted Single-Step Synthesis of Poly(lactic acid) by Direct Polycondensation of Lactic Acid. *Macromolecular Rapid Communications,* Vol. 28, No. 4 (February 2007), pp. 437-442, ISSN: 1521-3927

Nuchter, M.; Ondruschka, B.; Bonrath, W. & Gum, A. (2004). Microwave Assisted Synthesis - a Critical Technology Overview. *Green Chemistry,* Vol. 6, No. 3 (2004), pp. 128-141, ISSN: 1463-9270

Perreux, L. & Loupy, A. (2001). A Tentative Rationalization of Microwave Effects in Organic Synthesis According to the Reaction Medium, and Mechanistic Considerations. *Tetrahedron,* Vol. 57, No. 45 (November 2001), pp. 9199-9223, ISSN: 0040-4020

Polshettiwar, V.; Nadagouda, M. N. & Varma, R. S. (2009). Microwave-Assisted Chemistry: a Rapid and Sustainable Route to Synthesis of Organics and Nanomaterials. *Australian Journal of Chemistry,* Vol. 61, No. 1 (January 2009), pp. 16-26, ISSN: 0004-9425

Prasad, K.; Meena, R. & Siddhanta, A. K. (2006). Microwave-Induced Rapid One-Pot Synthesis of κ-Carrageennan-*g*-PMMA Copolymer by Potassium Persulphate Initiating System. *Journal of Applied Polymer Chemistry,* Vol. 101, No. 1 (July 2006), pp. 161-166, ISSN: 1097-4628

Razzaq, T. & Kappe, C. O. (2008). On the Energy Efficiency of Microwave-Assisted Organic Reactions. *ChemSusChem,* Vol. 1, No. 1-2 (February 2008), pp. 123-132, ISSN: 1864-564X

Riaz, U.; Ashraft, S. M. & Sharma, H. O. (2011). Mechanical, Morphological and Biodegradation Studies of Microwave Processed Nanostructured Blends of Some Bio-based Oil Epoxies with Poly(vinyl alcohol). *Polymer Degradation and Stability,* Vol.96, No. 1 (January 2011), pp. 33-42, ISSN: 0141-3910

Satgé, C.; Verneuil, B.; Branland, P.; Granet, R.; Krausz, P.; Rozier, P. & Petit, C. (2002). Rapid Homogeneous Esterification of Cellulose Induced by Microwave Irradiation. *Carbohydrate Polymers,* Vol. 49, No. 3 (August 2002), pp. 373-376, ISSN: 0144-8617

Sahu, A.; Goswami, P. & Bora, U. (2009). Microwave Mediated Rapid Synthesis of Chitosan. *Journal of Material Science: Materials in Medicine,* Vol. 20, No. 1 (January 2009), pp. 171-175, ISSN: 1573-4838

Sharma, A. K. & Mishra, A. K. (2010). Microwave Induced β-cyclodextrin Modification of Chitosan for Lead Sorption. *International Journal of Biological Macromolecules,* Vol. 47, No. 3 (October 2010), pp. 410-419, ISSN: 0141-8130

Singh, V.; Singh, S. K. & Maurya, S. (2010). Microwave Induced Poly(acrylic acid) Modification of *Cassia Javanica* Seed Gum for Efficient Hg(II) Removal from Solution. *Chemical Engineering Journal,* Vol. 160, No. 1 (May 2010), pp. 129-137, ISSN: 1385-8947

Singh, V.; Sharma, A. K. & Maurya, S. (2009). Efficient Cadmium (II) Removal from Aqueous Solution Using Microwave Synthesized Guar Gum-Graft-Poly(ethylacrylate). *Industrial & Engineering Chemistry Research,* Vol. 48, No. 10 (May 2009), pp. 4688-4696, ISSN: 0888-5885

Singh, V.; Tiwari, A.; Tripathi, D. N. & Sanghi, R. (2004). Microwave Assisted Synthesis of Guar-g-polyacrylamide. *Carbohydrate Polymers,* Vol. 58, No. 1 (October 2004), pp. 1-6, ISSN: 0144-8617

Singh, V.; Tiwari, A.; Tripathi, D. N. & Sanghi, R. (2004). Grafting of Polyacrylonitrile onto Guar Gum under Microwave Irradiation. *Journal of Applied Polymer Science,* Vol. 92, No. 3 (May 2004), pp. 1569-1575, ISSN: 0021-8995

Sinwell, S.; Schmidt, A. M. & Ritter, H. (2006). Direct Synthesis of (Meth-)acrylate Poly(ε-caprolactone) Macromonomers. *Journal of Macromolecular Science - Part A: Pure and Applied Chemistry,* Vol. 43, No. 3 (March 2006), pp. 469-476, ISSN: 1520-5738

Sivalingam, G.; Agarwal, N. & Madras, G. (2004). Kinetics of Microwave-Assisted Polymerization of ε-Caprolactone. *Journal of Applied Polymer Science,* Vol. 91, No. 3 (February 2004), pp. 1450-1456, ISSN: 0021-8995

Strauss, C. R. & Trainor, R. W. (1995). Developments in Microwave-assisted Organic Chemistry. *Australian Journal of Chemistry*, Vol. 48, No. 10 (October 1995), pp. 1665-1692, ISSN: 0004-9425

Tan, Y.; Cai, S.; Liao, L.; Wang, Q. & Liu, L. (2009). Microwave-Assisted Ring-Opening Polymerization of ε-Caprolactone in Presence of Hydrogen Phosphonates. *Polymer Journal*, Vol. 41, No. 10 (October 2009), pp. 849-854, ISSN: 0032-3896

Thostenson, E. T. & Chou, T. W. (1999). Microwave Processing: Fundamentals and Applications. *Composites – Part A: Applied Science and Manifacturing*, Vol. 30, No. 9 (September 1999), pp. 1055-1071, ISSN: 1359-835X

Tong, Z.; Peng, W.; Zhiqian, Z. & Baoxiu Z. (2004). Microwave Irradiation Copolymerization of Superabsorbents from Cornstarch and Sodium Acrylate. *Journal of Applied Polymer Science*, Vol. 95, No. 2 (January 2004), pp. 264-269, ISSN: 0021-8995

Torres, F. G.; Boccaccini, A. R. & Troncoso, O. P. (2007). Microwave Processing of Starch-based Porous Structures for Tissue Engineering Scaffolds. *Journal of Applied Polymer Science*, Vol. 103, No. 2 (January 2007), pp. 1332-1339, ISSN: 0021-8995

Velmathi, S.; Nagahata, R.; Sugiyama, J. & Takeuchi, K. (2005). A Rapid, Eco-Friendly Synthesis of Poly(butylene succinate) by a Direct Polyesterification under Microwave Irradiation. *Macromolecular Rapid Communications*, Vol. 26, No. 14 (July 2005), pp. 1163-1167, ISSN: 1521-3927

Vogel, B. M.; Mallapragada, S. K. & Narasimhan, B. (2003). Rapid Synthesis of Polyanhydrides by Microwave Polymerization. *Macromolecular Rapid Communications*, Vol. 25, No. 1 (January 2004), pp. 330-333, ISSN: 1521-3927

Wan, Z.; Xiong, Z.; Ren, H.; Huang, Y.; Liu, H.; Xiong, H.; Wu, Y. & Han, J. (2011). Graft Copolymerization of Methyl Methacrylate onto Bamboo Cellulose under Microwave Irradiation. *Carbohydrate Polymers*, Vol. 83, No. 1 (January 2011), pp. 264-269, ISSN: 0144-8617

Waugh, M. R. & Parkin, I. P. (2010). Enhanced Solubility and Covalent Solubilization of Single Walled Carbon Nanotubes via Atmospheric Pressure Microwave Reflux and the Subsequent Spray Coating of Transparent Conducting Thin Films. *Current Nanoscience*, Vol. 6, No. 3 (June 2010), pp. 232-242, ISSN: 1573-4137

Wiesbrock, F.; Hoogenboom, R. Abeln, C. H. & Schubert, U. S. (2004). Single-Mode Microwave Ovens as New Reaction Devices: Accelerating the Living Polymerization of 2-Ethyl-2-oxazoline. *Macromolecular Rapid Communications*, Vol. 25, No. 22 (November 2004), pp. 1895-1899, ISSN: 1521-3927

Wu, C. Y. & Benatar, A. (1997). Microwave Welding of High Density Polyethylene Using Intrinsically Conductive Polyaniline. *Polymer Engineering and Science*, Vol. 37, No. 4 (April 2007), pp. 738-743, ISSN: 1548-2634

Xu, Q.; Zhang, C.; Cai, S.; Zhu, P. & Liu, L. (2010). Large-Scale Microwave-Assisted Ring Opening Polymerization of ε-Caprolactone. *Journal of Industrial and Engineering Chemistry*, Vol. 16, No. 5 (September 2010), pp. 872-875, ISSN: 1226-086X

Yarlagadda, P. K. D. V. & Chai, T. C. (1998). An Investigation into Welding of Engineering Thermoplastics Using Focused Microwave Energy. *Journal of Materials Processing Technology*, Vol. 74, No. 1-3 (February 1998), pp. 199-212, ISSN: 0924-0136

Yu, Z. J.; Liu, L. J. & Zhuo, R. X. (2003). Microwave-Improved Polymerization of ε-Caprolactone Initiated by Carboxylic Acids. *Journal of Polymer Science - Part A: Polymer Chemistry,* Vol. 41, No. 1 (January 2003); pp. 13-21, ISSN: 1099-0518

Yu, Z. J. & Liu, L. J. (2004). Effect of Microwave Energy on Chain Propagation of Poly(ε-caprolactone) in Benzoic Acid Initiated Ring Opening Polymerization of ε-Caprolactone. *European Polymer Journal,* Vol. 40, No. 9 (September 2004), pp. 2213-2220, ISSN: 0014-3057

Yu, Z. J. & Liu, L. J. (2007). Biodegradable Poly(vinyl alcohol)-*graft*-poly(ε-caprlactone) Comb-like Polyesters: Microwave Synthesis and its Characterization. *Journal of Applied Polymer Science,* Vol. 104, No. 6 (June 2007), pp. 3973-3979, ISSN: 0021-8995

Zhang, C.; Liao, L. & Liu, L. (2004). Rapid Ring-Opening Polymerization of D,L-Lactide by microwaves. *Macromolecular Rapid Communications,* Vol. 25, No. 15 (August 2004), pp. 1402-1405, ISSN: 1521-3927

Zhang, C.; Liao, L. & Gong, S. (2007). Microwave-Assisted Synthesis of PLLA-PEG-PLLA Triblock Copolymers. *Macromolecular Rapid Communications,* Vol. 28, No. 4 (February 2007), pp. 422-427, ISSN: 1521-3927

Zhang, Y.; Wang, P.; Han, N. & Lei, H. (2009). Synthesis of Poly(D,L-Lactide) under Continuous Microwave Irradiation. *Shiyou Huagong/Petrochemical Technology,* Vol. 38, No. 8 (August 2009), pp. 861-865, ISSN: 1000-8144

Zhou, J.; Shi, C.; Mei, B.; Yuan, R. & Fu, Z. (2003). Research on the Technology and Mechanical Properties of the Microwave Processing of Polymers. *Journal of Materials Processing Technology,* Vol. 137, No. 1-3 (June 2003), pp. 156-158, ISSN: 0924-0136

Zhu, X.; Zhou, N.; He, X.; Cheng, Z. & Lu, J. (2003). Atom Transfert Radical Bulk Polymerization of Methyl Methacrylate under Microwave Irradiation. *Journal of Applied Polymer Science,* Vol. 88, No. 7 (May 2003), pp. 1787-1793, ISSN: 0021-8995

# 10

# Microwave Pyrolysis of Polymeric Materials

Andrea Undri, Luca Rosi, Marco Frediani and Piero Frediani
*University of Florence, Department of Chemistry, Firenze, Italy*

## 1. Introduction

The consumption of polymeric materials is growing ceaselessly in the world even in spite of the financial crisis.

World's plastics production in 2009 was $2.3 \cdot 10^8$ tons and in Europe it was $4.5 \cdot 10^7$ tons whose 54% is disposed as waste. The annual average production of tires in Europe is more than 2.5 $10^6$ tons. In the 2008 in Italy were produced $3.5 \cdot 10^6$ tons of plastics among which $4.1 \cdot 10^5$ tons of tires, and $1.5 \cdot 10^6$ of waste plastics tons were collected for disposal. (Chen et al., 2007; Federazione Imprese e Servizi, Unione Nazionale Imprese Recupero [FISE UNIRE], 2009; PlasticsEurope, 2010).

World rubber demand is foreseen to increase up to 4% annually to 26.5 million metric tons in 2011 (Freedonia, 2010). Therefore the disposal of waste polymers is a serious environmental problem against which public is becoming more aware. The interest of waste polymeric materials disposal is focused on new uses rather than land filling or incineration.

Regarding scrap tires, a strong attention has been paid over last years to the claims for their recycling or reprocessing. In consideration of their complex composition, slow degradation rate in landfill, high calorific value and shape hindrance they may be burned hardly but otherwise they cannot send to landfill anymore and an alternative methodology must be eligible in order to dispose scrap tires. European Directive No. 31/1999 states that the disposal of scrap tires in landfills is banned with the exclusion of bicycle tires and tires with an external diameter greater than 1400mm. Since July 2006 the ban has been extended also to shredded tires.

Waste plastics and tires are very attractive as a source of renewed raw materials and chemical substances. These products may be achieved by pyrolysis, heating usually in the absence, but sometimes in the presence, of an oxidative agent, and these processes may be viewed as a promising technology.

The pyrolysis of polymeric materials or plastic-containing wastes including scrap tires is a possible answer to the problem of their disposal because it let recover of gas, oil and solid able to be employed as a source of products and energy. Therefore the relevance of the pyrolysis processes of plastic waste has been growing.

A plethora of studies over the thermal degradation of polymeric materials are carried out using conventional heating method with internal or external heating source, under inert or oxidizing atmosphere. Generally the thermal decomposition needs operating temperature above 450°C (Kaminsky et al., 2004; Mastral et al., 2002; Whesterhout et al., 1998).

In this area also the interest to microwave heating technologies has become remarkable since they could represent a charming alternative to current technologies based on other conventional heating processes. Microwave heating provides a number of advantages with respect conventional heating methods, mainly referred to the rate and distribution of the heating. It is commonly thought that many limitations should occur nevertheless when switching microwave technology from a laboratory or pilot-scale process to a large scale plant. These limitations mainly concern the lack of data on raw materials amenability with microwave heating and therefore a proper set up of the experimental parameters of the process.

The main differences occurring between conventional and microwave heating are shown in Table 1.

| **Conventional Heating** | **Microwave heating** |
|---|---|
| Long reaction time (hours) | Short reaction time (minutes) |
| Hard transfer of heat, thermal conductivity of polymer is low | The low thermal conductivity of polymers may be overwhelmed, easy heating of polymers. |
| Heating efficiency is hardly obtained | High heating efficiency |
| Every fuel source may be employed | Electrical power is required |
| Additive are not required | Microwave absorber are required |

Table 1. Comparison between conventional and microwave heating technologies for pyrolysis of plastics.

In this chapter we report the state of art gathered from the scientific literature and patents in addition to our experience and results, concerning the pyrolysis process with microwave heating of polymers and complex plastics. Emphasis will be given to the influence of the main process variables on products obtained: apparatus set-up, temperature, heating rate, microwave absorber and so on. Strong attention will be devoted to the pyrolysis of scrap tires. Tires, due to their high carbon content (up to 30%) are suitable for direct microwave pyrolysis. They are able to absorb microwave and quickly turning it to heat (Meredith, 1998).

The chapter consists of two main sections. The first part is exclusively dedicated to microwave pyrolysis of tires. We report a review of literature and patents, we discuss the results on the bases of our own experience and we integrate them together with our recent achievements in this field. Within the products the liquid fraction is the most attractive due to its high content of valuable hydrocarbons like: benzene, toluene, xylenes and limonene. We thoroughly deal with the relation between layout technologies, operating parameters and products composition. We address in detail the problem to maximize the amount of liquid fraction and valuable hydrocarbons.

The second section is focused on a review of microwave heating pyrolysis of plastic materials. These polymers don't absorb microwave and in order to carry out the pyrolysis is necessary to mix them with a microwave absorbent. Typical absorbent are: coal or carbon-containing materials (for instance tires), metals (Hussain et al., 2010) or metal-oxide (Horikoshi et al., 2009). We report about microwave pyrolysis of the main waste polymeric materials avoiding to send them to landfill (Kaminsky, 1992): polyethylene (PE), polypropylene (PP), polystyrene (PS), poly(ethylene terephthalate) (PET) and polyvinylchloride (PVC). We present the processes proposed and the operative parameters employed in order to obtain renewed raw materials, refinery feedstock or energy source.

## 2. Tires pyrolysis

### 2.1 Why tires pyrolysis?

Usually the recycling policy try to maximize the recovery of materials and minimize the loss of material and energy stored up in wastes (Natural Resources Defense Council [NRDC], 2008). End life tire can be reused for reconstruction processes which magnify the material and energy recovery. However this procedure is limited by quality of waste tires, their deterioration and can be repeated no more than one or two times. Where reuse and remanufacturing are not possible, scrap tires, whole or chopped, can be used in engineering works for many applications, such as: roads (asphalt, where the granules improves the mechanical strength, reduces noise and eliminates the aquaplaning), street furniture (beds for curbs, bollards, bike paths, parking lots, play areas) and sports (football pitches in synthetic grass and sports flooring for athletic tracks).

Sure enough not the whole production of scrap tires can be reprocessed or reused indeed 19% of them is disposed in Italy with heat treatment processes mainly for the recovery of energy content (FISE UNIRE, 2009). The main technological processes available for heat treatments are listed below:

- incinerator (including municipal solid waste incinerators) (Sharma et al., 2000);
- fuel for rotary kilns or furnaces to produce cement or steam (Giugliano et al.; 1999);
- pyrolysis.

Just the last process mentioned, the pyrolysis, has the characteristics to transform waste polymeric materials in products suitable for energy production or petrochemical feedstock.

### 2.2 Microwave or non-microwave heating in pyrolysis processes: two different approaches

Pyrolysis is a thermal cracking process performed in inert atmosphere, such as nitrogen, helium or $CO_2$, and let to the total recovery of mass as a solid (non-volatile material), liquid (condensable fraction) and gaseous (non-condensable fraction) products (Kaminsky, 2006). These three products are always obtained regardless the apparatus, heating source, operating temperature and heating speed. The aim in all research is usually to enhance yields in liquid product and to control products characteristics. The pyrolysis is an endothermic process, typical energy consumption 4.0-5.7 MJ $Kg^{-1}$ (Piskorz et al., 1999), and the thermal energy required could be provided in several ways resembled in two main categories: conventional and non-conventional.

Conventional heating included heating sources internal (partial combustion of the load for instance) or external to the reactor: electrical resistance (Burrueco et al., 2005), flame (Williams et al., 1998) or microwave (Ludlow-Palafox & Chase, 2001). Among non-conventional heating may be included also plasma (Tang & Huang, 2004) and supercritical fluids (Chen et al., 1995). Up to now a wide number of apparatus and reaction conditions were experimented for conventional, non-microwave, heating pyrolysis. Alternative to this technology is microwave pyrolysis.

Only the pyrolysis process with conventional (external or internal) heating are up to now for large scale applications the only one employed.

Any apparatus, besides autoclave, is composed of few main parts: reactor, heat production and exchanger, collecting system.
The fundamental part of any apparatus is the reactor, where the heat is transferred from the source to the material. Many type of reactors are available, here are listed only the main classes: autoclave (de Marco Rodriguez et al., 2001), rotary kiln (Li et al., 2004), static bed reactor (Cunliffe & Williams, 1998; Burrueco et al., 2005) and fluidized bed reactor (Kaminsky & Mennerich, 2001, Aylòn et al., 2008; Aylòn et al., 2010). Their feature and main operating parameters are reported and compared in table 2.

| Reactor Type | Temperature range (°C) | Heating Rate (°C $min^{-1}$) | Main Feature |
|---|---|---|---|
| Autoclave | 300-700 | 15 | Products don't vent, poor thermal transfer |
| Rotary kiln | 450-650 | - | Fixed speed feeding, heat transfer improved by lowering tire size |
| Static bed reactor | 400-700 | 5-12 | Long reaction time, products vent from reactor |
| Fluidized bed reactor | 500-600 | - | Continuous process, optimal thermal transfer |

Table 2. Reactor employed in non-microwave pyrolysis: main features, operating temperature and heating rate.

In table 2 are also underlined the temperature range and heating rate tested. The temperature and the heating speed are parameters of paramount importance and together with apparatus affect the products composition. In order to simplify the approach we focus just on these three process variables. Other variables such as load size, apparatus set-up, pyrolysis running time and material flux inside the reactor are discussed afterwards.
The choice of one of these reactors is linked with two aims: effectiveness and manageable transfer of heat to the polymeric material and volatilize products from the reactor towards collecting system. It is essential an efficient heat transfer due to the poor thermal conductivity of tires and polymeric material. The possibility to control the heat transfer is fundamental to obtain uniform products.
Without taking care of the others variables just the fluidized bed reactor fulfills all the above requirements. The heating medium is usually quartz sand preheated and kept at the prefixed temperature while the polymeric material is continuously fed (Kaminsky & Mennerich, 2001). Nevertheless the process is sensitive to fibers and high amounts of metals and fillers (Kaminsky et al., 2004). It requires high operating cost for heating and complex feedstock preparation making this kind of reactor relatively expensive (Juma et al., 2006). For instance Kaminsky use 9 kg of quartz sand to treat up to 4 kg of chopped polymers (Kaminsky & Mennerich, 2001).
On the other side a simpler approach is possible when working with static bed reactor or autoclave in a batch process. The reactor is just heated together with the polymer to the required temperature. It is kept to this temperature until the end of the experiment (Cunliffe & Williams, 1998; Mastral et al., 2000; Gonzàlez et al., 2001; Burrueco et al., 2005). Anyway the heat transfer is not effective in this type of reactors and it can be improved by upgrading

the reactor with a continuous mixing of the load or with a spinning movement that is a rotary kiln (Li et al., 2004).

The heating rate and final temperature affect the properties and yields of the three products. It is deeply described in paragraph 2.3. Anyway different reactors may give different products even if the final or operating temperature is the same in each reactor.

An improving of the reactor technologies to effectively transfer the heat requires a control over the load size. A static bed reactor, autoclave, or rotary kiln doesn't need any particular load size, just small enough to be filled during the experiments. In the fluidized bed reactor usually a size of 1-2 mm (Kaminsky & Mennerich, 2001), 2 mm (Aylón et al., 2008) or no more than 5 mm (Aylón et al., 2010) is required.

The limitations displayed above are held back controlling the other experiment's variables: pyrolysis running time, apparatus set-up and material flux. Where a good heat transfer is achieved shorter reaction time is required. Kamisky reports 230 minutes to pyrolyze 4 kg with its fluidized bed reactor (Kaminsky & Mennerich, 2001) while Berrueco's process requires up to 300 minutes to pyrolyze 0.3 kg with its static-bed reactor (Berrueco et al., 2005).

Another way to overwhelm the not optimal heat transfer is the control of the flux of material and consequently the apparatus set up. Except for autoclave, where all the products are forced inside the reactor, all the others described reactors require a gas circulating system to remove the pyrolyzed products at the desired time.

Again a longer residence time of the volatilized products is used when the heat transfer is not optimal and it lets a more efficient cracking. Static-bed reactor requires up to 120 s of residence time of the gas in the reactor and this is achieved by a controlled nitrogen flux (Berrueco et al., 2005) or apparatus set-up (Cunliffe & Williams, 1998). The residence time of the gas in the fluidized-bed reactor is less than 3 s and it is achieved with strong flux of nitrogen or other inert gas (Kaminsky & Mennerich, 2001).

This description of main reactor technologies available is far away to be exhaustive; it wants only to give an idea of the possible approaches to the pyrolysis of tires and plastics.

When we switch to a microwave heating system the apparatus set up is almost the same: efficiently and manageable transfer of heat to obtain products with desired proprieties. Only the reactor is the feature which must be changed.

The scientific literature is extremely poor with respect to the patent literature. Unfortunately for scientific spread the patents don't give extensive information concerning the products achievable by the described apparatus. Anyway it is possible to understand the approach to the microwave pyrolysis of tire and plastics.

The goal in each patent is to provide a simple and economically convenient approach to the pyrolysis of both tires and plastics. It is achieved by treating the polymeric substrate in a static bed-type reactor. Holland (Holland, 1995) uses a static-bed reactor filled with carbonaceous materials (tires or coal) between 400 °C and 800°C to pyrolyze any kind of polymers. More recent patents showing similar approach, which always resemble a static bed reactor, are reported by many authors (Holland, 1992; Parker, 1992; Johnson et al., 1996; Pringle, 2006; Pringle, 2007; Kasin, 2009). Just the feeding system and the reactor design diversify the several patents. They are planned to be employed for industrial application. Strong attention is paid to productivity and energy efficiency.

Recently a more complete microwave pyrolysis plant was proposed from Scandinavian Biofuel Company (Scandinavian Biofuel Company [SBC], 2011). All the products were evaluated as source of electric energy and after subtraction of the energy required for the process give a production of 1.55 MWh/ton from tires or 3.98 MWh/1 ton from plastics. The hypothetical working capacity for SBC plant is reported in table 3.

| Feedstock | Plastics | Tires |
|---|---|---|
| Capacity/yr | 25,000 tons | 25,000 tons |
| Water Content | 5% | 1% |
| Net capacity/yr | 25,000 tons | 20,000 tons |
| Output oil | 12,500 tons | 5,600 tons |
| Output gas | 10,000 tons | 3,000 tons |
| Output carbon | 1,200 tons | 11,000 tons |
| Electricity output | 103.4 GWh | 44.5 GWh |
| Electricity use | 3.79 GWh | 4.57 GWh |
| Electricity sales | 99.65 GWh | 39.89 GWh |

Table 3. Project examples for electricity production from SBC plant.

From the few scientific articles available up to now for microwave pyrolysis of polymeric materials two kind of apparatus came out: static bed batch reactor (Hussain et al., 2010) and fluidized bed reactor (Ludlow-Palafox & Chase, 2001).

The static bed reactor used by Hussain et al. is as simple as possible. The microwave absorbent is an iron mesh mixed with the polymer (just polystyrene in this article). The products are collected by two cold traps.

Ludlow-Palafox & Chase use a reactor resembling the fluidized bed reactor of Kaminsky, where the heat carrier is carbon instead of quartz sand (Kaminisky & Mannerich, 2001). The temperature is risen up and kept to the prefixed temperature before adding the polymers. Small amount of polymers is dropped at time intervals directly into the reactor, 50g of polymers for 1000g of heated carbon, to prevent large temperature fluctuation. It is required a small amount of time (120 s) to complete the pyrolysis in these conditions. Anyway with this productivity to pyrolyze 4 kg, as reported by Kaminsky & Mannerich with their fluidized bed reactor, is possible to extrapolate a reaction time of 120 minutes instead of 230 minutes for fluidized bed reactor. There's a huge difference. Furthermore the full potentiality of microwave oven is not used.

In our experiments we use a static bed reactor approach in a batch process to simplify most of the apparatus and to minimize the energy consumption.

The system is composed by a microwave oven, a heat exchanging pipes and collecting flasks. The volatilized material is moved towards the condensing traps by the products formed in the course of the pyrolysis and the partial condensation of gases. In this way the gases vented out are not diluted by transport gas. Up to now the experiments were carried out by heating the chopped tires and plastics from room temperature to the complete pyrolysis of material. The pyrolysis starts usually 20 seconds after switching on microwave. The experiments are conducted using an increasing microwave power. In this way only the energy necessary for the pyrolysis itself is supplied inside the oven.

In the best conditions tested until now we are able to pyrolyze 0.4 Kg of tires in 14 minutes. Here is possible to extrapolate 140 minutes for 4 kg of tires. It is a big improvement from the static bed reactor and slightly better than fluidized bed reactor from Ludlow-Palafox & Chase. Also in this case the full potential of the microwave oven is not used.

### 2.2.1 The problem of high temperature measurement when switching on microwave

When the microwave is not employed the temperature is easily measured by thermocouples. When microwave is turned on a metal thermocouple can't be used anymore. It absorbs the microwave radiation and provides a skewed temperature.

Up to now are available two technological solutions to investigate the high temperature during the microwave pyrolysis process: infrared and fiber optic thermometer. The infrared thermometer is a contactless thermometer which can work properly in a wide range of temperature. It measures the radiation emitted in the infrared region of the spectrum which is linked to the temperature. It consists in a lens to focus the infrared energy onto the detector which converts the energy to an electric signal that may be displayed in units of temperature after calibration. The main limitations of this instrument are its calibration and the medium value of the temperature of objects present. It is available at a relatively low price. The common fiber optic thermometer commercially available is not suited for high temperature detection. They can be used in a temperature range up to 400°C, over this temperature they melt. A quite new technology in optic fiber is now accessible based on sapphire crystalline fiber (Djeu, 2000). It can work up to temperature of 950°C, and it is perfectly suitable for temperature detection in a microwave pyrolysis experiment (Micromaterials Inc., 2010). The main advantage of this fiber optic probe is the detection of temperature in a single position where the probe is placed. The infrared thermometer on the other side averages the temperature over a large area and this could lead significant errors. We have to remember that the heating is not homogeneously distributed but localized on the microwave absorbent that are mixed with the polymeric materials.

## 2.3 Conventional tires pyrolysis: products characteristics

The three products, solid, liquid and gas, have been fully characterized and the results reported in many papers on non-microwave pyrolysis. Due to this reason it is really interesting to review them for a better understanding of microwave influence on products characteristics.

Tires have a variable composition, depending on brand, dimensions and use. In tables 4 and 5 are reported the average composition of truck and car tires (Rubber Manufacturers Association [RMA], 2011).

| Tires for | Polisoprene 1,4-cis (wt%) | Synthetic Polymers (wt%) | Carbon Black (wt%) | Steel (wt%) | Sulfur (wt%) | Textile, additive (wt%) |
|---|---|---|---|---|---|---|
| Car | 14 | 27 | 28 | 14-15 | 1.5 | 16-17 |
| Truck | 27 | 14 | 28 | 14-15 | 2.5 | 16-17 |

Table 4. Average composition of truck and car tires. Synthetic polymers include: polybutadiene (BR, butadien rubber), styrene-butadiene copolymers (SBR, styrene-butadiene rubber) and EPDM rubber (ethylene propylene diene Monomer (M-class) rubber).

Even if tire composition is variable, yields obtained from the pyrolysis are not strongly affected by it (Kyari et al., 2005). Variation in products characteristics are appreciated only when significant differences between tires occurred. Their characteristics are mainly affected by temperature, heating speed and feeding size as reported in paragraph 2.2.

| Element | % |
|---|---|
| C | 80.1 |
| H | 7.0 |
| N | 0.4 |
| S | 1.5 |
| O | 3.0 |
| Ash | 8.0 |

Table 5. Average elemental composition (CHNSO) of tires.

The main gas components are: $H_2$, $H_2S$, $CO$, $CO_2$, $CH_4$, $C_2H_4$, $C_2H_6$, $C_3H_6$, $C_3H_8$ and other organic substances in lower amount. It has a high gross calorific value of 68-84 MJ $m^{-3}$ (de Marco Rodriguez et al., 2001) when it is rich in hydrocarbons and a lower one, 30-40 MJ $m^{-3}$ (Williams et al., 1990), when depleted in organic compounds. It may be used as an energy source for sustaining the pyrolysis process itself (Williams et al. 1998; Kyari et al., 2005).

The liquid, or pyrolysis oil, is the most scientific and economical interesting product due to its high gross calorific value and content of valuable hydrocarbons such as: benzene, toluene, xylenes, limonene and so on. It can be employed as energy source, refinery feedstock or reservoir for aromatics and limonene (Laresgoiti et al., 2004).

The solid is mainly made of carbon and non volatilizable material: steel and inorganic additives (de Marco Rodriguez et al., 2001). The steel can be easily removed with electro-magnets. The carbonaceous fraction is employed up to now as fillers in tires manufacturing (de Marco Rodriguez et al., 2001), activated carbon production (Mui et al., 2004) and smokeless fuel (Dìez et al., 2004).

The yields for the three products are strictly connected with the process variables as introduced in paragraph 2.2.

In table 6 the yields obtained by various authors in different reaction conditions are reported.

A comparative study is not easy to carry out. The data reported in table 5 are not congruent among them due to the different reactors and conditions employed. Anyway a trend in all of them may be easily obtained. There's a critical temperature above which the pyrolysis is completed and the solid yield doesn't significative change. In any study it is around 500°C.

When the temperature is rising over 500°C, while the solid product remain almost constant, the yield of liquid product goes down due a stronger cracking which improve the yield of non-condensable fraction.

The yield of liquid is lower when the residence time of liquid in the oven increase. The autoclave, where the pyrolysis products don't vent out, the yield of non-condensable fraction is three times higher than the one achieved in a static bed reactor. In a fluidized bed reactor with a residence time of 3 seconds gives similar gas yield than a static bed reactor where the residence time is 120 seconds. When the reaction time increases the cracking of liquid products remarkable rises up to 3 times.

| Author | Max Temp (°C) | Heating speed (°C/min) | Running time (min) | Yields | | | Comments |
|---|---|---|---|---|---|---|---|
| | | | | S | L | G | |
| Burreco et al., 2005 | 400 | NA | 240 | 64.0 | 30.0 | 2.4 | Static-bed batch reactor |
| | 500 | NA | 240 | 52.7 | 39.9 | 3.6 | |
| | 550 | NA | 240 | 52.5 | 39.1 | 3.6 | |
| | 700 | NA | 240 | 51.3 | 42.8 | 4.4 | |
| de Marco Rodiguez et al., 2001 | 400 | 15 | 55* | 55.9 | 24.8 | 19.3 | Autoclave |
| | 500 | 15 | 62* | 44.8 | 38.0 | 17.2 | |
| | 600 | 15 | 68* | 44.2 | 38.2 | 17.5 | |
| | 700 | 15 | 75* | 43.7 | 38.5 | 17.8 | |
| Gonzàlez et al., 2001 | 400 | - | 30 | 55.1 | 42.9 | 2.0 | Static-bed batch reactor, sample 4g. |
| | 500 | - | 30 | 38.6 | 55.4 | 6.0 | |
| | 600 | - | 30 | 37.0 | 52.2 | 10.8 | |
| | 700 | - | 30 | 36.7 | 36.6 | 26.7 | |
| Aylòn et al., 2010 | 600 | - | NA | 40.6 | 41.5 | 17.9 | Moving bed reactor, without steel |
| | 700 | - | NA | 39.0 | 31.3 | 29.7 | |
| | 800 | - | NA | 41.0 | 27.5 | 31.5 | |
| Kaminsky & Mannerich, 2001 | 500 | - | 230 | 30 | 65 | 5 | Moving bed reactor, tires composition simulate |
| | 550 | - | 220 | 34 | 57 | 9 | |
| | 600 | - | 240 | 40 | 51 | 9 | |
| Cunliffe & Williams, 1998 | 450 | 5 | 150 | 37.4 | 58.1 | 4.5 | Static-bed batch reactor, load pre-heated at 150°C |
| | 500 | 5 | 160 | 38.3 | 56.2 | 5.5 | |
| | 560 | 5 | 172 | 38.1 | 55.4 | 6.5 | |
| | 600 | 5 | 180 | 38.0 | 53.1 | 8.9 | |
| Li et al., 2004 | 450 | - | 200-250 | 43.9 | 43.0 | 13.1 | Rotary kiln, 50 kg processed |
| | 500 | - | 200-250 | 41.3 | 45.1 | 13.6 | |
| | 550 | - | 200-250 | 39.9 | 44.6 | 15.5 | |
| | 600 | - | 200-250 | 39.3 | 42.7 | 18.0 | |

Table 6. Yields and parameters of non-microwave tires pyrolysis reported by various authors. S: solid; L: liquid; G: gas. *Estimated. NA: not available.

### 2.3.1 Non-condensable fraction (gas)

Each apparatus is provided with a condensation system, in order to collect the liquid products. In order to maximize the yield of the liquid products different heat exchanging systems were tested and this is an extra variable affecting the gas composition in each report.

However the non-condensable fraction contains permanent gas (such as: $H_2$, $H_2S$, CO, $CO_2$ and $CH_4$) and organic substances with high vapor pressure difficult to condensate.

In table 7 is summarized the amount of the main compounds identified in each gaseous mixture reported by various authors.

| Author | Max Temp (°C) | Compound (vol. %) | | | | | | | | | |
|---|---|---|---|---|---|---|---|---|---|---|---|
| | | $H_2$ | $CO_x$ | $CH_4$ | $C_2H_6$ | $C_2H_4$ | $C_3$ | $C_4$ | $C_5$ | $C_6$ | $H_2S$ |
| Burreco et al., 2005 | 400 | 2.6 | 2.5 | 1.0 | 0.4 | 0.3 | 0.5 | 3.2 | 0.0 | 0.3 | - |
| | 500 | 14.2 | 1.3 | 4.3 | 1.5 | 0.7 | 1.3 | 2.3 | 0.0 | 0.4 | - |
| | 550 | 17.9 | 1.1 | 5.6 | 1.8 | 0.8 | 1.5 | 2.0 | 0.0 | 0.3 | - |
| | 700 | 10.1 | 1.6 | 5.4 | 1.6 | 1.2 | 1.6 | 2.5 | 0.0 | 0.5 | - |
| de Marco Rodiguez et al., 2001 | 400 | - | 14.9 | 4.4 | 4.5 | 4.3 | 8.5 | 36.9 | 16.5 | 7.3 | 2.6 |
| | 500 | - | 14.2 | 19.8 | 9.1 | 9.4 | 10.8 | 21.3 | 7.6 | 2.8 | 5.1 |
| | 600 | - | 15.3 | 20.0 | 9.0 | 9.7 | 10.6 | 21.9 | 7.4 | 2.5 | 3.6 |
| | 700 | - | 21.8 | 20.6 | 8.1 | 8.9 | 7.7 | 19.8 | 6.7 | 2.5 | 3.9 |
| Gonzàlez et al., 2001* | 400 | 36.7 | 50.6 | 12.7 | 0.0 | 0.0 | - | - | - | - | - |
| | 500 | 19.8 | 34.4 | 22.9 | 15.3 | 7.6 | - | - | - | - | - |
| | 600 | 16.5 | 20.4 | 28.5 | 28.5 | 6.1 | - | - | - | - | - |
| | 700 | 18.7 | 7.1 | 34.7 | 32.5 | 7.1 | - | - | - | - | - |
| Kaminsky & Mannerich, 2001 ** | 500 | 0.1 | 12.5 | 3.0 | 4.6 | 3.1 | 10.5 | 63.2 | - | - | 3.0 |
| | 550 | 0.2 | 7.5 | 3.8 | 5.4 | 5.1 | 13.7 | 62.5 | - | - | 1.8 |
| | 600 | 0.2 | 8.7 | 4.1 | 5.1 | 8.0 | 16.4 | 55.2 | - | - | 2.2 |
| Dìez et al., 2004 | 350 | 24 | 3.3 | 20 | 29 | | 12 | 5.7 | 3 | 1 | - |
| | 450 | 30 | 3.0 | 24 | 26 | | 9 | 4.3 | 2 | 0.2 | - |
| | 550 | 40 | 3.0 | 26 | 20 | | 6 | 2.8 | 1.1 | 0.1 | - |

Table 7. Main compounds present in the gas. * Values are extrapolated (Pressure of 25 atm) where they are reported as mol kg$^{-1}$ of scrap tire. ** Values are calculated where are expressed as mass percentage.

The inorganic substances, $H_2S$, $CO_x$ and $H_2$, are connected to three tire components. The sulfur compounds come from the vulcanization process of the polymeric tire constituents. A variable amount of sulfur is used to improve the rubber elasticity. In the gases only hydrogen sulfide is found as sulfur containing compound, any organic sulfur containing products are not found. Carbon dioxide and carbon monoxide find their origin from two sources: metal carbonates and oxygen containing inorganic substances. Both of them are additives in tire formulation. Hydrogen is a common cracking product. What happen in a pyrolysis reactor is close to a thermal cracking in a petroleum refinery. The organic substances find their origin from the random cracking of polymers. The growing presence of olefin instead of their paraffinic equivalents is linked to a working temperature in the range 400-700°C where alkenes are thermodynamically more stable than alkane.

### 2.3.2 Condensable fraction (liquid)

The condensable fraction, also known as liquid fraction or pyrolysis oil, is the most interesting and studied product. The maximum yield is obtained working with a pyrolysis temperature between 450 and 550 °C as shown in table 5, it varies between 30% and 58% depending if steel cords are considered in calculation. Its color is dull dark-brown and it has a bad-smelling flavor. It has a complex composition which is usually studied using a gas chromatograph coupled with a mass detector. An enormous amount of substances are identified, more than 300, but only few compounds are present in an amount greater than 1% (Laresgoiti et al., 2004).

In table 8 are reported the top 12 substances identified inside the condensable fraction from some authors.

| Author | Max T (°C) | Compound | | | | | | | | | | | |
|---|---|---|---|---|---|---|---|---|---|---|---|---|---|
| | | Benzene | Toluene | Ethylbenzene | Xylenes | Styrene | 2-Ethyltoluene | α–methyl styrene | Limonene | 1H-Indene | Naphthalene | Benzothiazole | 1,6-dimethyl naphthalene |
| Laresgaoiti et al., 2004 | 400 | 0.65 | 1.00 | 1.07 | 0.69 | 0.36 | 0.15 | 3.22 | 0.17 | 0.18 | 0.53 | 0.25 | 0.65 |
| | 500 | 0.94 | 1.00 | 1.25 | 0.88 | 0.40 | 0.11 | 1.84 | 0.32 | 0.26 | 0.33 | 0.30 | 0.94 |
| | 600 | 2.34 | 1.00 | 1.09 | 1.03 | 0.16 | 0.11 | 1.70 | 0.34 | 0.34 | 0.39 | 0.36 | 2.34 |
| | 700 | 1.46 | 1.00 | 1.33 | 1.01 | 0.50 | 1.26 | 2.30 | 0.42 | 0.41 | 0.53 | 0.50 | 1.46 |
| de Marco Rodiguez et al., 2001* | 400 | - | - | - | - | - | - | - | - | - | - | 2.10 | - |
| | 500 | - | - | - | - | - | - | - | - | - | - | 1.80 | - |
| | 600 | - | - | - | - | - | - | - | - | - | - | 2.00 | - |
| | 700 | - | - | - | - | - | - | - | - | - | - | 1.80 | - |
| Kyari et al., 2005* | 500 | 0.85 | 1.00 | 0.11 | 0.52 | 1.91 | 0.27 | 3.34 | 0.16 | 0.05 | 0.17 | 0.28 | 0.85 |
| Kaminsky & Mannerich, 2001 | 500 | 0.73 | 1.00 | 0.21 | 1.60 | - | 0.57 | - | 0.08 | 0.07 | 0.60 | - | 0.73 |
| | 550 | 1.00 | 1.00 | 0.38 | 0.87 | - | 0.37 | - | 0.09 | 0.08 | 0.40 | - | 1.00 |
| | 600 | 1.24 | 1.00 | 0.32 | 1.82 | - | 0.44 | - | 0.18 | 0.15 | 0.21 | - | 1.24 |

Table 8. The 12 main substances present in the liquid fraction. The data reported are normalized to toluene. * It is the percentage over total liquid fraction.

The liquid fraction is a mixture of organic compounds from $C_4$ to $C_{24}$, mainly olefins, single ring aromatics and limonene. Within the minor substances strong attention should be devoted to three classes of compounds always present: polycyclic aromatics hydrocarbons (PAHs), sulfur containing products and other heteroatoms containing substances.

In figures 1-4 are reported the mechanism proposed for the degradation of polymeric materials (Cunliffe & Williams, 1998; Mastral et al., 2000).

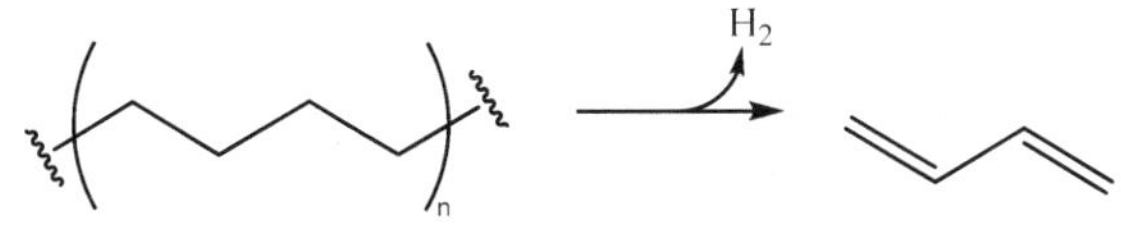

Fig. 1 Cracking of linear polyolefins.

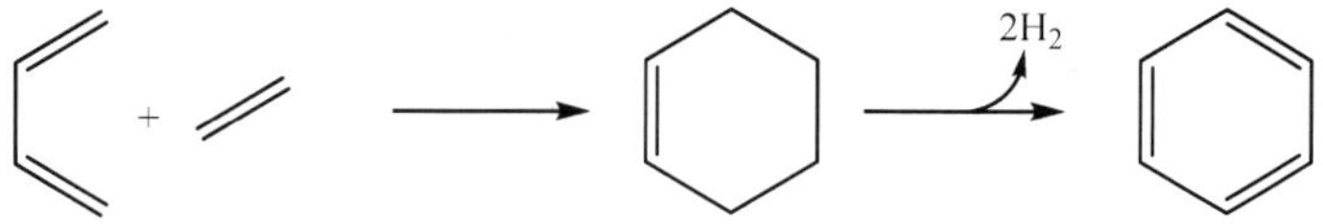

Fig. 2. Diels-Alder reaction followed by dehydrogenation reaction that yield benzene.

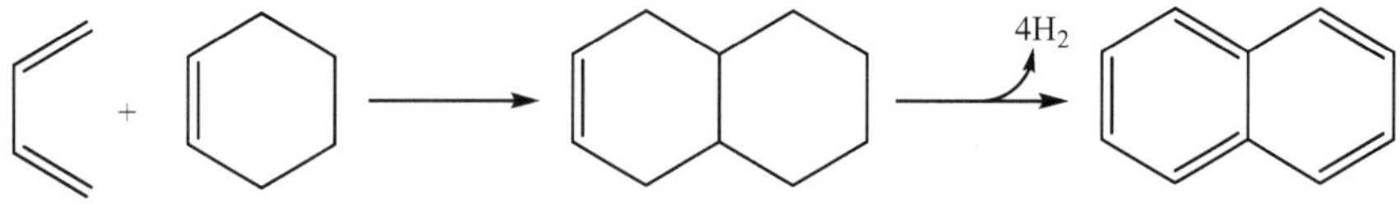

Fig. 3. PAHs via Diels-Alder and dehydrogenation reactions.

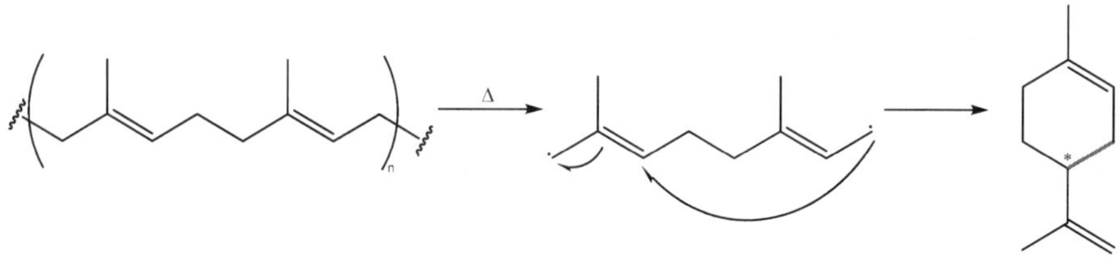

Fig. 4. Radical synthesis of rac-limonene from polyisoprene

These reactions explain the presence of huge amount of $H_2$, aromatics and limonene. The decrease of limonene when rising the reaction temperature can be connected with cracking and especially dehydrogenation reaction. Other reactive compounds may be involved in analogous reactions.

Benzothiazole derivates are present in tires and unsubstitued benzothiazole is widely found in condensable products. Other hetero compounds are also formed due to the high content of sulfur compounds in tire.

Anyway the sulfur content in the liquid is too high to let its use directly as fuel. This oil need refinery treatments or must be blended with refinery oil before any use as fuel.

The gas chromatographic analysis is just one way to characterize the liquid fraction. Other analysis becomes notable when the liquid is compared to petrochemical fuel. The simplest way to compare the liquid fraction with a petrochemical product is distillation. It let to split the liquid fraction in two common petroleum fractions: gasoline and diesel oil. Gasoline distillation temperature is in the range between 30 and 210°C and only 20% of liquid products are within this range. The typical interval of distillation temperature for diesel oil is between 150 and 370°C and 60% of pyrolysis oil distillate in this range. This product must be characterized as a fuel by the gross calorific value. It is between 43 and 44 MJ $kg^{-1}$, higher than commercial oil (de Marco Rodriguez et al., 2001; Lasgaroiti et al., 2004).

### 2.3.3 Non-volatile fraction (solid)

The non-volatile fraction, or simply solid, contains steel cords, carbon, sulfur and other metals containing compounds. In table 9 is reported the average composition of a non-volatile fraction (Zabariotou et al., 2004; Pantea et al., 2003).

We have also identified by ICP-MS analysis the presence of others metal in minor amounts such as, in order of abundance: Co, Pb, Cu, Mn, Ni, Sn, Cr, Cd, V, As and V.

| Element | Abundance (wt%) |
|---|---|
| C | 71.5 |
| O | 13.3 |
| Fe | 5.4 |
| S | 2.8 |
| Zn | 2.3 |
| Ca | 1.3 |
| Al | 0.3 |

Table 9. Elemental composition of a non-volatile fraction (solid).

The steel cords are not scratched at pyrolysis temperature and they may be separated by an electromagnet. The remaining solid product is investigated and may be employed as active carbon in grain or pellets (Mui et al., 2004; Pantea et al., 2003). It is obtained by treating the raw carbonaceous non-volatile product at temperature between 800°C and 1000°C in the presence of an activating agent such as $H_2O$ or $CO_2$. The active carbon obtained in this way has high active surface similar to the commercial product.

The carbon may be used as sorbent for organics (phenols and colorants), metals, chlorinated compounds and gases (butane for instance). Anyway the polluting metals present inside the active carbon (Cr, Cd) may be released and any application must be accurately monitored. Carbon may be also employed as a source of energy by combustion process.

## 2.4 Microwave tires pyrolysis

Up to now the microwave pyrolysis are exploited in laboratory and pilot plant scale. Anyway the greatest difference, if correlated to thermal heating, dwells in uniform and efficient heating. We carry out experiments using a microwave oven that assure uniform distribution of the microwaves inside the chamber and equipped with the possibility to control the power of microwave supply. In this way it is possible to modulate the speed of thermal degradation and the yield of products together with their characteristics.

Yields of the liquids are in the range between 37-40%, similar to those reported for non-microwave pyrolysis (steel cords are counted) when the optimal conditions are achieved, otherwise fewer yields are obtained.

In table 10 are listed the yield of five representative experiments.

| MW power (kW) | Time (min) | Solid (wt%) | Liquid (wt%) | Gas (wt%) |
|---|---|---|---|---|
| 6 | 13 | 50.10 | 26.72 | 23.18 |
| 3 | 39 | 50.45 | 37.21 | 12.34 |
| 1.5 | 100 | 69.38 | 18.09 | 12.53 |
| 1.5 to 4.5 | 76 | 49.65 | 38.76 | 11.59 |
| 1.5 to 6 | 59 | 48.36 | 36.35 | 15.29 |

Table 10. Microwave pyrolysis: correlation among power, time and yields.

The cracking is already effective when half of the maximum power is employed. If the power is lower than 1.5 kW, in our oven, it is not possible to complete the pyrolysis itself. Anyway the properties of collected products vary amazingly.

The properties of non-volatile fraction anyway are slightly affected by the microwave power and the product has close characteristics to that one obtained from a non-microwave oven.

The composition of non-condensable organic fraction changes as function of microwave power and the results are summarized in table 11.

| Substance | MW power (kW) | | | | |
|---|---|---|---|---|---|
| | 6 | 3 | 1.5 | 1.5 to 4.5 | 1.5 to 6 |
| Hydrocarbons C1 + C2 | 46.54 | 76.79 | 70.29 | 27.39 | 29.22 |
| Hydrocarbons C4 | 25.29 | 11.60 | 17.30 | 40.44 | 38.26 |
| Hydrocarbons C5 | 1.20 | 0.75 | 0.94 | 1.79 | 2.07 |
| 2-methylbutadiene | 10.23 | 3.60 | 5.18 | - | 10.88 |

Table 11. Main substances identified in non-condensable product.

High microwave power increases the production of volatile materials which flow quickly out of the reactor and they are not subsequently cracked. A slow heating is not good enough to break the gaseous molecules while a mid-power heating is suited to crack the polymers macromolecules and let them to stay for a long time in the reactor to be degraded to simple molecules such as methane, ethane and ethene. Minor amount of organic sulfur containing substances like ethylene sulfide are present in gases.

Among inorganic substances remarkable is the presence of $H_2$ which rises up to 36% when the microwave power is 3 kW. This result is not easy comparable with non-microwave results: just in few experiments where the residence time of gases inside the reactor is long similar results are achieved.

The condensable fraction (liquid) is deeply affected by microwave power. Its appearance switches from a dark dull liquid when the full power is used to a yellow clear solution when 1.5 kW is used. Decreasing the power also the viscosity is lower.

Anyway the GC/MS analysis gives a better idea how much the liquid product changes its properties.

The main substances identified and their relative abundances are reported in table 12.

| Compound | Abundance % | | | | |
|---|---|---|---|---|---|
| | 6 (kW) | 3 (kW) | 1.5 (kW) | 1.5to4.5 (kW) | 1.5to6 (kW) |
| 2-butene | 1.63 | 1.49 | 1.05 | 1.21 | 0.81 |
| 1,3-pentadiene | 3.18 | 1.63 | 1.49 | 1.70 | 0.31 |
| 3-methyl-1-esene | 2.09 | 1.57 | 1.40 | 1.63 | 0.64 |
| Benzene | 3.91 | 1.71 | 2.78 | 2.26 | 1.66 |
| Toluene | 4.55 | 2.87 | 3.52 | 3.11 | 2.51 |
| Xylenes | 6.92 | 5.22 | 7.48 | 7.08 | 3.43 |
| Styrene | 3.06 | 1.39 | 1.45 | 1.85 | 0.76 |
| Trimethylbenzenes | 3.23 | 3.03 | 4.57 | 4.54 | 2.53 |
| α–methylstyrene | 0.62 | 0.39 | - | 1.33 | 0.24 |
| 1-ethyl-3-methylbenzene | 0.99 | 1.01 | 1.27 | 1.41 | 0.78 |
| 1,2,4,5-tetramethylbenzene | 0.76 | 1.40 | 2.10 | 2.41 | 1.19 |
| Limonene | 3.72 | 3.43 | 2.49 | 3.49 | 0.82 |
| 1H-Indene | 0.87 | 0.38 | 0.28 | 0.50 | 0.46 |
| Naphthalene | 0.97 | 0.43 | - | 0.20 | 0.77 |
| Benzothiazole | 0.59 | 0.57 | 0.25 | 0.32 | 0.58 |

Table 12. Main substances in liquid product, obtained using different microwave power.

A fast pyrolysis with a power of 6 kW yields a huge amount of aromatics (up to 20%). Decreasing the microwave power the aromatic yield decreases. Probably the microwave energy is not enough to perform dehydrogenation reactions.
The reduction of microwave power leads a change in other properties of liquid such as: viscosity, density and fraction of distilled product.
Viscosity and density are directly connected with microwave power. The density changes among 0.800 g $cm^{-3}$ and 0.900 g $cm^{-3}$, while the viscosity is in the range 0.7 - 2.6 cPs.
The fraction of product distilling at temperature lower than 210°C changes deeply when the power is lowered as shown in table 13.

| MW Power (kW) | Distillate under 210°C (%) |
|---|---|
| 6 | 22.93 |
| 3 | 37.61 |
| 1.5 | 74.61 |

Table 13. The distillation of condensable fraction. Only the amount of distillate under 210°C is reported.

The trend is almost linear and interesting results are achieved. In non-microwave pyrolysis only 20% of liquid fraction is distillable under 210°C.
Using a microwave heating is possible to manage properly the yields and proprieties of condensable and non-condensable products.

## 3. Microwave pyrolysis of polymeric materials

As well as tires pyrolysis the microwave pyrolysis of single polymeric materials has been investigated by many authors (Kaminsky & Zorriqueta, 2007; Arandes et al., 2008; Kim, 2001; Yoshioka et al., 2004; de la Puente et al., 1998). Five main categories of polymers were tested: polyethylene (PE), polypropylene (PP), polystyrene (PS), poly(ethylene terephthalate) (PET) and polyvinylchloride (PVC). All of these polymers are not able to absorb microwave and turn it into heat. So a microwave absorbent must be employed to carry out the pyrolysis, as reported in a previous paragraph. Carbon powder (to avoid unwanted interference) or chopped tires has been employed as microwave absorbent.
The microwave influence on the products yields is poor but it is strong on the properties of liquid fraction and this influence is reported in the following paragraphs.
The non-microwave pyrolysis of the above mentioned polymers has been deeply studied and characterized. In the first instance the microwave pyrolysis seemed to be more suited for multi materials pyrolysis such as: Tetra Pak® or laminates (Ludlow-Palafox & Chase, 2001).
Usually the pyrolysis of a pure plastic material yields just a liquid and a gas product. In few experiments is reported the growing up of a solid product and usually low attention has been paid to this aspect. In our apparatus a low amount of non-volatilized fraction is always found when the polymers are pyrolyzed in the presence of coal, presumably due to a strong localized heating. In table 14 are listed some experiments with relative yields for the pyrolysis of single plastics.

| MW power range (kW) | Polymer | Time (min) | Absorbent-polymer ratio | Solid (wt%) | Liquid (wt%) | Gas (wt%) |
|---|---|---|---|---|---|---|
| 1.8 to 3* | HDPE | 33 | 2:1 | 42.49 | 44.98 | 12.09 |
| 3 | HDPE | 75 | 1:2 | 0.40 | 83.92 | 15.68 |
| 1.2 to 6* | HDPE | 265 | 2:1 | 35.52 | 45.83 | 18.65 |
| 1.8 to 3* | PP | 39 | 2:1 | 35.29 | 54.00 | 10.72 |
| 3 | PP | 68 | 1:2 | 15.89 | 70.82 | 13.29 |
| 1.8 to 3* | PVC | 71 | 2:1 | 53.16 | 17.70 | 29.14 |
| 1.8 to 3 | PVC | 47 | 1:1 | 13.92 | 10.25 | 75.83 |
| 3 | PVC | 21 | 1:2 | 14.69 | 3.44 | 81.87 |
| 1.8 to 3* | PET | 40 | 2.5:1 | 38.20 | 35.32 | 26.48 |
| 1.2 to 3* | PET | 207 | 4.2:1 | 48.29 | 26.34 | 31.99 |
| 3 to 6 | PS | 59 | 1:2 | 6.83 | 89.25 | 8.92 |

Table 14. Microwave pyrolysis: correlation between microwave power and yield. * Tires instead of coal as microwave absorber.

### 3.1 PE microwave pyrolysis: products characteristics

The polyethylene (PE) is the most important and used plastic. It is a thermoplastic polymer, chemically inert and is used in an enormous range of applications. It is produced as a linear High Density (HDPE) or a branched Low Density (LDPE) polyethylene.

The thermal degradation of HDPE starts over 380°C (Bockhorn et al., 1998). When HDPE is heated to a temperature below 730°C it forms mainly a wax (a semi-solid product) and a liquid. Together with the increasing of the temperature the amount of wax decreases in advantages to liquid and gas products. At 730°C there is no further formation of wax, just liquid and gas (Mastral et al., 2002). Over 800°C the conversion of the polymer in gases is almost total: 91.2% at 850°C (Westerhout et al., 1998).

The cracking of the polymers chain leads to a large distribution of numerous linear hydrocarbons having different length. This distribution is a function of microwave power and working temperature. At high cracking temperature the product is liquid at room temperature otherwise it is solid.

The gas is composed of $H_2$, $CH_4$, $C_2H_6$, $C_2H_4$, $C_3H_8$, and $C_3H_6$. The liquid and wax are made of alkanes, from $C_{11}$ to $C_{57}$, and their equivalent 1-alkene and 1,3-dialkene (Williams & Williams, 1999a).

Anyway too high temperature causes the formation of aromatics with the reaction path proposed in figure 3. Together with high temperature also long residence time promotes the formation of aromatic compounds.

The pyrolysis of HDPE using microwave heating requires HDPE flakes mixed with a microwave absorbent such as coal or chopped tires. Different microwave power gives completely different products. In table 14 are listed the yields in solid, liquid and gas using different microwave power and absorber.

When microwave power of 3 kW is used a wax is largely produced and the amount of a liquid product inside the wax for analysis is hardly recovered. Anyway when milder reaction conditions are used, lowering the microwave power, a liquid product is collected. It has been obtained using 1.8 - 1.2 kW, but in this condition it is not possible to complete the pyrolysis without the erogation of a higher power of microwave in a subsequent step. By the

simple modulation of the microwave power is not possible to obtain, without other treatments or apparatus, only a liquid product.
The liquid fraction, analyzed via GC/MS, contains linear saturated hydrocarbons and their terminal unsaturated equivalent from $C_6$ to $C_{32}$.
The only way to obtain a liquid product instead of wax is the cracking of the wax in a second reactor, refluxing them again in the pyrolysis reactor or using a catalytic system which resemble a refinery unit (Arandes et al., 2008).

### 3.2 PP microwave pyrolysis: products characteristics

The polypropylene (PP) is the second most used polyolefin in the world due to the low-cost of monomer production together with flexible properties.
The degradation mechanism yields a chain cracking to a wax and a liquid. They are composed of hydrocarbons from $C_{11}$ to $C_{57}$ (Westerhout et al., 1998). Rising the pyrolysis temperature the wax yield decreases. Anyway these conditions lead to the formation of aromatics due to further reactions of hydrocarbons obtained in the first instance (see scheme reported in figure 3).
The PP is more unstable than HDPE to thermal degradation due the presence of a large extent of tertiary carbons giving more stable radicals with respect to primary or secondary radicals formed in the pyrolysis of PE. The pyrolysis of PP may be carried out at 500°C instead of 700°C as required for HDPE (Kaminsky & Zorriqueta, 2007)
An analogous effect is shown when the pyrolysis is performed using a microwave oven. The products are liquid at room temperature independent from the microwave power and the microwave absorbent: coal or chopped tires. In table 14 are reported two experiments performed at different microwave power.
The high reactivity of PP leads often to the formation of a solid product.
The liquid fraction contains organic compounds from $C_6$ to $C_{30}$. In table 15 are reported the main components identified by GC/MS analysis in the low boiling fraction (below 145°C) of the liquid product obtained using 3 kW as microwave power source.

| Compounds | Composition (%) |
|---|---|
| 4-methylhepthane | 3.78 |
| 2,6-dimethyl-2-octene | 3.69 |
| 1,3,5-trimethylcyclohexane | 4.88 |
| 3,7-dimethyl-1-octene | 57.3 |
| 1,2,4-trimethylcyclohexane | 5.17 |
| 3,5,5-trimethylcyclohexane | 3.45 |
| 1-nonene | 18.93 |
| Total | 97.20 |

Table 15. Main compounds present in the low boiling fraction of the microwave pyrolysis of PP using a power source of 3 kW.

All the substances formed by PP degradation are methyl derivates. For instance 3,7-dimethyl-1-octene is very close to the repeating unit of PP as shown in figure 5.
Anyway few aromatic substances are present in liquid products when pure PP is pyrolyzed using a microwave oven with respect to the compounds present in a thermal pyrolysis of PP (Kaminsky, 2006). This is connected with a different way of heat transfer. The heat provided by microwave oven is quickly employed to break the polymers avoiding subsequent reactions.

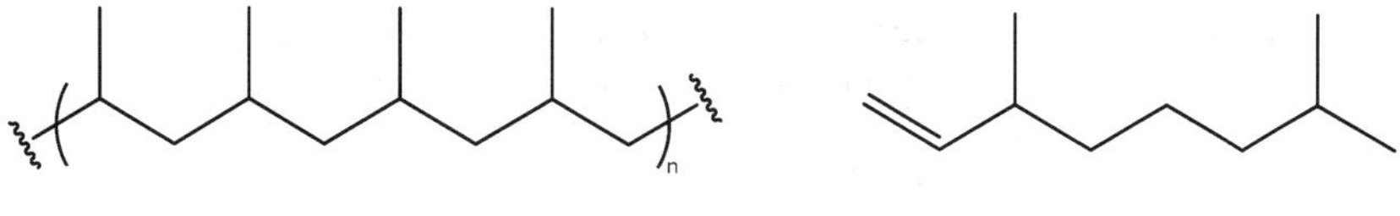

PP 3,7-dimethyl-1-octene

Fig. 5. A molecule formed in the liquid from pyrolysis of PP.

### 3.3 PVC microwave pyrolysis: products characteristics

The polyvinylchloride (PVC) holds a unique position among commercial polymers. It is cheap and used in a wide range of applications due to its high versatility.

Usually it is added with stabilizers, lubricating, plasticizers, fillers and so on. To avoid interference due to large amount of additives present, the microwave pyrolysis of PVC has been carried out on pure PVC.

The thermal degradation of PVC takes place in two-steps process: dehydrochlorination followed by residual chain cracking. The elimination of hydrogen chloride from the chain back bone is fast and complete. It happens at temperature among 300°C and 330°C (Bockhorn et al., 1998; Kim, 2001). The reaction scheme is reported in figure 6.

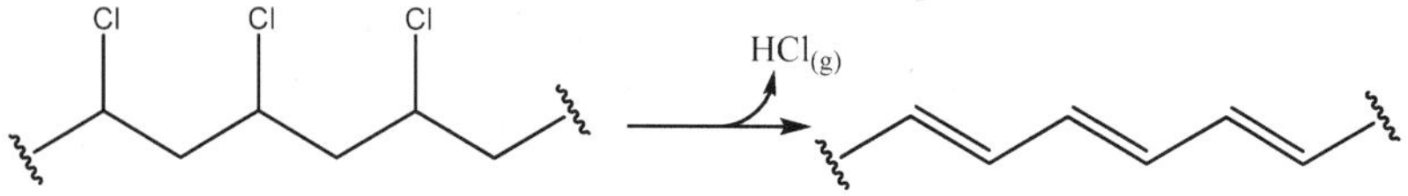

Cl Cl Cl $HCl_{(g)}$

Fig. 6. Dehydrochlorination of PVC.

The gases are mainly composed of HCl, ethylene and propylene. The liquid fraction contains aromatic compounds such as benzene, toluene and styrene (Saeed et al., 2004).

The microwave pyrolysis of PVC follows the same two steps of thermal pyrolysis. The polymer quickly eliminates HCl that has been absorbed in NaOH traps after the collecting flasks. Up to 97% of chlorine in PVC chain is found in the traps. Few chlorine organic compounds are present in liquid products. In table 14 are reported the PVC pyrolysis at different microwave power.

As reported in the literature for thermal pyrolysis also in the microwave pyrolysis the main products present in liquid fraction are aromatic compounds, the main components of the liquid are listed in table 16.

| Compounds | Composition (%) |
|---|---|
| Benzene | 18.56 |
| Toluene | 8.73 |
| 1,3-dimethyl-benzene | 4.25 |
| 1,4-dimethyl-benzene | 2.49 |
| Styrene | 0.46 |
| 1,2-dimethyl-benzene | 3.44 |
| Naphthalene | 4.01 |
| Total | 41.94 |

Table 16. Main aromatic compounds formed in PVC/coal pyrolysis.

The formation of aromatics in high yield may be attributed to Diels-Alder type reactions involving the poly-unsaturated chain formed after dehydrochlorination of PVC. This liquid may be a source of interesting compounds such benzene, toluene and xylenes.

### 3.4 PET microwave pyrolysis: products characteristics

The poly(ethylene terephthalate) (PET) is one of the most important polyester (together with polybutylenetherephatlate, PBT) and it is used to obtain: fiber textiles, carpets, medical accessories, belts for cars, electronics and items for cars, photographic film, magnetic tapes for audio and video, packaging, bottles and so on.

The gases obtained in PET pyrolysis are mainly $CO_2$ (up to 22.71%) and CO (up to 13.29%) due to decarboxylation reactions involving the esters group present in the polymeric chain (Williams & Williams, 1999b).

Oxygenated compounds, like benzoic acid or other compounds soluble in aqueous NaOH, are largely found. Anyway while rising the temperature from 510°C to 610°C the concentration of oxygenated compounds decreases from 45 wt% to 29% (Yoshioka et al., 2004).

A lot of aromatics and oxygenated compounds are found in the liquid fraction, all of them coming from the therephtalic moiety present in the polymer. Their concentrations increase when rising the pyrolysis temperature.

The PET microwave pyrolysis shows very different results from the non-microwave pyrolysis. The most significant difference comes from high amount of acetaldehyde and benzene. In table 14 are reported the yields of two PET pyrolysis where chopped tires where used as microwave absorbent.

The high amount of gas is linked with the $CO_2$ and CO formation. The main substances present in the liquid fraction are reported in table 17; for the pyrolysis with increasing power from 1.2 to 3 kW two liquid fractions where collected when the power is changed.

| Compounds | Composition % | | |
|---|---|---|---|
| | 1.8 to 3 (kW) | 1.2 to 3 (kW) | |
| | | Fraction 1 | Fraction 2 |
| Benzene | 32.03 | 5.80 | 14.26 |
| Toluene | 9.40 | 4.37 | 9.57 |
| 1,2-dimethylbenzene | 4.29 | 3.13 | 6.04 |
| 1,3-dimethylbenzene | 3.70 | 2.31 | 5.25 |
| Styrene | 3.43 | - | 1.07 |
| Benzaldehyde | - | 1.19 | - |
| 1,2,3-trimethylbenzene | 4.71 | 2.54 | 4.06 |
| Limonene | 14.51 | 2.31 | - |
| 1H-indene | 2.44 | - | - |
| Pentylbenzene | 2.10 | - | - |
| Benzoic Acid | 5.15 | 1.14 | - |
| Biphenyl | 3.66 | - | 3.02 |
| Total | 85.42 | 22.79 | 43.27 |

Table 17. The main products present in the liquid from microwave pyrolysis of PET.

The acetaldehyde was identified by trapping it with (2,4-dinitrophenyl)hydrazine. The hydrazone was characterized via $^1$H-NMR.
The formation of acetaldehyde is not reported for non-microwave pyrolysis. It can be accounted as a microwave heating prerogative. The heat is provided from microwave absorbent, coal or tires, in close contact with the polymer. This heat is straight shifted from the coal to the surrounding macromolecules.
In figure 7 is reported the steps proposed for acetaldehyde formation.

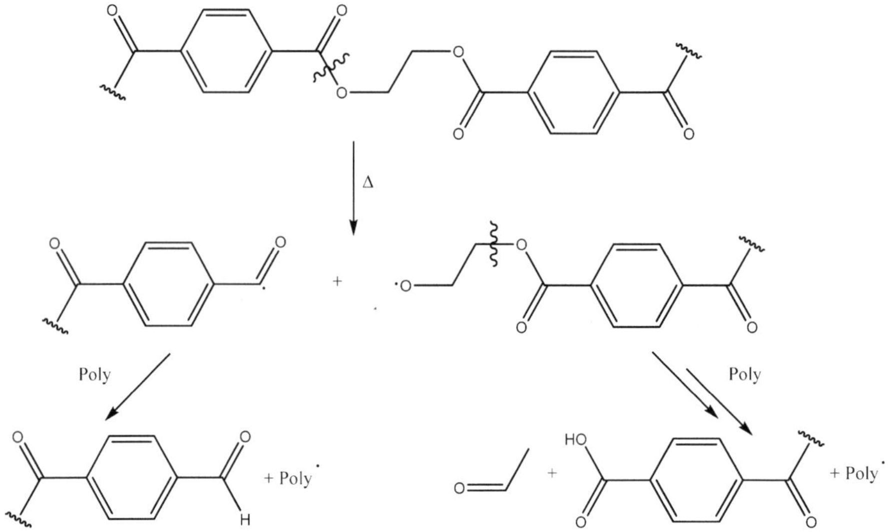

Fig. 7. Suggested reactions involved in microwave pyrolysis of PET.

### 3.5 PS microwave pyrolysis: products characteristics

The polystyrene (PS) is produced and sold as an amorphous polymer or as co-polymers with monomers such as acrylonitrile and butadiene.
The thermal pyrolysis of PS yields the formation of just a liquid product. A limited amount of gas is also collected (Williams & Williams, 1999b; Williams & Bagri, 2004). In the liquid product styrene is found up to 76.2 wt%. The remaining liquid fraction contains other aromatics such as: benzene, toluene, xylenes, ethylbenzene, cumene and α-methylstyrene.
The gas fraction contains: $H_2$, $CH_4$, $C_2H_6$, $C_2H_4$, $C_3H_8$, $C_3H_6$ and $C_4H_8$ (Williams & Williams, 1999b).
The PS is easy cracked also via microwave heating in the presence of a microwave absorber like with the thermal cracking. The pyrolysis leads to the complete depolymerization of PS to its monomer and other aromatics. The main products present in the liquid fraction are reported in table 18.
The yields are strongly different from a non-microwave pyrolysis (de la Puente & Sedran, 1998). The amount of styrene is lower than 50% but the amount of toluene and other aromatic compounds are higher with respect to non-microwave pyrolysis. These results

may be explained only by a different way in which the heat is transferred from the absorber to the polymer.

| Compound | Composition (%) |
|---|---|
| Benzene | 1.62 |
| Toluene | 8.99 |
| 1,2-dimethyl-benzene | 8.70 |
| Styrene | 47.49 |
| α-methyl-styrene | 13.39 |
| 4-penthyl-benzene | 5.06 |
| Total | 85.25 |

Table 18. The main compounds present in liquid fraction from microwave pyrolysis of PS.

## 4. Conclusion

Microwaves are a precious tool and they may be employed in pyrolysis process of plastics, such as HDPE, PP, PVC, PET and PS, or complex polymeric materials such as tires or mixed waste plastics. The use of a microwave heating source open new paths in waste disposal via pyrolysis, where the yields and products proprieties may be controlled just by the power of microwave it is employed. The problem of poor thermal conductivity of polymers is overwhelmed and a simpler apparatus may be employed like static bed reactors. Together with a simple apparatus the reaction time could be strongly reduced from hours to few minutes.

In addition no pretreatments are required for the polymeric waste with the exception of the addition of a microwave absorbent. Another of the big advantage in tires pyrolysis with regard to thermal pyrolysis, for instance, is the possibility to heat not chopped materials.

The microwave pyrolysis results as an optimal way to transform simple or complex plastic wastes or more generally plastic containing wastes in new products such as fuels, oil, important hydrocarbons such as BTX or to obtain gas pyrolysis non-contaminated with combustion gas and nitrogen present in the air employed for combustion and consequently the gases obtain show a high calorific value.

## 5. Acknowledgment

This work was partially supported by CAF Scarl - Florence. The authors want to thanks Dr. Silvio Occhialini for suggestions and Mr. Maurizio Passaponti for his glassware masterpiece.

## 6. Reference

Arandes, J.M.; Torre, I.; Azkoiti, M.J.; Castaño, P.; Bilbao, J. & de Lasa, H. (2008). Effect of catalyst properties on the cracking of polypropylene pyrolysis waxes under FCC conditions, *Catalysis Today*, Vol.133-135, April-June 2008, pp. 413-419, ISSN 0920-5861.

Aylón, E.; Fernàndez-Colino, A.; Navarro, M. V.; Murillo, R.; Garcìa, T. & Mastral A. M. (2008), Waste Tire Pyrolysis: Comparison between Fixed Bed Reactor and Moving Bed Reactor, *Industrial & Engineering Chemistry Research*, Vol.47., No.12, May 2008, pp. 4029–4033, ISSN 1520-5045.

Aylón, E.; Fernàndez-Colino, A.; Murillo, R.; Navarro, M. V.; Garcìa, T. & Mastral A. M. (2008), Valorisation of waste tyre by pyrolysis in a moving bed reactor, *Waste Management*, Vol.30, No.7, July 2010, pp. 1220–1224, ISSN 0956-053X.

Berrueco, C.; Esperanza, E.; Mastral, F.J.; Ceamanos, J. & Garcìa-Bacaicoa, P. (2005). Pyrolysis of waste tyres in an atmospheric static-bed batch reactor: Analysis of the gases obtained, *Journal of Analytical and Applied Pyrolysis*, Vol.74, No.1-2, August 2005, pp. 245-253, ISSN 0165-2370.

Bockhorn, H.; Hornung, A. & Hornung, U. (1998). Stepwise pyrolysis for raw material recovery from plastic waste, *Journal of Analytical and Applied Pyrolysis*, Vol.46, No.1, June 1998, pp. 1-13, ISSN 0165-2370.

Chen, D.T.; Perman, C.A.; Riechert, M.E. & Hoven, J. (1995). Depolymerization of tire and natural rubber using supercritical fluids, *Journal of Hazardous Materials*, Vol.41, No.1, November 1995, pp. 53-60, ISSN 0304-3894.

Chen, S.-J.; Sua, H.-B.; Chang, Juu-En; Leeb, W.-J.; Huanga, K.-L.; Hsieha, L.-T.; Huang, Y.-C.; Lind, W.-Y. & Lin, C.-C. (2007). Emissions of polycyclic aromatic hydrocarbons (PAHs) from the pyrolysis of scrap tires, *Atmospheric Environment*, Vol.41, No.6, February 2007, pp. 1209-1220, ISSN 1352-2310.

Cunliffe, A.M. & Williams, P.T. (1998). Composition of oils derived from the batch pyrolysis of tyres, *Journal of Analytical and Applied Pyrolysis*, Vol.44, No.2, January 1998, pp. 131-152, ISSN 0165-2370.

de la Puente, G. & Sedran, U. (1998). Recycling polystyrene into fuels by means of FCC: performance of various acidic catalysts, *Applied Catalysis B: Environmental*, Vol.19, No.3-4, December 1998, pp. 305-311, ISSN 0926-3373.

de Marco Rodriguez, I.; Laresgoiti, M.F.; Cabrero, M.A.; Torres, A.; Chomòn, M.J. & B. Caballero. (2001). Pyrolysis of scrap tyres, *Fuel Processing Technology*, Vol.72, No.1, August 2001, pp. 9-22, ISSN 0378-3820.

Dìez, C.; Martínez, O.; Calvo, L. F.; Cara, J. & Morán, A. (2004). Pyrolysis of tyres. Influence of the final temperature of the process on emissions and the calorific value of the products recovered, *Waste Management*, Vol.24, No.5, January 2004, pp. 463-469, ISSN 0956-053X.

Djeu, N.I. (2000). Fiber-optic high temperature sensor, *United States Patent*, April 2000, US6045259.

Federazione Imprese e Servizi, Unione Nazionale Imprese Recupero [FISE UNIRE]. (2009). L'Italia del recupero - Rapporto FISE UNIRE sul riciclaggio dei rifiuti, 10° edizione, Rimini, Italy.

Freedonia. (2010). World Rubber & Tire to 2013 - Demand and Sales Forecasts, Market Share, Market Size, Market Leaders, in: *Freedonia Industry Research*, 15.02.2011, Available from: http://www.freedoniagroup.com/DocumentDetails.aspx?ReferrerId=FG-01&studyid=2575.

Giugliano, M.; Cernuschi, S.; Ghezzi, U. & Grosso, M. (1999). Experimental Evaluation of Waste Tires Utilization in Cement Kilns, *Journal of the Air & Waste Management Association*, Vol.49, December 1999, pp. 1405-1414, ISSN 1047-3289.

Holland, K.M. (1992). Process of destructive distillation of organic material, *United States Patent*, January 1992, US005084141A.

Holland, K.M. (1995). Apparatus for Waste Pyrolysis, *United States Patent*, February 1995, US005387321A.

Horikoshi, S.; Matsubara, A.; Takayama, S.; Sato, M.; Sakai, F.; Kajitani, M.; Abe, M. & Serpone N. (2009). Characterization of microwave effects on metal-oxide materials: Zinc oxide and titanium dioxide, *Applied Catalysis B: Environmental*, Vol.91, No.1-2, September 2009, pp. 362-367, ISSN 0926-3373.

Hussain, Z.; Khan,K.M. & Hussain, K. (2010). Microwave–metal interaction pyrolysis of polystyrene, *Journal of Analytical and Applied Pyrolysis*, Vol.89, No.1, September 2010, pp. 39-43, ISSN 0165-2370.

Johnson, A.C.; Lauf, R.J.; Bible, D.W. & Markunas, R.J. (1996). Apparatus and method for microwave processing of materials, *United States Patent*, May 1996, US005521360A.

Juma, M.; Koreňovà, Z.; Markoš, J.; Annus, J. & Jelemenský, L. (2006). Pyrolysis and combustion of scrap tire, *Petroleum & Coal*, Vol.48, No.1, 2006, pp 15-26, ISSN: 1335-3055.

Kaminsky, W. (1992). Plastics, Recycling, In: *Ullmann's Encyclopedia of Industrial Chemistry*, pp. 57-73, Wiley-VCH, ISBN 3-527-30385-5, Weinheim, Germany

Kaminsky, W. & Mennerich, C. (2001). Pyrolysis of synthetic tire rubber in a fluidised-bed reactor to yield 1,3-butadiene, styrene and carbon black, *Journal of Analytical and Applied Pyrolysis*, Vol.58-59, No.1, April 2001, pp. 803-811, ISSN 0165-2370.

Kaminsky, W.; Predel, M. & Sadiki, A. (2004). Feedstock recycling of polymers by pyrolysis in a fluidised bed, *Polymer Degradation and Stability*, Vol.85, No.3, September 2004, pp. 1045-1050, ISSN 0141-3910.

Kaminsky, W. (2006). *Feedstock Recycling and Pyrolysis of Waste Plastics*, John Wiley & Sons, ISBN 0-470-02152-7.

Kaminsky, W. & Zorriqueta, I.-J.N. (2007). Catalytical and thermal pyrolysis of polyolefins, *Journal of Analytical and Applied Pyrolysis*, Vol.79, No.1-2, May 2007, pp. 368-374, ISSN 0165-2370.

Kasin, K.I. (2009). Microwave gasification, pyrolysis and recycling of waste and other organic materials, *United States Patent Application Publication*, January 2009, US2009000938A1.

Kim, S. (2001). Pyrolysis kinetics of waste PVC pipes, *Waste Management*, Vol.21, No.7, 2001, pp. 609-616, ISSN 0165-2370.

Kyari, M.; Cunliffe, A. & Williams, P.T. (2005). Characterization of Oils, Gases, and Char in Relation to the Pyrolysis of Different Brands of Scrap Automotive Tires, *Energy & Fuels*, Vol.19, No.3, April 2005, pp. 1165-1173, ISSN 1520-5029.

Laresgoiti, M. F.; Caballero, B. M.; de Marco, I.; Torres, A.; Cabrero, M. A. & Chomón M. J. (2004). Characterization of the liquid products obtained in tyre pyrolysis, *Journal of Analytical and Applied Pyrolysis*, Vol.71, No.2, June 2004, pp. 917-934, ISSN 0165-2370.

Li, S.-Q.; Yao, Q.; Chi, Y.; Yan, J.-H. & Cen, K.-F. (2004). Pilot-Scale Pyrolysis of Scrap Tires in a Continuous Rotary Kiln Reactor, *Industrial & Engineering Chemistry Research*, Vol.43., No.17, June 2004, pp. 5133-5145, ISSN 1520-5045.

Ludlow-Palafox, C. & Chase, H.A. (2001). Microwave-Induced Pyrolysis of Plastic Waste, *Industrial & Engineering Chemistry Research*, Vol.40., No.22, October 2001, pp. 4749-4756, ISSN 1520-5045.

Mastral, A.M.; Murillo, R.; Callén, M. S.; Garcìa, T. & Snape, C. E. (2000). Influence of Process Variables on Oils from Tire Pyrolysis and Hydropyrolysis in a Swept Fixed Bed Reactor, *Energy & Fuels*, Vol.14, No.4, July/August 2000, pp. 739-744, ISSN 1520-5029.

Mastral, F.J.; Esperanza, E.; Garcìa, P. & Juste, M. (2002). Pyrolysis of high-density polyethylene in a fluidised bed reactor. Influence of the temperature and residence time, *Journal of Analytical and Applied Pyrolysis*, Vol.63, No.1, March 2002, pp. 1-15, ISSN 0165-2370.

Meredith, R. (1998). *Engineers' handbook of industrial microwave heating*, The Institution of Electrical Engineers, ISBN 0852969163, London, United Kindom.

Micromaterials Inc., Tampa, FL, U.S.A. (2010). Fiber Optic Thermometer- Up to 950°C, In: *Products*, 25.02.2011, Available from:
http://www.micromaterialsinc.com/specsSensor.html.

Mui, E.L.K.; Ko, D.C.K. & McKay G. (2004). Production of active carbons from waste tyres–a review, *Carbon*, Vol.42, No.14, 2004, pp. 2789-2805, ISSN 0008-6223.

Natural Resources Defense Council [NRDC]. (February 2008). The 3r's Still Rule, In: *This Green Life*, 20.02.2011, Available from:
http://www.nrdc.org/thisgreenlife/0802.asp.

Pantea, D.; Darmstadt, H.; Kaliaguine, S. & Roy, C. (2003). Heat-treatment of carbon blacks obtained by pyrolysis of used tires. Effect on the surface chemistry, porosity and electrical conductivity, *Journal of Analytical and Applied Pyrolysis*, Vol.67, No.1, March 2003, pp. 55-76, ISSN 0165-2370.

Parker, T.H. (1992). Apparatus for pyrolysis of tires and waste. *United States Patent*, December 1992, US005167772A.

Piskorz, J.; Majerski, P.; Radlein, D.; Wik, T. & Scott, D. S. (1999). Recovery of Carbon Black from Scrap Rubber, *Energy & Fuels*, Vol.13, No.3, March 1999, pp. 544-551, ISSN 1520-5029.

Plastics Europe. (2010). Plastics – the Facts 2010 An analysis of European plastics production, demand and recovery for 2009, In: *Plastic Market*, 09.03.2011, Available from:
http://www.plasticseurope.org/documents/document/20101028135906-final_plasticsthefacts_26102010_lr.pdf.

Pringle, F.G. (2007). Microwave-based recovery of hydrocarbons and fossil fuels, *United States Patent Application Publication*, June 2007, US20070131591A1.

Pringle, J.A. (2006). Microwave pyrolysis apparatus for waste tires, *United States Patent*, September 2006, US007101464B1.

Rubber Manufacturers Association [RMA]. (2011). Scrap Tire Characteristics, In: *Scrap Tires and the Environment*, 28.02.2011, Available from
http://www.rma.org/scrap_tires/scrap_tire_markets/scrap_tire_characteristics/

Saeed, L.; Tohka, A.; Haapala, M. & Zevenhoven, R. (2004). Pyrolysis and combustion of PVC, PVC-wood and PVC-coal mixtures in a two-stage fluidized bed process, *Fuel Processing Technology*, Vol.85, No.14, September 2004, pp. 1565–1583, ISSN 0378-3820.

Scandinavian Biofuel Company [SBC]. (2011). Microwave Assisted Pyrolysis, In: *Microwave pyrolysis*, 09.03.2011, Available from:
http://www.sbiofuel.com/pyrolysis.html.

Sharma, V.K.; Mincarini, M.; Fortuna, I F.; Cognini, F. & Cornacchia G. (2000). Disposal of waste tyres for energy recovery and safe environment—review, *Energy Applied Energy*, Vol.65, No.1-4, April 2000, pp. 381-394, ISSN 0306-2619.

Tang, L. & Huang H. (2004). An investigation of sulfur distribution during thermal plasma pyrolysis of used tires, *Journal of Analytical and Applied Pyrolysis*, Vol.72, No.1, August 2004, pp. 35-40, ISSN 0165-2370.

Westerhout, R.W.J.; Waanders, J.; Kuipers, J. A. M. & van Swaaij, W. P. M. (1998). Development of a Continuous Rotating Cone Reactor Pilot Plant for the Pyrolysis of Polyethene and Polypropene, *Industrial & Engineering Chemistry Research*, Vol.37, No.6, May 1998, pp. 2316-2322, ISSN 1520-5045.

Williams, P.T.; Besler, S & Taylor, D.T. (1990). The pyrolysis of scrap automotive tyres: The influence of temperature and heating rate on product composition, *Fuel*, Vol.60, No.12, December 1990, pp. 1474-1482, ISSN 0016-2361.

Williams, P.T.; Bottrill, R.P. & Cunliffe, A.M. (1998). Combustion of Tyre Pyrolysis Oil, *Process Safety and Environmental Protection*, Vol.76, No.4, November 1998, pp. 291-301, ISSN 0957-5820.

Williams, P.T. & Williams, E.A. (1999a). Fluidised bed pyrolysis of low density polyethylene to produce petrochemical feedstock, *Journal of Analytical and Applied Pyrolysis*, Vol.51, No.1-2, July 1999, pp. 107-126, ISSN 0165-2370.

Williams, P.T. & Williams, E.A. (1999b). Interaction of Plastics in Mixed-Plastics Pyrolysis, *Energy & Fuels*, Vol.13, No.1, 1999, pp. 188-196, ISSN 1520-5029.

Williams, P.T. & Bagri, R. (2004). Hydrocarbon gases and oils from the recycling of polystyrene waste by catalytic pyrolysis, *International Journal of Energy Research*, Vol.28, No.1, January 2004, pp. 31-44, ISSN 0363-907X.

Yoshioka, T.; Grause, G.; Eger, C.; Kaminsky, W. & Okuwaki, A. (2004). Pyrolysis of poly(ethylene terephthalate) in a fluidised bed plant, *Polymer Degradation and Stability*, Vol.86, No.3, December 2004, pp. 499-504, ISSN 0141-3910.

Zabaniotou, A.; Madau, P.; Oudenne, P.D.; Jung, C.G.; Delplancke, M.-P. & Fontana, A. (2004). Active carbon production from used tire in two-stage procedure: industrial pyrolysis and bench scale activation with $H_2O$–$CO_2$ mixture, *Journal of Analytical and Applied Pyrolysis*, Vol.72, No.2, November 2004, pp. 289-297, ISSN 0165-2370.

# Part 5

## Synthesis

# 11

# Synthesis of SiC Powders and Whiskers by Microwave Heating

Jianlei Kuang, Fu Wang, Qiang Wang,
Matiullah Khan and Wenbin Cao
*University of Science and Technology Beijing, Beijing,
China*

## 1. Introduction

Silicon carbide (SiC) has been widely used in industry as both structural and functional materials due to its high hardness, strength, good wear resistance, light weight, excellent resistance to oxidation and corrosion, lower thermal expansion coefficient and high thermal conductivity. The main method for the production of SiC powders (SiC*p*) in industry is known as the Acheson process, in which SiC ingots are synthesized at about 2200 °C through the carbothermal reduction of quartz sand by coke lasting for a few days in general(Krstic, 1992), following with cracking, grinding, and meshing, etc.. It is energy-intensive and difficult to produce fine powders with a narrowed size distribution and controllable morphology. This process has been modified as so called rapid carbothermal reduction (RCR) to produce fine and pure SiC powders(Setiowati & Kimura, 1997). Unlike conventional carbothermal reduction process, the aerosol consisting of ultrafine carbon and silica precursors have been used as starting materials in the RCR process(Weimer et al., 1991; Weimer et al., 1993; Weimer et al., 1994), so that the carbothermal reduction can be completed in a few seconds through the reactor tube by coupling the thermal radiations and the fast kinetics. It indicates the probability that SiC could be synthesized in a very short time with appropriate methods.

Therefore, new methods, such as sol-gel(Hatakeyama & Kanzaki, 1990; Seog & Kim, 1993), plasma(Guo et al., 1995), CVD(Klein et al., 1998; Halamka et al., 2003), and laser(Li et al., 1994) were attempted during the last few decades to synthesize fine SiC powders with narrow particle size distribution and high purity. However, these processes have not satisfied the industry requirement of the quick production of ultra-fine SiC powders and whiskers due to its complex process, high cost and less productivity.

Compared with that of the other methods, the microwave heating has been demonstrated to be an attracting method in synthesis of ceramic materials(Rao et al., 1999) because of its rapid and bulk heating, selective coupling, improvement of heat transfer, enhanced reaction kinetics, acceleration of the reaction rate and these all results in quick synthesis of ultra-fine SiC powders(Satapathy et al., 2005). The basic idea behind microwave heating is that, it transfers heat from inner to surface of materials and as a result the temperature of the inner part will be higher than that of the other parts and the first the reaction will begin in the inner core. Thus a significant reduction in the reaction temperature and thermal processing time will be achieved, which is energy-effective and economical.

SiC whiskers (SiC*w*) are widely used in reinforcing metallic matrix and ceramic matrix composites. At present, solid phase synthesis method is used for the industrial production of SiC*w*, by using Fe, Ni and NaF etc. as catalyst via VLS mechanism. However this method will introduce metallic impurities which lead to lower the quality of SiC*w* by forming stacking fault, dislocation, twin boundary, rough surfaces and nodular structure. In addition, CVD, which use gaseous precursor or template (carbon nanotube) to prepare SiC*w*, is also a common method; however, low yield and complex equipment are still the major disadvantage. As the inner part of the reactants will be firstly heated to the reaction temperature under the microwave irradiation, the reactions begin at the inner will increase the concentration of gaseous phase reactants, and benefit the synthesis of SiC*w* by VS or VLS mechanisms.

In this chapter, the effect of the silicon sources and carbon sources, reaction temperature and holding time on the synthesis of SiC powders and whiskers have been discussed and reviewed mainly from our published papers and dissertations (Kuang et al, 2011; F. Wang, 2008a; F. Wang et al, 2008b, 2009; Q. Wang 2009;) and the cited works from the other groups..

## 2. Synthesis of SiC powders

Different carbon sources were selected to investigate their effects on the particle size and morphologies of the synthesized SiC powders, detail of which is mentioned below.

### 2.1 Experimental

Silicon powders and different carbons (their chemical and physical properties are listed in Table 1) were mixed as starting materials at the Si/C mol-ratio of 1/1. The mixtures were well homogenized by ball-milling in a polyethylene milling jar with alumina balls, ratio of balls to powders was 3/1, with the ball milling time of 4 h, and alcohol was used as the dispersant. The prepared mixtures were dehydrated at 110 °C for 24 h. The obtained powders were charged into alumina crucible, and then heated in a 2.45 GHz microwave filed in flowing Ar atmosphere with pressure of 0.1 MPa. The schematic diagram of heating system is shown in Fig. 1. The output microwave power is adjusted automatically ranged from 0~4 kW. The as-prepared products were decarbonized at 650 °C for 2 h.

| powder | element (wt %) | | | | | | | $O_2$ | BET | grain size |
|---|---|---|---|---|---|---|---|---|---|---|
| | Si | Fe | Al | Ca | Sn | C | S | (wt %) | ($m^2/g$) | (μm) |
| silicon | 99.1 | 0.18 | 0.11 | 0.04 | 0.20 | - | - | 2.17 | 1.32 | < 44 μm |
| activated carbon | - | - | - | - | - | >99.5 | - | - | - | 1~30 μm |
| graphite | - | - | - | - | - | >99.5 | - | - | - | 1~20 μm |
| carbon black | - | - | - | - | - | 96.2 | 3.8 | 6.28 | 76 | 0.5 μm |

Table 1. Chemical composition and physical properties of silicon and carbon black

### 2.2 Synthesis of $SiC_p$ by silicon and activated carbon powders

Silicon powders and activated carbon are used as starting materials to synthesize SiC powders whose detail is given in the incoming lines.

The SEM images of the starting activated carbon, silicon powders and their mixtures after ball milling are shown in Fig. 2. Activated carbon particles are irregular in shape and wide in particle size distribution. The size was distributed from about 1 to 30 μm. While that of the silicon particles with sharp angle is about 50 μm. After ball milling, the shape of the mixture powders is still irregular with sharp angles and edges and its size has been reduced to less than 20 μm.

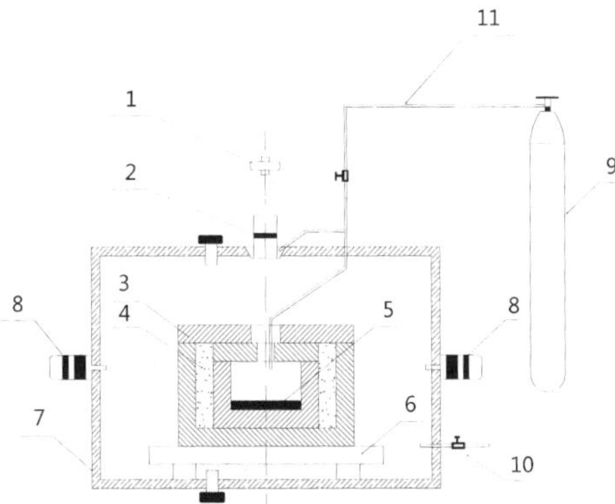


Fig. 1. Schematic diagram of the microwave furnace. 1- infrared radiation thermometer; 2- silica glass; 3- porous alumina crucible; 4- alumina fiber; 5- reactants; 6- alumina plate; 7- stainless steel chamber; 8- microwave generator; 9- argon gas tank; 10- outlet valve; 11- inlet valve.

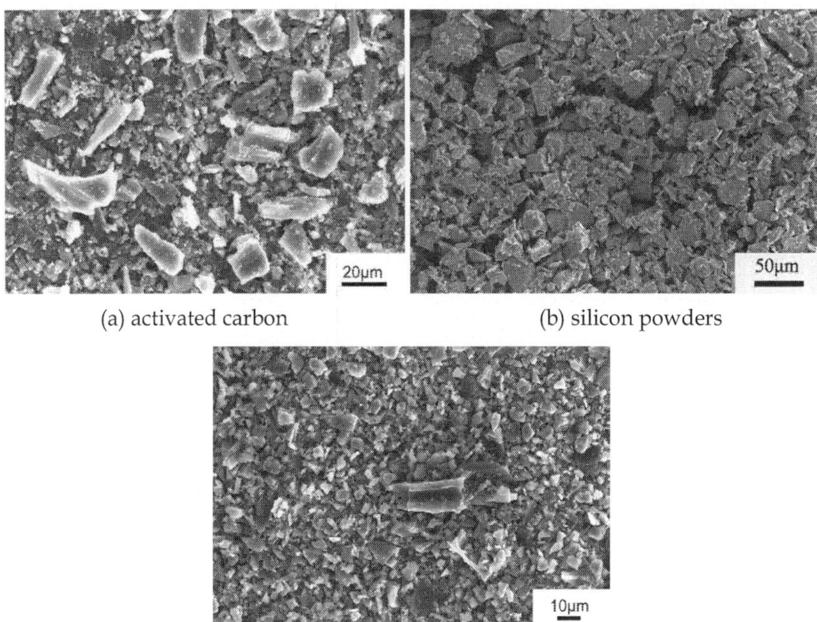


(a) activated carbon (b) silicon powders

(c) activated carbon mixed with silicon powders by ball milling

Fig. 2. SEM images of the starting materials

The heating temperature has been varied from 900 to 1200 °C (100 °C interval) and holding for 5 to 30 min. The curve of the heating temperature vs. heating time is shown in Fig. 3.

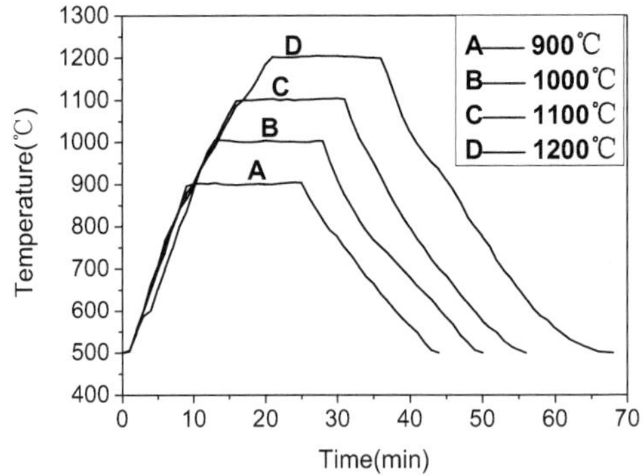


Fig. 3. Heating temperature (°C) vs. time (min)

### 2.2.1 Identification of the phase compositions of the synthesized products

Schematic composition structure of the synthesized product is illustrated in Fig. 4. After completion of the reaction, the color of the top and inner parts of the synthesized product was changed into dark green (part A in Fig. 4), however that of the parts closed to the wall and bottom of the crucible was still black (part B in Fig. 4).

The X-ray diffraction peaks reveal that the black parts are unreacted silicon and carbon black. Microwave heats from inside of the reactants and is driven to the surface. This is reverse of the conventional heating, where the surface of the reactants is heated first (Metaxas, 1991; Shawn et al., 2008). Heat loss will happen at the sidewall and the bottom of the crucible due to the heat transfer, which further decreases the heating temperature and leads to the uncompleted reaction. However, the heat will transfer upwards and thus the reaction temperature difference between that of the top part and the inner part is very small. Then the reaction on the top surface also can complete. Thus, only the inner part of the products was used for further characterization in the later experiment.

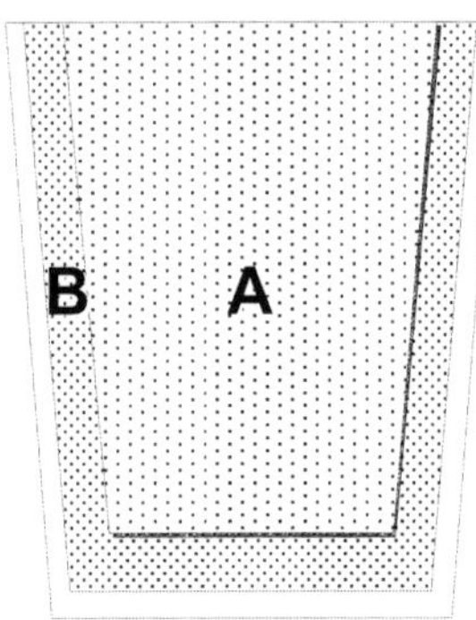


Fig. 4. Schematic composition structure of the product synthesized in the crucible

Figure 5 shows the XRD patterns of the products synthesized at different heating temperatures holding for 15 min. In the case of 900 °C, β-SiC was formed but the reactant Si still could be identified in the products due to the uncompleted reaction. The content of β-SiC in the products was increased with the increase of the heating temperature while that of the unreacted Si decreased. When the heating temperature was increased to 1200 °C, pure β-SiC powders were synthesized successfully.

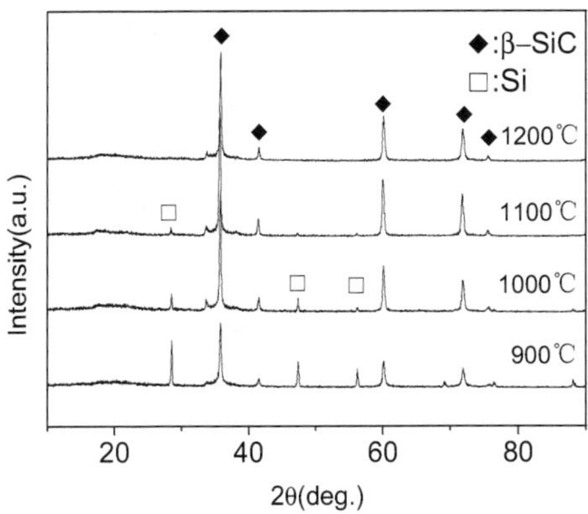


Fig. 5. XRD patterns of the products synthesized at different heating temperatures holding for 15 min

The effect of holding time on the synthesis of SiC at 1200 °C was investigated. The XRD patterns of the products synthesized at 1200 °C holding for different time are shown in Fig. 6. When the holding time was set as only 5 min, no unreacted Si could be detected in the products, which means the reaction has proceeded completely. The crystallinity and particle size will be further increase with the increase of holding time according to the peak intensity. It can be concluded that pure phase β-SiC could be synthesized at 1200 °C holding for only 5 min.

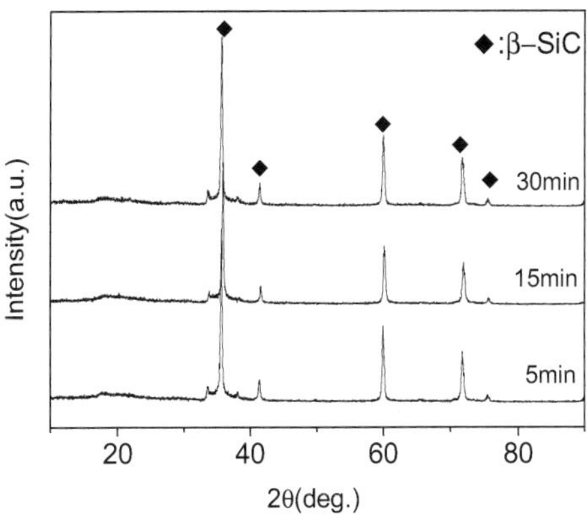


Fig. 6. XRD patterns of the products synthesized at 1200 °C holding for different time

### 2.2.2 Effect of reaction temperature on the morphology of the synthesized products

The effect of reaction temperature on the morphology of the products has been investigated. The SEM images of the products prepared at different reaction temperatures holding for 15 min are shown in Fig. 7. In the case of 900 °C, similar morphology with that of the starting materials can be seen in the products due to the uncompleted reaction. There are still some unreacted silicon and carbon particles, which is consistent with that from the X-ray diffraction analysis. When the reaction temperature was increased to above 1000 °C, there were large number of agglomerated SiC particles in the products and a little unreacted silicon particles could be found. Agglomerated pure SiC particles with size of 1~10 μm have been synthesized when the reaction temperature was increased to 1200 °C.

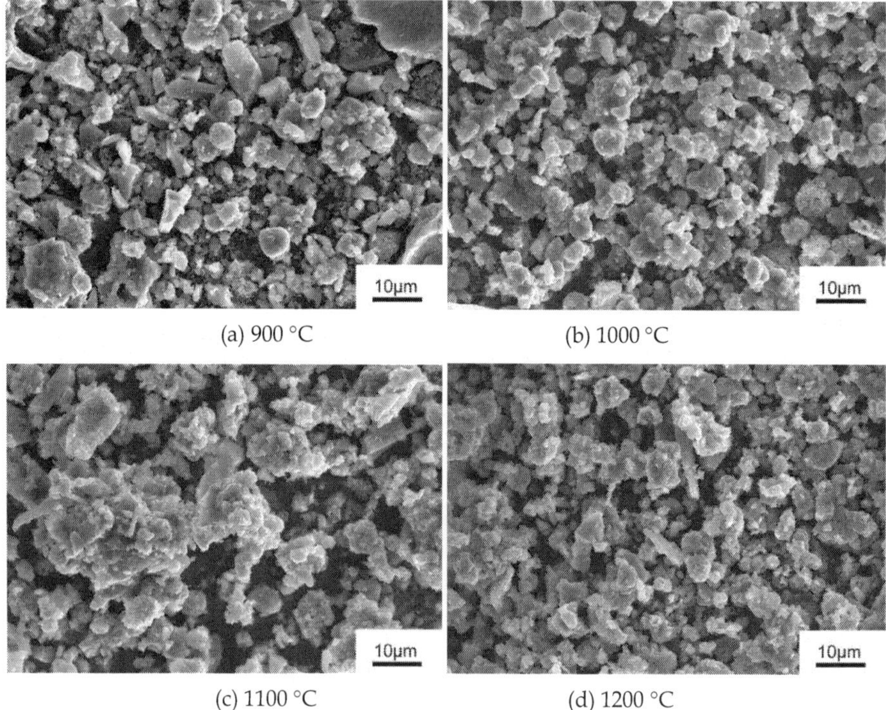


(a) 900 °C (b) 1000 °C

(c) 1100 °C (d) 1200 °C

Fig. 7. SEM images of products synthesized at different heating temperatures holding for 15 min

### 2.2.3 Effect of heating time on the morphology of the synthesized products

The SEM images of the products prepared at 1200 °C holding for different time are shown in Fig. 8. The shape of the SiC particles tends to be equiaxial with the increase in the holding time. When holding for 30 min, equiaxial particles with size of about 1~10 μm were synthesized.

(a) 5 min (b) 15 min

(c) 30 min

Fig. 8. SEM images of the products synthesized at 1200 °C holding for different time

### 2.3 Synthesis of SiC*p* by using silicon and graphite powders

Figure 9 shows the SEM images of the graphite and the as-milled mixture powders. The average size of the original graphite particles is 1~20 μm and that of the mixture was decreased to about ~10 μm after ball milling. Both the original and the ball-milled particles have sharp angles and edges due to their brittle properties.

(a) (b)

Fig. 9. SEM images of (a) graphite and (b) mixtures of graphite and silicon powders

### 2.3.1 Identification of phase compositions of the synthesized products

The XRD patterns of the products prepared at various heating temperatures holding for 15 min are shown in Fig. 10. All the peaks are corresponding to those of the β-SiC. Pure phase β-SiC can be synthesized when the heating temperature was as low as 900 °C. While with the further increase of the heating temperature, the peak intensity of the synthesized SiC increased, which was assigned to the increase in grain size and crystallinity. It was found that the heating temperature has positive impact on the reaction. Higher heating temperature benefits the progress of reaction. Figure 11 shows the XRD patterns of products reacted at 1200 °C holding for different heating time. The results show that even the holding time was as less as 5 min, the reactants could be completely converted into pure β-SiC. And the further increase in the holding time gives the increase in the crystallinity and grain size.

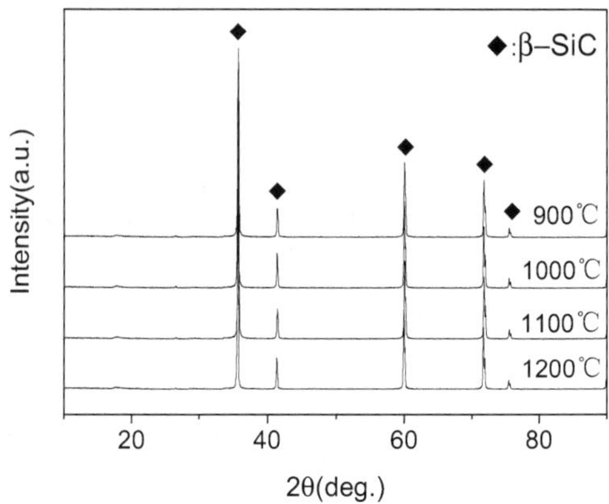


Fig. 10. XRD patterns of the products synthesized at different heating temperatures holding for 15 min

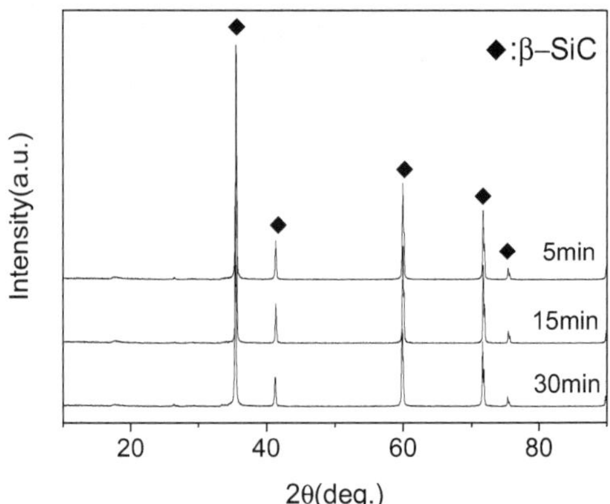


Fig. 11. XRD patterns of the products synthesized at 1200 °C holding for different time

### 2.3.2 Effect of reaction temperature on the morphology of the synthesized products

The SEM images of the products prepared at different heating temperatures holding for 15 min are shown in Fig. 12. It is clear that the particle size of the synthesized SiC increased with the increase of the heating temperature. The synthesized particles tend to be equiaxial and have a narrower size distribution with the increase of the heating temperature. In the case of 1200 °C, equiaxed SiC powders with averaged size of about 5 μm were prepared.

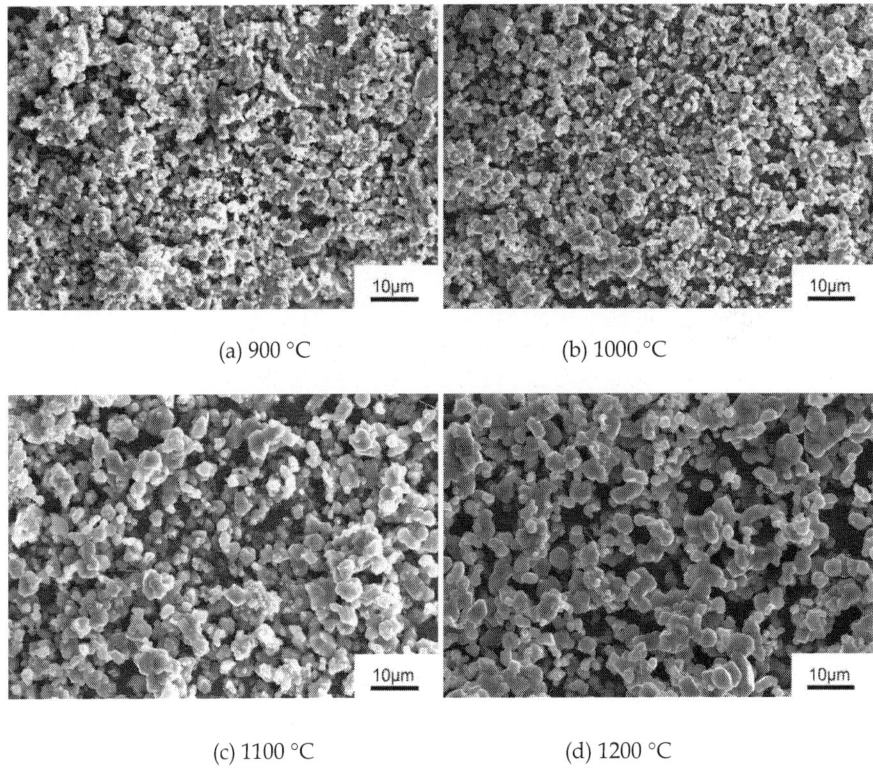


(a) 900 °C (b) 1000 °C

(c) 1100 °C (d) 1200 °C

Fig. 12. SEM images of the products synthesized at different heating temperatures holding for 15 min

### 2.3.3 Effect of heating time on the morphology of the synthesized products

The SEM images of products prepared at 1200 °C holding for different time are shown in Fig. 13. Uniform equiaxial SiC powders with increased particle size and narrower size distribution have been synthesized with the increase of the holding time. In the case of holding for 30 min, equiaxed SiC powders with a size of about 5~10 μm were synthesized.

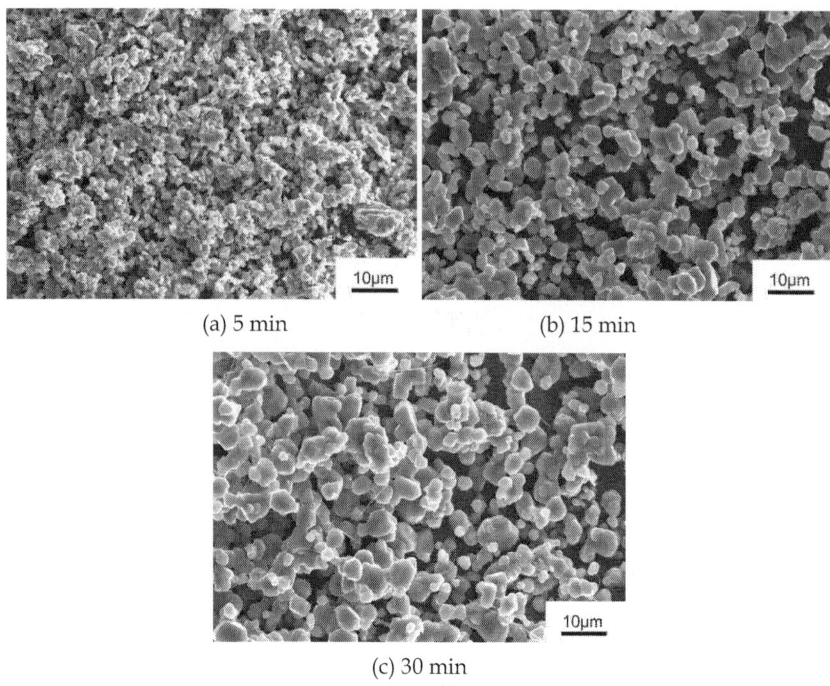


(a) 5 min (b) 15 min

(c) 30 min

Fig. 13. SEM images of the products synthesized at 1200 °C holding for different time

### 2.4 Synthesis of SiC*p* using silicon powders and carbon black

The SEM images of the carbon black and the mixtures with silicon after ball milling are indicated in Fig. 14. The averaged particle size of the carbon black is about 500 nm. After ball milling, the averaged particle size of the silicon in the mixtures has been decreased to about 10 μm. The silicon particles, which still have sharp angles, have been dispersed homogeneously in the carbon black after ball milling.

(a) (b)

Fig. 14. SEM images of (a) carbon black and (b) mixtures of carbon black and silicon powders

#### 2.4.1 Identification of phase compositions of the synthesized products

Figure 15 shows the XRD patterns of the products synthesized at various heating temperatures holding for 30 min. β-SiC starts to form at the heating temperature of 900 °C. There are still some unreacted silicon can be detected in the products. When it was increased to 1100 °C, all the diffraction peaks are indexed to β-SiC, which means that pure phase β-SiC was prepared at 1100 °C for 30 min. Further increasing the heating temperature to 1200 °C, the increased intensity of the diffraction peaks was attributed to the grain growth and improved crystallinity.

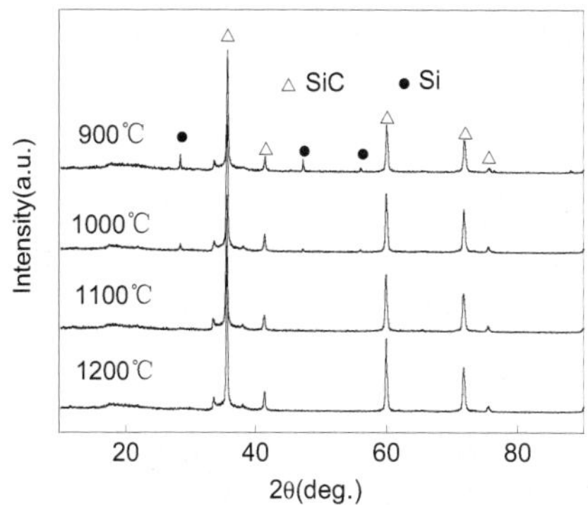


Fig. 15. XRD patterns of the products synthesized at various heating temperatures holding for 30 min

XRD patterns of the products synthesized at 1200 °C holding for various time are shown in Fig. 16. The major composition in the product was β-SiC and a small amount of unreacted silicon was also detected when heated for 5 min. Pure phase SiC was obtained when the holding time was set as 15 min and 30 min.

Thus it is concluded that both increase in heating temperature and holding time can promote the formation of β-SiC.

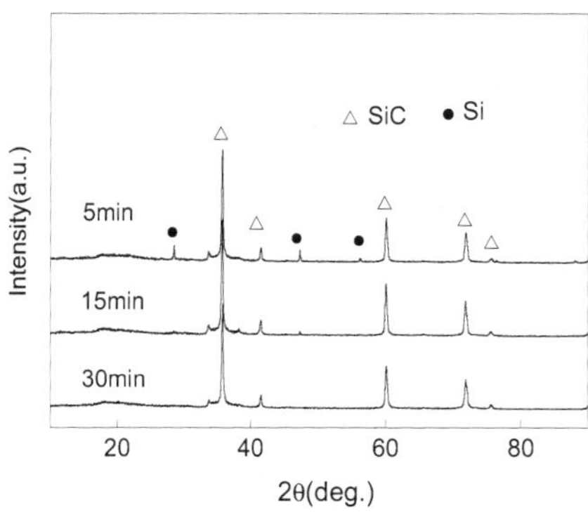


Fig. 16. XRD patterns of the products heated at 1200 °C for various holding time

#### 2.4.2 Effect of reaction temperature on the morphology of the synthesized products

Figure 17 shows that the SEM images of the products synthesized at various heating temperatures holding for 30 min. It is clear that all the products were composed of particles with irregular shapes. In the case of 900 and 1000 °C, the particle size was about several hundred nanometers. With the increase in heating temperature to 1100°C an obvious grain growth was found, further increase to 1200°C gives particles having size of 0.5~2 μm.

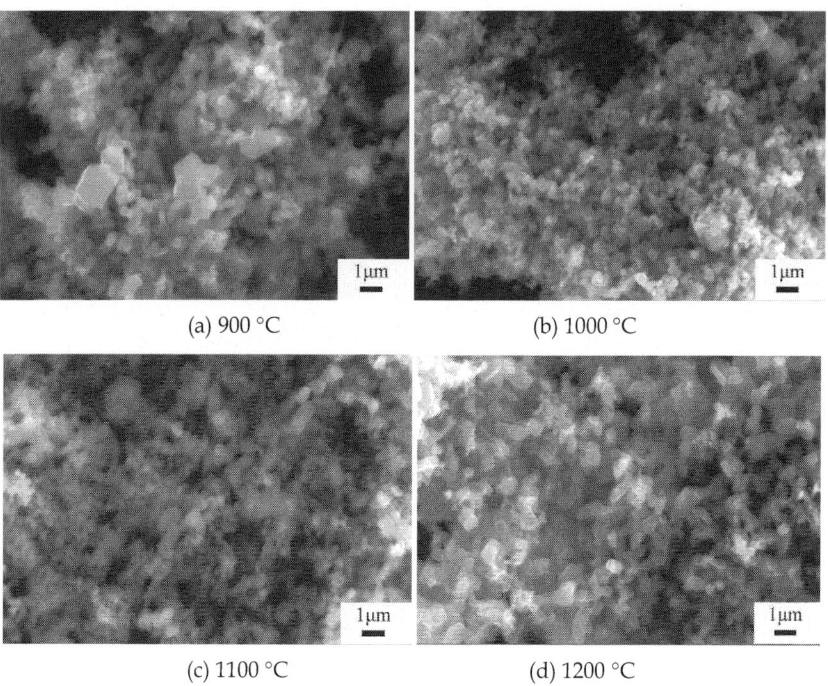


(a) 900 °C (b) 1000 °C

(c) 1100 °C (d) 1200 °C

Fig. 17. SEM images of the products synthesized at various heating temperatures holding for 15 min

#### 2.4.3 Effect of the heating time on the morphology of the synthesized products

Figure 18 shows the SEM images of the products prepared at 1200 °C for various holding time. It can be seen that when the holding time was increased from 5 min to 30 min the particle size of the products rapidly increased from 0.1~1 μm to 1~3 μm, and became narrower in particle size distribution. The above results indicate that increasing heating temperature and extending holding time can also promote the growth of SiC particles.

Under high-temperature SiC sublimates to form Si, $Si_2C$ and $SiC_2$ gas, and the crystal growth depends on the coacervation of these gaseous phases. These gas phases transfer from high-temperature part to low-temperature part and then condense on the surface of SiC crystal; or directly condense on the nearby larger crystal and promote the growth of crystal. The relationship between saturated vapor pressure, surface tension and radius of curvature is shown in Eq. (1) (Young-Laplace equation). The smaller crystal has the larger saturated vapor pressure which means that under the same vapor pressure, it is saturated for the large crystal, but it must be under saturated for the smaller one. Based on this phenomenon, with the increase in the reaction time, the smaller crystal will evaporate continuously until it disappears, however the larger crystal will continue to grow to form large crystal(Tairov, 1995).

$$P = 2\gamma / R \tag{1}$$

(a) 5 min (b) 15 min

(c) 30 min

Fig. 18. SEM images of the products synthesized at 1200 °C holding for various time

#### 2.4.4 Effect of the silicon particle size on the morphology of the synthesized products

There are two mechanisms about the effect of particle size of the reactants on the morphology of the synthesized SiC. The first one is dissolution-precipitation mechanism which considers that the morphology of the synthesized SiC is determined by that of silicon(Yamada et al., 1989), and the second is so called diffusion mechanism which believes that the morphology of the SiC is controlled by carbon(Pampuch et al., 1987; Narayan et al., 1994).

Silicon powder with particle size of ~44 μm and 15 μm were tried with the same kind of carbon black to synthesize SiC powders at 1200 °C for holding 15 min. SEM images of the silicon powders, carbon black (500 nm) and the corresponding products are shown in Fig. 19. It is clear that the particle size of the as-prepared SiC was about 1~2 μm in both cases, which suggests that the morphology of the SiC was dominated by that of carbon by diffusion mechanism as the size was much more close to that of the carbon black.

(a) ~44 µm (b) ~2 µm (c) ~15 µm (d) ~2 µm

Fig. 19. SEM images of (a) ~44 µm and (c) ~ 15 µm silicon powders; products prepared by (b) ~44 µm and (d) ~15 µm silicon powders and acetylene carbon black at 1200 °C for 15 min

### 2.5 Synthesis of SiC by quartzite and coke powders

In order to further reduce the cost of the synthesis of SiC, quartzite and coke were selected as raw materials whose chemical compositions are listed in Table 2 and Table 3, respectively. The proportion of reactant is strictly based on Eq. (2). There were two kinds of particle size of coke, less than 20 µm and 100~200 µm, each one mixed with ~1 µm quartzite, and then pressed under 15 MPa to form compacts. The heating temperature was varied from 1300 to 1600 °C, holding time changed from 15 to 60 min and the heating rate was set as 50 °C/min.

| Composition | $SiO_2$ | $Fe_2O_3$ | $Al_2O_3$ | CaO | MgO | Burn reduces |
|---|---|---|---|---|---|---|
| Content (%) | >98.8 | <0.3 | 0.24 | <0.16 | <0.2 | <0.2 |

Table 2. Chemical composition of quartzite

| Composition | Fixed carbon | Ash | Volatile | Water |
|---|---|---|---|---|
| Content (%) | 82.5 | 15.7 | 1.1 | 0.8 |

Table 3. Chemical composition of coke

$$SiO_2(s) + 3C(s) = SiC(s) + 2CO(g) \quad (2)$$

### 2.5.1 Identification of phase compositions

The coke with the particle size of less than 20 μm was selected to investigate the effect of reaction temperature and holding time on the phase compositions. Figure 20 shows the XRD patterns of products heated at various temperatures holding for 30 min. In the case of 1300 °C, $SiO_2$ is the major composition and very small amount of β-SiC could be identified; with the increase in reaction temperature, the relative amount of β-SiC increased, and that of $SiO_2$ decreased; when the reaction temperature was 1600 °C holding for 30 min, pure phase β-SiC was synthesized.

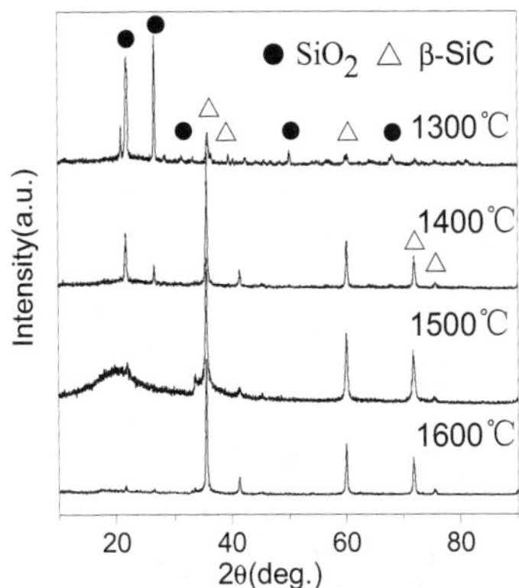


Fig. 20. XRD patterns of products synthesized at various temperatures holding for 30 min

Holding time is one of the most important factors which affect the conversion of the SiC. Different holding times have been tried to investigate their effect on the progress of reaction. XRD patterns of the products synthesized at 1500 °C holding for various time are shown in Fig. 21.

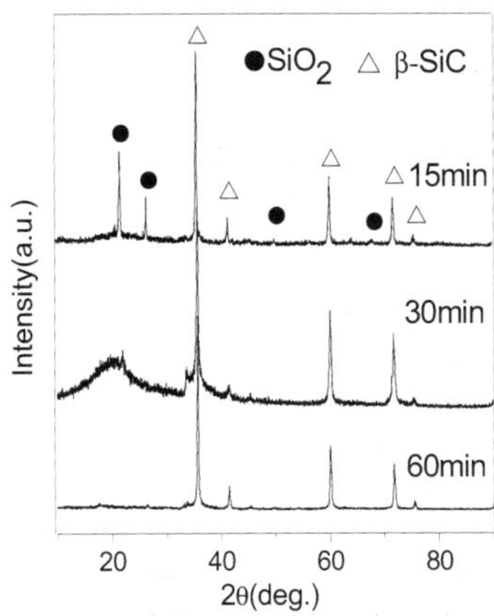


Fig. 21. XRD patterns of products synthesized at 1500 °C holding for various time

It is clear that when the holding time was set as 15 or 30 min, there are still some unreacted $SiO_2$ could be detected in the products. But the content of $SiO_2$ was decreased with the increase of the holding time, which suggests that the holding time should be further increased. At the case when the holding time was increased to 60 min, pure β-SiC has been synthesized by XRD. This heating temperature and holding time are remarkably lower than that of the conventional Acheson process.

### 2.5.2 Effect of coke particle size on the morphology of SiC

The effect of the particle size of coke on the morphology of the synthesized products has been investigated. Figure 22 shows SEM images of coke particles with size of less than 20 μm and 100~200 μm, and the products prepared by using them with ~1 μm quartzite heated at 1600 °C for holding 30 min. The synthesized SiC has almost the similar morphology and particle size as those of the coke. Size distribution of SiC powders prepared from coke with various particle size is shown in Table 4. The D50 value of SiC is very close to that of the corresponding coke, which was considered that the morphology and size of the SiC were controlled by those of coke through the diffusion mechanism(Pampuch et al., 1987).

Thus it can be concluded that in the microwave synthesis process the morphology and particle size of the synthesized SiC*p* could be easily controlled by adjusting the corresponding coke size.

(a) ~20 μm coke (b) ~20 μm SiC

(c) 100~200 μm coke (d) 100~200 μm SiC

Fig. 22. SEM images of (a) 20 μm and (c) 100~200 μm coke powders; products prepared by (b) 20 μm and (d) 100~ 200 μm silicon powders and carbon black at 1600 °C for 30 min

| No. | Temperature/ °C | Holding time/min | D50 of coke/μm | D50 of product/μm |
|---|---|---|---|---|
| 1 | 1600 | 30 | 13.795 | 8.604 |
| 2 | 1600 | 30 | 154.235 | 156.907 |

Table 4. Size distribution of SiC prepared with various grain size of coke

Except our works, the data from the researches on the microwave synthesis of SiC*p* by using other silicon and carbon sources are listed in Table 5. These works also reveal that microwave heating can lower the synthesis temperature and reduce the reaction time.

| silicon source | Carbon source | Form temperature | Complete reaction | products | morphology |
|---|---|---|---|---|---|
| silicon powder[1] | activated charcoal | < 977 °C | - | β-SiC | - |
| precipitated silica[2] | char | 1180 °C for 20 min | - | β-SiC and a little α-SiC | 3~80 nm |
| | resin | | | | |
| | carbon black | | | | |
| silicon powder[3] | amorphous carbon powder | 900 °C for 30 min | 1200 °C for 15 min | agglomerated β-SiC particles | ~20 μm |
| silica gel[4] | graphite and carbon black | 1225 °C | 1454 °C | equiaxed β-SiC particles | 13.6 and 58.2 nm |
| silicon powder[5] | carbon Black | - | 1200 °C for 15 min | homogeneous β-SiC particles | 1 μm |

Table 5. Synthesis of SiC*p* by microwave heating. 1-(Ramesh et al., 1994); 2-(Dai et al., 1997b); 3-(Satapathy et al., 2005); 4-(Ebadzadeh & Marzban-Rad, 2009); 5-(Jin et al., 2010).

## 3. Preparation of SiC whiskers

### 3.1 Preparation of SiC*w* by silica-sol and activated carbon

As shown in Fig. 23, the temperature of the inner part of the reactants is always higher than that of other parts under microwave irradiation, which means the reaction will be initiated

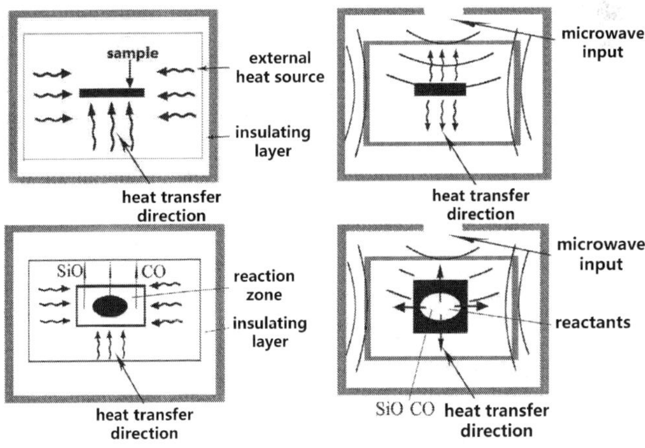


Fig. 23. Conventional heating vs. microwave heating

firstly inside the reactants, results in the reduce of the loss of the produced gaseous intermediates. Then the gaseous intermediates could be maintained with a higher concentration in the inner part, which benefits the synthesis of SiC*w* by VS or VLS mechanisms and the increase of the SiC*w* productivity.

This section will discuss the synthesis of SiC*w* by using silica-sol as silicon source and activated carbon (or carbon black) as carbon source by microwave heating without the use of any catalysts, such as Fe, Ni, NaF, etc..

### 3.2 Experimental

Due to the ultrafine silica particles dispersed in silica sol could be coated well on the surface of the carbon particles, thus the silica and carbon could be uniformly mixed. So silica sol is considered as one of the excellent silica sources for the synthesis of SiC*w*. The content, particle size of $SiO_2$ in silica sol is about 30% and 10~20 nm, respectively. Certain amount of silica sol and activated carbon were mixed and stirred for 2 h as starting materials. The Si/C molar ratio was set as 1/4. 1 *wt*% sodium hexametaphosphate was used as the dispersant for better mixing. The prepared mixtures were dehydrated at 110 °C for 24h, and then crushed and screen separated into three kinds of powders with size distribution of 200-300 μm, 30-80 μm and 1-10 μm. The powders were charged into alumina crucibles and heated in a microwave furnace. The heating temperature varied from 1200 to 1600 °C (100 °C interval), while holding time changed from 5 to 60 min. The heating temperature vs. heating time is shown in Fig. 24. The as-prepared products were decarbonized in air at 650 °C for 2 h.

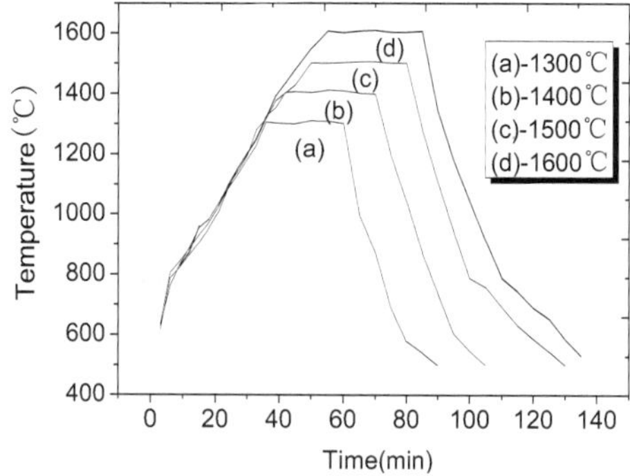


Fig. 24. Heating temperature (°C) vs. heating time (min)

### 3.3 Identification of phase compositions

The 1-10 μm mixtures were used to investigate the effect of processing parameters on the phase compositions of the synthesized products. The XRD patterns of different parts of the product in the crucible heated at 1500 °C for 15 min were shown in Fig. 25. The patterns show that different parts of the products were composed with different compositions. The bottom part consists of unreacted $SiO_2$ and very small amount of β-SiC; the inner part consists of pure phase β-SiC without any $SiO_2$; while the top part consists of β-SiC and very small amount of $SiO_2$. This phenomenon is similar with that of the synthesis of SiC*p* as mentioned before. Thus, only the inner parts of the products synthesized under different conditions were selected for further characterization in the followed investigates.

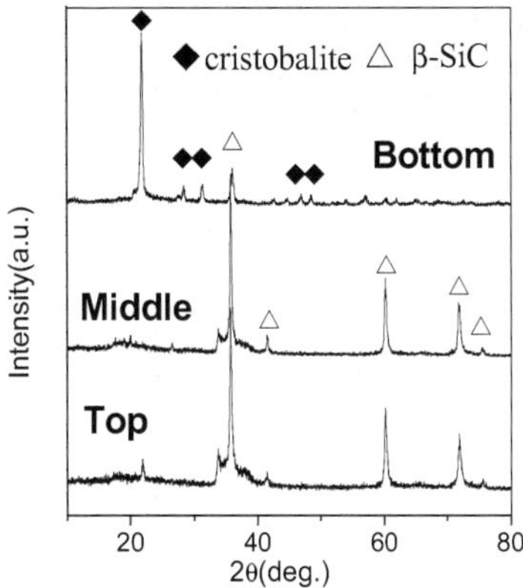


Fig. 25. X-ray diffraction patterns of the different parts of the products synthesized at 1500 °C for 15 min

The X-ray diffraction patterns of the products heated at different temperatures for 15 min are shown in Fig. 26. In the case of 1200 °C, $SiO_2$ is the main composition phase and very small amount of β-SiC could be detected, which shows that the reaction rate is slow and it is hard to synthesize SiC at this reaction temperature by microwave heating. The peak intensity of β-SiC increased and that of $SiO_2$ decreased with the increase of reaction temperature. Pure β-SiC could be prepared at 1500 °C holding for 15 min. While when the reaction temperature was increased to 1600 °C, the crystallinity of β-SiC has been remarkably increased.

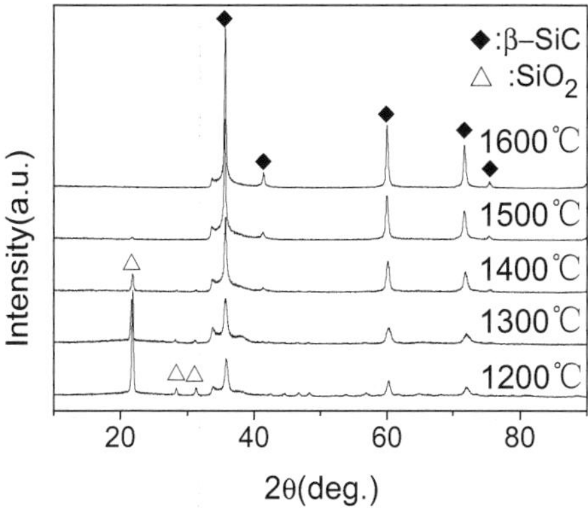


Fig. 26. XRD patterns of the products synthesized at various temperatures holding for 15 min

The effect of the holding time on the synthesis of SiC at 1500 °C has been investigated. Figure 27 shows the phase compositions of the products heated at 1500 °C holding for 5, 10, 15, and 30 min, respectively. The content of $SiO_2$ decreased with the increase of the holding time. It is impossible to complete the reaction when holding for less than 15 min. The pure phase β-SiC could be synthesized when holding for at least 15 min. The further increase of the holding time will improve the crystallinity.

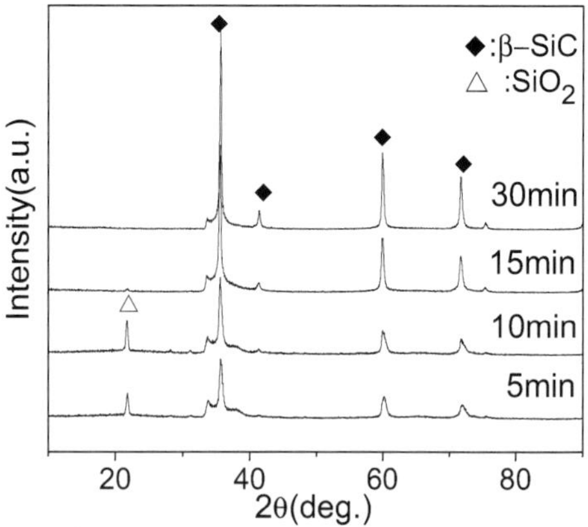


Fig. 27. XRD patterns of the products synthesized at 1500 °C holding for different time

It can be concluded from the above results that carbothermal reduction could be promoted by increasing either the heating temperature or the holding time.

### 3.4 Effect of reaction temperature on the morphology of the products

Figure 28 shows the SEM images of the products synthesized at various temperatures holding for 15 min. It is clear that there was only very small amount of SiC*w* could be observed in the images of the product synthesized at 1200 °C. When the heating temperature was increased from 1200 °C to 1500 °C, the amount of SiC*w* increased gradually. However at the case of 1600 °C, the product is almost composed of SiC particles. This could be explained by the VS formation mechanism of SiC*w*. In the carbothermal reduction the formation of SiC mainly depends on Eq. (3)~(5). According to Eq. (3), SiO and CO will be formed firstly, and then the formed SiO will react with C or CO to form SiC. The products synthesized through Eq. (4) will mainly be SiC particles, and the products synthesized through the reaction Eq. (5) will primarily be composed of SiC whiskers. When the heating temperature is higher than 1300 °C and the partial pressure of CO is higher than 0.027 MPa, the reaction Eq. (5) will dominate the process to synthesize SiC*w*. When the reaction temperature is too high, violent reaction rate will result in excessive production of SiO and CO, thus SiC will grow in several directions to form SiC particles. So, 1500 °C is the advisable reaction temperature while the temperatures over that are disadvantage for the synthesis of SiC*w* via microwave heating (Sharma et al., 1984; Biernacki & Wotzak, 1989; Chrysanthou et al., 1991).

$$SiO_2(s) + C(s) = SiO(g) + CO(g) \tag{3}$$

$$SiO(g) + 2C(s) = SiC(s) + CO(g) \tag{4}$$

$$SiO(g) + 3CO(g) = SiC(s) + 2CO_2(g) \tag{5}$$

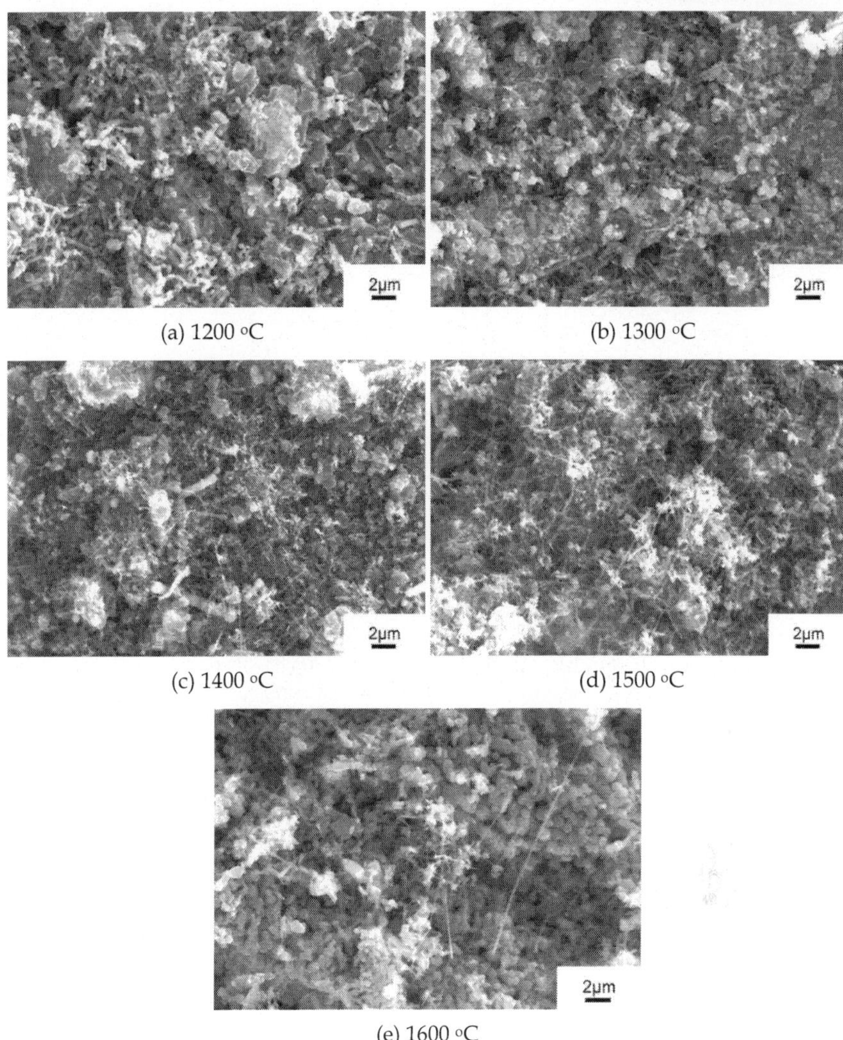


(a) 1200 °C (b) 1300 °C

(c) 1400 °C (d) 1500 °C

(e) 1600 °C

Fig. 28. SEM images of products synthesized at different heating temperatures holding for 15 min

### 3.5 Effect of reaction time on the morphology of the products

Figure 29 shows the SEM images of the inner part of the products synthesized at 1500 °C for various holding time. The images show that SiC*w* could be synthesized in all cases, and the amount of SiC*w* increased with the increase in holding time. When the holding time was set as 30 min, SiC*w* has the maximum aspect ratio of 100:1. The results reveal that increasing the holding time is benefit for the growth of SiC*w*.

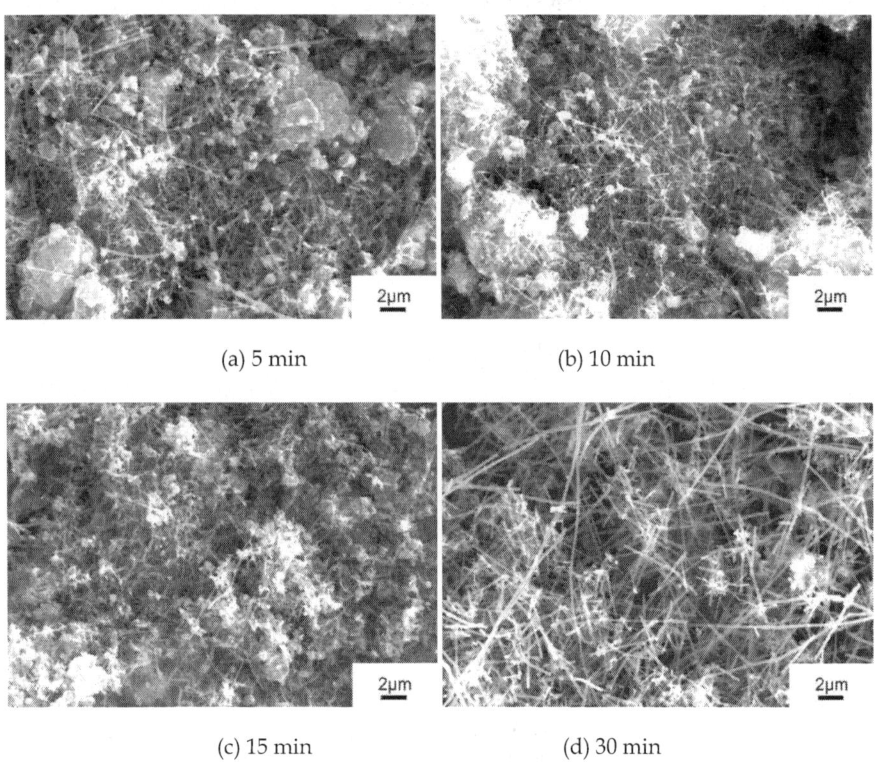


(a) 5 min (b) 10 min

(c) 15 min (d) 30 min

Fig. 29. SEM images of products heated at 1500 °C holding for various time

### 3.6 Effect of reaction temperature and holding time on the yield of SiC*w*

The yield of SiC*w* synthesized at various temperatures and time were measured and listed in Table 6. With the same holding time of 30 min, the yield of SiC*w* prepared at 1300 and 1400 °C are very low because lower heating temperature inhibits the progress of the reaction. The yield reached 56.9% when the temperature was increased to 1500 °C and it will hit 60.1 wt% if the holding time was extended to 60 min. While when the heating temperature was further increased to 1600 °C, the yield was decreased to 28.7wt% due to the formation of equiaxial SiC*p*. On the other hand, with the decrease in the particle size of raw materials, the yield of SiC*w* was increased which is shown in Table 6. The results indicate that, the product synthesized at 1500 °C for 60 min has the maximum yield of SiC*w*, but holding for 30 min is more economic in our cases.

Figure 30 shows the SEM images of the products prepared at 1500 °C for 30 min, by using raw materials with particle size of 30~80 and 200~300 μm, respectively. It can be seen that there were many large grains of SiC have been synthesized in the products, and no SiC*w* could be observed on the surface of them. As the particle of raw materials is relatively large, it is hard to form SiO and CO and put the formation of SiC*w* at a disadvantage.

| Heating temperature / °C | Holding time / min. | Particle size of raw materials / μm | Yield of SiC*w* / weight % |
|---|---|---|---|
| 1300 | 30 | 1-10 | 4.8 |
| 1400 | 30 | 1-10 | 7.1 |
| 1500 | 30 | 1-10 | 56.9 |
| 1600 | 30 | 1-10 | 28.7 |
| 1500 | 15 | 1-10 | 30.3 |
| 1500 | 60 | 1-10 | 60.1 |
| 1500 | 30 | 30-80 | 34.6 |
| 1500 | 30 | 200-300 | 28.3 |

Table 6. The yield of SiC*w* prepared at various conditions

(a)30~80 μm (b) 200~300 μm

Fig. 30. SEM image of products prepared by using different grain size of raw materials

### 3.7 Effect of carbon particle size on the morphology of the products

In order to investigate the effect of particle size of carbon on the synthesis of SiC*w*, the carbon black (500 nm) was selected instead of activated carbon as carbon source, mixed with silica sol to prepare SiC*w*. The preparation process is the same with that when using activated carbon. The phase compositions and morphology of product synthesized at 1500 °C for 30 min are shown in Fig. 31. The obtained product was β-SiC however its morphology was different from that prepared from the activated carbon-silica sol system. The product is composed of equiaxed particles of SiC with a narrowed particle size distribution, and SiC*w* could not be observed by SEM.

The carbothermal reduction process between the nano $SiO_2$ and the carbon black could be illustrated by Fig. 32 according to the literature (Wang et al., 1992). Firstly, SiO and CO were formed in the carbothermal reduction, and the formation of SiC was mainly depended on the gas-solid reaction between SiO and C, because its Gibbs free energy was lower than that of the reaction between SiO and CO. SiC formed on the surface of the carbon particles and the formed SiC layer hindered the diffusion of carbon atom and then the reaction rate between SiO and C was decreased. When the carbon particle was ultrafine (500 nm) the formed SiC layer was also thin, thus the carbon atom could diffuse through this layer and react with SiO to form SiC, thus, the morphology of as-prepared SiC is similar to carbon particle. As the size of carbon particle was several microns, thicker SiC layer would form, which hindered the diffusion of carbon atom and the reaction between SiO and C. In this case the formation of SiC was depending mainly on the gas-solid reaction between SiO and CO. Generally the product of this reaction was believed to synthesize one-dimension SiC, such as whisker, nanowire, and nanorod by VS mechanism.

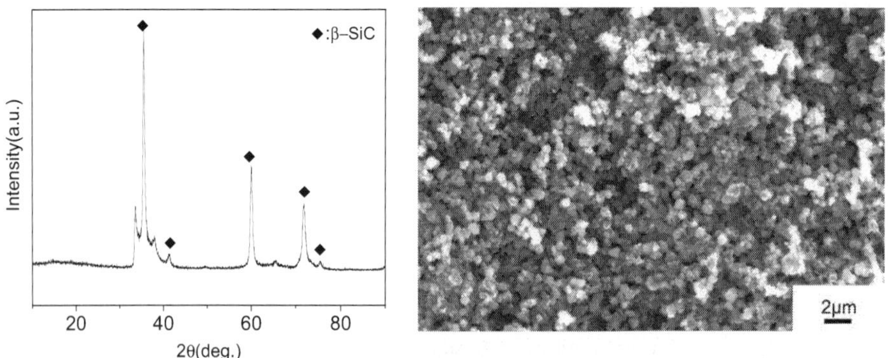


Fig. 31. XRD pattern (left) and SEM image (right) the product synthesized at 1500 °C for 30 min by using carbon black as carbon source

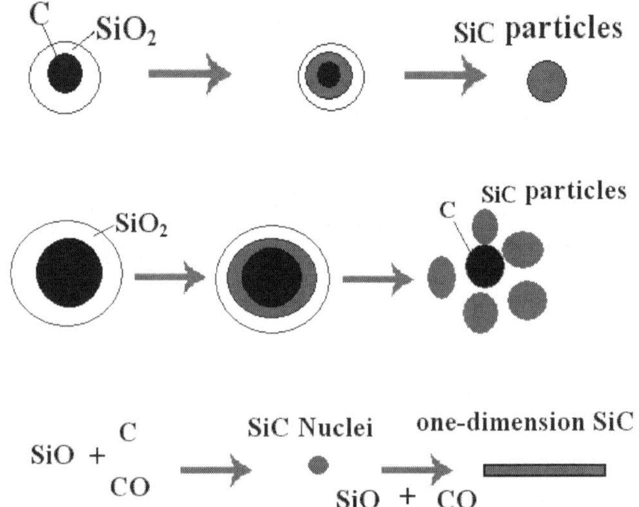


Fig. 32. Schematic diagram of carbothermal reduction

Except for silica sol-activated carbon and silica sol-carbon black systems we used, other works that using different silicon and carbon source to prepare SiC*w* by microwave heating have been published, and some of them are listed in Table 7. No data on yield of SiC*w* has been mentioned in their papers.

| silicon source | carbon source | preparation method | products | morphology |
|---|---|---|---|---|
| precipitated silica[1] | char carbon black | 1180, 1240 and 1350 °C for 20 min | β-SiC and a little α-SiC | *D*: 20 ~ 50nm *L*: several microns |
| silicon powder[2] | phenolic resin | 1300~1400 °C for 0.5-2h | SiC | *D*: 20 ~ 100nm *L*: tens microns |
| tetraethylorthosilicate (TEOS) [3] | sucrose | > 1300 °C for 30 min | 6H-SiC | *D*: 5~200 nm *L*: tens to hundreds of microns |
| quartz plate[4] | methane | 1050~1200 °C for 60 min | β-SiC | *D*: 100nm *L*: several tens of microns |

Table 7. Synthesis of SiC*w* by microwave heating. 1-(Dai et al., 1997a); 2-(Lu et al., 2007); 3-(Wei et al., 2008); 4-(Fu et al., 2009).

**3.8 Effect of packing density of the reactants on the morphology of the products**

It is thought that the growth of SiC*w* by VLS or VS mechanism will firstly start in the free space in the reactants. So the porosity in the reactants will greatly affect the morphology and productivity of SiC*w*.

The mixture powders of silica sol and activated carbon were well homogenized by ball-milling for 15min, 30min, 1h, 2h, 4h and 8h, which resulting in six different powder mixtures with different particle size. Each powder mixtures (30 g) was loose packed in crucible and heated at 1500 °C for 30 min in flowing Ar atmosphere with pressure of 0.1 MPa.

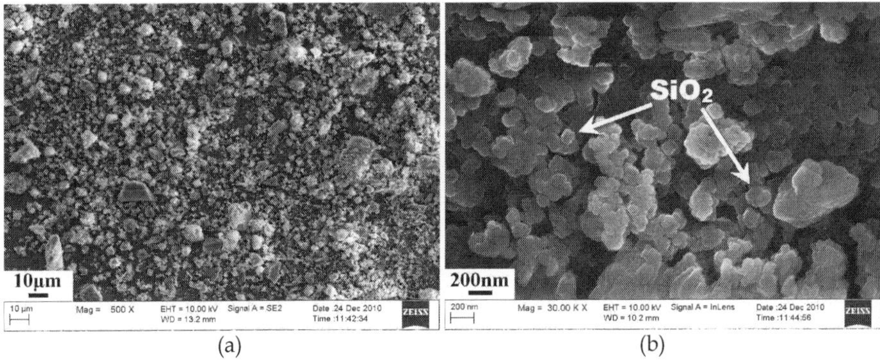


(a) (b)

Fig. 33. SEM images of (a) powder mixtures after 8 h-ball-milling; (b) nano-$SiO_2$ coated on the surface of activated carbon (local magnified image of Fig. 33(a))

Figure 33 shows the SEM images of powder mixtures after 8 h-ball-milling and its local magnified image. It can be seen that the nano-$SiO_2$ was coated uniformly on the surface of activated carbon. The relationship between the particle size of raw material powders and ball milling time is indicated in Fig 34(a). AC represents activated carbon. When the ball milling time was increased from 15 min to 8 h, the D10, D50 and D90 of powders decreased from 5.1, 14.1 and 30.2 μm to 2.1, 4.4 and 9.0 μm respectively. Figure 34(b) shows the loose packing density and the corresponding porosity of the powder mixtures after different ball milling time, which were calculated based on the Eq. (6) and (7).

$$p = 1 - (V_P/V) \tag{6}$$

$$\rho = W_P/V \tag{7}$$

The symbols p, $V_P$, V, ρ, $W_P$ represent the porosity, true volume, total volume, packing density, and weight of raw material powders respectively. The value of V and $W_P$ were measured in our experiment; $V_P$ was calculated by the density and weight of $SiO_2$ (2.33 $g/cm^3$) and activated carbon (2.2 $g/cm^3$). With the increase in ball milling time, porosity of powders increased from 80.2% to 89.2%, and loose packing density decreased from 0.44 $g/cm^3$ to 0.24 $g/cm^3$.

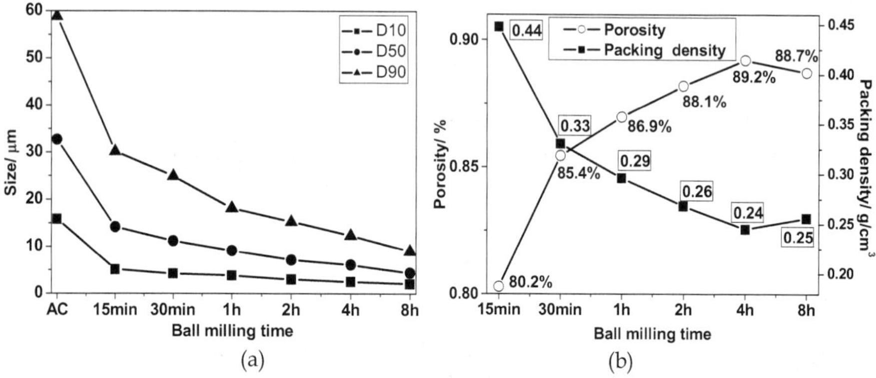


Fig. 34. (a) particle size (b) porosity and loose packing density of powders vs. ball milling time

Figure 35 shows the SEM images of the synthesized SiC whiskers. When the loose packing density was 0.44 and 0.33 $g/cm^3$ (Fig. 35 a,b), two kinds of whiskers, flexural one and straight one, have been synthesized. The diameter of the whiskers was 100-300 nm, and the whiskers ranged in length from 1 μm to up to several microns. At the case of 0.29 $g/cm^3$ (Fig. 35 c), flexural whiskers were almost disappeared, but still were about several microns in length. As the packing density was less than 0.26 $g/cm^3$ (Fig. 35 d,e,f), the morphologies of the SiC whiskers look quite different compared with that of the previously synthesized samples. SiC whiskers were straight and dispersal and the length of them were increased to more than 10 microns. But the diameter of the SiC whiskers was almost the same as that of the previous samples.

The variation in the morphologies of SiC whiskers from flexural to straight with the decrease of packing density of the raw materials could be explained by the different volume of the growing space. When the growing space was very small, the straight growth of whiskers was limited; whiskers would have a curving or circular movement during the growing process. Straight whiskers could be synthesized as a result having enough growing space. This tendency was consistent with the change of the loose packing density. The concentration of gaseous reactants should be also considered(Fu et al., 2005). When the packing density increases and the porosity decreases, the relative concentration of the gaseous phase will increase and then the whiskers will grow in several directions results in the formation of flexural and shorter whiskers.

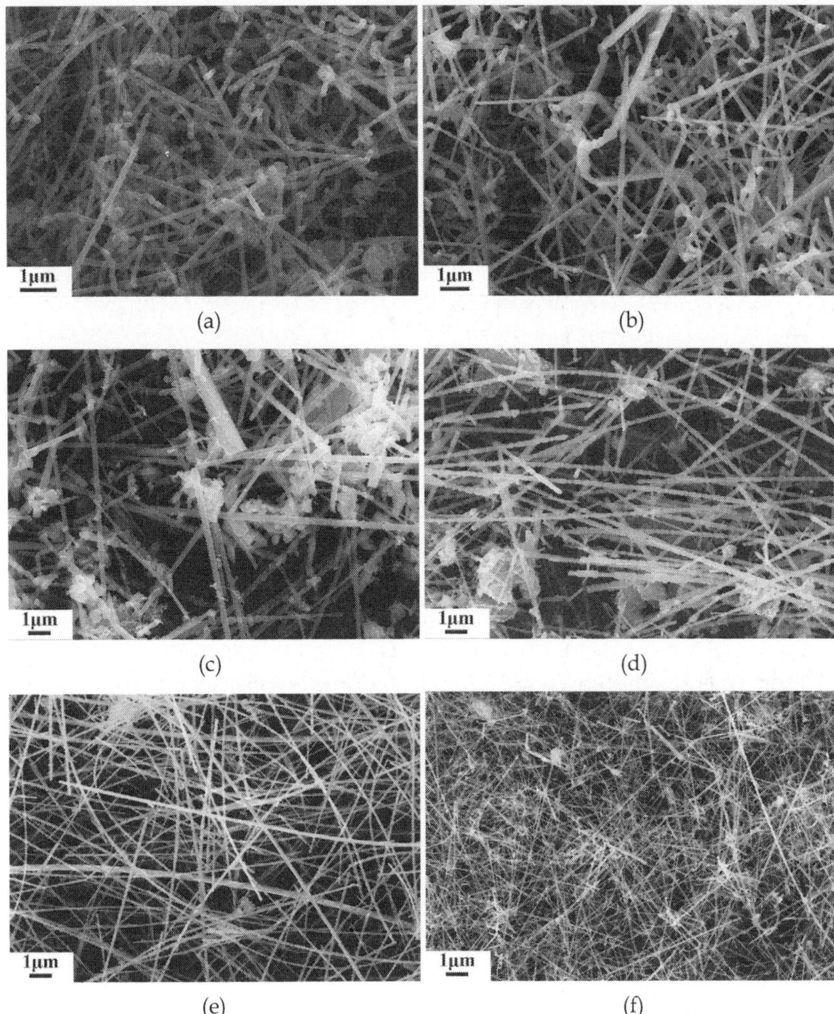


(a) (b) (c) (d) (e) (f)

Fig. 35. SEM images of samples synthesized at 1500 °C for 30 min with the loose packing density(g/cm$^3$) of (a) 0.44; (b) 0.33; (c) 0.29 (d) 0.26; (e) 0.24; (f) 0.25

## 4. Summary

Pure β-SiC powders were synthesized by microwave heating at the range of 900 °C~1600 °C and holding for 5~60 min through the reaction between different silicon and carbon sources.

Agglomerated particles of SiC have been synthesized at 1200 °C by activated carbon-silicon. While in graphite-silicon system, the β-SiC can be synthesized at 900 °C and the problem of agglomeration of the synthesized products can be partly solved. The uniform equiaxial SiC particles with a narrow particle size distribution were synthesized at 1200 °C by using carbon black and silicon. The full conversion in the coke-silica system can be achieved at 1600 °C for holding 30 min, during which the content of SiC in the reacted powders was 98.1%. The size and morphology of the synthesized SiC particles are primarily dominated by carbon sources.

SiC whiskers were prepared from silica-sol and activated carbon by microwave heating. The effect of reaction temperature, holding time and loose compact density of the starting materials on the phase composition, morphology and yield of the product have been investigated. The yield of SiCw increase with the increase of the heating temperature and the maximum of that is 60.1 wt% when heated at 1500 °C holding for 60 min. After that, the yield will decrease with the further increase of heating temperature. The shape of the SiC*w* was found to vary from flexural to straight with the decrease of packing density of the raw materials which was explained on the basis of different volume of the growing space. Larger growing space will benefit the growth of straight whiskers while in the case of very small growing space curved or circular whiskers will be obtained.

## 5. Acknowledgement

The authors appreciate the financial support by the National Natural Science Foundation of China under Grant No. 50472094.

## 6. References

Biernacki, J. J. & Wotzak, G. P. (1989). Stoichiometry of the C + $SiO_2$ Reaction. *Journal of the American Ceramic Society*, Vol. 72, No. 1, pp. 122-129, ISSN 1551-2916

Chrysanthou, A., Grieveson, P. & Jha, A. (1991). Formation of silicon carbide whiskers and their microstructure. *Journal of Materials Science*, Vol. 26, No. 13, pp. 3463-3476, ISSN 0022-2461

Dai, C., Zhang, X. & Zhang, J. (1997a). Microwave Synthesis of Ultrafine Silicon Carbide Whiskers. *Journal of the American Ceramic Society*, Vol. 80, No. 5, pp. 1274-1276, ISSN 1551-2916

Dai, C., Zhang, X. & Zhang, J. (1997b). The synthesis of ultrafine SiC powder by the microwave heating technique. *Journal of Materials Science*, Vol. 32, No. 9, pp. 2469-2472, ISSN 0022-2461

Ebadzadeh, T. & Marzban-Rad, E. (2009). Microwave hybrid synthesis of silicon carbide nanopowders. *Materials Characterization*, Vol. 60, No. 1, pp. 69-72, ISSN 1044-5803

Fu, D., Zeng, X. & Zou, J. (2009). In situ synthesis and photoluminescence of SiC nanowires by microwave-assisted pyrolysis of methane. *Journal of Alloys and Compounds*, Vol. 486, No. 1-2, pp. 406-409, ISSN 0925-8388

Fu, Q., Li, H. & Shi, X. (2005). Microstructure and growth mechanism of SiC whiskers on carbon/carbon composites prepared by CVD. *Materials Letters*, Vol. 59, No. 19-20, pp. 2593-2597, ISSN 0167-577X

Guo, J. Y., Gitzhofer, F. & Boulos, M. I. (1995). Induction plasma synthesis of ultrafine SiC powders from silicon and $CH_4$. *Journal of Materials Science*, Vol. 30, No. 22, pp. 5589-5599, IS SN 0022-2461

Halamka, M., Kavecky, S. & Docekal, B. (2003). Synthesis of high purity $Si_3N_4$ and SiC powders by CVD method. *CERAMICS - SILIKATY*, Vol. 47, No. 3, pp. 88-93, IS SN 0862-5468

Hatakeyama, F. & Kanzaki, S. (1990). Synthesis of Monodispersed Spherical β-Silicon Carbide Powder by a Sol-Gel Process. *Journal of The American Ceramic Society*, Vol. 73, No. 7, pp. 2107-2110, IS SN 1551-2916

Jin, H. B., Cao, M. S. & Zhou, W. (2010). Microwave synthesis of Al-doped SiC powders and study of their dielectric properties. *Materials Research Bulletin*, Vol. 45, No. 2, pp. 247-250, IS SN 0025-5408

Klein, S., Winterer, M. & Hahn, H. (1998). Reduced-Pressure Chemical Vapor Synthesis of Nanocrystalline Silicon Carbide Powders. *Chemical Vapor Deposition*, Vol. 4, No. 4, pp. 143-149, IS SN 1521-3862

Krstic, V. D. (1992). Production of Fine, High-Purity Beta Silicon Carbide Powders. *Journal of The American Ceramic Society*, Vol. 75, No. 1, pp. 170-174, IS SN 1551-2916

Kuang, J. L., Li, L. & Cao, W. B. (2011). Preparation of SiC Whiskers by Microwave Heating. *Materials Science Forum*, In press, ISSN 0255-5476

Li, Y. L., Liang, Y. & Zheng, F. (1994). Carbon Dioxide laser Synthesis of Ultrafine Silicon Carbide Powders from Diethoxydimethylsilane. *Journal of The American Ceramic Society*, Vol. 77, No. 6, pp. 1662-1664, IS SN 1551-2916

Lu, B., Liu, J., Zhu, H. (2007). SiC Nanowires Synthesized by Microwave Heating. *Materials Science Forum*, Vol. 561-565, No. pp. 1413-1416, IS SN 1662-9752

Metaxas, A. C. (1991). Microwave heating. *Power Engineering Journal*, Vol. 5, No. 5, pp. 237-247, IS SN 0950-3366

Narayan, J., Raghunathan, R. & Chowdhury, R. (1994). Mechanism of combustion synthesis of silicon carbide. *Journal of Applied Physics*, Vol. 75, No. 11, pp. 7252-7257, IS SN 0021-8979

Pampuch, R., Stobierski, L. & Lis, J. (1987). Solid combustion synthesis of β SiC powders. *Materials Research Bulletin*, Vol. 22, No. 9, pp. 1225-1231, IS SN 0025-5408

Ramesh, P. D., Vaidhyanathan, B. & Ganguli, M. (1994). Synthesis of β-SiC powder by use of microwave radiation. *Journal of Materials Research*, Vol. 9, No. 12, pp. 3025-3027, IS SN 2044-5326

Rao, K. J., Vaidhyanathan, B. & Ganguli, M. (1999). Synthesis of Inorganic Solids Using Microwaves. *Chemistry of Materials*, Vol. 11, No. 4, pp. 882-895, IS SN 0897-4756

Satapathy, L. N., Ramesh, P. D. & Agrawal, D. (2005). Microwave synthesis of phase-pure, fine silicon carbide powder. *Materials Research Bulletin*, Vol. 40, No. 10, pp. 1871-1882, IS SN 0025-5408

Seog, I. & Kim, C. (1993). Preparation of monodispersed spherical silicon carbide by the sol-gel method. *Journal of Materials Science*, Vol. 28, No. 12, pp. 3277-3282, IS SN 0022-2461

Setiowati, U. & Kimura, S. (1997). Silicon Carbide Powder Synthesis from Silicon Monoxide and Methane. *Journal of The American Ceramic Society*, Vol. 80, No. 3, pp. 757-760, IS SN 1551-2916

Sharma, N. K., Williams, W. S. & Zangvil, A. (1984). Formation and Structure of Silicon Carbide Whiskers from Rice Hulls. *Journal of the American Ceramic Society*, Vol. 67, No. 11, pp. 715-720, IS SN 1551-2916

Shawn, A., Holly, S. & Morgana, F. (2008). Microwave Heating Technologies. *Heat Treating Progress*, Vol. 8, No. 3, pp. 39-42, IS SN 1536-2558

Tairov, Y. M. (1995). Growth of bulk SiC. *Materials Science and Engineering B*, Vol. 29, No. 1-3, pp. 83-89, IS SN 0921-5107

Wang, L., Wada, H. & Allard, L. F. (1992). Synthesis and characterization of SiC whiskers. *Journal of Materials Research*, Vol. 7, No. 1, pp. 148-163, IS SN 0884-2914

Wang, F., Wang, Q., Cao, W. B. (2009). Microwave Synthesis of SiC Powders with Coke and Quartzite. *Journal of Materials Engineering*, No. 7, pp. 32-35, ISSN 1001-4381

Wang, F. (2008a). Study on preparation of β-SiC powders. Beijing: University of Science & Technology Beijing

Wang, F., Sun, J. L., Cao, W. B. (2008b). Microwave synthesis of silicon carbide whisker. *Refractories*, Vol. 42, No. 5, pp. 357-361, ISSN 1001-1935

Wang, Q. (2009). Research on the synthesis of silicon carbide powders by microwave heating method. Beijing: University of Science & Technology Beijing

Wei, G., Qin, W. & Wang, G. (2008). The synthesis and ultraviolet photoluminescence of 6H–SiC nanowires by microwave method. *Journal of Physics D: Applied Physics*, Vol. 41, No. 23, pp. 235102, IS SN 0022-3727

Weimer, A. W., Moore, W. G. & Rafaniello, W. (1994). Process for Preparing Silicon Carbide, U.S. Patent 5,340,417

Weimer, A. W., Nilsen, K. J. & Cochran, G. A. (1993). Kinetics of carbothermal reduction synthesis of beta silicon carbide. *AIChE Journal*, Vol. 39, No. 3, pp. 493-503, IS SN 1547-5905

Weimer, A. W., Roach, R. P. & Haney, C. N. (1991). Rapid carbothermal reduction of boron oxide in a graphite transport reactor. *AIChE Journal*, Vol. 37, No. 5, pp. 759-768, IS SN 1547-5905

Yamada, O., Hirao, K. & Koizumi, M. (1989). Combustion Synthesis of Silicon Carbide in Nitrogen Atmosphere. *Journal of the American Ceramic Society*, Vol. 72, No. 9, pp. 1735-1738, IS SN 1551-2916

# 12

# Microwave Synthesis of Core-Shell Structured Biocompatible Magnetic Nanohybrids in Aqueous Medium

Ling Hu[1], Aurélien Percheron[1],
Denis Chaumont[1] and Claire-Hélène Brachais[2]
[1]*Laboratoire Interdisciplinaire Carnot de Bourgogne, Université de Bourgogne*
[2]*Institut de Chimie Moléculaire de l'Université de Bourgogne,*
*Université de Bourgogne,*
*France*

## 1. Introduction

In the past decade, biocompatible magnetic nanohybrids, i.e. materials consisting of an inorganic core encapsulated by a biocompatible polymeric corona, went throw various developments in biomedical applications especially in the fields of diagnosis and therapy. Numerous descriptions of their syntheses can be found in the literature (Zhang et al., 2002; Flesch et al., 2004; Fan et al., 2007; etc). These two-steps protocols often describe the use of organic or aqueous solvents, classical thermal heating, long time reaction as well as fastidious exchange and drying steps. In recent years, microwave heating has been proven to be a very original technology for nanoparticles synthesis due to its almost instantaneous "in core" heating of materials in a homogeneous and selective way. As a consequence, this technology allows an interesting control over crystallization rate and size of the nanomaterials (Bellon et al., 2001; Michel et al., 2001; etc). The formation of magnetic nanoparticles is usually realized in aqueous medium. However, the functionalization of magnetic nanoparticles is mainly conducted in organic solvent. For further biomedical applications, it seems favorable to realize these different grafting steps maintaining initial aqueous medium, thus avoiding the solvent exchanges and the particles drying. In this context, we can summarize the objective of this chapter as the study of new methods to obtain core-shell structured nanohybrids in aqueous medium from iron oxide nanoparticles and water-soluble biocompatible polymers. Is it possible to obtain iron oxide colloidal suspension and well-defined nanohybrids with simplified protocols and relatively short microwave heating time? This chapter is divided is three part, consisting for the first one, of a general presentation of biocompatible magnetic nanohybrids (properties, applications and synthesis). The second part deals with the principles of microwave heating and the description of microwave-assisted synthesis of inorganic nanoparticles. Finally, the last part is dedicated to the use of microwave heating towards magnetic nanohybrid synthesis compared to classical thermal heating process with a focus on nanoparticles characterization (morphology, size and grafted amount of polymer).

## 2. Biocompatible magnetic nanohybrids

### 2.1 Definition and properties

Nanoscience is an interdisciplinary knowledge-generating domain which has the potential to create many advances in diagnosis and therapy. Nanomaterials such as magnetic nanoparticles and more especially iron oxide nanoparticles like magnetite $Fe_3O_4$ and maghemite $\gamma$- $Fe_2O_3$, present three very interesting properties for in *vivo* applications, which depend on their magnetic capacities. The most simple and natural property is their movement in a predetermined direction like a magnet when they are in a magnetic field (Gupta & Curtis, 2004). These particles can also produce a strong signal during a scan using a magnetic field, in RMI analysis for example (Baghi et al., 2005; Schultz et al. 1999). The third property is their superparamagnetic behaviour, *i.e.* they have not persistent magnetic field when no magnetic field is applied. In this case, the temperature of the particles increases when they are subjected to a varying magnetic field with a suitable frequency (Flesch et al., 2005a). Furthermore, these nanoparticles can be easily recycled by human cells using the classic metabolic pathways of iron treatment (Mowat, 2007). They thus provide new properties compared to conventional drugs. Their applications will be more detailed in Part 2.2.

For their utilization in diagnosis and therapy, nanoparticles can be injected intravenously before being delivered by the blood circulation to the treated area. The nanoparticles can be also injected directly to the treated area. In the first case, after their injection in the blood circulation, the nanoparticles are quickly covered by plasma protein. This process is the first implementation of the immune system and makes the nanoparticles recognizable by macrophages (Davis, 1997). Nanoparticles capture and elimination by macrophages strongly limit their lifetime in the body and therefore their medical uses. An efficient way to overcome this problem is to make discrete nanoparticles by grafting a biocompatible polymer layer on their surfaces (Figure 1). The major difficulty is to find a polymer which is perfectly compatible with the human body and can be easily grafted onto an inorganic material. Biocompatibility is one of the most important characteristics of a biomedical polymer material which can provide a function with an appropriate response without any

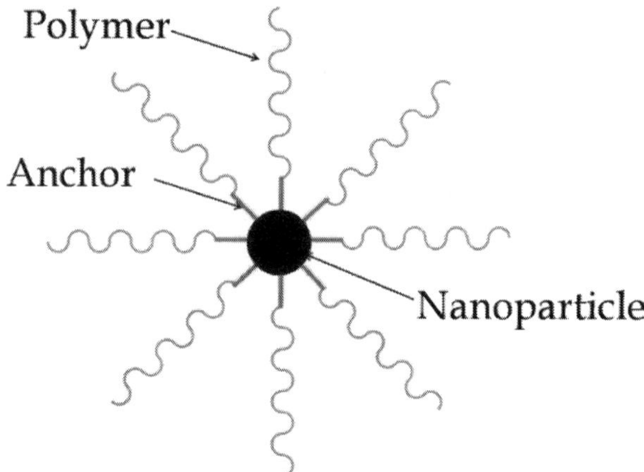


Fig. 1. Schematic representation of a core-shell nanohybrid

adverse effects on the biological environment. The biocompatibility of polymers depends largely on their interactions with the tissues during their whole stay in the human body. The surface composition and the surface state of a biomaterial are two important factors to evaluate its biocompatibility. Usually, the polymers used to increase half life in the human body are hydrophilic polymers. Indeed, the grafting of hydrophilic polymer onto inorganic nanoparticles avoids the process of recovering and decreases their rate of elimination (Lu et al., 1999; Gaur et al., 2000). It is not the same case for the hydrophobic particle surfaces which are easily covered by blood plasma components and removed rapidly from the blood circulation (Gupta & Curtis, 2004).

A second essential point of the utilization of magnetic nanoparticles for diagnosis end therapy is their suspension stability. In the case of nanohybrids, the stabilization is ensured by the repulsive forces between interacting polymers attached to the particle surfaces and prohibiting the inter-particle contact. The total potential energy curve of this particle-polymer system presents a potential barrier. When the polymer chains interpenetrate, there is a loss of configuration entropy which increases the free energy of the system and then the repulsive forces between the chains can take place. The physical barrier also helps to reduce the Van der Waals forces between particles by keeping them at a certain distance. The presence of polymer on the nanoparticle surfaces gives thus better dispersion stability. Table 1 gives some examples of polymers which can be used in biomedical applications.

| Polymer | Advantages | Reference |
|---|---|---|
| PEG | Improves the circulation time of nanoparticles in the blood and the efficiency of internalization of nanoparticles | Zhang et al., 2002; Gupta & Curtis, 2004 |
| PVP | Stabilizes the colloidal solution and improves the circulation time of nanoparticles in the blood. | D'Souza et al., 2004 |
| PVA | Stabilizes the colloidal solution | Shan et al., 2003 |
| PAA | Increases the stability and biocompatibility of nanoparticles | Changez et al., 2003 |
| PCL | Used as biomaterials for prostheses, sutures and controlled release systems | Désévaux, 2002 |
| PMMA | Used for prosthetic constraint but not easily sterilizable (by alcohol for example) | Bonnans-Plaisance et al., 2001 |
| PVC | Widely used in medicine but not in prosthesis, because of the plasticizer, which diffuses into the blood | Drakakis et al., 2008 |
| PTFE | Very hydrophobic, flexible, non-reactive Adapted to prostheses of blood vessels / arteries | Chen et al., 2003 |
| Dextran | Stabilizes the colloidal solution and improves the circulation time of nanoparticles in the blood | Berry et al., 2003 |
| Chitosan | Used as a no viral vector for gene therapy | Khor & Lim, 2003 |

Table 1. Some examples of biocompatible polymers used in medical application.
(PEG: poly(ethyleneglycol); PVP: poly(vinylpyrrolidone); PVA: poly(vinylalcohol); PAA: poly(acrylicacid); PCL: poly(ε-caprolactone); PMMA: poly(methylmethacrylate); PVC: poly(vinylchloride); PET: poly(ethyleneterephtalate); PTFE: poly(tetrafluorethylene))

## 2.2 Potential therapeutic applications

Different applications of nanohybrids in the medical world depend directly on their properties regarding to Part 2.1 (controlled movement in a magnetic field; temperature rise in an alternating magnetic field; strong magnetic response in medical image analysis). Their principal applications can be resumed in three classes and are described below:

- contrast agent for Magnetic Resonance Imaging (MRI);
- mediators of magnetic hyperthermia treatment;
- nanovectors for drug delivery by magnetic targeting.

**Contrast enhancement of Magnetic Resonance Images**

Magnetic Resonance Imaging (MRI) is a non-invasive imaging technique widely used in clinical for determining the functional parameters associated or not with pathologies. The use of contrast agents is necessary to target specifically certain biological mechanisms. It is well known that the introduction of magnetic nanoparticles in tissues allows obtaining a strong signal from MRI scanner. Magnetic nanoparticles offer many advantages (high specific magnetic moment, biocompatibility, functionalizable surface etc.) compared with the classical gadolinium contrast agents (Kubaska et al., 2001).

A wide variety of nanohybrids was already studied, differing in particle size or type of cover material. Particle size influences their physico-chemical and pharmacokinetic properties (Brigger et al., 2002). Lee *et al.* presented the synthesis and characterization of nanoparticles of maghemite $\gamma$-$Fe_2O_3$ coated with poly(vinylpyrrolidone) (PVP) (Lee et al., 2006). MTT cytotoxicity test, which is a rapid counting method of living cells, revealed the excellent biocompatibility of grafted nanoparticles. Observation of MRI images of distribution of grafted nanoparticles into the rabbit marginal vein showed a darker color of the liver after particles injection. Thus, the significant improvement in the detection of liver lesions by MRI using magnetic nanohybrids as contrast agent is proved.

**Targeted cancer cell destruction with magnetic hyperthermia**

The magnetic hyperthermia, a well-known therapy for cancer treatment, exposes the human cancer tissues to an alternating magnetic field which can not be absorbed by living tissues although can be applied within the whole human body. When magnetic particles are subjected to an alternating magnetic field, heat energy is produced due to the magnetic hysteresis loss. The amount of heat energy produced depends on the nature of magnetic materials and the properties of magnetic field. This magnetic therapy is based on the fact that cancer cells can be destroyed at a temperature around 43 °C (Berry et al. 2003), whereas normal cells can survive at higher temperatures.

Some experiments were performed on animals or cancer cells in order to detect the therapeutic effects of magnetic nanohybrids on different types of tumours. Wada *et al.* have proven the effectiveness of magnetic nanohybrids (magnetite $Fe_3O_4$-dextran) (MD) for hyperthermia against cancer of the mouth (Wada et al., 2003). A suspension of MD was injected into tumours of the tongue and the tongue was heated to 43-45 °C by applying a magnetic field of 500 kHz. Owing to this technique, the inhibition of tumour growth is more effective than that of untreated tumours.

**Magnetic targeting for site-specific drug delivery**

Magnetic targeting drug delivery is an interesting technique for administration of pharmacologically active molecules. The objective is to improve their therapeutic index. The term vector refers to the molecular structure or the colloidal compounds used to carry or transport molecule of biological interest into cells. The role of the vector is both to protect the active molecules against the chemical or enzymatic degradation occurring in vivo and also to transport them to their sites of action.

Plassat et al. (2008) showed that the liposomes, sterically stabilized by hydrophilic chains of polyethylene glycol and encapsulating superparamagnetic maghemite $\gamma$-$Fe_2O_3$ nanocrystals, can be exploited to target breast cancer tumours. This exploitation is possible by applying an external magnetic field gradient at the region of interest, and by associating the nanohybrids to an antiestrogen (RU 58668), which is effective to stop the growth of estrogen-dependent tumours (Maillard et al., 2006).

### 2.3 Synthesis of biocompatible magnetic nanohybrids

#### 2.3.1 Preparation methods in the literature

The use of an anchor is necessary to attach the organic component (the polymer) to the inorganic component (the nanoparticle) (Figure 1). The role of this anchor is to create a strong bond such as an ionic or a covalent bond between the two components to avoid desorption (Jo & Park, 2000). Macromolecules ionically attached to the particles surface are very sensitive to the dissolution and / or the depletion. The conditions of utilization of this type of nanohybrids would be rather limited for medical applications, since the biological environment could separate the polymer from the nanoparticle (Delair et al., 2003). In addition, the anchor also acts as a spacer between nanoparticle core and polymer chains to facilitate the approach of the reactive polymer to the nanoparticle.

The core-shell structured nanohybrid materials can be prepared following two approaches: *grafting from* and *grafting to*, considering the polymers already formed or not (Figure 2). Table 2 shows various examples of the synthesis of nanohybrids using these two methods.

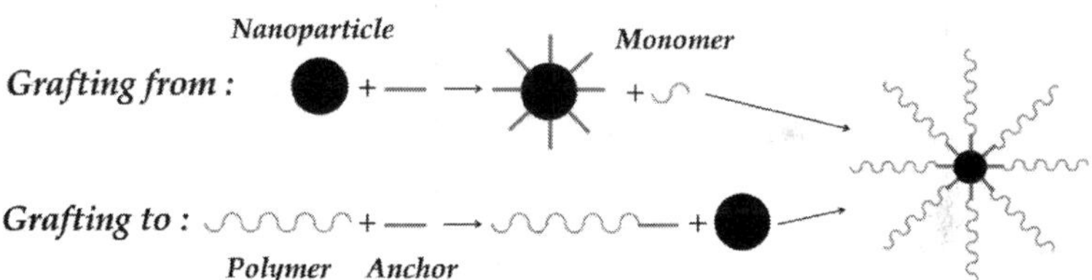


Fig. 2. Presentation of two principal grafting methods: *grafting from* and *grafting to*

**Grafting from**

*Grafting from* includes monomer adsorption onto particles followed by a subsequent polymerization which can be catalyzed by an initiator promoting the process (Ninjbadgar et al., 2004). This approach proceeds in two steps: the anchoring on the particle surface and the polymerization of monomer from the anchor.

**Grafting to**

In the *Grafting to* process, the polymer is modified to react with reactive groups of the nanoparticles surface (Brachais et al. 2010; Hu et al. 2008). In this case, the grafting is preformed either by placing an anchoring system on the particle surface or by a prior functionalization of polymer. The length of grafted chains on the nanoparticles is determined by the choice of the starting polymer.

| Protocol | Core | Shell | Solvent | Anchor | Synthesis parameters | Grafted amount [a] | Reference |
|---|---|---|---|---|---|---|---|
| *Grafting from* | $\gamma$-$Fe_2O_3$ | PCL | toluene | AAPM | 50 °C, under $N_2$, 48 h | 10 % | Flesch et al., 2005b |
| | $\gamma$-$Fe_2O_3$ | PMMA | DMF | CDS | 70 °C, 10 h, under $N_2$ | - | Ninjbadgar et al., 2004 |
| | $\gamma$-$Fe_2O_3$ | PEG | water | MPS | 70 °C, under $N_2$, 23 h | 20 % | Flesch et al., 2005c |
| | $Fe_3O_4$ | PDMAEMA | DMF | BIB | 90 °C, 24 h, under Ar | 14 % | Zhou et al., 2009 |
| | $Fe_3O_4$ | SPAN | HCl 0,2 M aqueous solution | APTMS | 26 h, under $N_2$ | 12 % | Reddy et al., 2007 |
| | $Fe_3O_4$ | P(PEGA) | THF | CPA | 30 °C, 18 h | 34 % | Fan et al., 2007 |
| | $NiFe_2O_4$ | PMAA | water | SDS | 70 °C, under Ar, 2 h | - | Rana et al., 2007 |
| *Grafting to* | $\gamma$-$Fe_2O_3$ | dextran | water | APS | 28 h, under $N_2$ | - | Mornet et al., 2005 |
| | $\gamma$-$Fe_2O_3$ | PCL | DMF | IPTS | 120 °C, under $N_2$, 22 h | 15 % | Flesch et al., 2004 |
| | $\gamma$-$Fe_2O_3$ | PVI | toluene/ methanol | MPS | 48 h, under $N_2$ | - | Takafuji et al., 2004 |
| | $\gamma$-$Fe_2O_3$ | PANI | toluene | Oleic acid | 24 h, under $N_2$ | 6 % | Dallas et al., 2006 |
| | $Fe_3O_4$ | *m*PEG-PCL | water | Oleic acid | $T_{room}$, 24 h, under $N_2$ | 12 % | Meerod et al., 2008 |
| | $Fe_3O_4$ | PEG | water | APS | 60 °C, 4 h, under $N_2$ | - | Zhang et al. 2002 |
| | $NiFe_2O_4$ | PVA | water | - | 25 °C, 15 h, under $N_2$ | - | Sindhu et al., 2006 |

[a] determined by thermogravimetric analysis

Table 2. Synthesis of nanohybrids using "grafting from" and "grafting to" methods.
(PCL: poly(ε-caprolactone); AAPM: N-(2-aminoethyl)-3-aminopropyltrimethoxysilane;
PMMA: poly(methylmethacrylate); DMF: N'N-dimethylformamide;
CDS: (chloromethyl)phenylethyl-dimethylchlorosilane; PEG: polyethylene glycol;
MPS: methacryloxypropyltrimethoxysilane;
PDMAEMA: poly(2-(dimethylamino)ethylmethacrylate);
BIB: N-bromoisobutyrate acid; SPAN: poly(aniline-co-aminobenzenesulfonic acid);
APTMS: aminopropyltrimethoxysilane;
P(PEGMA): poly(poly(ethyleneglycol) monomethacrylate); THF: tetrahydrofurane;
CPA: chloropropionic acid; PMAA: poly(methacrylicacid); SDS: sodium dodecylsulfate;
APS: 3-aminopropyltrimethoxysilane; IPTS: 3-isocyanatopropyltriethoxysilane;
PVI: poly(1-vinylimidazole); PANI: poly(aniline); PVA: poly(vinyl alcohol))

**Comparison between *grafting from* and *grafting to***

The literature shows that *grafting from* process allows reaching higher contents of grafted polymers whereas *grafting to* process allows a good control over the length of the polymeric chains. Furthermore, *grafting to* method can be applied to all types of polymers possessing reactive ends. The length of polymer chain is known and well controlled. The only parameter to control is the grafting rate. Thus nanocomposites obtained by this method can be better characterized.

### 2.3.2 Reaction mechanisms proposed in the literature

In this part, we present the various steps proposed in the literature concerning the grafting reaction mechanism of polyethylene glycol (PEG) onto an inorganic core (iron oxide nanoparticle) using 3-isocyanatopropyltriethoxysilane (IPTS) as the anchor. PEG has been used widely for surface modification because of its unique properties such as hydrophilicity, flexibility, nontoxicity and nonimmunogenecity (Jo & Park, 2000).

**Formation of a urethane bond**

In the case of anchoring of polymer chains, Kim *et al.* (2005) studied the functionalization of polyethylene glycol (PEG) with 3-isocyanatopropyltriethoxysilane (IPTS) (Figure 3), using dibutyl tin dilaurate (DBTL) as catalyst in tetrahydrofuran (THF).

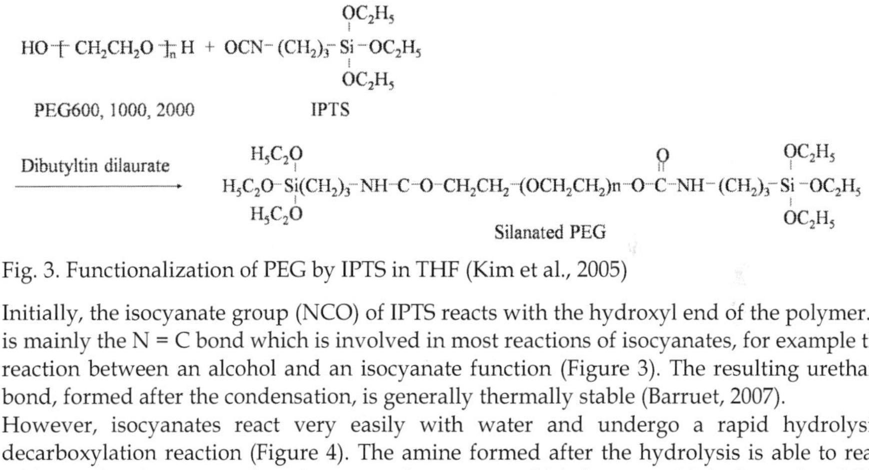

$$HO-[CH_2CH_2O]_n-H + OCN-(CH_2)_3-Si(OC_2H_5)_3$$

PEG600, 1000, 2000 IPTS

$$\xrightarrow{\text{Dibutyltin dilaurate}} (H_5C_2O)_3Si(CH_2)_3-NH-\overset{O}{\overset{\|}{C}}-O-CH_2CH_2-(OCH_2CH_2)n-O-\overset{O}{\overset{\|}{C}}-NH-(CH_2)_3-Si(OC_2H_5)_3$$

Silanated PEG

Fig. 3. Functionalization of PEG by IPTS in THF (Kim et al., 2005)

Initially, the isocyanate group (NCO) of IPTS reacts with the hydroxyl end of the polymer. It is mainly the N = C bond which is involved in most reactions of isocyanates, for example the reaction between an alcohol and an isocyanate function (Figure 3). The resulting urethane bond, formed after the condensation, is generally thermally stable (Barruet, 2007).
However, isocyanates react very easily with water and undergo a rapid hydrolysis-decarboxylation reaction (Figure 4). The amine formed after the hydrolysis is able to react with another isocyanate function to produce urea, which has very high thermal stability preventing possible subsequent substitution (Figure 5) (Barruet, 2007). Thus it is important to avoid the presence of humidity in the reaction medium during the grafting step using isocyanate functions.

$$R-N=C=O + H_2O \longrightarrow R-\overset{H}{\overset{|}{N}}-\overset{O}{\overset{\|}{C}}-O-H \longrightarrow R-NH_2 + CO_2$$

Fig. 4. Decarboxylation of isocyanates in water (Barruet, 2007)

$$R-NH_2 + R'-N=C=O \longrightarrow R'-NH-C(=O)-NH-R$$

Fig. 5. Formation of urea after the reaction between an amine and an isocyanate (Barruet, 2007)

**Grafting of alkoxysilane on the nanoparticles**

Alkoxysilane functions introduced during the anchoring are then used for the grafting on the nanoparticles. According to Zhang et al. (2002), the grafting of PEG on the magnetite nanoparticles takes place by direct reaction between surface hydroxyl groups (-OH) of magnetite nanoparticles and anchor siloxane functions of polymer. PEG is grafted on the surface of magnetite by forming a covalent Fe-O-Si bond (Figure 6).

$$\bullet\!-OH + Si(OCH_2CH_3)_3-PEG \longrightarrow \bullet\!-O-Si(OC_2H_5)_2-PEG + CH_3CH_2OH$$

Fig. 6. Grafting reaction of functionalized PEG with magnetite nanoparticle (Zhang et al., 2002)

**Different catalytic media for the grafting process of alkoxysilanes**

It is possible to perform the grafting process under acidic, basic catalytic media or with organometallic catalyst, for catalyzing the initiation of condensation between the hydroxyl groups of particles and the silanol groups.

**Acid catalysis**

Alkoxysilane hydrolysis in acidic medium is based on an electrophilic substitution mechanism leading to the formation of silanol (Grignon-Dubois et al., 1980). The following condensation between silanol and surface hydroxyl functions of the nanoparticles lays essentially on an equilibrium between the protons and the silanol functions forming the temporary charged species (Figure 7). These positively charged species strongly attract the surface hydroxyl groups of nanoparticles.

$$-Si-OH + H^+ \longrightarrow -Si-O^+H_2$$

Fig. 7. Temporary formation of positively charged species in acidic medium

**Basic catalysis**

The basic hydrolysis is based on a nucleophilic substitution mechanism (Buurma et al., 2003). An ionic equilibrium is established between the silanols and the hydroxyde ions, leading to the formation of silanolate anions (Figure 8).

$$-\overset{|}{\underset{|}{Si}}-OH + OH^{-} \longrightarrow -\overset{|}{\underset{|}{Si}}-O^{-} + H_2O$$

Fig. 8. Formation of silanolate anions in basic medium

**Organometallic catalysis**

Tin salts such as dibutyltin dilaurate (DBTL) are widely used as catalysts for reaction between the isocyanate and alcohol functions. The reaction mechanisms were well defined by Houghton & Mulvaney (1996) and Brosse et al. (1995). Moreover, it was reported that the effect of DBTL is also to reduce the activation energy involved in the hydrolysis step of alkoxysilanes and to catalyze the condensation between silanol and hydroxyl functions (Kelnar & Schätz, 1993; Narkis et al., 1985).

### 2.4 Conclusion

Nanohybrid particles combine the advantages of their two components: magnetic core (magnetic hyperthermia, strong signal in magnetic field and magnetic vectorization) and polymer corona (biocompatibility). These combinations are essential for medical uses. The preparation of magnetic nanoparticles is generally carried out in aqueous medium whereas the functionalization of the polymer is commonly performed in organic medium. However, the colloidal suspensions formed are widely used for medical diagnosis and pharmaceutical applications in aqueous phase. For those applications, a particle drying followed by a solvent exchange are necessary before the grafting of the polymer. It is important to note that the agglomeration of nanoparticles is easily formed during the drying process. This problem has already been described by several authors (Reddy et al., 2007; Ramirez & Landfester, 2003; Sreeja & Joy, 2007). In this case, the grafting of polymers is performed on particle clusters instead of nanoparticles. Therefore it seems advantageous to carry out the synthesis by avoiding the solvent exchanges. Furthermore the preparation time is quite long (until several hours) because of the separate steps as shown in Table 2. Thus the finding of an appropriate protocol for the synthesis of nanohybrids remains still important and necessary.

## 3. Microwave synthesis

### 3.1 Principles of microwave heating

The microwave region of the electromagnetic spectrum lies between infrared and radio frequencies. Their frequency ranges from 30 GHz to 300 GHz and the wavelengh from 1cm to 1m. Equation 1 shows the relation between the energy E, the frequency $v$ or the wavelength λ of the radiation:

$$E = h \cdot v = \frac{h \cdot c}{\lambda} \tag{1}$$

where h is the Planck constant and c is the speed of light. The microwave photon energy is around $10^{-5}$ eV. This value is lower than the mean kinetic energy of Brownian motion (0.017 eV), the hydrogen bonds (0.04 to 0.44 eV) and the covalent bonds (5 eV) (Michel, 2003a). Consequently, microwave field cannot be the origin of bond breaking.

The physical origin of microwave heating comes from the ability of some materials to convert electromagnetic energy into heat energy. Indeed, the electric component of microwave field can induces a polarization of the charge carriers of the material. In the case of materials with dielectric losses, the polarization inversion process is slower than the electric field inversion of the applied electric field. There is therefore the apparition of a phase shift between the both. This phase shift gives rise to the energy conversion. A part of electromagnetic energy is stored in the material and the other part is converted into heat energy. It is possible to quantify this part of energy by the dielectric permittivity $\tilde{\varepsilon}$ (a physical complex data) defined by Equation 2.

$$\tilde{\varepsilon} = \varepsilon^{'} - j\varepsilon^{''} = \varepsilon_0\varepsilon_r^{'} - j\varepsilon_0\varepsilon_r^{''} \quad (2)$$

where $\varepsilon_0$ is the dielectric permittivity in vacuum, $\varepsilon'$ is the real part of dielectric permittivity, $\varepsilon''$ is the imaginary part of dielectric permittivity, $\varepsilon_r'$ is the real part of relative dielectric permittivity, and $\varepsilon_r''$ is the imaginary part of relative dielectric permittivity. The real part corresponds to the electromagnetic energy stored and the imaginary part corresponds to the heating conversion. Then, the power dissipation $P_{diss}$ within the material is proportional to $\varepsilon^{''}$, according to Equation 3.

$$P_{diss} = \iiint_{material} \omega\varepsilon^{''} |E|^2 \, dV \quad (3)$$

where $\omega$ is the electric field pulsation and E is the electric field amplitude. It is interesting to note that the dielectric loss changes with temperature. Two cases may arise: a) the temperature dependence of dielectric loss is positive, and in that case there is an accelerating heating, which can lead to thermal runaway; b) this dependence is negative and the microwave heating is then self-regulated.

In conclusion, the material heating under microwave irradiation is dependent on three parameters: the material dielectric loss, the frequency and the amplitude of electric field. The dielectric parameter limits the material class that could be used, the frequency is fixed by the microwave device (usually 2,45GHz), the intensity (square of the amplitude) or the power delivered is chosen by the user.

### 3.2 Characteristics of microwave heating

One of the most important features of microwave heating is the material volumetric heating (core heating), contrary to the classical heater devices working by convection and conduction. This property is the direct consequence of phase shift between electric field and dipolar moment of charge carriers. Thus the entire material volume is heated at the same time, contrary to the case of classical heating producing a spatial distribution of heat. Furthermore, the microwave heating kinetic could be very fast, several degrees by second. If the power applied is strong, about kilowatts over several tens of milliliters, the volumetric heating induces very high temperature within the material.

Pinto-Gateau (1995) showed the influence of the position and the distribution of electric field on the temperature evolution of the material heated. She showed also the importance of a rigorous control of the dimension and the position of the material in the microwave applicator. The simulations and experiments of microwave heating in the case of water-filled tubes indicated that:

- Temperature increases more quickly when the electric field is orthogonal to the main material axe.

- When water volume and power puissance are constant, heating kinetic depends on the tube diameter.

Consequently, a microwave heating system must allow to position as precisely as possible the material regarding to the electric field, and ensure the repeatability of experiments.

### 3.3 Microwave heating devices

#### 3.3.1 Generalities about microwave heating systems

Figure 9 shows the main elements of a classical microwave heating system: the generator, the insulator, the waveguide (or coupler) and the applicator containing the reactor and the chemical precursors.

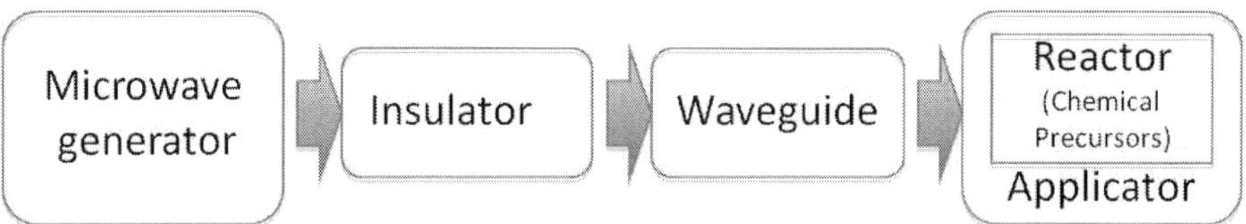


Fig. 9. Presentation of main elements of a classical microwave heating device

The microwave generator produces the microwave energy from the alternative electric network. It is composed of a magnetron. The most common commercial devices produce waves at 2.45GHz and power of about 1 or 2kW. The isolator protects the generator from any potentially destructive wave returns to the magnetron. It has a ferrite circulator which deflects the reflected fields on an absorbent filler, usually a water load. The transfer of energy between the generator and the applicator is provided by the waveguide (or coupler).

The applicator must transfer the energy under the best possible conditions to the material to be heated. Several types of applicator are available and depend on the desired applications. Progressive wave applicators are made of rectangular waveguides, split at the maximum electric field position and where the material to be heated circulates. The resonant applicators (or the resonant cavities) allow the optimum transfer of energy to the sample because of the containment of the microwave field. The condition of resonance is that the cavity dimensions are equal to an integer number of half-wavelengths. These cavities are qualified "single mode" or "multi mode" depending on their spatial distribution modes of energy. Some cavities, equipped with a piston, are tunable: the action on the piston causes a change in the length of the cavity and thus of the resonant frequency.

This microwave heating devices, we used, are specially designed by our research team in University of Burgundy

#### 3.3.2 The laboratory microwave refluxing (open reactor system)

The laboratory microwave refluxing device we used, is composed by:

- A generator MSE (Microwave System Energy), output frequency: 2.45 GHz; continuous and adjustable output power from 0 to 2 kW.
- An insulator (for generator protection).
- A waveguide to transfer of microwave energy between the generator and the refluxing reactor. The waveguide is equipped with a piston: the action on the piston causes a change in the length of the cavity and thus of the resonant frequency.

- The quartz tube of the refluxing reactor is placed on the hole in the middle of the waveguide. The quartz tube containing the chemical precursors is topped with a classical reflux condenser for chemistry.

The advantages of a microwave refluxing heating device are numerous:

- Because of the core heating, there are no conduction or convection phenomena. The entire system volume is heated at the same time.
- The microwave heating is selective: its effectiveness depends on the dielectric losses of the reagents.
- The absence of thermal inertia of microwave reactor, owing to internal sources, allows the optimal temperature control.
- The energy is focused on the reactor and then provides a high power density and high temperature rise.

### 3.3.3 The laboratory microwave autoclave reactor (closed reactor system)

The laboratory microwave autoclave reactor we used, is composed by:

- A generator MSE (Microwave System Energy), output frequency: 2.45 GHz; continuous and adjustable output power from 0 to 2 kW.
- An insulator (for generator protection) and a waveguide to transfer of microwave energy between the generator and the rectangular resonant cavity. The waveguide with radiating slots has a tuning plunger.
- A rectangular resonant cavity which contains the autoclave reactor placed in its middle.
- The autoclave is connected to pressure monitoring system. It is also possible to have an *in situ* temperature measurement using optic thermometry. In addition, the inert gas such as argon can be introduced into the reactor to work under inert atmosphere and to control the risk of explosion with flammable solvents, oxygen and possible electric sparks. The autoclave reactor body is composed of a polymer material, the polyetherimide (PEI). This material was chosen because of its transparency to microwaves which allows the full transfer of microwave energy to the reaction medium. The second reason is its usability under severe operating conditions of temperature and pressure induced by the microwave field. The PEI presents a good mechanical strength at high temperature. Finally, the material chosen must withstand to chemical species involved during synthesis. Inside the reactor, a container in Teflon® which is transparent to microwaves and chemically inert, contains the reaction medium. The Teflon® container can hold up to 20 mL of precursor solution.

The laboratory microwave autoclave reactor combines the advantages of a classical microwave heating device (see above) and the advantage of an autoclave reactor: the synthesis under pressure and at high temperatures (12 bars, 180 °C). Then, the laboratory microwave autoclave reactor allows a very fast temperature rise (around 5 $°C.s^{-1}$). This advantage gives very interesting perspectives due to the kinetic which is in the same order of magnitude than the reaction time. Stuerga & Gaillard (1996) showed that the control of thermal paths allows the selective syntheses, and the heating rate can promote one reaction path among all the concurrent processes. The real-time control of pressure inside the reactor allows to adjust the generator power during synthesis to maintain the pressure at a setpoint (Bellon, 2000).

The advantages of the utilization of microwave heating in a closed reactor device (laboratory microwave autoclave reactor) are numerous. However, the use of certain compounds (such as glycols, ethers and alkanes) in a closed reactor could damage the system. In this case, the laboratory microwave refluxing device is used instead of the autoclave system. This open system provides the possibility to work safely in solution.

### 3.4 Microwave-assisted synthesis

We have seen the usefulness of microwave heating in comparison with other heating methods. During the synthesis in solution, this gives to particles synthesized under microwave field, some remarkable properties.

#### 3.4.1 Solvent effects on microwave synthesis

**Aqueous medium**

Water is a solvent with high dipolar moment (1.84 Debye) and a strong dielectric constant ($\varepsilon_r^{'} = 78.5$). Furthermore water has strong dielectric loses at 2.45 GHz ($\varepsilon_r^{''} = 11$ at 20℃). These properties are sufficient to ensure a good conversion of electromagnetic energy into heat energy.

The synthesis in aqueous solution could be classified into two categories: the synthesis of nanoparticles by thermal initiation and the synthesis by basic initiation:

- Thermal initiation: In this case, the increase of temperature initiates the condensation process and thus the nanoparticle formation. The initiation step, more specially the nucleation step, is controlled by the heating kinetic. Microwave heating allows to separate the nucleation and growth steps. It must be noted that if microwave heating system is associated to an autoclave, the reaction temperature is higher than that in atmospheric pressure.
- Basic initiation: In the second case, in a basic medium, the initiation starts by forming a precipitate when the solutions are mixed. These precipitates are then placed in a microwave field. In this case, microwave heating controls the growth and the evolution of the precipitate instead of the initiation process (Daichuan et al., 1995).

**Organic solvents**

Simple alcohols are widely used in microwave synthesis. They are more employed than polyols due to their low toxicity. The alcoholic synthesis avoids the oxidation contrary to the aqueous solvent (Thanh et al., 2000; Cirera et al., 2000). The alcohol/water mixture presents a lower dielectric constant, and thus decreases the solubility of salts and starts the precipitation earlier (Choi & Kim, 1999).

Glycols, especially ethylene glycol, are used for the synthesis of metal nanoparticles by the reduction of a hydroxide, a sulfate or a nitrate species. This process, called "polyol process" in the literature, is based on redox reactions and leads to finely divided metal powders (Komarneni et al., 2002; Harpeness & Gedanken, 2005; Grisaru et al., 2003). Under conventional heating, this process is very long (several hours) and their transposition under microwave heating can divide the reaction time by 10.

#### 3.4.2 Synthesis of nanoparticles by microwave heating

Recent works of our laboratory showed that the singular and original microwave heating induces some interesting properties to the particles obtained by microwave heating (Bellon et al., 2001; Michel et al., 2001, 2003b; Combemale et al., 2005; Bousquet-Berthelin et al., 2008).

The effects of the core heating, of the high power focused, of the non-diffusional heating process, of the fast temperature rise, are:

- quick synthesis
- high crystallinity of the oxides
- composition of polycation oxide is controlled
- crystalline phase is controlled
- size and size distribution controlled
- morphology controlled
- chemical selectivity

We proved the possibility to realize magnetic nanoparticles by microwave heating (Caillot et al., 2002; Caillot et al. 2004; Michel, 2003b). Microwave heating allows to accelerate chemical reactions, differents from the classical ones. Main advantage is its almost instantaneous "in core" heating of materials in a homogeneous and selective manner which results in a reduction of the processing time and energy cost.

Sreeja & Joy (2007) synthesized maghemite nanoparticle of 10 nm from solution of $FeSO_4$ and $FelCl_3$ with MARS5 CEM system (max power of 1200W) in 25 min. Kulikov et al. (2003) prepared maghemite nanoparticles of 5 nm by microwave heating of $Fe(NO_3)_3$ in aqueous solution. The system was domestical microwave (2450 MHz, 850 W). These examples show that microwave heating allows to decrease the reaction time compared to classical heating (Ge et al., 2006).

## 4. Contribution of microwave heating towards the magnetic nanohybrids synthesis

As previously stated in paragraph 2.3, magnetic nanohybrids are commonly synthesized in two steps under classical heating. The first step is the preparation of magnetic nanoparticles, and the second step is the grafting of polymers onto the nanoparticles. The major problems imposed by these protocols are the obligation of a solvent exchange and the long reaction times (Table 2).

We developed a new approach of the magnetic nanohybride aqueous synthesis: a microwave-assisted one-step grafting of the polymer on the maghemite nanoparticles during their formation. mPEG was used because it has only one alcohol end function which does not allow the intraparticle grafting or the interparticle grafting (Werne & Patten, 2001).

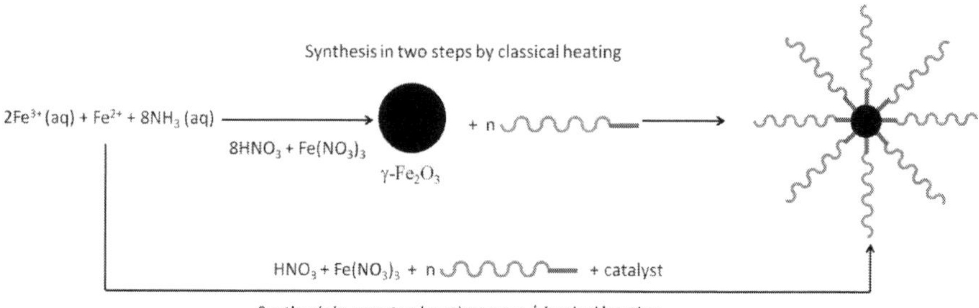


Fig. 10. Representation of preparations of magnetic nanohybrids in one and two steps by classical and microwave synthesis

A comparison will be made between this microwave treatment with those prepared using classical heating mode (Figure 10) according to several criteria: particle size and crystallinity, reaction time and grafted amount of polymer on the nanoparticles.

### 4.1 Experimental part

**Synthesis of silanated mPEG (mPEG-Si)**

The trialkoxysilylated mPEG (mPEG-Si) was prepared by direct coupling of mPEG with IPTS through a urethane bond. IPTS was chosen as coupling agent to allow the grafting of mPEG chains at the surface of maghemite nanoparticles.
40 mL of anhydrous THF is added to 16 g of dried mPEG under nitrogen. Then, 5.5 mL (0.022 mol) of IPTS and 0.56 ml ($0.9.10^{-3}$ mol) of DBTL is added to the mPEG solution. The mixture is stirred continuously for 48 h under nitrogen. After the reaction, the silanated mPEG is precipitated into hexane (400 mL) and dried in vacuum at room temperature for 24 h (yield = 71 %).
According to the work published by Jo & Park (2000), the 300 MHz spectrum shows six sets of peaks:
$^1$H-NMR ($CDCl_3$, 300 MHz): $\delta$ = 0.55 (t, 2H, $J$ = 8.3 Hz, -$CH_2Si$), 1.16 (t, 9H, $J$ = 6.9 Hz, -$OCH_2CH_3$), 1.49−1.59 (m, 2H, -$CH_2CH_2CH_2Si$), 3.31−3.82 (m, 4nH, -$(CH_2CH_2O)_n$-), 4.14 (t, 2H, $J$ = 4.7 Hz, -$CH_2OC(O)NH$-), 4.94 (s-large, 1H, -OC(O)NH-). These results show that the functionalization step is complete at the end of the reaction with IPTS which is used in excess.

**mPEG grafting onto maghemite nanoparticles**

In the following procedures (Figure 10), the grafting of the maghemite nanoparticles is performed in water with a typical w/w (Si-mPEG/$\gamma$-$Fe_2O_3$) ratio of 10. DBTL is added as a catalyst. After the reactions, the excess of Si-mPEG was removed by 10 centrifugation-dispersion cycles in distilled water.

**Two-step grafting procedure by classical heating (Sample 1)**

The maghemite nanoparticles are synthesized in advance according to the procedure developed by Massart (1981). Briefly, 0.0019 mol of $FeCl_2$ and 0.0038 mol of $FeCl_3$ are mixed at 80 °C with 100 mL of distilled water. 10 minutes of agitation is carried out after adding 4.5 mL of 28 % ammonia to the reaction mixture and magnetite black particles precipitate. Magnetite particles are oxidized with 7.5 mL of $HNO_3$ 2 M for 10 min and then with 7.5 mL of $Fe(NO_3)_3$ 0.33 M solution at 80 °C for 30 min. 1 mL of freshly prepared ferrofluid reacts with 0.10 g of Si-mPEG at 80 °C for one hour.

**One-step grafting procedure by classical heating (Sample 2)**

0.00228 mol of $FeCl_2$, 0.00038 mol of $FeCl_3$ and 0.00013 mol of Si-mPEG are dissolved at 80 °C in distilled water. 10 minutes of agitation is carried out after adding 0.45 mL of 28 % ammonia to the reaction mixture and magnetite black particles precipitate. Nanoparticles are oxidized with 0.75 mL of $HNO_3$ 2 M for 10 min and then with 0.75 mL of $Fe(NO_3)_3$ 0.33 M solution at 80 °C for 30 min.

**One-step grafting procedure by microwave heating (Sample 3)**

For this one-step synthesis, we didn't choose the laboratory microwave autoclave reactor for 3 reasons: (i) There is a possibility that the polymer clogs the system during the reaction; (b) the PEG does not support the operating temperature close to 180 °C in the laboratory microwave autoclave reactor and (c) the grafting in alcoholic medium is not proved. Thus the functionalization is performed by the microwave (2.45 GHz) refluxing device.

The power used was 400 W, higher microwave power might cause the difficult control of reaction system. 0.00228 mol of $FeCl_2$, 0.00038 mol of $FeCl_3$ and 0.00013 mol of Si-mPEG are dissolved in water, then the reaction mixture is submitted to a microwave irradiation power of 400 W for 10 min after adding 0.45 mL of 28 % ammonia and magnetite black particles precipitate. They are oxidized with 0.75 mL of $HNO_3$ 2 M and 0.75 mL of $Fe(NO_3)_3$ 0.33 M solution under microwave irradiation for 10 min additional treatment.

The experimental conditions of these syntheses are summarized in Table 3. The results show that maghemite nanoparticles can be synthesized by the microwave heating method in an attractive shorter time.

| Sample | Grafting protocol | Heating mode | Total reaction time (min) |
|---|---|---|---|
| 1 | 2 steps | Classical | 110 |
| 2 | 1 step | Classical | 50 |
| 3 | 1 step | Microwave | 20 |

Table 3. Reaction conditions of different nanohybrids synthesis

### 4.2 Characterization of the grafted nanoparticles

**Structural characterization**

In the case of one step grafting procedure, the grafting step occurs at the same time as the formation of the nanoparticles. The diffractogrammes imply that the nanoparticles have a cubic crystal structure (Figure 11).

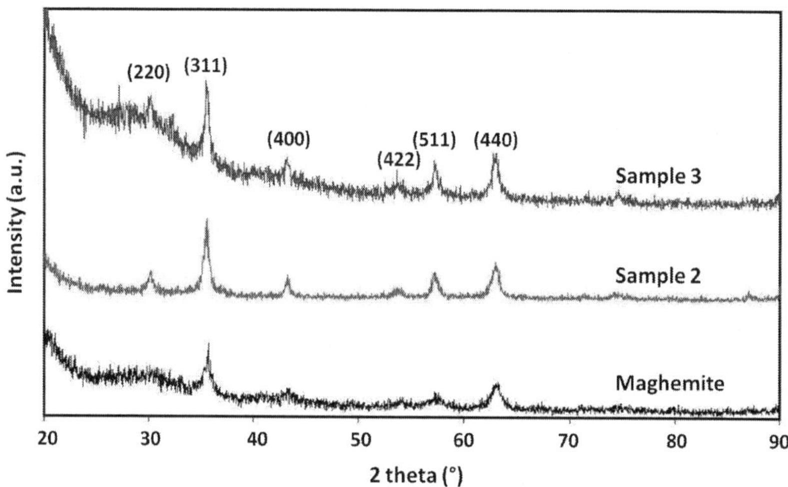


Fig. 11. Diffractograms of ungrafted and grafted maghemite nanoparticles

Raman spectra allow us to confirm that these samples contain pure maghemite nanoparticles, by the presence of characteristic bands of maghemite phase around 350, 500 and 700 cm$^{-1}$ (Figure 12). Nakashima et al. (2001) showed that the relative intensities of the bands are related to the crystallinity of the particles. The intensity of the bands of particles grafted by microwave reflux heating is stronger than those of particles grafted by conventional heating, which means that the microwave reflux heating tends to improve the crystallinity of the particles. The microwave system reflux can induce a very rapid heating rate throughout the system and lead to the instantaneous nucleation of monodisperse particles.

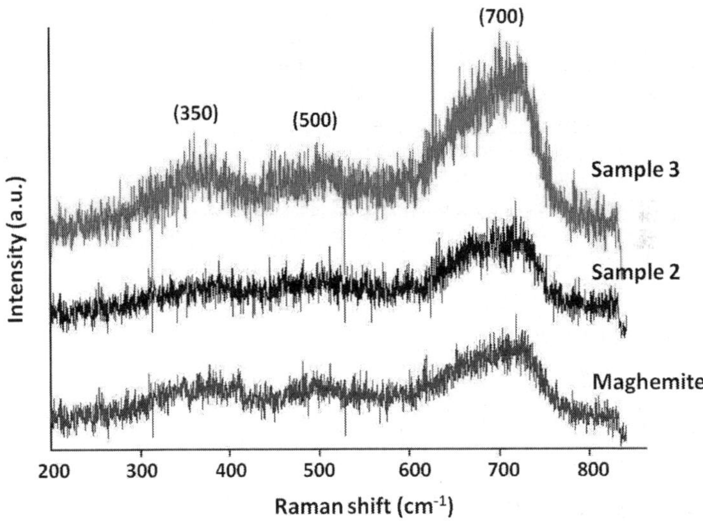


Fig. 12. Raman spectra of ungrafted and grafted maghemite nanoparticles

**Particle morphology and size**

The particle morphology is studied initially by Transmission Electron Microscopy (TEM). Nanosized almost spherical nanoparticles are observed (Figure 13). Table 4 reveals a good agreement with the average particle sizes determined by XRD profile analysis. The presence of the macromolecular chain is not visible on the images of grafted maghemite but the pictures reveal that the nanoparticles can be isolated.

| Sample | $d_{DRX}$ (nm) | $d_{MET}$ (nm) |
|---|---|---|
| Maghemite | 14 ± 2 | 12 ± 1 |
| 1 | - | 12 ± 1 |
| 2 | 20 ± 3 | 20 ± 1 |
| 3 | 18 ± 3 | 21 ± 1 |

Table 4. Particle sizes determined by XRD and TEM

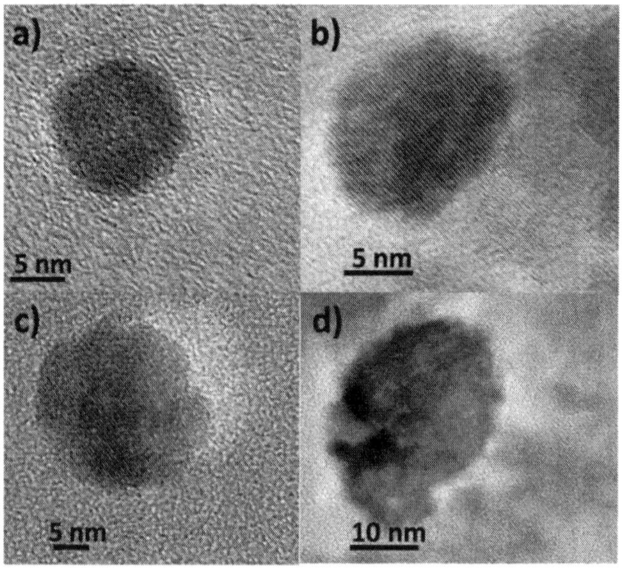


Fig. 13. TEM images of ungrafted (a) and grafted maghemite nanoparticles: sample 1 (b), sample 2 (c) and sample 3 (d)

**Surface properties**

Zeta potential of ungrafted maghemite is 42.3 mV (Table 5) which signifies that the solution is stable. The protonated hydroxyl groups present on their surface ensured efficient electrostatic stabilization of the iron oxide particles. The decrease of surface charge of the grafted samples showed successful PEG grafting. To explain this phenomenon, two reasons are suggested: neutralization of the positive charge of nanoparticles by a covalent bond with mPEG-Si or screening effect of the polymer covering the surface which hides the positive charge (Lu et al., 2007).

| Sample | Zeta Potential (mV) |
|---|---|
| Maghemite | +42.3 ± 2,1 |
| 1 | -0.1 ± 0,1 |
| 2 | +0.2 ± 0,1 |
| 3 | +0.1 ± 0,1 |

Table 5. Zeta potential of ungrafted and grafted maghemite nanoparticles

**Chemical bonds in the nanohybrids**

The comparison of FTIR spectra of the ungrafted and grafted nanoparticles, in the 400 - 2000 cm-1 range, is shown in Figure 14. The peaks at 631 and 567 $cm^{-1}$ in the spectrum of ungrafted particles are assigned to the Fe-O bond vibration of maghemite particles (Li et al., 2004). The peak at 1384 $cm^{-1}$ is attributed to the $v$(N-O) deformation from the surface nitrates (Mornet et al., 2005).

The spectrum of Si-mPEG is presented for the reference. The peak at 1715 $cm^{-1}$ is attributed to the vibration of the urethane carbonyle bond $v(C=O)$ (Kim et al., 2005). The presence of absorption peak at 1531 $cm^{-1}$ reveals the presence of urethane bond $\delta(NH)$ (Schoonover et al., 2007). The peaks at 1466, 1348 and 1279 $cm^{-1}$ are attributed to the vibration of $-CH_2$ (Flesch et al., 2005a). The signal at 1240 $cm^{-1}$ is attributed to the vibration of $v(C\text{-}O)$ (Zhang et al., 2008). The band around 1105 $cm^{-1}$ is due to the $v(Si\text{-}O)$ vibration (Ma et al., 2003). The bands at 961 and 839 $cm^{-1}$ correspond to the ethoxy groups $-OC_2H_5$ (Steitz et al., 2007). The absorption band at 943 $cm^{-1}$ belongs to -CH out-of-plane bending vibration of PEG (Zhang et al., 2002).

The characteristic peaks of mPEG-Si appear on the spectra of grafted maghemite, besides the Fe-O signals of maghemite: $v(C=O)$ at 1695 $cm^{-1}$, $\delta(NH)$ at 1539 $cm^{-1}$, $\delta(CH_2)$ at 1452 $cm^{-1}$, $T(CH_2)$ at 1348 $cm^{-1}$, $\omega(CH_2)$ at 1285 $cm^{-1}$, $v(C\text{-}O)$ at 1246 $cm^{-1}$ and $\omega(\text{-}CH)$ at 943 $cm^{-1}$. All these vibrations confirm that Si-mPEG is well grafted at the surface of maghemite particles. The absorption band at 1624 $cm^{-1}$ refers to the vibration of remaining $H_2O$ on the sample (Mena Duran et al., 2007). The Si-mPEG is adsorbed on the maghemite surface by a Fe-O-Si bond. However, in the literature, this band is situated around 584 $cm^{-1}$ and therefore overlaps with the Fe-O vibration of maghemite (Yamaura et al., 2004). The vibration of the carbonyle bond is shifted from 1715 to 1695 $cm^{-1}$ after grafting, since this bond is weakened by hydrogen bonds that are formed between carbonyl and surface hydroxy groups (Posthumus et al., 2004). The disappearance of the characteristic peaks for the ethoxy groups at 961 and 839 $cm^{-1}$ shown in the spectrum of silanated m-PEG, indicates that hydrolysis

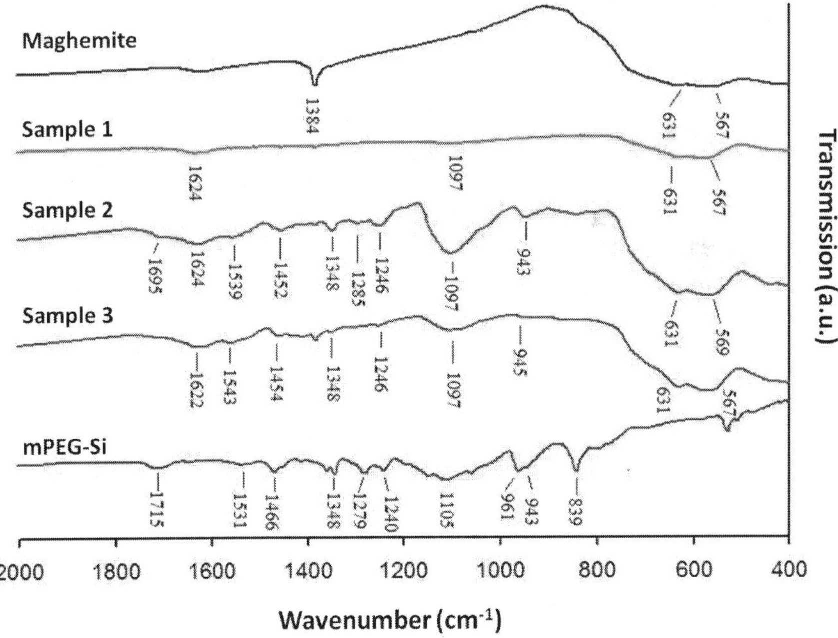


Fig. 14. FTIR spectra of ungrafted and grafted maghemite nanoparticles and spectrum of mPEG-Si shown for reference

took place. The large band around 1097 $cm^{-1}$ is due to the *υ*(Si-O-Si) vibration which shows that the triethoxysilane groups have self condensed to form a polysiloxane film on the maghemite surface. The organosilane is hydrolyzed in aqueous solution, alkoxide groups -$OC_2H_5$ are replaced by hydroxyl groups -OH to form reactive silanol groups which condense with other silanol groups to produce siloxane bonds Si-O-Si. The presence of the group Si-O-Si confirmed that the last mechanism took place in our case.

**Grafted amounts of polymers**

The grafted amount of mPEG-Si determined by thermogravimetric analysis (TGA) reveals the effectiveness of mPEG-Si grafting whatever the grafting methods used (Table 6). It has been calculated from the difference in losses between the grafted maghemite and the ungrafted maghemite in the temperature range between 30 and 850 °C.

| Sample | Heating energy | Total reaction time (min) | Weight loss (%) | Difference in weight loss (%) |
|---|---|---|---|---|
| Maghemite | Classical | 50 | 6 | 0 |
| 1 | Classical | 110 | 16 | 10 |
| 2 | Classical | 50 | 21 | 15 |
| 3 | Microwave | 20 | 18 | 12 |

Table 6. Grafted amount of samples 1, 2 and 3

Weight losses of 15 % and 12 % of grafted mPEG-Si at the nanoparticles surface were determined for the two samples prepared by the one step procedure. These values are slightly superior to the one obtained by the two steps procedure (10 %). However, the one step procedures take less time to reach this grafted amount. Especially, the sample 3 prepared by the microwave refluxing allows to reduce 82 % of the reaction time. Compared to the classical two steps method, microwave heating provides the best grafting rate of mPEG onto nanoparticles. Ge et al. (2006) has shown the same tendency in his work. These results show that the microwave heating can increase the grafting rate over the conventional method.

### 4.3 Conclusion

In this part, we studied the preparation of magnetic nanohybrids by microwave heating, compared with the classical system. One two-step and two one-step procedures in terms of PEG grafting onto maghemite nanoparticles in aqueous medium were compared. In all cases, the formation of pure polycrystalline maghemite nanoparticles was proved by XRD and Raman analysis, the PEG grafting was evidenced by FTIR spectroscopy and by means of zeta potential measurements, and the grafted amount was determined by TGA. The main difference between microwave heating and conventional heating is a consequent time saving. It is obvious that one-step microwave heating procedure gives the same amount of PEG grafting by using less reaction time (82 % shorter) than the other two methods. Moreover, Raman studies reveal that microwave treatment tends to result in better crystallized particles. Microwave heating provides a simple and convenient way for the PEG grafting that could potentially be useful in the field of magnetic nanohybrids.

## 5. Conclusion

The magnetic nanohybrids (magnetic core/biocompatible polymeric corona) are increasingly used in medical imaging applications, for drug delivery, for hyperthermia treatment...

While conventional methods for the synthesis of nanohybrids have the disadvantages of changing solvent and to be long, we have proposed simplistic approaches. The use of polymer functionalized with an anchor allows to achieve grafting directly in aqueous medium containing magnetic oxide nanoparticles. Another improvement is to grow the magnetic nanoparticles in the presence of polymer chains linked with an anchor: this procedure guarantees the non-agglomeration of nanoparticles and the grafting of individual particles. Finally we take advantage of microwave heating (core heating, high heating power, high speed of temperature rise, uniform heating ...) for the synthesis of nanohybrids in aqueous medium. The microwave was used to obtain magnetic particles of pure $\gamma$-$Fe_2O_3$, with a good crystallinity, of monodisperse sizes (21 nm) and a grafting amount of PEG identical to that obtained by the conventional methods but with a synthesis time considerably shorter, at least 5 times faster.

This approach can be applied to other hybrid systems and for many other applications: in plant biology, in cosmetics, in cements, in paints, in formulation...

## 6. References


Baghi, M., Mack, M. G., Hambek, M., Rieger, J., Vogl, T., Gstoettner, W. & Knecht, R. (2005). *The efficacy of MRI with ultrasmall superparamagnetic iron oxide particles (USPIO) in head and neck cancers*, Anticancer Research, Vol. 25, No. 5, pp. 3665-3670

Barruet, J. (2007). *Polymères réactifs à base d'isocyanates bloqués: développement de méthodologies de synthèse pour la bioconjugaison*, PhD thesis, University of Paris 12

Bellon, K. (2000). *Elaboration de sols et de poudres nanométriques par hydrolyse force microonde. Applications aux oxydes de fer (III) et de zirconium (IV)*, PhD thesis, University of Burgundy

Bellon, K., Chaumont, D. & Stuerga, D. (2001). *Flash synthesis of zirconia nanoparticles by microwave forced hydrolysis*, Journal of Materials Research, Vol. 16, pp. 2619-2622

Berry, C. C., Wells, S., Charles, S. & Curtis, A. S. G. (2003). *Dextran and albumin derivatised iron oxide nanoparticles: influence on fibroblasts in vitro*, Biomaterials, Vol. 24, pp. 4551-4557

Bonnans-Plaissance, C., Levesque, G. & Toulin, V. (2001). *Functionalization of PS and PMMA by a chain transfer agent and study of their degradation, Reactive & Functional Polymers*, Vol. 47, pp. 77-85

Bousquet-Berthelin, C., Chaumont, D. & Stuerga, D. (2008). *Flash microwave synthesis of trevorite nanoparticles*, Journal of Solid State Chemistry, Vol. 181, pp. 616-622

Brachais, C.-H., Hu, L., Hach, D., Chaumont, D., Percheron, A. & Couvercelle, J.-P. (2010). *Aqueous chemical grafting of modified-PEG onto maghemite nanoparticles: Influence of grafting conditions*, e-polymers

Brigger, I., Dubernet, C. & Couvreur, P. (2002). *Nanoparticles in cancer therapy and diagnosis*, Advanced Drug Delivery Reviews, Vol. 54, pp. 631-651

Brosse, C., Hamdaoui, A. E., Soutif, J. C. & Brosse, J. C. (1995). *Synthèse de polyuréthanes utilisables pour des techniques de microencapsulation –I. Modification chimique d'alcool polyvinylique par des isocyanates d'alkyle*, European Polymer Journal, Vol. 31, pp. 425-429

Buurma, N. J., Blandamer, M. J. & Engberts, B. F. N. (2003). *General-base catalyzed hydrolysis and nucleophilic substitution of activated amides in aqueous solutions*, Journal of Physical Organic Chemistry, Vol. 16, pp. 438-449

Caillot, T., Aymes, D., Stuerga, D., Viart, N. & Pourroy, G. (2002). *Microwave flash synthesis of iron and magnetite particles by disproportionation of ferrous alcoholic solutions*, Journal of Materials Science, Vol. 37, pp. 1-6

Caillot, T., Pourroy, G. & Stuerga, D. (2004). *Microwave hydrothermal flash synthesis of nanocomposites Fe-Co alloy/cobalt ferrite*, Journal of Solid State Chemistry, Vol. 177, pp. 3843-3848

Changez, M., Burugapalli, K., Koul, V. & Choudhary, V. (2003). *The effect of composition of poly(acrylic acid)-gelatin hydrogel on gentamicin sulphate release: in vitro*, Biomaterials, Vol. 24, pp. 527-536

Chen, M., Zamora, P. O., Som, P., Pena, L. A. & Osaki, S. (2003) *Cell attachment and biocompatibility of polytetra uoroethylene (PTFE) treated with glow-discharge plasma of mixed ammonia and oxygen*, Journal of Biomaterials Science, Polymer Edition, Vol. 14, pp. 917-935

Choi, J. Y. & Kim, D. K. (1999). *Preparation of monodisperse ans spherical powders by heating of alcohol-aqueous salt solutions*, Journal of Sol-Gel Science and Technology, Vol. 15, pp. 231-241

Cirera, A., Vila, A. & Dieguez, A. (2000). *Microwave processing for the low cost, mass production of undoped and in situ catalytic doped nanosized $SnO_2$ gas sensor powders*, Sensors and Actuators, Vol. 64, pp. 65-69

Combemale, L., Caboche, G, Stuerga, D & Chaumont, D. (2005). *Microwave synthesis of yttria stabilized zirconia*, Materials Research Bulletin, Vol. 40, pp. 529-536

Daichuan, D., Pinjie, H. & Shushan, D. (1995). *Preparation of uniform β-FeO(OH) colloidal particles by hydrolysis of ferric salts under microwave irradiation*, Materials Research Bulletin, Vol. 30, pp. 537-541

Dallas, P., Moutis, N., Devlin, E., Niarchos, D. & Petridis, D. (2006). *Characterization, electrical and magnetic properties of polyaniline/maghemite nanocomposites*, Nanotechnology, Vol. 17, pp. 5019-5026

Davis, S. S. (1997). *Biomedical applications of nanotechnology- implication for drug targeting and gene therapy*, TIBTECH, Vol. 15, pp. 217-224.

Delair, T., Elaissari, A., Perrin, A. & Mandrand, B. (2003). *Les polymères de synthèse, supports du diagnostic médical*, pp. 79-84

Désévaux, C. (2002). *Développement d'un implant solide biodegradable à base d'amidon réticulé à teneur élevée en amylose pour la libération contrôlée d'un principe actif*, PhD thesis, University of Montréal

Drakakis, E. M., Toumazou, C. & Cass, A. E. G. (2008). *Biocompatible ion selective electrode for monitoring metabolic activity during the growth and cultivation of human cells*, Biosensors and Bioelectronics, Vol. 24, pp. 435-441

D'Souza, A. J. M., Schowen, R. L. & Topp, E. M. (2004). *Polyvinylpyrrolidone-drug conjugate: synthesis and release mechanism*, Journal of Controlled Release, Vol. 94, pp. 91-100

Fan, Q. L., Neoh, K. G., Kang, E. T., Shuter, B. & Wang, S. C. (2007). *Solvent-free atom transfer radical polymerization for the preparation of poly(poly(ethyleneglycol) monomethacrylate)-*

*grafted $Fe_3O_4$ nanoparticles: synthesis, characterization and cellular uptake*, Biomaterials, Vol. 28, pp. 5426-5436

Flesch, C., Delaite, C., Dumas, P., Bourgeat-Lami, E. & Duguet, E. (2004). *Grafting of poly(ε-caprolactone) onto maghemite nanoparticles*, Journal of Polymer Science: Part A: Polymer Chemistry, Vol. 42, pp. 6011-6020

Flesch, C., Jourbert, M., Bourgeat-Lami, E., Mornet, S., Duguet, E., Delaite, C. & Dumas, P. (2005a). *Organosilane-modified maghemite nanoparticles and their use as co-initiator in the ring-opening polymerization of ε-caprolactone*, Colloids and Surfaces A: Physicochemical and Engineering Aspects, Vol. 262, pp. 150-157

Flesch, C., Bourgeat-Lami, E., Mornet, S., Duguet, E., Delaite, C. & Dumas, P. (2005b). *Synthesis of colloidal superparamagnetic nanocomposites by grafting poly(ε-caprolactone) from the surface of organosilane-modified maghemite nanoparticles*, Journal of Polymer Science: Part A: Polymer Chemistry, Vol. 43, pp. 3221-3231

Flesch, C., Unterfinger, Y., Bourgeat-Lami, E., Duguet, E., Delaite, C. & Dumas, P. (2005c). *Poly(ethylene glycol) surface coated magnetic particles*, Macromolecular Rapid Communications, Vol. 26, pp. 1494-1498

Gaur, U., Sahoo, S. K., De, T. K., Ghosh, P. C., Maitra, A. & Ghosh, P. (2000). *Biodistribution of fluoresceinated dextran using novel nanoparticles evading reticuloendothelial system*, International Journal of Pharmaceutics, Vol. 202, pp. 1-10

Ge, H. C., Pang, W. & Luo, D. (2006). *Graft copolymerization of chitosan with acrylic acid under microwave irradiation and its water absorbency*, Carbohydrate Polymers, Vol. 66, pp. 372-378

Grignon-Dubois, M., Dunoguès, J. & Calas, R. (1980). *Influence de la taille du cycle sur la substitution électrophile des cycloalkyltriméthylsilanes. Synthèse de cycloalkyldiméthylfluorosilanes*, Canadian Journal of Chemistry, Vol. 58, pp. 291-295

Grisaru, H., Palchik, O., Gedanken, A., Palchik, V., Slifkin, M. A. & Weiss, A. M. (2003). *Microwave-assisted polyol synthesis of $CuInTe_2$ and $CuInSe_2$ nanoparticles*, Inorganic Chemistry, Vol. 42, pp. 7148-7155

Gupta, A. K. & Curtis, A. S. G. (2004). *Surface modified superparamagnetic nanoparticles for drug delivery: interaction studies with human fibroblasts in culture*, Journal of Materials Science: Materials in Medicine, Vol. 15, pp. 493-496

Harpeness, R. & Gedanken, A. (2005). *The microwave-assisted polyol synthesis of nanosized hard magnetic material, FePt*, Journal of Materials Chemistry, Vol. 15, pp. 698-702

Houghton, R. P. & Mulvaney, A. W. (1996). *Mechanism of tin(IV)-catalyzed urethane formation*, Journal of Organometallic Chemistry, Vol. 518, pp. 21-27

Hu, L., Hach, D., Chaumont, D., Brachais, C.-H. & Couvercelle, J.-P. (2008). *One step grafting of monomethoxy poly(ethylene glycol) during synthesis of maghemite nanoparticles in aqueous medium*, Colloids and Surfaces A: physicochemical and engineering, Vol. 330, pp. 1-7

Jo, S. B. & Park, K. N. (2000). *Surface modification using silanated poly(ethylene glycol)s*, Biomaterials, Vol. 21, pp. 605-616

Kelnar, I. & Schätz, M. (1993). *Silane cross-linking of PVC. I. Grafting of mercaptoalkylalkoxysilanes onto PVC: properties of the grafted and cross-linked product*, Journal of Applied Polymer Science, Vol. 48, pp. 657-668

Khor, E. & Lim, L. Y. (2003). *Implantable applications of chitin and chitosan*, Biomaterials, Vol. 24, pp. 2339-2349

Kim, H., Lim, C. & Hong, S. I. (2005). *Gas permeation properties of organic-inorganic hybrid membranes prepared from hydroxyl-terminated polyester and 3-isocyanatopropyltriethoxysilane*, Journal of Sol-Gel Science and Technology, Vol. 36, pp. 213-221

Komarneni, S., Li, D. & Newalkar, B. (2002). *Microwave-polyol process for Pt and Ag nanoparticles*, Langmuir, Vol. 18, pp. 5959-5962

Kubaska, S., Sahani, D. V., Saini, S., Hahn, P. F. & Halpern, E. (2001). *Dual contrast enhanced magnetic resonance imaging of the liver with superparamagnetic iron oxide followed by gadolinium for lesion detection and characterization*, Clinical Radiology, Vol. 56, pp. 410-415

Kulikov, F. A., Vanetsev, A. S., Muav'eva, G. P., Il'inskii, A. L., Oleinikov, N. N. Tret'yakov, Y. D. (2003). *Microwave synthesis of* $\gamma$-$Fe_2O_3$, Inorganic Materials, Vol. 39, pp. 1074-1075

Lee, H. Y., Lim, N. H., Seo, J. A., Yuk, S. H., Kwak, B. K., Khang, G., Lee, H. B. & Cho, S. H. (2006). *Preparation and magnetic resonance imaging effect of polyvinylpyrrolidone-coated iron oxide nanoparticles*, Journal of Biomedical Materials Research Part B: Applied Biomaterials, Vol. 79, pp. 142-150

Li, J., Zeng, H., Sun, S., Liu, P. & Wang, L. (2004). *Analyzing the structure of CoFe-*$Fe_3O_4$ *core-shell nanoparticles by electron imaging and diffraction*, Journal of Physical Chemistry B, Vol. 108, pp. 14005-14008

Lu, A. H., Salabas, E. L. & Schüth, F. (2007). *Magnetic nanoparticles: synthesis, protection, functionalization, and application*, Angewandte Chemie International Edition, Vol. 46, pp. 1222-1244

Lu, W. G., Yang, D. Q., Sun, Y., Guo, Y., Xie, S. P. & Li, H. L. (1999). *Preparation and structural characterization of nanostructured iron oxide thin films,* Applied Surface Science, Vol. 147, pp. 39-43

Ma, M., Zhang, Y., Yu, W., Shen, H. Y., Zhang, H. Q. & Gu, N. (2003). *Preparation and characterization of magnetite nanoparticles coated by amino silane*, Colloids and Surfaces A : physicochemical and engineering aspects, Vol. 212, pp. 219-226

Maillard, S., Gauduchon, J., Marsaud, V., Gouilleux, F., Connault, E., Opolon, P., Fattal, E., Sola, B. & Renoir, J. M. (2006). *Improved antitumoral properties of pure antiestrogen RU 58668-loaded liposomes in multiple myeloma*, Journal of Steroid Biochemistry & Molecular Biology, Vol. 100, pp. 67-78

Massart, R. (1981). *Preparation of aqueous magnetic liquids in alkaline and acidic media*, IEEE Transactions on magnetics, Vol. 17, pp. 1247-1248

Meerod, S., Tumcharem, G., Wichai, U. & Rutnakornpituk, M. (2008). *Magnetite nanoparticles stabilized with polymeric bilayer of poly(ethylene glycol) methyl ether-poly(ε-caprolactone) copolymers*, Polymer, Vol. 49, pp. 3950-3956

Mena-Duran, C. J., Sun Kou, M. R., Lopez, T., Azamar-Barrios, J. A., Aguilar, D. H., Dominguez, M. I., Odriozola, J. A. & Quintana, P. (2007). *Nitrate removal using natural clays modified by acid thermoactivation*, Applied Surface Science, Vol. 253, pp. 5762-5766

Michel, E. (2003a). *Thermohydrolyse micro-onde: des nanoparticles aux films minces. Applications à* $SnO_2$ *et* $TiO_2$ *rutile et anatase*, PhD thesis, University of Burgundy, France

Michel, E., Sturga, D. & Chaumont, D. (2001). *Microwave flash synthesis of tin dioxide sols from tin chloride aqueous solutions*, Journal of Material Science Letters, Vol. 20, pp. 1593-1595

Michel, E., Chaumont, D. & Stuerga, D. (2003b). *$SnO_2$ thin films prepared by dip-coating from microwave synthesized colloidal suspensions*, Journal of Colloid and Interface Science, Vol. 257, pp. 258-262

Mornet, S., Portier, J. & Duguet, E. (2005). *A method for synthesis and functionalization of ultrasmall superparamagnetic covalent carriers based on maghemite and dextran*, Journal of Magnetism and Magnetic Materials, Vol. 293, pp. 127-134

Mowat, P. (2007). *IRM cellulaire de lymphocytes marques par des nanoparticules d'oxydes de fer: application au diagnostic en cancerologie*, PhD thesis, University of Angers

Nakashima, S., Nakatake, Y., Ishida, Y., Talkahashi, T. & Okumura, H. (2001). *Detection of defects in SiC crystalline films by Raman scattering*, Physica B, Vol. 308-310, pp. 684-686

Narkis, M., Tzur, A. & Vaxman, A. (1985). *Some properties of silane-grafted moisture-crosslinked polyethylene*, Polymer Engineering and Science, Vol. 25, pp. 857-862

Ninjbadgar, T., Yamamoto, S. & Fukuda, T. (2004). *Synthesis and magnetic properties of the $\gamma$-$Fe_2O_3$/poly-(methyl methacrylate)-core/shell nanoparticles*, Solid State Science, Vol. 6, pp. 879-885

Pinto-Gateau, N. (1995). *Chauffage microonde et résonance dimensionnelle. Des concepts aux applications en géométrie cylindrique*, PhD thesis, University of Burgundy, France

Plassat, V., Martina, M. S., Barratt, G., Marsaud, V., Ménager, C., Renoir, J. M. & Lesieur, S. (2008). *Liposomes superparamagnétiques PEG-yles pour l'IRM et la thérapie anticancéreuse: pharmacocinétique, biodistribution et étude in-vitro sur cellules MCF-7*, Fourth meeting of the Conference of "Vectorisation de molécules actives et ciblage biologique" (Bordeaux, France)

Posthumus, W., Magusin, P. C. M. M., Brokken-Zijp, J. C. M., Tinnemans, A. H. A. & Van der Linde, R. (2004). *Surface modification of oxidic nanoparticles using 3-methacryloxypropyltrimethoxysilane*, Journal of Colloid and Interface Science, Vol. 269, pp. 109-116

Ramirez, L. P. & Landfester, K. (2003). *Magnetic polystyrene nanoparticles with a high magnetite content obtained by miniemulsion processes*, Macromolecular Chemistry and Physics, Vol. 204, pp. 22-31

Rana, S., Gallo, A., Srivastava, R. S. & Misra, R. D. K. (2007). *On the suitability of nanocrystalline ferrites as a magnetic carrier for drug delivery: functionalization, conjugation and drug release kinetics*, Acta Biomaterialia, Vol. 3, pp. 233-242

Reddy, K. R., Lee, K. P., Gopalan, A. I. & Kang, H. D. (2007). *Organosilane modified magnetite nanoparticles/poly(aniline-co-o/m-aminobenzenesulfonic acid) composites: synthesis and characterization*, Reactive & Functional Polymers, Vol. 67, pp. 943-954

Schoonover, J. R., Steckle Jr., W. P., Cox, J. D., Johnston, C. T., Wang, Y., Gillikin, A. M. & Palmer, R. A. ((2007). *Humidity-dependent dynamic infrared linear dichroism study of a poly(ester urethane)*, Spectrochimica Acta Part A, Vol. 67, pp. 208-213

Schultz, J. F., Bell, J. D., Goldstein, R. M., Kuhn, J. A. & McCarty, T. M. (1999). *Hepatic tumor imaging using iron oxide MRI: comparison with computed tomography, clinical impact, and cost analysis*, Annals of Surgical Oncology, Vol. 6, No. 7, pp. 691-698

Shan, G. B., Xing, J. M., Luo, M. F., Liu, H. Z. & Chen, J. Y. (2003) *Immobilization of pseudomonas delafieldii with magnetic polyvinyl alcohol beads and its application in biodesulfurization*, Biotechnology Letters, Vol. 25, pp. 1977-1981

Sindhu, S., Jegadesan, S., Parthiban, A. & Valiyaveettil, S. (2006). *Synthesis and characterization of ferrite nanocomposite spheres from hydroxylated polymers*, Journal of Magnetism and Magnetic Materials, Vol. 296, pp. 104-113

Sreeja, V. & Joy, P. A. (2007). *Microwave-hydrothermal synthesis of γ-$Fe_2O_3$ nanoparticles and their magnetic properties*, Materials Research Bulletin, Vol. 42, pp. 1570-1576

Steitz, B., Hofmann, H. & Petri-Fink, A. (2007). *Production and biofunctionalization of magnetic nanobeads for magnetic separation of messenger RNA*, Biophysical Reviews and Letters, Vol. 2, pp. 109-122

Stuerga, D. & Gaillard, P. (1996). *Microwave heating as a new way to induce localized enhancements of reaction rate, non-isothermal and heterogeneous kinetics*, Tetrahedron, Vol. 52, pp. 5505-5510

Takafuji, M., Ide, S., Ihara, H. & Xu, Z. (2004). *Preparation of poly(1-vinylimidazole)-grafted magnetic nanoparticles and their application for removal of metal ions*, Chemistry of Materials, Vol. 16, pp. 1977-1983

Thanh, S. P., Gaslain, F. & Leblanc, M. (2000). *Rapid synthesis of hybrid fluorides by microwave heating*, Journal of Fluorine Chemistry, Vol. 101, pp. 161-163

Wada, S., Tazawa, K., Furuta, I. & Nagae, H. (2003). *Antitumor effect of new local hyperthermia using dextran magnetite complex in hamster tongue carcinoma*, Oral Diseases, Vol. 9, pp. 218-223

Werne, T. V. & Patten, T. E. (2001). *Atom transfer radical polymerization from nanoparticles: a tool for the preparation of well-defined hybrid nanostructures and for understanding the chemistry of controlled/living radical polymerizations from surfaces*, Journal of American Chemical Society, Vol. 123, pp. 7497-7505

Yamaura, M., Camilo, R. L., Sampaio, L. C., Macedo, M. A., Nakamura, M. & Toma, H. E. (2004). *Preparation and characterization of (3-aminopropyl) triethoxysilane-coated magnetite nanoparticles*, Journal of Magnetism and Magnetic Materials, Vol. 279, pp. 210-217

Zhang, J., Rana, S., Srivastava, R. S. & Misra, R. D. (2008). *On the chemical synthesis and drug delivery response of folate receptor-activated, polyethylene glycol-functionalized magnetite nanoparticles*, Acta Biomaterialia, Vol. 4, pp. 40-48

Zhang, Y., Kohler, N. & Zhang, M. Q. (2002). *Surface modification of superparamagnetic magnetite nanoparticles and their uptake*, Biomaterials, Vol. 23, pp. 1553-1561

Zhou, L. L., Yuan, J. Y., Yuan, W. Z., Sui, X. F., Wu, S. Z., Li, Z. L. & Shen, D. Z. (2009). *Synthesis, characterization, and controllable drug release of pH-sensitive hybrid magnetic nanoparticles*, Journal of Magnetism and Magnetic Materials, Vol. 321, pp. 2799-2804

Michel, E., Stuerga, D. & Chaumont, D. (2001). *Microwave flash synthesis of tin dioxide sols from tin chloride aqueous solutions*, Journal of Material Science Letters, Vol. 20, pp. 1593-1595

# 13

# Synthesis of Carbon-Based Materials by Microwave Hydrothermal Processing

Marcela Guiotoku[1], Claudia M. B. F. Maia[1], Carlos R. Rambo[2] and Dachamir Hotza[3]
*[1]National Forestry Research (EMBRAPA)*
*Estrada da Ribeira, km 111, Colombo-PR,*
*[2]Department of Electrical Engineering (EEL);*
*[3]Department of Chemical Engineering (EQA),*
*Federal University of Santa Catarina (UFSC), Florianópolis - SC,*
*Brazil*

## 1. Introduction

The conventional methods of converting biomass into renewable energy are based on thermal, biochemical or physical processes. Carbonization is one of the possible thermochemical conversion of biomass into energy, where a solid residue known as charcoal is produced through a slow process of partial thermal decomposition of wood in the absence or controlled presence of oxygen (Bridgwater, 2003).

Each temperature range in carbonization is responsible for a product or byproduct. Both temperature and the raw material influence the quality of the obtained charcoal and they play a key role in several reactions that occur in the carbonization process used to produce compounds with different physical and chemical properties.

In Brazil, the conventional charcoal production consists in cycles of 8-10 days to produce 20 $m^3$ of charcoal, depending on the oven type. Furthermore, the carbonization process generates byproducts such as carbon dioxide (60%), carbon monoxide (30%) among others (Trugilho & Silva, 2001).

These facts raise concern about global warming and environmentally friendly processes of energy production, and the need for research that aims to develop clean technology or green chemistry, new energy production routes that reduce waste and pollutants.

The use of renewable energy has been widely discussed as an alternative to fossil fuels. The biomass, consisting mainly of agricultural and forestry waste, can be regarded as a renewable energy source with potential to supply the global energy demands. Moreover, the use of biomass contributes to reduce the greenhouse effect.

The discussion on biofuels and renewable energy sources became a relevant topic both in academia and industry. The development of new technologies that leads to new and more products with specific applications and economically feasible is a challenge for scientific and technological advances.

A great contribution to environmentally sustainable processes was the development of a carbonization method that uses aqueous media, the hydrothermal carbonization (HTC). Hydrothermal carbonization is a thermochemical process for biomass conversion to produce a solid material, named "hydrochar". In this chapter, authors feature the use of microwave energy in hydrothermal carbonization of lignocellulosic materials as an innovative process and discuss the potential of this technique in biochar production.

## 2. Hydrothermal carbonization

There are many methods to produce advanced materials. One of them is hydrothermal processing. This technique enables the production of complex materials with interesting physicochemical properties. A wide range of materials such as metals, oxides, hydroxides, silicates, carbonates, phosphates and sulphates are being produced by this technique as nanosctructured particules (nanotubes, nanowires, nanospheres). It is also a method used to produce carbonaceous materials with both $sp^2$ and $sp^3$ hybridization type.

According to Yoshimura and Byrappa (2008), a hydrothermal process can be defined as any homogeneous or heterogeneous chemical reaction in the presence of solvent (whether aqueous or non-aqueous) above room temperature and at pressure greater than 1 atm in a closed system. As previously mentioned, the hydrothermal carbonization (HTC) is a thermochemical conversion process of biomass or lignocellulosic raw materials that yields a solid product, known as hydrochar. HTC has been widely used to simulate the coalification in laboratory (Funke and Ziegler, 2010). Due to increasing demand for efficient biomass conversion technologies, hydrothermal carbonization has attracted much attention as a promising large scale apllication.

Several studies have already been published reporting on hydrothermal environment to carbonize materials like cellulose (Sevilla & Fuertes, 2009; Inoue et al., 2008), switchgrass and corn stover (Kumar et al., 2011), wood (Liu et al., 2010), microalgae (Heilmann et al., 2010), swine-manure (Cao et al., 2010) and sugars such as xylose (Ryu et al., 2010), glucose (Mi et al., 2008), sucrose and starch (Sevilla & Fuertes, 2009).

Basically, the method consists of heating the biomass in the presence of a catalyst in a closed vessel under pressure, at temperatures ranging from 180 to 300°C, with reaction times between 1 and 48 h. Thus, the hydrothermal carbonization allows the use of stored energy in biomass more efficiently. Figure 1 shows a simplified scheme of energy comparison of HTC with the most common methods for processing biomass (adapted from Titirici et al., 2007).

Theoretically, 15% of the energy stored in biomass is already lost when the carbohydrates are converted into alcohol, for example, and two of six carbon atoms are released as $CO_2$, generating a carbon efficiency of 0.66 or 60%. The carbon conversion efficiency (CE) can be defined as the amount of carbon derived from biomass, which remains linked to the final product after procesing. In the anaerobic conversion, about 18% of energy is lost and 50% of carbon is released as $CO_2$ (CE=50%). In the HTC process, the carbon efficiency is very close to 100, *i.e.* almost the carbon from biomass is converted into carbonized material, without generating CO and $CO_2$ (Titirici et al., 2007).

HTC is one of the most advanced technologies to convert biomass and waste with high moisture levels, because it eliminates the drying step. In addition, HTC efficiently decomposes the carbohydrates in biomass, such as cellulose and hemicellulose by hydrolysis, to produce sugars and other decomposition byproducts, *i.e.* organic acids and aldehydes (Mochidzuki et al., 2003; Fujino et al., 2002).

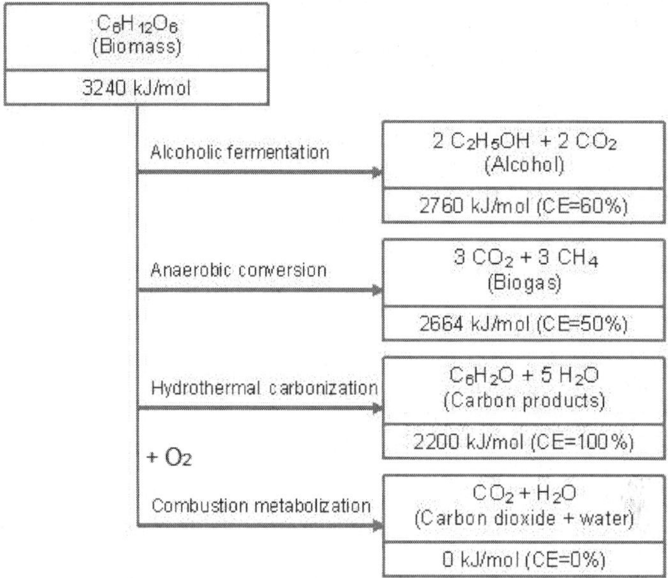


Fig. 1. Simplified diagram comparing different processing of biomass conversion.

Water acts both as reagent and reactive environment, which aids the hydrolysis, depolymerization, dehydration and decarboxylation reactions. However, these conventional hydrothermal processes need special systems that support pressure, temperature (usually an autoclave with pressure safety device), and reaction times ranging from hours to days, making them expensive and time consuming. Another inconvenience is the low selectivity, generating by products in the same reaction.

## 3. Microwave-assisted hydrothermal carbonization

### 3.1 Microwave in materials processing

Microwaves are electromagnetic waves with frequencies ranging from 300 MHz to 300 GHz and wavelengths between 1 m and 1 mm. Originally, microwaves were used in telecommunications, such as radar and telephone technology. Only during World War II, Percy Spencer discovered that microwaves had the ability to heat food, based on the fact that the wave energy in this range of electromagnetic spectrum matches the energy of the rotational movement of some dipolar molecules like water, fats and sugars (Spencer, 1941). Thereafter, microwaves have been used in several applications like as civil aviation radars, cellphones and especially in domestic ovens.

From the physicochemical point of view, heating by microwave radiation is a result of the interaction of electromagnetic wave with the electric dipole of the molecule.

Food heating and cooking in a microwave oven, for example, takes place because food contains water, which is formed by polar molecules able to align with the microwave electric field, as shown in Figure 2 (adapted from Titirici et al., 2007).

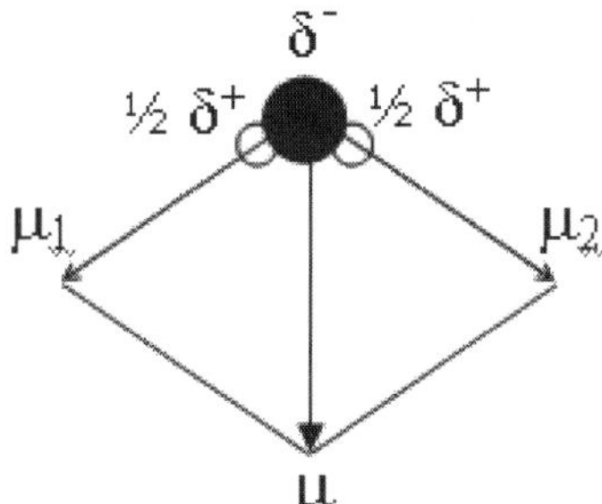


Fig. 2. Electric dipole of the water molecule. δ- and δ+ are the negative and positive partial charges of oxygen and hydrogen, respectively; μ1 and μ2 are the dipole moments of the water molecule and μ matches to the resulting dipole moment.

In chemistry and materials science fields, application of microwave technology has attracted special interest in the synthesis of organic and inorganic compounds and heat treatment of materials (drying and sintering).

Microwave radiation has been used in material processing and synthesis reactions, such as the synthesis of advanced ceramic materials (Keyson et al., 2006), carbides (Rambo et al., 1999), oxide nanoparticles (Palchik et al., 2000), synthesis from catalytic reactions (Balalaie et al., 2000) and recycling plastics (Lulow-Palafox & Chase, 2001) in addition to use in organic synthesis with organic solvents (Nücher et al., 2004) and in the absence of solvents (Caddick, 1995).

In contrast to conventional ovens, the material processed in a microwave oven interacts with electromagnetic radiation and not with the radiant energy. Due to the heat being generated by the material itself in its bulk, the heat reaches the entire volume and can be much faster and selective. These characteristics, when properly monitored, result in a homogeneous material, with faster production, while providing a significant reduction in energy losses (Clarck & Sutton, 1996).

All this set of benefits even leads to the greatest: the economic. Furthermore, the use of microwaves in a process is classified as clean technology, following the global trend towards the use of alternative environmentally friendly methods.

The use of microwaves on hydrothermal carbonization contributed to simplify and accelerate the process, since the reaction time was decreased compared to conventional hydrothermal carbonization and the obtained products are usually homogeneous. The main difference between the conventional and microwave heating is how the heat is generated. In the conventional process, energy is transfrerred to material by convention, conduction and radiation of heat from the material surface.

On the other hand, microwave energy is derived directly from the material by molecular interactions with electromagnetic waves. The material is processed rapidly, with selective power, homogeneous heating and energy conservation. These factors contribute to the improvement of the properties of the final products and enable the synthesis of new materials that could not be obtained by conventional methods (Yang et al., 2002).

The use of microwaves in the carbonization process have been intensively investigated over the last decade with different raw materials like wood (Wan et al., 2009; Miura et al., 2000), corn stover (Wan et al., 2009), grass (Orozco et al., 2007) in addition to some sugars

like as glucose and fructose (Qi et al., 2008) and activated carbons (Deng et al., 2010; Franca et al., 2010, Xin-hui et al., 2011)
The innovative process discussed in this chapter depicts the use of microwave heating in the biomass hydrothermal carbonization, which is named microwave-assisted hydrothermal carbonization (MAHC) (Guiotoku et al., 2009).
The method is based on previous work conducted by Antonietti and co-workers (2006), and uses pine sawdust and α-cellulose as raw material, water (hydrothermal environment), mild temperature (200°C) and citric acid as catalyst. The experiments were carried out using a microwave oven Millestone (Ethos Plus) suitable for sample digestion. The device frequency used was 2.45 GHz, 12.25 cm wavelength and 1000 W power supply. Figure 3 illustrates a simplified scheme of the microwave oven chamber (adapted from Walter et al., 1997).

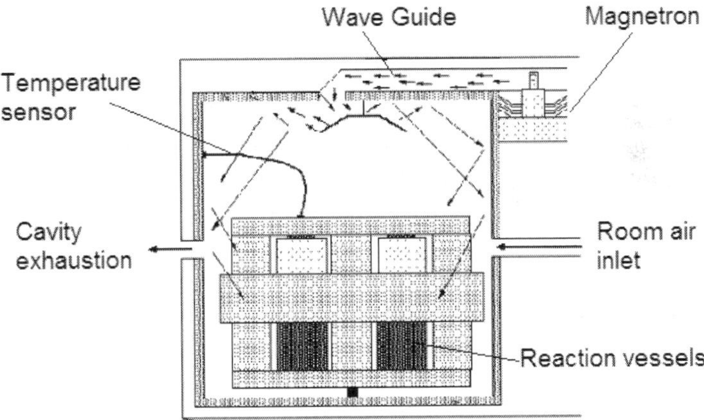


Fig. 3. Typical laboratory cavity-type microwave system . TFM$^{TM}$, a modified polytetrafluorethylene, reaction closed vessels with 100 mL volume were used in the carbonization process. These vessels withstand temperatures up to 300°C and the pressure inside may range from 80 to 100 bar.

Different catalyst concentrations were tested with pine sawdust (PS) as raw material. Figure 4 illustrates the experiments (Guiotoku, 2008).

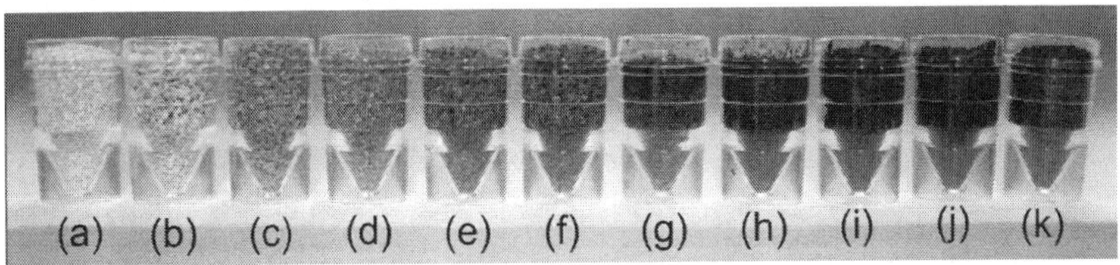


Fig. 4. Pine sawdust samples (a) in natura; hydrothermally carbonized during 30 minutes with different catalyst concentration: (b) 0.00; (c) 0.10; (d) 0.15; (e) 0.25; (f) 0.5; (g) 1.0; (h) 1.5; (i) 2.0; (j) 2.5 and (k) 3.0 mol·L$^{-1}$.

The color in the carbonized pine sawdust samples depends on the catalyst concentration. As the acid concentration increases, the samples become darker. The catalyst concentration chosen to conduct the experiments was 1.5 mol·L$^{-1}$. The changes caused during the cellulose heating in water are predominantly determined by hydrolytic reactions and especially in this case, by acid hydrolysis reactions.
The key factor that influences the acid hydrolysis reaction is the way in which it takes place, since the most important reaction in cellulose occurs in heterogeneous phase: solid polysaccharide in acid media.

### 3.1.1 Energy assessment

When subjected to carbonization or conventional pyrolysis lignocellulosic material generally undergoes thermal decomposition of their components - hemicellulose, cellulose and lignin - under inert or oxygen restricted atmospheres. Each component volatilizes more intensively at specific ranges of temperature: hemicellulose at 200 - 300°C, cellulose at 240 - 350°C and lignin at 350 - 500°C. The decomposition products are basically oil, gas and solid products (charcoal).
In order to evaluate the energetic potential of pine sawdust (PS) and α-cellulose (C), proximate analysis and gross calorific value were carried out as shown in Table 1 (Guiotoku, 2008). The samples were hydrothermally carbonized in a microwave oven at 200°C for 60, 120 and 240 min with 10 mL of 1.5 mol.L$^{-1}$ catalyst solution.

| Sample | Time (min) | % VM[a] | % Ash | %FC[b] | HGC[c] (MJ/kg) |
|---|---|---|---|---|---|
| PS60 | 60 | 61.41 | 0.20 | 38.39 | 22.60 |
| PS120 | 120 | 55.14 | 0.20 | 44.66 | 24.22 |
| PS240 | 240 | 50.94 | 0.21 | 48.85 | 25.42 |
| C60 | 60 | 53.37 | 0.25 | 46.38 | 21.62 |
| C120 | 120 | 51.36 | 0.28 | 48.36 | 24.91 |
| C240 | 240 | 48.05 | 0.21 | 51.74 | 27.12 |

[a]Volatile matter; [b]Fixed carbon; [c]High gross calorific value.

Table 1. Proximate analysis and calorific values of pine sawdust (PS) and α-cellulose (C) samples carbonized in MAHC for 60, 120 and 240 min.

In the proximate analysis properties such as volatile matter, ash and fixed carbon were determined. There is a small decrease in volatile matter with increasing carbonization time, due to volatilization of low weight molecules during longer reaction times. The fixed carbon is intrinsically linked to high gross calorific values and their values increases with the MAHC reactions times, indicating that more carbonized material was generated.
Figure 5 shows the low gross calorific values (LGC) for some solid and liquid fuels. Hachured bars correspond to the MAHC materials (adapted from Ministério de Minas e Energia, 2007). LGC values may be calculated from equation (1):

$$HGC = LGC + m\,(c \cdot \Delta T + L) \quad (1)$$

where HGC = high gross calorific value; m = water mass combustion; c = water specific heat; ΔT = difference between ambient and equilibrium temperature before condensation, and L = condensation latent heat of water vapor.

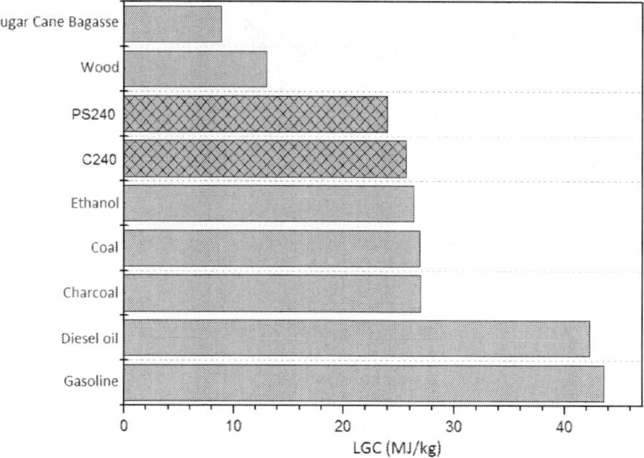


Fig. 5. Low gross calorific values for solid and liquid fuels. P240 and C240 are the MAHC samples.

Gross calorific values of MAHC products are comparable to a fossil fuel (coal), charcoal and ethanol. However, gross calorifc values found for petroleum-based fuels, such as gasoline and diesel oil are 48% higher than the carbonaceous materials obtained in MAHC process.
Thermogravimetric analysis (TGA) provides information about the thermal stability of carbonized materials. Figure 6 shows the weight loss curves for the pine and α-cellulose samples under argon atmosphere.

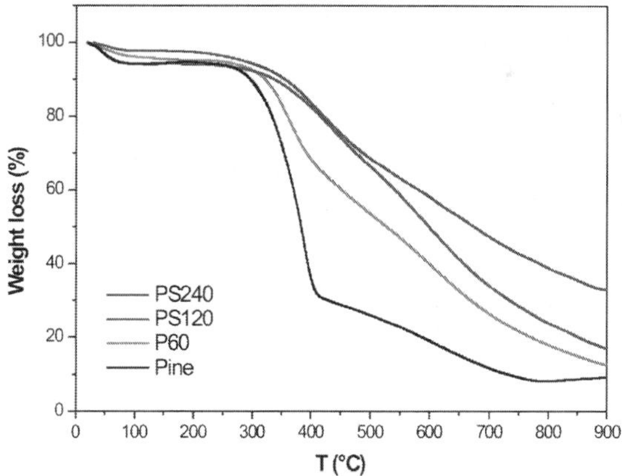


Fig. 6. Thermogravimetric curves of pine sawdust samples *in natura* and MAHC pine sawdust samples (PS) carbonized in microwave for 60, 120 and 240 min.

TG curves for pine samples reveal three weight loss stages, the first matches the moisture loss and occurs at 95 to 110°C, the second stage at 300 to 470°C is assigned to the cellulose

thermal decomposition with cellulose macromolecule breakdown. The third weight loss at 470 to 740°C, is assigned to thermal degradation of lignin. The α-cellulose samples showed similar trends.
Thermogravimetric analysis showed that the carbonization time was not enough to promote complete carbonization. These results were confirmed by elemental analysis of MAHC charcoals and raw materials (Table 2) (Guiotoku et al., 2009). Elemental analysis data from α-cellulose carbonized in a tubular furnace at 900°C, for 3 h, under nitrogen atmosphere, was used as reference material to complete carbonization.

| Time | C (wt.%) | H (wt.%) | N (wt.%) | O[a] (wt.%) | Atomic ratio | |
|---|---|---|---|---|---|---|
| | | | | | H/C | O/C |
| *Pine sawdust* | 45.45±0.06 | 6.22±0.09 | 0.02±0.01 | 48.31±0.16 | 1.16 | 0.79 |
| 0 | 60.01±0.15 | 5.51±0.00 | 0.02±0.00 | 34.46±0.16 | 1.10 | 0.43 |
| 60 | 64.74±0.15 | 5.29±0.18 | 0.04±0.01 | 29.93±0.35 | 0.98 | 0.35 |
| 120 | 63.54±0.12 | 5.19±0.10 | 0.71±0.01 | 30.56±0.20 | 0.98 | 0.36 |
| *α-cellulose* | | | | | | |
| 0 | 40.5±0.15 | 6.43±0.02 | 0.09±0.06 | 52.98±0.11 | 1.90 | 0.97 |
| 60 | 63.11±0.08 | 4.74±0.007 | 0.26±0.05 | 31.89±0.13 | 0.90 | 0.38 |
| 120 | 63.63±0.00 | 4.64±0.03 | 0.06±0.02 | 31.67±0.06 | 0.87 | 0.37 |
| 240 | 63.75±0.09 | 4.50±0.04 | 0.46±0.01 | 32.19±0.10 | 0.85 | 0.37 |
| *α-cel charcoal* | 91.08±0.12 | 1.33±0.05 | 0 | 7.59±0.08 | 0.17 | 0.06 |

Table 2. Elemental analysis for pine sawdust and α-cellulose at different times of MAHC and α-cellulose charcoal (Guiotoku et al., 2009). [a] The oxygen content was determined by mathematical difference [100% - (C% + H% + N%)].

Compared with their respective raw materials, the samples subjected to the microwave-assisted hydrothermal carbonization process had their carbon content increased by 40% and 57% for pine sawdust and α-cellulose, respectively. In contrast, the amount of O and H decreased in all carbonized materials, which suggests aromatization process.
In order to evaluate qualitatively the carbonization process, H/C and O/C molar ratios were plotted using the van Krevelen diagram, which provides information about the changes in chemical structure after carbonization, as seen in Figure 7. The van Krevelen diagram is widely used in study and classification of kerogen, which is a mixture of organic chemistry compounds modified by geological actions that originates the most of fossil fuels such as oil, gas and coal. This classification is basically provided by analysis of H/C and O/C molar ratios (van Krevelen, 1950).
In the diagram, H/C and O/C ratios of carbonized materials decrease when compared to their natural samples, suggesting that changes in the materials were taking place. The loss of H and O occurred by dehydrogenation, deoxygenation and dehydration processes.
However, as seen in TGA curves, the products are not completely carbonized, since they are in the center of diagram, between the raw material and α-cellulose charcoal. Such partially carbonized materials are commonly obtained by hydrothermal carbonization and their chemical structure can be described as amorphous aromatic carbon -OH and -COOH substituted (Titirici et al., 2007a).

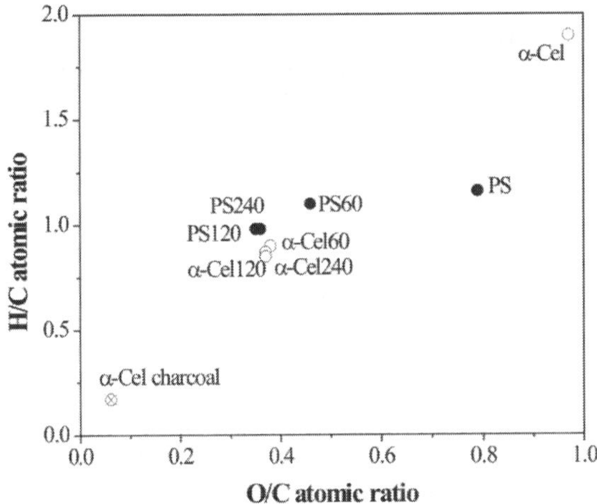


Fig. 7. The van Krevelen diagram for pine sawdust (•), α-cellulose (○) and α-cellulose charcoal (⊗) (Guiotoku et al., 2009).

SEM (Scanning Electron Microscopy) micrographs of pine sawdust and α-cellulose both *in natura* and in carbonized form are depicted in Figure 8.

Fig. 8. SEM micrographs of pine sawdust (A) and α-cellulose raw material (C); hydrothermally carbonized in microwave oven pine sawdust (B) and α-cellulose (D) (Guiotoku et al., 2009).

Carbonized pine sawdust apparently maintained its micro-morphological features after hydrothermal carbonization (Fig. 8B). In contrast, the carbonized α-cellulose is characterized by a noticeable morphological change after the MAHC process (Fig. 8 C and D). α-cellulose exhibits a fibrous aspect (Fig. 8C). After carbonization the fibers changed to spherical particles, with sizes of aproximately 1.5 μm in diameter (Fig. 8D).

### 3.1.2 Chemical characterization

To better understand the molecules breakdown mechanism under microwave-assisted hydrothermal carbonization, Pyrolysis Gas Chromatography Mass Spectrometry (Py-GC-MS) analysis were carried out in order to identify and quantify the carbonized compounds. A list of compounds identified by Py-GC-MS and total ion current (TIC) is shown in Table 3. Carbonized products were divided into five classes: (A) alkyl furans, (B) oxygen-functionalized furans, (C) benzenoids, (D) benzofurans and (E) unknown compounds. Only compounds with total percentage area higher than 1.0% had their spectra analyzed (Guiotoku et al., submitted).

Eight chromathografic peaks, 27.2% of total area, did not provide pure compounds. For this reason their mass spectra were not useful for interpretation. Major classes identified in the hydrochar were benzofurans followed by oxygen-functionalized furans, which correspond to 32.1% and 24.3%, respectively. The minor classes were alkyl furans and benzenoids.

| RT (min) | Compound | Class | MW | Fragment ions | % |
|---|---|---|---|---|---|
| 2.18 | 2,5-dimethyl-furan | A | 96 | 43, 53, 81 | 2.45 |
| 2.73 | *impure* | ? | - | - | 1.33 |
| 3.20 | methyl-benzene | C | 92 | 65, 91 | 1.29 |
| 3.85 | 2,3,5-trimethyl-furan | A | 110 | 43, 67, 95 | 1.12 |
| 4.44 | ?-methyl-phenol | C | 108 | 77, 107 | 1.07 |
| 4.51 | 2-ethyl-furan | A | 96 | 53, 81 | 1.82 |
| 5.22 | *impure* | ? | - | - | 2.65 |
| 6.03 | 2-acetyl-furan | B | 110 | 67, 95 | 3.1 |
| 6.75 | 2,3-dihydro-furfuryl-ethanol | B | 114 | 53, 68, 96, 113 | 21.25 |
| 7.06 | *impure* | ? | - | - | 2.01 |
| 7.56 | trimethyl-benzene | C | 120 | 77, 105 | 2.52 |
| 8.10 | *impure* | ? | - | - | 3.42 |
| 8.39 | *impure* | ? | - | - | 5.57 |
| 8.96 | *impure* | ? | - | - | 2.13 |
| 9.49 | *impure* | D | - | - | 7.38 |
| 9.55 | methyl-benzofuran | D | 132 | 77, 103 | 2.57 |
| 15.24 | methyl-2-ethyl-benzofuran (probably) | D | 160 | 77,103, 131 | 10.82 |
| 18.70 | *impure* | ? | - | - | 2.73 |
| 20.81 | dimethyl-2,3dihydro-2-acetyl-benzofuran | D | 190 | 91, 119, 147, 175 | 18.71 |

Table 3. Identified compounds and their relative distribution in Py-GC-MS data of microwave assisted hydrothermal product of cellulose (hydrochar).

The most pyrolytic furan-like and benzenoids compounds present in the hydrochar are detected in Py-GC-MS analysis of raw materials and carbonized cellulose, A and B classes (Pastorova et al., 1994). However, levoglucosan, the main product detected by pyrolysis analysis of raw cellulose, and other sugar markers (*e.g.* pyranones) were not identified in the hydrochar, indicating that all original material was modified during the hydrothermal treatment.

Furthermore, condensed aromatic compounds (*e.g.* naphtalene and phenantrene) usually present in the macromolecular structure of carbonized cellulose were also not detected. Therefore, apparently the hydrochar material did not preserve the oligosaccharide structure

of the started material in spite of low temperature used, and no strong evidence of the cellulose charcoal was detected in the final hydrochar product.
The solid state $^{13}C$ NMR spectrum of hydrochar is showed in Figure 9 and is characterized by the presence of two broadened peaks centered at 25 and 125 ppm.

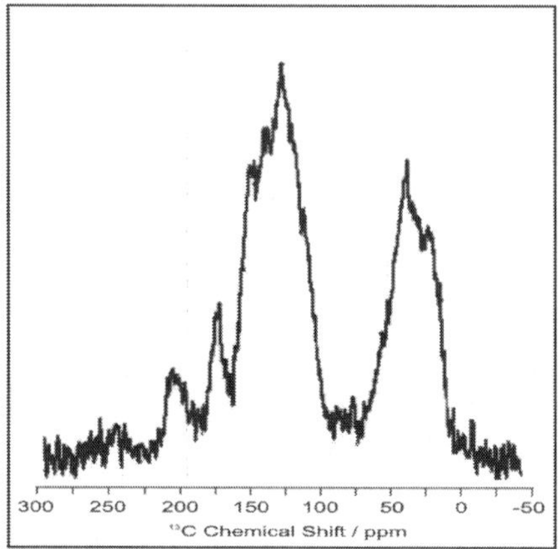


Fig. 9. Solid state $^{13}C$ NMR spectrum of hydrothermally carbonized on microwave cellulose (hydrochar) (Guiotoku et al., submitted).

Contribution of aliphatic groups can be observed between 10 and 60 ppm, with methylene resonance contribution ranging from 20 to 40 ppm. The peak at 125 ppm is typical of charred residues due to polycyclic aromatic structures. Other peaks of low intensities appear at 175 and 205 ppm, and they are consistent with the presence of carboxyl and carbonyl functional groups, respectively (*i.e* aliphatic carboxylic acids - 175 ppm and aldehydes and ketones - 205 ppm).
Results of the NMR and Py-GC-MS analyses confirmed that the hydrochar product contains an intermediate polymer composed of mainly furan elements. A strong cellulose signal is also absent in the NMR spectrum, since signals around 75 ppm, indicative of the presence of hydroxylated methylene gourps, were not prominent. The same results were observed in the pyrolysis analysis.
Interestingly, NMR spectra provided information on the nature of aromatic carbons (129 ppm), which may be interpreted as a graphite-like structure (Jiang et al., 2002). Py-GC–MS analysis provided a more detailed view of the aromatic structure, suggesting that aromatic monomers of hydrochar polymer are built on benzofuran-like components.
According to a study about hydrochar production from cellulose, the carbon-rich production from cellulose takes place via: (i) hydrolysis; (ii) dehydration and fragmentation of sugars; (iii) polymerization or sugar derivatives condensation, and (iv) aromatization (Sevilla & Fuertes, 2009). Similar mechanism should occur in hydrochar production by microwave hydrothermal carbonization, leading to partial oxygenated and aromatic products.
Hydrothermal carbonization is a novel route to produce carbon-rich material, and its use as a soil additive should be considered, mainly due to its specific properties like a functionalized chemical nature structure (hydrochar). In contrast to several biochar

agronomic studies (Sohi et al., 2010 and references therein), only preliminary studies using hydrochar have been carrying out.

## 4. MAHC and biochar

Biochar is any source of biomass previously heated under low or absence of oxygen supply - a carbonization process - whose purpose is to be applied to soil in order to improve its agronomic and environmental quality. This process results in a very stable carbon-rich material not only capable of improving physical and chemical soil properties, and therefore soil productivity, but also of increasing soil carbon storage on large scale and for a long period of time (Sohi et al., 2010).

Biochar application to soil involves a combination of biological and physico-chemical routes to reduce atmospheric greenhouse gas levels, but its behavior in soil needs to be deeply evaluated for large scale use (Hammond et al., 2011). Nevertheless, there are few alternatives of carbon-neutral or carbon negative technologies, with great economic viability and low cost, leading to significant changes in the global carbon balance, such as that can be expected by adopting the currently technology known as biochar.

Usually, biochar is obtained from the partial transformation of biomass in to coal-like material by thermal decomposition under low oxygen atmosphere, resulting in partial degradation of cellulose into smaller molecules. This material is characterized by condensed aromatic structures with no functional groups (*e.g.* alcohol and organic acids), which are very important for reactivity and soil fertility when used as an additive. Therefore, methods for production and conversion of lignocellulosic biomass waste into highly functionalized biochar have emerged as a major challenge.

Soil fertility, especially in tropical soils, is substantially influenced by soil organic matter contents and its physical and chemical properties. Soil organic matter (SOM) is a complex mixture of organic compounds, of plants and animals origin, in gradual decomposition stage due to chemical or biological transformation. Pyrogenic carbon is the most recalcitrant fraction of SOM and, according to Seiler & Crutzen (1980), it can be described as "a continuum form partly charred plant material, through char and charcoal, to soot and graphite particles without distinct".

Pyrogenic carbon is derived from partial carbonization mainly of lignocellulosic materials resulting in different sizes of condensed polyaromatic units, hydrogen and oxygen deficient. Soot particles that condense from gas phase typically develop as concentric shells of graphene stacks ("onion-like" structure) (Preston & Schmidt, 2006). Pyrogenic carbon is highly resistant to thermal, chemical and photo degradation, excellent characteristics for soil carbon sequestration (Glaser et al., 2001; Masiello, 2004). However, partial oxidation of peripheral aromatic units may produce carboxylic groups that elevate the total acidity of SOM and, therefore, the cation exchange capacity (CEC) and soil fertility (Novotny et al., 2009). This effect is demonstrated in Figure 10 where the $^{13}C$ NMR spectra of charcoal before and after chemical oxidation is shown (Novotny et al., 2007).

The use of biochar as a soil conditioner and several scientific publications can be found from the beginning of the 20$^{th}$ century. Nevertheless, recent studies about Amazonian Dark Earths, known as *Terras Pretas de Índios* (TPI), renewed the scientific interest on pyrogenic carbon in soils. TPI are anthropogenic dark earths which surface horizons of variable depth enriched in organic matter (OM), pottery shards lithic artifacts and other evidences of human activity (Kämpf et al., 2009).

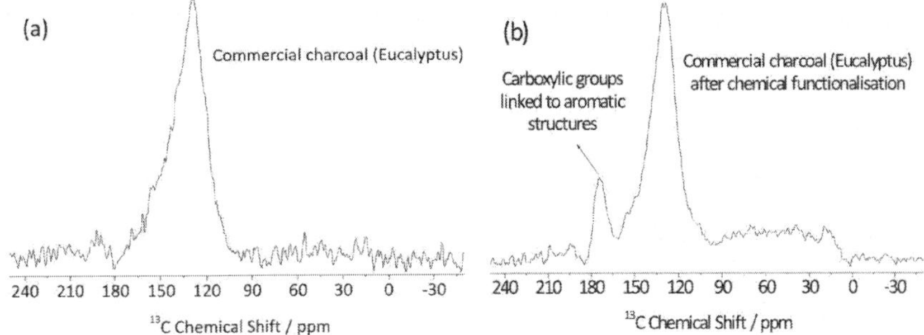


Fig. 10. Solid state $^{13}$C NMR spectra of charcoal (a) before and (b) after chemical oxidation (Novotny et al., 2007).

These archeological Amazonian sites are highly fertile, very unusual among Amazonian soils, typically of low fertility, highly weathered and very acid. TPI soils have carbon content up to 150 g·kg$^{-1}$ of soil, compared with 20 to 30 g·kg$^{-1}$ in adjacent soils (Novotny et al., 2009). This additional carbon is present mainly in the pyrogenic form, which is more stable than other carbons forms (Glaser et al., 2001), distributed along the whole soil organic profile, which can reach up to 200 cm deep, averaging from 40 to 50 cm, while surrounding soils are limited from 10 to 20 cm. Therefore, carbon stocks in TPI can be an order of magnitude higher than in surrounding soils.

Humic acids extracted from TPI and analyzed by $^{13}$C NMR spectroscopy, revealed that this fraction is rich in condensed aromatic structures and functionalized with carboxylic groups linked directly to the recalcitrant polyaromatic rings. In contrast, humic acids from adjacent non-anthropogenic soils have a higher content of labile compounds, shuch as carbohydrates and aminoacids. Once in soil biochar increases the pyrogenic fraction and the stable carbon pool. Besides carbon sequestration, biochar can provide other benefits such as: enhancing productivity (from 0 to 300%); decreasing emission of methane and nitrous oxide (up to 50% estimations); decreasing nutrients leaching and enhancing water holding capacity (Gaunt & Lehmann, 2007).

The success in obtaining functionalized materials by MAHC leads to new perspectives for the application of spherical carbonaceous particles, as soil amendments. Although soil pyrogenic carbon is rich in condensed polyaromatic compounds of low H/C and O/C ratios, chemical characteristics which provide high recalcitrant, resilience and resistance to thermal, chemical and photo degradation (Seiler & Crutzen, 1980), as a soil organic matter component, the presence of functionalized groups as carboxyls and hydroxyls is desirable to improve cationic exchange capacity (CEC). Charcoal is an efficient material for carbon sequestration (half-life ranging from decades to millennia), but it does not contain carboxylic groups that could improve its reactivity and CEC. Functionalization of the aryl backbone can be naturally obtained, slowly, or through chemical, thermal or biological oxidation like MAHC.

Hydrothermal carbonization produces carbonaceous materials under milder conditions than other carbonization processes. The structure of the hydrochar produced is basically an aromatic nucleus with hanging functional groups such as hydroxyl (-OH), carboxyl (-OOH) and carboxylic acids (-COOH), making it hydrophilic and less graphitized as compared with conventional charcoal (Rillig et al., 2010).

Conventional pyrolysis results in graphite-like structures with low O/C and H/C ratios and, therefore, few functional organic groups. Due to its polar structure, hydrochar is especially interesting as a soil conditioner, since it can play an active role in soil CEC and reactivity.

Biochar has been considered a very promising technology around the world mainly because it perfectly fits in the bioenergy sector as a solution for mitigation of greenhouse gas emissions. Scientific evidences indicate that biochar improves soil biodiversity, enhances soil performance, and reduces its susceptibility to weathering and its fertilization input. With the additional suppression of trace gas emission, a biochar-to-soil strategy based on agricultural waste streams offers a sustainable, carbon negative energy production by cutting farming´s carbon footprint, and offering to the society a rare win-win option for combating climate change (Sohi et al., 2010).

Some of the benefits of biochar with respect to biomass are low moisture retention, high gross calorific value (and consequent reduction in transportation costs), ease of handling, it non perishablity and long term storage ability. The acceleration of biomass carbonization by a factor of $10^6$ to $10^9$ by microwave hydrothermal reaction can be considered promising technology.

Despite these advantages of MAHC, further studies are still needed, in order to evaluate the viability of the wide range of possible applications. Considering its potential, the microwave-assisted hydrothermal carbonization is a new thermochemical conversion technology that should be seriously considered among biomass conversion technologies for the near future.

## 5. Research challenges and perspectives

Brazil is now the world's largest exporter not only of coffee, sugar, orange juice and tobacco but also of ethanol, beef and chicken, and the second-largest source of soybean products (The Economist, 2010). With this huge agricultural production, the first challenge is exactly the large variety of biomass source available to char. Each biomass source is physically and chemically different and has specific ideal conditions to char, and different soil responses. Besides, the large climate variation of the country also contributes to the variety of biomass and soil. With all these possibilities, it is necessary to prioritize the biochar research on certain biomass sources. At the moment, the natural trend seems to point to Brazilian energetic matrix, which is based in renewable sources (45% compared with a world average of 15%), from which 27.2% is supplied by biomass (Lora & Andrade, 2009). The Brazilian agro-energy chain involves 121.15 million tons of bagasse from sugar cane industry, although others sources of wasted biomass are currently in the biochar research agenda. As for MAHC procedures, the challenge is even larger since as a biochar making technique, is barely studied.

## 6. Conclusions

The purpose of this chapter was to demonstrate how the use of microwaves in an already well established process such as the hydrothermal, can produce materials with several interesting physico-chemical properties. As this process has been recently developed, there are no sufficient studies for material comparison and the discussed results during the chapter was based on the work of the authors research group. However, the authors hope that the disclosure of this book may inspire further studies and applications for microwaves on carbonaceous materials in all fields of knowledge, such as materials engeneering, chemistry, physics, biology, agronomy and others.

## 7. Acknowledgment

The authors thank Prof. Dr. Adilson Curtius and Prof. Dr. Vera Azzolin Frescura Bascuñan from Federal University of Santa Catarina for their valuable contribution during the microwave-assisted hydrothermal carbonization experiments, and to CNPq for financial support.

## 8. References

Antonietti, M. (2006) Magic coal from steam cooker. *Max Planck Research*, Vol. 3, pp. 20-22, ISSN 1616-4172

Balalaie, S., Arabanian, A., Hashtroudi, M. S. (2000) Zeolite HY and silica gel as new efficient heterogeneous catalysis for the synthesis of triarylimidazoles under microwave irradiation. *Monatshefte für Chemie*, Vol. 131, pp. 945-948, ISSN 1434-4475

Bridgwater, A.V. (2003). Renewable fuels and chemicals by thermal processing of biomass. *Chemical Engeneering Journal*, Vol. 91, pp. 87-102, ISSN 1385-8947

Caddick S. (1995). Microwave assisted organic reactions. *Tetrahedron*, Vol. 51, No. 38, pp. 10403-10432, ISSN 0040-4020

Cao, X.; Ro, K. S.; Chappell, M.; Li, Y.; Mao, J. (2011). Chemical structures of swine-manure chars produced under different carbonization conditions investigated by advanced solid state $^{13}C$ nuclear magnetic resonance (NMR) spectroscopy. *Energy & Fuels*, Vol. 25, pp. 388-397, ISSN 1520-5029

Clarck, E. & Sutton, W. H. (1996). Microwave processing of materials. *Annual Reviews of Material Science*, Vol. 26, pp. 299-331, ISSN 084-6600

Deng, H.; Li, G.; Yanga, H.; Tanga, J.; Tanga, J. (2010). Preparation of activated carbons from cotton stalk by microwave assisted KOH and $K_2CO_3$ activation. *Chemical Engineering Journal*, Vol. 163, pp. 373–381, ISSN 1385-8947

Franca, A. F.; Oliveira, L. S.; Nunes, A. A.; Alves, C. C. O. (2010). Microwave assisted thermal treatment of defective coffee beans press cake for the production of adsorbents. *Bioresource Technology*, Vol. 101, pp. 1068–1074, ISSN 09608524

Fujino, T.; Moreno, J. M. C.; Swamy, S.; Hirose, T., Yoshimura, M. (2002). Phase and structural change of carbonized wood materials by hydrothermal treatment. *Solid State Ionics*, Vol. 151, pp. 197-203, ISSN 0167-2738

Funke, A. & Ziegler, F. (2010). Hydrothermal carbonization of biomass: A summary and discussion of chemical mechanisms for process engineering. *Biofuels, Bioproducts and Biorefining*, Vol. 4, pp. 160-177, ISSN 1932-1031

Gaunt, J. L. & Lehmann, J. (2007). Abstracts of the International Biochar Innitiative 2007 Conference, Terrigal, Australia.

Glaser, B., Haumaier, L., Guggenberger, G., Zech, W. (2001). The "Terra Preta" phenomenon: a model for sustainable agriculture in the humic tropics. *Naturwissenschaften*, Vol. 88, pp. 37-41, ISSN 0028-1042

Guiotoku, M. (2008). Universidade Federal de Santa Catarina. Programa de Pós-graduação em Ciência e Engenharia de Materiais. Obtenção de partículas submicrométricas de carbono a partir da celulose utilizando carbonização hidrotérmica por microondas. Tese (Doutorado) Florianópolis, Santa Catarina, Brasil.

Guiotoku, M., Rambo, C. R., Hansel, F. A., Magalhães, W. L. E., Hotza, D. (2009). Microwave-assisted hydrothermal carbonization of lignocellulosic materials. *Materials Letters*, Vol. 63, pp. 2707-2709, ISSN 0167-577X

Guiotoku, M., Hansel, F. A., Novotny, E. H., Maia, C. M. B. F., Molecular and morphology characterization of hydrochar produced by microwave-assisted hydrothermal carbonization of cellulose. *Pesquisa Agropecuária Brasileira*, submitted.

Hammond, J., Shacley, S., Sohi, S., Brownsort, P. (2011). Prospective life cycle carbon abatement for pyrolysis biochar systems in the UK. *Energy Police*, Vol. 39, pp. 2646-2655, ISSN 0301-4215.

Heilmann, S. M.; Davis, H. T.; Jader, L. R.; Lefebure, P. A.; Sadowsky, M. J.; Schendel, F. J.; Keitz, M. G.; Valentas, K. J. (2010). Hydrothermal carbonization of microalgae. *Biomass & Bioenergy*, Vol. 34, pp. 875-882, ISSN 0961-9534

Inoue, S.; Uno, S.; Minowa, T. (2008). Carbonization of cellulose using the hydrothermal method. *Journal of Chemical Engeneering of Japan*, Vol. 41, No. 3, pp. 210-215, ISSN 0021-9592

Kämpf, N., Woods, W.I., Sombroek, W., Kern, D.C., Cunha, T.J.F. (2003). *Classification of Amazonian dark earths and other ancient anthropic soils*. In: Amazonian dark earths. Origin, properties, management. Lehmann, J., Kern, D.C., Glaser, B., Woods, W.I. (Eds.), Kluwer Academic Publichers: Dordrecht, pp. 77-104, ISBN-10 1402018398

Keyson, D. Longo, E., Vasconcelos, J. S., Varela, J. A., Éber, S., DerMaderosian, A. (2006) Síntese e processamento de cerâmicas em forno de microondas doméstico. *Cerâmica*, Vol. 52, pp. 50-56, ISSN 0366-6913

Kumar, S.: Kothari, U.; Kong, L.; Lee, Y. Y.; Gupta, R. B. (2011). Hydrothermal pretreatment of switchgrass and corn stover for production of ethanol and carbon microshperes. *Biomass and Bioenergy*, Vol. 35, pp. 956-968, ISSN 0961-9534

Liu, Z.; Zhang, F. S.; Wu, J. (2010). Characterization and application of chars produced from pinewood pyrolysis and hydrothermal treatment. *Fuel*, Vol. 89, pp. 510-514, ISSN 0016-2361

Lora, E. S. & Andrade, R. V. (2009). Biomass as energy source in Brazil. *Renewable and Sustainable Energy Reviews*, Vol. 13, pp. 777-788, ISSN 1364-0321

Lulow-Palafox, c., Chase, H. A. (2001) Microwave induced pyrolysis of plastic wastes. *Industrial & Engeneering Chemistry Research*, Vol 4., pp. 4749-4756, ISSN 1520-5045

Masiello, C. A. (2004). New directions in black carbon organic geochemistry. *Marine Chemistry*, Vol. 202, pp. 201-213, ISSN 0304-4203

Mi, Y.; Hu, W.; Dan, Y.; Liu, Y. (2008). Synthesis of carbon micro-spheres by a glucose hydrothermal method. *Materials Letters*, Vol. 62, pp. 1194-1196, ISSN 0167-577X

Ministério de Minas e Energia, Empresa de Pesquisa Energética. (2007). *Balanço Energético Nacional (BEN) 2007: ano base 2006*. Rio de Janeiro, Brasil, ISS 0101-6636,

Miura, M.; Kaga, H.; Tanaka, S.; Takahashi, K.; Ando, K. (2000). Rapid microwave pyrolysys of wood. *Journal of Chemical Engeneering of Japan*, Vol. 33, No.2, pp. 299-302, ISSN 0021-9592

Mochidzuki, K.; Sakoda, A.; Suzuki, M. (2003). Liquid-phase thermogravimetric measurement of reaction kinetics of the conversion of biomass wastes in pressurized hot water. *Advances in Environmental Research*, Vol. 7, pp. 421-428, ISSN 1093-7927

Novotny, E. H., Azevedo, E. R., Bonagamba, T. J., Cunha, T. J. F., Madari, B. E., Benites, V. M. Hayes, M. H. B. (2007). Studies on the composition of humic acids from Amazonian Dark Earth Soils. *Environmental Science & Technology*. Vol. 41, pp. 400-405, ISSN 0013-936X

Novotny, E. H., Azevedo, E.R., Bonagamba, T.J., Cunha, T.J.F., Madari, B.E., Benites, V.M., Hayes, M.H.B. (2009). In *Amazonian Dark Earths: Wim Sombroek's Vision*. Woods, W.I., Teixeira, W.G., Lehmann, J., Steiner, C., WinklerPrins, A., Rebellato, L., eds., Springer Science, Heidelberg, Ch. 21. ISBN 978-1-4020-9030-1

Novotny, E. H., Hayes, M.H.B., Madari, B.E., Bonagamba, T.J., Azevedo, E.R., Souza, A.A., Song, G., Nogueira, C.M., Mangrich, A.S. (2009a). Lessons from the Terra Preta de Índios of the Amazon Region for the Utilisation of charcoal for soil amendment. *Journal of Brazilian Chemical Society*, Vol. 20, pp. 1003-1010, ISSN 0103-5053

Nücher, M. Ondruschka, B., Bonrath, W., Gum, A. (2004). Microwave assisted sythesis – a critical technology overview. *Green Chemistry*, Vol. 6, pp. 126-141, ISSN 1463-9270

Orozco, A.; Ahmad, M.; Rooney, D.; Walker, G. (2007). Dilute acid hydrolysis of cellulose and cellulosic bio-waste using a microwave system. *Process Safety and Environmental Protecttion*, Vol. 85, No. B5, pp. 446-449, ISSN 0957-5820

Qi, X.; Watanabe, M.; Aida, T. M.; Smith Jr., R. L. (2008). Catalytical conversion of fructose and glucose into 5-hydroxymethylfurfural in hot compressed water by microwave heating. *Catalysis Communications*, Vol. 9, pp. 2244-2249, ISSN 1556-7367

Preston, C. M. & Schmidt, M. W. (2006). Black (pyrogenic) carbon in boral forests: a syntehis of current knowledge and uncertainties. *Biogeosciences Discussion*, Vol. 3, pp. 211-271, ISSN 1810-6285

Palchik, O. P., Zhu, J., Gedanken, A. (2000) A microwave assisted preparation of binary oxide nanoparticles. *Journal of Material Chemistry*, Vol. 10, pp 1251-1254, ISSN 0959-9428

Pastorova, I., Botto, R. E., Ariz, P. W., Boon, J. J. (1994). Cellulose char structure: a combined analytical Py-GC-MS, FTIR and NMR study. *Carbohydrate Research*. Vol. 262, pp. 27-47, ISSN 0008-6215

Rambo, C. R., Martinelli, J. R., Bressiani, A. H. A. (1999) Synthesis of SiC and cristobalite from rice husk by microwave heating. *Advanced Powder Technology*, Vol. 299, No. 3, pp. 63-66, ISSN 0921-8831

Rillig, M. C., Wagner, M., Salem, M., Antunes, P.M., George, C., Ramke, H.-G., Titirici, M.-M., Antonietti, M. (2010). Material derived from hydrothermal carbonization: Effects on plant growth and arbuscular mycorrhiza. *Applied Soil Ecology*, Vol. 45, pp. 238-242, ISSN 0929-1393

Ryu, J.; S.,Y. W.; S., D. J.; Ahn, D. J. (2010). Hydrothermal preparation of carbon microspheres from mono-saccharides and phenolic compunds. *Carbon*, Vol. 48, pp. 1990-1998, ISSN 0008-6223

Seiler, W. & Crutzen, P. J. (1980). Estimates of gross and net fluxes of carbon between the biosphere and the atmosphere from biomass burning. *Climate Change*, Vol.2, pp. 207-247, ISSN 1573-1480

Sevilla, M. & Fuertes, A. B. (2009). The production of carbon materials by hydrothermal carbonization of cellulose. *Carbon*, Vol. 47, pp. 2281-2289, ISSN 0008-6223

Sevilla, M. & Fuertes, A. B. (2009a). Chemical and structural properties of carbonaceous products obtained by hydrothermal carbonization of saccharides. *Chemistry A European Journal*, Vol. 15, pp. 4195-4203, ISSN 1521-3765

Sohi, S. P., Krull, E., Lopez-Capel, E., Bol, R. (2010). A Review of biochar and its use and function in soil. *Advances in Agronomy*, Vol. 105, Chap. 2, pp. 47-82, ISBN 0065-2113

Spencer, P. L. Hall of fame - inventor profile. High Efficiency Magnetron. Patent Number 2,408,235. Avaiable from: http://invent.org/hall_of_fame/136.html

The Economist, (2010). *It´s only natural*, Publisher: *The Economist Newspaper Limited*, London, UK.

Titirici, M. M.; Thomas, A.; Antonietti, M. (2007). Back in black: hydrothermal carbonization of plant material as an efficient chemical process to treat the CO2 problem? *New Journal of Chemistry*, Vol. 31, pp. 787-789, ISSN 1144-0546

Titirici, M. M., Thomas, A., Yu, S. H., Müller, J., Antonietti, M. (2007a). A direct synthesis of meosporous carbons with bicontinuous pore morphology from crude plant material by hydrothermal carbonization. *Chemistry of Materials*, Vol. 19, pp.4205-4212, ISSN 1520-5002

Trugilho, P. F. & Silva, D. A. (2001). Influência da temperatura final de carbonização nas características físicas e químicas do carvão vegetal de jatobá (Hilmenea courbaril L.). *Scientia Agraria*, Vol. 2, No. 1, pp. 39-44, ISSN 1983-2443

Van Krevelen, D. W. (1950). Graphical-statistical method for the study of structure and reaction process of coal. *Fuel*, Vol. 29, pp. 269-284. ISSN 0016-2361

Xin-hui, D.; Srinivasakannan, C.; Li-bo, Z.; Zheng-yong, Z. (2011). Preparation of activated carbon from Jatropha hull with microwave heating: Optimization using response surface methodology. *Fuel Processing Technology*. Vol. 92, pp. 394–400, ISSN 0378-3820

Yang, K. S.; Yoon, Y. J.; Lee, M. S.; Lee, W. J., Kim, J. H. (2002). Further carbonization of anisotropic and isotropic pitch-based carbons by microwave irradiation. *Carbon*, Vol. 40, pp. 897-903, ISSN 0008-6223

Yoshimura, M. & Byrappa, K. (2008). Hydrothermal processing of materials: past, present and future. *Journal of Materials Science*, Vol. 48, pp.2085-2103, ISSN 0022-2461

Walter, P. J., Chalk, S., Kingston, H. M., (1997). *Overview of Microwave-assisted sample Preparation*, In: Microwave Enhanced Chemistry Fundamentals Sample Preparation, and Applications, Kingston, H. M., Haswell, S. J., Eds. American Chemical Society. Washington: D.C., pp. 55-222, ISBN-10 0841233756

Wan, Y.; Chen, P.; Zhang, B.; Yang, C.; Liu, Y.; Lin, X.; Ruan, R. (2009). Microwave-assisted pyrolisys of biomass: catalysts to improve product selectivity. *Journal of Analytical Applied Pyrolysis*, Vol. 86, pp. 161-167, ISSN 0165-2370

# 14

# Sucrose Chemistry: Fast and Efficient Microwave-Assisted Protocols for the Generation of Sucrose-Containing Monomer Libraries

M. Teresa Barros[1], Krasimira T. Petrova[1] and Paula Correia-da-Silva[1,2]
[1]*REQUIMTE, CQFB, Departamento de Química, Faculdade de Ciências e Tecnologia, Universidade Nova de Lisboa*
[2]*Instituto Superior de Ciências da Saúde Egas Moniz*
*Portugal*

## 1. Introduction

Sucrose is a carbohydrate feedstock of low molecular weight which is ubiquitous in its availability and is of relatively low cost (Lichtenthaler and Mondel, 1997). The potential value of sucrose as a raw material has been recognized for many years and has been the subject of considerable research. The quest for novel, sustainable, but also structurally robust materials is gaining momentum as the pressure on our environment is building up and the progressive changeover of the chemical industry to renewable feedstock for their raw materials emerges as an inevitable necessity. Although extensive work has been done on the synthesis of glycopolymers, more research efforts are needed to bridge the conceptual, technological and economical gap between fossil hydrocarbons and renewable carbohydrates (Khan, 1984).

Sucrose is a particularly appropriate material for use in the formation of many products of commercial significance produced currently from petroleum-based materials, because: it is a naturally occurring, relatively abundant renewable resource; it is polyfunctional, with three reactive primary hydroxyl groups that can be readily and selectively derivatized; it is a 1-2′ linked disaccharide and therefore a nonreducing sugar and thus it does not have the potential for the wide variety of reactions that reducing sugars have; it is a naturally occurring carbohydrate, therefore products based on it are potentially biocompatible and biodegradable (Davis and Fairbanks, 2002).

As part of a program directed toward the valorization of sucrose by incorporating it into polymers, its conversion to natural compounds, and its possible application as a chiral auxiliary in selective asymmetric cycloaddition (Lichtenthaler and Peters, 2004), we needed to obtain derivatives of sucrose which could be functionalized easily and selectively at the primary hydroxyl groups (Fig. 1) (Jarosz and Mach, 2002). For this, the development of short and easy routes for the preparation of larger quantities of these target molecules was necessary.

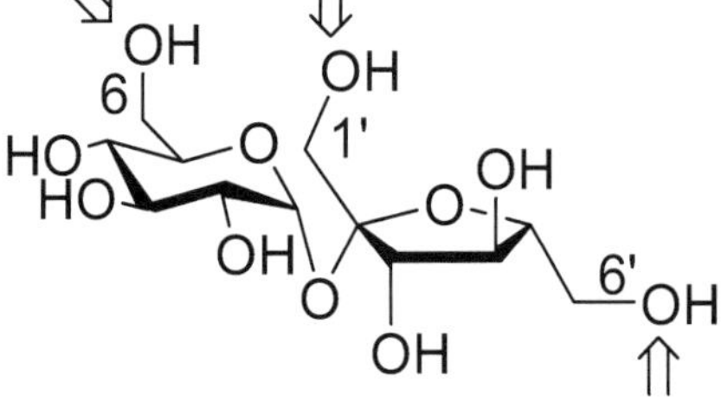

Fig. 1. The primary hydroxyl groups in the sucrose molecule can be selectively functionalized.

A high degree of regioselectivity can be achieved by the enzymatic approach (Patil et al., 1991; Potier et al., 2000), although some disadvantages and limitations can be found in the use of enzymes (low stability of the enzymes in organic solvents or the correct choice of reagents for introducing the unsaturation, (Chen and Park, 2000).

The functionalization of carbohydrates normally involves various protection-deprotection steps, which take a lot of experimental time (Jarosz and Mach, 2002). In order to access selectively specific positions with nonselective reagents, the first step would always be blocking of one, two, or all three primary positions of sucrose with bulky substituents. The groups most utilized for this purpose are triphenylmethyl- (trityl-) (Mach et al., 2001) and *tert*-butyldiphenylsilyl- (TBDPS) chlorides. The advantage of using trityl chloride over TBDPSCl is that it is a relatively cheap reagent, but unfortunately, it is not selective for mono-protection. Thus, the first step in many protection-deprotection sequences to obtain valuable monofunctionalized compounds often consists of a regioselective silylation of the 6′-hydroxyl group of the sucrose (Barros et al., 2000a; Barros et al., 2004b; Crucho et al., 2008) using 1.1 equiv. of *tert*-butyldiphenylchlorosilane (TBDPSCl) in dry pyridine. The methods have been described for sucrose first by Karl et al.,(Karl et al., 1982) to obtain a 49 % yield, and later (Barros et al., 2004b) optimized to 85 % by closely following the reaction by TLC and stopping it when di-*O*-TBDPS sucrose first appeared as a less polar product. A similar strategy has been applied to obtain 6,6′-di-*O*-TBDPS-sucrose, where 2.2 eq TBDPSCl were used (Karl et al., 1982). With the 6 and 6′ positions blocked, regioselective access to the 1′-hydroxyl was gained, and the compound 1′-bromo-1′-deoxy-6,6′-di-*O*-TBDPS-sucrose has been reported (Andrade et al., 2007).

The first publications from our group on this topic (Barros et al., 2000a; Barros et al., 2000b) were directed to selective protection-deprotection of silyl ethers of sucrose, based on the fact that the primary alcohol groups are more reactive than the secondary and that between the primary hydroxyls there is a relatively well defined reactivity (Barros et al., 2001). Thus, 6′- or 1′-OH free heptabenzoyl sucrose derivatives (Fig. 2 and 6) have been obtained via a short sequence (three to five steps from sucrose) with good overall yields. The flexibility of this pathway allowed the facile exchange of the protecting groups, as has been shown by the preparation of the heptaacetate and heptabenzyl homologues, and made possible the selective preparation of several monohydroxy sucrose esters or ethers (Fig. 2).

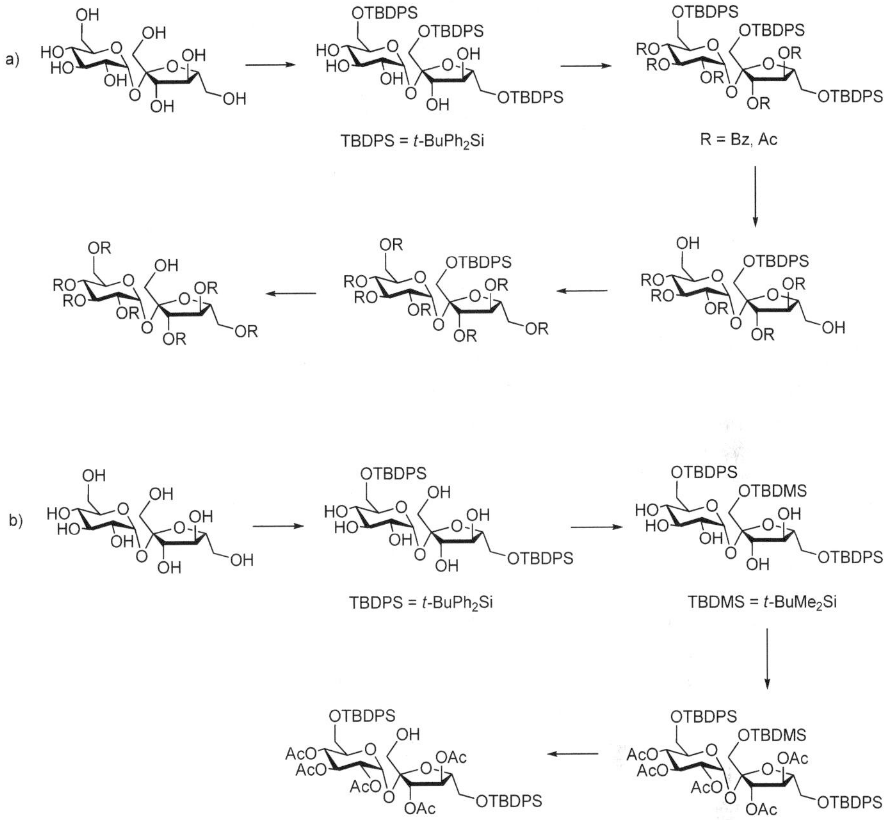

TBDPS = $t$-$BuPh_2Si$

R = Bz, Ac

TBDPS = $t$-$BuPh_2Si$

TBDMS = $t$-$BuMe_2Si$

Fig. 2. Access to the 1′-position of the fructose moiety: a) via selective *O*-desilylation, 5 steps, 66 % overall yield (Barros et al., 2000b); b) via selective deprotection of *tert*-butyldimethylsilyl ether, 30 % overall yield (Barros et al., 2000a).

The potential applications of these compounds are numerous and thus several types of reactions involving the primary positions have been studied and various novel compounds prepared - by galloylation, three monogalloyl sucroses with antioxidant activity were prepared (Barros et al., 2000a); *O*-silyl protected sugars have been converted into their corresponding formates using the Vilmeier-Haack complex (Andrade and Barros, 2004); using Wittig olefination, branched-chain olefinic compounds have been obtained (Andrade et al., 2007), and their application as chiral auxiliaries in 1,3-dipolar cycloaddition to azides has been reported (Andrade and Barros, 2009). With the aim of developing cleaner and faster technologies, the latter two methods have been carried out using microwave-assisted synthesis (Fig. 3, 4).

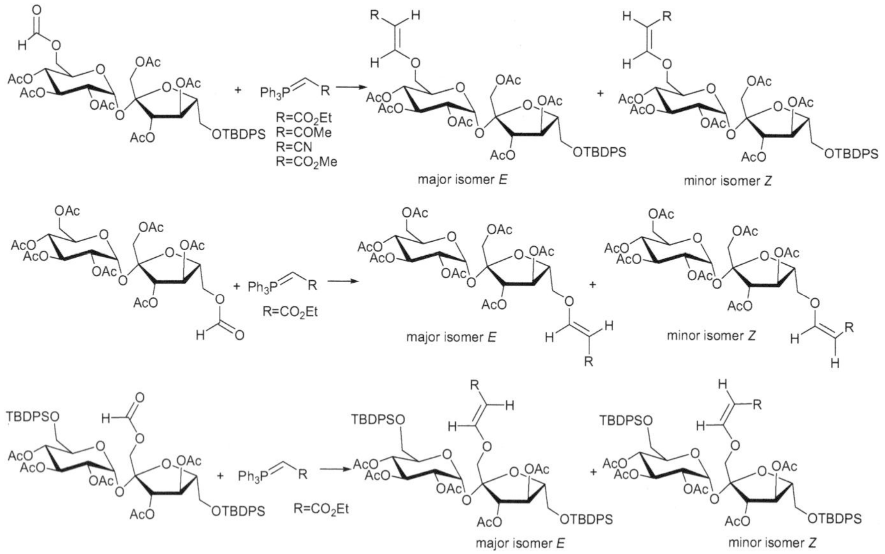

Fig. 3. Microwave assisted selective synthesis of sucrose derivatives containing unsaturated systems by Wittig reaction.

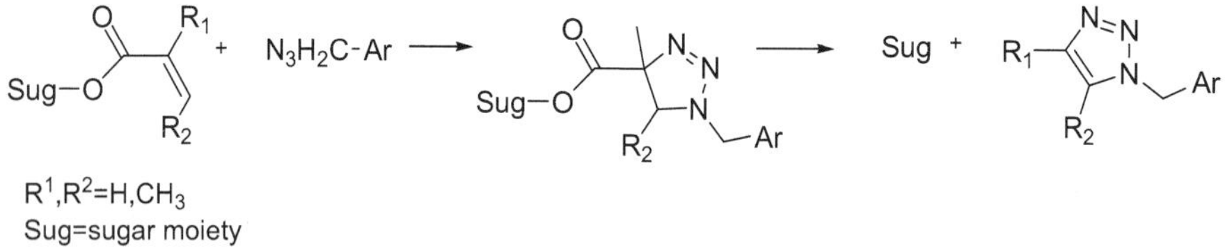

Fig. 4. Microwave assisted 1,3-dipolar cycloaddition of vinylic sugars with aryl azides.

Petrochemical derived polymers (such as polystyrene), functionalized with sugar, to form biodegradable polymers is a recently discovered application of a sugar linked synthetic polymer (Galgali et al., 2002). The class of sugar based polymers, generally known as poly(vinylsaccharide)s, has also been investigated for a variety of applications, particularly in the biomedical field (Carneiro et al., 2001; Kobayashi et al., 1985). The most widely used method for the synthesis of poly (vinylsaccharide)s was based on the free radical polymerizations of vinyl sugars (Klein et al., 1990). An extensive review of the preparation and applications of this type of polymers is available (Varma et al., 2004). Kobayashi (Kobayashi, 2001) has described various applications of glyconjugate polymers in the biological and biomedical fields. Synthetic carbohydrate-based polymers having pendant sugar residues are of great interest, not only as simplified models for biopolymers bearing oligosacharides, but also as artificial glycoconjugates in biochemistry and medicine.

The introduction of sugars into polymeric molecules can bestow new properties, such as increased polarity, chirality, biodegradability, and biocompatibility. Sucrose-containing polymers, having a polyvinyl backbone and pendant sucrose moieties, have been obtained by polymerization or copolymerization of sucrose derivatives - esters, ethers, and acetals, bearing a carbon-carbon double bond (Fig. 5), (Fanton et al., 1992; Ferreira et al., 2000; Jhurry et al., 1992; Patil et al., 1991). The monomers have been prepared either by multistep synthesis, leading to defined compounds and subsequently a well-defined polymerization processes (Fig. 6,7), (Barros et al., 2004b; Barros and Petrova, 2009; Barros et al., 2007; Barros and Sineriz, 2002; Crucho et al., 2008), or by direct functionalization of unprotected sucrose, leading to mixtures of isomers and therefore to more complex polymers (Barros et al., 2010b).
A number of monomers has been synthesized from sucrose during the last decades, as presented in Fig. 5. Nonselective derivatization of free sucrose to provide "statistical" mixtures that might find industrial applications were not of interest since undefined polymeric structures result. Also, the di- and tri- substituted derivatives were not of interest, as they result in cross-linked networks.
The first selectively obtained monomer described was the substituted derivative at the three primary positions, then there appeared substituted or hydrolized isopropylidene derivatives with free or acetylated hydroxyl groups, which were also a mixture of 2 isomers (Fanton et al., 1992). To the best of our knowledge, mono-unsaturated sucrose esters at the 1′-position were selectively obtained only with the aid of enzymes (Fig. 5) (Dordick et al., 1994; Potier et al., 2000).

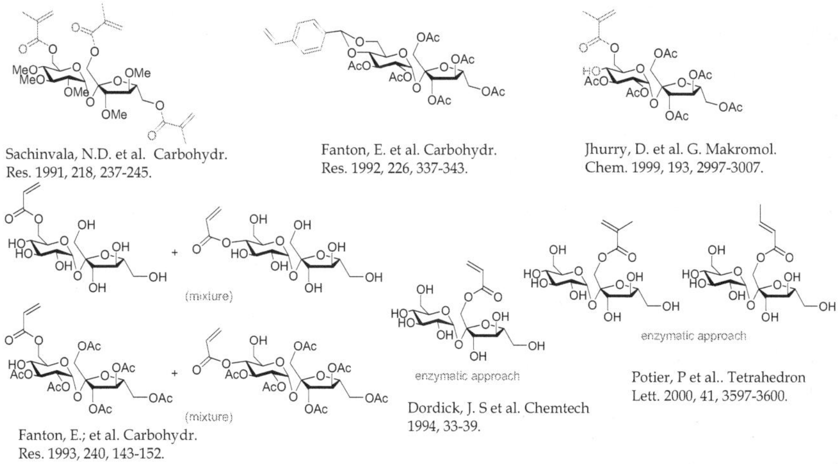

Sachinvala, N.D. et al. Carbohydr. Res. 1991, 218, 237-245.

Fanton, E. et al. Carbohydr. Res. 1992, 226, 337-343.

Jhurry, D. et al. G. Makromol. Chem. 1999, 193, 2997-3007.

(mixture)

(mixture)

Fanton, E.; et al. Carbohydr. Res. 1993, 240, 143-152.

enzymatic approach

Dordick, J. S et al. Chemtech 1994, 33-39.

enzymatic approach

Potier, P et al.. Tetrahedron Lett. 2000, 41, 3597-3600.

Fig. 5. Structures of some selectively obtained sucrose monomers.

The direct transformation of unprotected sucrose in the context of the preparation of derivatives of industrial interest is a challenging task (Queneau et al., 2004). In general, it is accepted that bulky substituents like *tert*-butyl-di-phenylsilyl (TBDPS), are introduced at the primary positions in the order 6-OH ≈ 6′-OH > 1′-OH (Lichtenthaler and Peters, 2004). Since sucrose has eight chemically active hydroxyl groups, regioselective derivatization is important in the selective synthesis of sucrose-containing linear polymers (Barros and Petrova, 2009; Barros et al., 2007; Barros et al., 2010a) and other new compounds. The route to selective derivatization of the 6'-position of the sucrose, developed in our laboratory (Barros et al., 2004a; Barros et al., 2000a), allowed us to obtain the fully benzylated sucrose with only the 6′-hydroxyl unprotected **4** (Fig. 6) in three steps from sucrose and 58 % overall yield. We have prepared several unsaturated sucrose esters and ethers from this intermediate (Fig. 7). These monomers could be converted into pure linear polymers, avoiding the formation of mixtures of di- and higher substituted unsaturated esters, which results in cross-linked polymers (Chen and Park, 2000). Saccharide containing synthetic polymers have attracted great attention because of their potential as biotechnological, pharmacological, and medical materials (Varma et al., 2004).

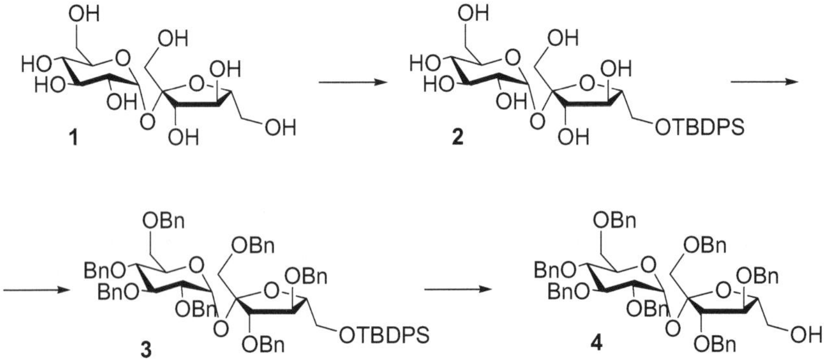

Fig. 6. Protection-deprotection sequence from sucrose.

Microwave-enhanced synthesis has been extended to almost all areas of chemistry (Lidstrom et al., 2001) with the exception of carbohydrate chemistry which has suffered a certain delay, as is testified by the limited number of applications (Corsaro et al., 2004; Soderberg et al., 2001). In particular, it had previously not been applied to sucrose chemistry, and the general opinion is that the method is hampered by competitive degradation of sucrose because of its thermal instability (Queneau et al., 2007). Herein, a method to overcome these limitations and to apply highly efficient and fast synthetic protocols for the synthesis of a series of sugar monomers under microwave irradiation is presented. These alternative protocols allowed a significant decrease of the reaction time to achieve useful sucrose derivatives, compared to other known routes (Barros et al., 2000a; Barros and Sineriz, 2002; Crucho et al., 2008). In all cases, a comparison was made between results obtained with conventional and microwave-assisted methods.

The key point to successful synthesis under microwave irradiation is to use proper equipment, especially designed for chemical laboratories. Monomodal microwave equipment has overcome the uncertainties associated with domestic microwave ovens, as it offers much more precise control over conditions of temperature and pressure than any previous technology and the software provides simplified process monitoring and control, which results in accurate, reproducible reaction conditions (www.milestonesrl.com). The energy transfer in a microwave-assisted reaction is incredibly quick, and only by programming temperature control the decomposition of the substrates has been avoided and comparatively high yields have been obtained in short reaction times. In this method, as the temperature reaches the input value, the power is reduced so that the reaction mixture does not exceed the set point. It then stays at a lower level in order to maintain the set temperature throughout the entire reaction.

The methods used to produce and purify carbohydrate derivatives are often tedious and complex, which makes the selective synthesis of sucrose derivatives laborious and new, simpler procedures are needed. In the development of new carbohydrates or in their transformations there is a need for faster and cleaner methods which can be provided by microwave heating. The library of microwave assisted protocols, presented in this chapter, allows significant reduction of time and energy and potential automatization of tedious multi-step synthesis.

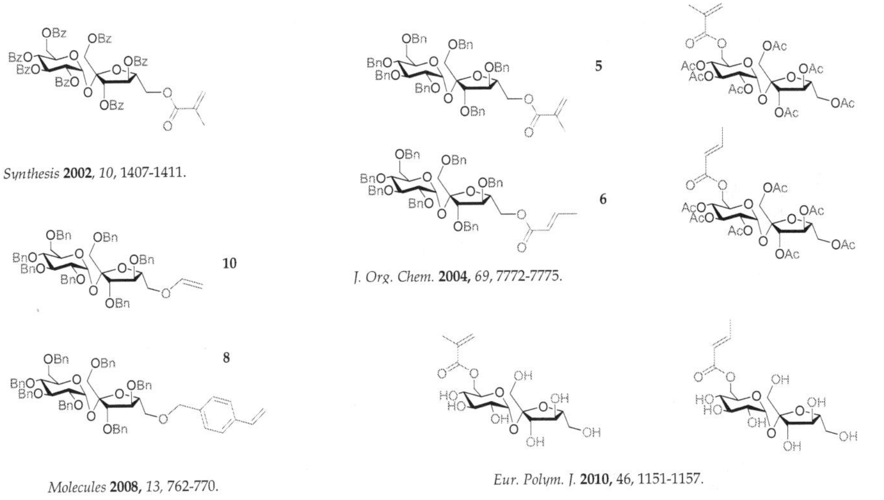

*Synthesis* **2002,** *10,* 1407-1411.

*J. Org. Chem.* **2004,** *69,* 7772-7775.

*Molecules* **2008,** *13,* 762-770.

*Eur. Polym. J.* **2010,** 46, 1151-1157.

Fig. 7. Structures of some sucrose monomers synthesized in our group.

## 2. Methods

### 2.1 General

Reagents and solvents were purified by standard procedures (Perrin et al., 1980). Starting compound **4** (1′,2,3,3′,4,4′,6-hepta-*O*-benzyl sucrose) has been synthesized as previously reported (Barros et al., 2004b). NMR spectra were recorded at 400 MHz in $CDCl_3$, or $D_2O$, with chemical shift values (δ) in ppm downfield from TMS (0 ppm) or the solvent residual peak of $D_2O$ (4.79 ppm) as internal standard. Optical rotations were measured at 20 °C on an AA-1000 polarimeter (0.5 dm cell) at 589 nm. The concentrations (c) are expressed in g/$10^{-2}$ mL. FTIR spectra were recorded on Perkin-Elmer Spectrum BX apparatus in KBr dispersions. The yields are all isolated yields after silica gel chromatography. The reactions under microwave irradiation were performed using a monomodal microwave reactor MicroSynth Labstation (MileStone, USA)(www.milestonesrl.com) in open flasks equipped with temperature control sensor and magnetic stirring.

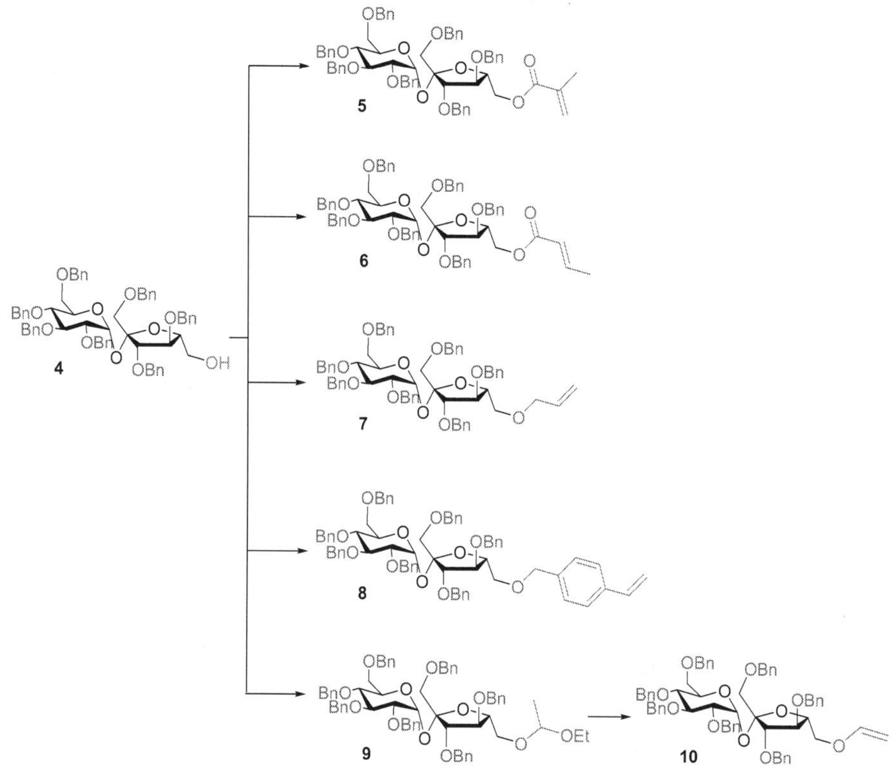


Fig. 8. Synthesis of benzylated sucrose monomers from **4**.

**1′,2,3,3′,4,4′,6-Hepta-*O*-benzyl-6′-*O*-methacryloyl-sucrose (5) and 1′,2,3,3′,4,4′,6-hepta-*O*-benzyl-6′-*O*-crotonyl-sucrose (6)**

To a 0.1 M solution of **4** (500 mg, 0.514 mmol) in anhyd. $CH_2Cl_2$, $Et_3N$ (130 mg, 1.286 mmol) and a catalytic amount of 4-DMAP was added. The mixture was cooled to 0 °C, and then 0.5 M solution of crotonic/ methacrylic anhydride (1.2 equiv.) in anhyd. $CH_2Cl_2$ was added. The reaction mixture was stirred at 0 °C for 10 min, and then placed in the microwave cavity and subjected to microwave irradiation (max 300W at constant temperature 35 °C) for 10 min. The mixture was diluted with more $CH_2Cl_2$ (40 ml) and washed with aq. 1.0 N HCl (8 ml), satd. aq. $NaHCO_3$, and dist. $H_2O$. The organic layer was dried ($Na_2SO_4$), and the solvent evaporated. Purification by flash column chromatography, eluent hexane-ethyl acetate, 5:1, yielded 0.273 g (51 %) of **5** and 0.348 g (65 %) of **6**.

**5:** $[\alpha]_D^{20}$ +46.9(*c* 1.2, $CHCl_3$); $^1H$ NMR ($CDCl_3$): δ 7.37-7.13 (m, 35H), 6.15 (d, *J* 3.6 Hz, 1H), 5.55 (d, *J* 3.6 Hz, 1H), 5.72 (d, *J* 3.7 Hz, 1H), 4.83 (d, *J* 11.0 Hz, 1H), 4.79 (d, *J* 11.0 Hz, 1H), 4.72-4.33 (m, 14H), 4.18 (d, *J* 12.0 Hz, 1H), 4.11 (dd, *J* 11.2, 5.5 Hz, 1H), 4.09-4.05 (m, 1H), 3.93 (t, *J* 9.2 Hz, 1H), 3.75-3.62 (m, 3H), 3.55-3.47 (m, 2H), 2.00 (s, 3H). Anal. Calcd for $C_{65}H_{68}O_{12}$: C, 74.98, H, 6.58. Found: C, 74.69; H, 6.88.

**6:** $[\alpha]_D^{20}$ +48.1 (*c* 1.3, $CHCl_3$); $^1H$ NMR ($CDCl_3$): δ 7.37-7.13 (m, 35H), 6.90 (dq, *J* 15.6, 6.8 Hz, 1H), 5.80 (d, *J* 15.6 Hz, 1H), 5.72 (d, *J* 3.7Hz, 1H), 4.83 (d, *J* 11.0 Hz, 1H), 4.79 (d, *J* 11.0 Hz, 1H), 4.72-4.33 (m, 14H), 4.18 (d, *J* 12.0 Hz, 1H), 4.11 (dd, *J* 11.2, 5.5 Hz, 1H), 4.09-4.05 (m, 1H), 3.93 (t, *J* 9.2 Hz, 1H), 3.75-3.62 (m, 3H), 3.55-3.47 (m, 2H), 1.75 (s, 3H). Anal. Calcd for $C_{65}H_{68}O_{12}$: C, 74.98; H, 6.58. Found: C, 74.74; H, 6.66.

**1′,2,3,3′,4,4′,6-hepta-*O*-benzyl-6′-*O*-allyl sucrose (7)**

**Conventional synthesis**

A solution of 1′,2,3,3′,4,4′,6-hepta-*O*-benzyl sucrose (**4**, 500 mg, 0.514 mmol) in DMF (6 mL) was added dropwise to a slurry of NaH (60% dispersion in mineral oil, 0.031 g, 0.77 mmol) in DMF (10 mL). The mixture was cooled to 0 °C in an ice bath and stirred with exclusion of moisture for 30 min. Allyl bromide (0.124 g, 1.03 mmol) was then added dropwise, the ice bath was removed and the reaction monitored by TLC (hexane-ethyl acetate, 5:1). When there was no more of the initial compound (2-3 hours), excess of the hydride was destroyed by careful addition of water, and the mixture was partitioned between water and ether (50 mL each). The organic phase was washed with water, dried, concentrated and the product **7** was isolated by column chromatography (5:1 hexane–EtOAc) as a colourless oil (0.312 g, 60%), which was in all aspects identical to the product obtained bellow.

**Microwave synthesis**

A solution of 1′,2,3,3′,4,4′,6-hepta-*O*-benzyl sucrose (**4**, 500 mg, 0.514 mmol) in DMF (6 mL) was added dropwise to a slurry of NaH (60% dispersion in mineral oil, 0.031 g, 0.77 mmol) in DMF (10 mL). The mixture was cooled to 0 °C in an ice bath and stirred with exclusion of moisture for 30 min, allyl bromide (0.124 g, 1.03 mmol) was then added dropwise, and the mixture was subjected to microwave irradiation (max 300W at constant temperature 145 °C) for 5 min. Excess of the hydride was destroyed by careful addition of water, the mixture was partitioned between water and ether (50 mL each), the organic phase washed with water, dried, concentrated and the product **7** was isolated by column chromatography (5:1 hexane–EtOAc) as a colourless oil (0.307 g, 59%), $[\alpha]_D^{20}$ +29.9 (*c* 1.1, $CHCl_3$). $^1H$ NMR ($CDCl_3$): δ 7.32 (m, 30H, Ar-H), 5.93 (m, 2H, H-1, *C*H-allyl group), 5.30 (dd, 2H, $J_{HA,HB}$ 17.3 Hz, $CH_2$-allyl group), 5.18 (t, 4H, $J_{3',4'}$ 10.8 Hz, H-3′, H-4′, Ar-$CH_2$), 4.94 (t, 3H, $J_{3,4}$ 11.1 Hz, H-4, Ar-$CH_2$), 4.82 (dd, 3H, $J_{6A,6B}$ 10.9 Hz, H-6, Ar-$CH_2$), 4.72 (d, 1H, *J* 10.8 Hz, Ar-$CH_2$), 4.61 (d, 1H, *J* 10.8 Hz, Ar-$CH_2$), 4.44 (d, 4H, $J_{6'A,6'B}$ 7.7 Hz, H-6′, Ar-$CH_2$), 4.14 (dd, 1H, $J_{1,2}$ 5.7 Hz, $J_{2,3}$ 12.7 Hz, H-2), 4.03 (dd, 2H, $J_{CH2,CH}$ 5.4 Hz, $J_{CH2A,B}$ 11.2 Hz, $O\underline{CH_2}CH{=}CH_2$), 3.70 (t, 1H, $J_{2,3,4}$ 10.4 Hz, H-3), 3.61 (m, 5H, H-1′, H-5 Ar-$CH_2$), 3.46 (m, 2H, H-5′, Ar-$CH_2$). $^{13}C$ NMR ($CDCl_3$): δ 138.6, 138.4, 138.2 ($C_q$ benzyl groups), 134.0 ($\underline{C}H$ allyl group), 128.3, 128.2, 127.9, 127.8, 127.7, 127.6, 127.5, 127.4 ($C_{Ar}$), 117.0 ($\underline{C}H_2$, allyl group), 108.2 (C-2′), 102.7 (C-1), 84.7, 82.3, 77.9, 77.3, 77.0, 76.7 (C-2,3,3′,4,4′,5,5′), 75.7, 75.0, 74.8, 72.5, 70.3, 68.9 (C-1′,6,6′, $O\underline{C}H_2Ph$, $O\underline{C}H_2CH{=}CH_2$). Anal. Calcd for $C_{64}H_{68}O_{11}$: C, 75.87; H, 6.78. Found: C, 75.71; H, 6.86.

**1′,2,3,3′,4,4′,6-hepta-*O*-benzyl-6′-*O*-vinylbenzyl sucrose (8)**

To a solution of **4** (0.5 g, 0.514 mmol) in DMF (12.5 mL) was added NaH (0.07 g, 1.0 eq) at 0°C. After 20 min, 4-vinylbenzyl chloride (0.09 mL, 1.0 eq) was added, the ice-bath removed, and the reaction mixture was subjected to microwave irradiation (max 300W at constant temperature 145 °C) for 10 min. The reaction mixture was poured into cold $H_2O$ (50 ml). The product was extracted with diethyl ether (4x30 ml), and the combined organic layers were washed with $H_2O$ (2x20 ml), dried over $Na_2SO_4$ and concentrated. The residue was purified by flash column chromatography on silica gel, eluent hex/EtOAc 3:1. Compound **8** was obtained as a light yellow oil (0.280 g, 50%); R*f* 0.7 (hex/AcOEt 3:1); $^1$H NMR ($CDCl_3$): δ 7.28-7.12 (m, 39H, Ar-H), 6.65 (dd, 1H, *J* 13.2, 8.2 Hz, PhCH=), 5.71- 5.66 (m, 2H, H-1, =$CH_{2trans}$), 5.20 (dd, 1H, *J* 10.8, 2.11 Hz, =$CH_{2cis}$), 4.86-4.35 (m, 16 H, H-5′, $CH_2$-Ph), 4.22-3.84 (m, 4H, H-6′$_a$, H-6′$_b$, H-5, H-4′, H-3), 3.71-3.25 (m, 7H, H-2, H-6$_a$, H-6$_b$, H-1′$_a$, H-1′$_b$, H-4, H-3′). $^{13}$C NMR ($CDCl_3$): δ 138.9 (Ar-C), 138.6 (Ar-C), 138.2 (Ar-C), 138.0 (Ar-C), 137.9 (PhCH=), 136.8 (Ar-C), 136.5 (Ar-C), 128.3 (Ar-CH), 128.3 (Ar-CH), 128.0 (Ar-CH), 127.9 (Ar-CH), 127.7 (Ar-CH), 127.6 (Ar-CH), 127.5 (Ar-CH), 126,2 (Ar), 113.7 ($CH_2$=), 104.6 (C-2′), 89.9 (C-1), 83.9 (C-5′), 82.4 ($CH_2$-Ph), 81.9 (C-3), 81.9 (C-4′), 79.8 (C-3′), 79.6 (C-5), 77.6 (C-2), 75.5 ($CH_2$-Ph), 74.8 ($CH_2$-Ph), 73.4 ($CH_2$-Ph), 73.0 ($CH_2$-Ph), 72.5 ($CH_2$-Ph), 72.2 ($CH_2$-Ph), 71.4 (C-6′), 71.2 (C-6), 70.6 (C-4), 68.4 (C-1′). Anal. Calcd for $C_{70}H_{72}O_{11}$: C, 77.18; H, 6.66. Found: C, 76.91; H, 6.78.

**1′,2,3,3′,4,4′,6-hepta-*O*-benzyl-6′-*O*-(1-ethoxy)ethyl sucrose (9)**

To a solution of **4** (1.00 g; 1.03 mmol) in dry $CH_2Cl_2$ (10 mL) were added PPTS (10 mg) and ethylvinyl ether (0.074 g; 1.55 mmol). The reaction mixture was subjected to microwave irradiation (max 300W at constant temperature 35 °C) for 10 min. $NaHCO_3$ (1 g) was added, stirred for 10 min more, filtered, concentrated and purified by flash column chromatography on silica gel 60 (0.04-0.06 mm), eluent: hex/EtOAc 3:1. Compound **9** was obtained as a light yellow oil (0.580 g, 54%); R*f* 0.40 (hex/EtOAc 3:1); IR: $\nu_{max}$ ($CH_2Cl_2$) 3088, 3040, 3030, 2976, 2897, 2867, 2000-1600, 1496, 1454, 1363, 1266, 1208, 1082, 1028, 736, 698 cm$^{-1}$; $^1$H NMR ($CDCl_3$): δ 9.79 (qd, 1H, *J* 2.13 Hz, CH), 7.33-7.20 (m, 32H, Ar-H), 7.12-7.10 (m, 2H, Ar-H), 5.54 (d, 1H, *J* 2.5 Hz, H-1), 4.92-4.28 (m, 16H, $CH_2$-Ph, H-5′ and H-3), 4.09-4.06 (m, 1H, H-6), 4.03-3.97 (m, 2H, H-4′ and H-3′), 3.87-3.80 (m, 1H, H-6′), 3.73-3.63 (m, 4H, H-2, H-1′ and $OCH_2CH_3$), 3.61-3.51 (m, 4H, H-5, H-1′, H-6′ and H-4), 3.48 (d, 1H, *J* 3.03 Hz, H-6), 2.20 (d, 1H, *J* 2.13 Hz, $CHCH_3$), 1.27-1.19 (m, 4H, $CH_2CH_3$). $^{13}$C NMR($CDCl_3$): δ 128.3 (Ph), 128.0 (Ph), 127.9 (Ph), 127.8 (Ph), 127.7 (Ph), 127.5 (Ph), 103.8 (C-2′), 91.1 (C-1), 83.6 (C-5′), 81.8 (C-3′), 81.2 (C-4′), 79.5 (C-3), 79.4 (C-5), 77.3 (C-2), 75.6 ($CH_2$-Ph), 74.9 ($CH_2$-Ph), 73.5 ($CH_2$-Ph), 73.4 ($CH_2$-Ph), 73.3 ($CH_2$-Ph), 72.9 ($CH_2$-Ph), 72.5 ($CH_2$-Ph), 71.3 (C-4), 71.2 (C-6′), 67.9 (C-1′), 61.2 (C-6), 60.6 ($OCH_2CH_3$), 20.0 ($O_2$CHCH3), 15.3 ($OCH_2CH_3$). Anal. Calcd for $C_{65}H_{72}O_{12}$: C, 74.69; H, 6.94. Found: C, 74.57; H, 7.08.

**1′,2,3,3′,4,4′,6-hepta-*O*-benzyl-6′-*O*-vinyl sucrose (10)**

To a solution of **9** (0.580 g, 0.555 mmol) in dry $CH_2Cl_2$ (7 mL) were added $Et_3N$ (0.067 g, 0.666 mmol) and TMSOTf (trimethylsilyl trifluoromethanesulfonate) (0.148 g, 0.666 mmol) at 0°C. After 10 min it was allowed to warm to r.t. and subjected to microwave irradiation (max 300W at constant temperature 145 °C) for 5 min. The reaction mixture was neutralised with 1N NaOH (4 mL), and extracted with diethyl ether (3 x 20 mL). After drying and concentrating, it was purified by flash column chromatography on silica gel 60 (0.04-0.06 mm), eluent: hex/EtOAc 3:1. Compound **10** was obtained as a light yellow oil (0.166 g, 30%); R*f* 0.5 (Hex/AcOEt 3:1); IR:$\nu_{max}$ ($CH_2Cl_2$) 3088, 3064, 3030. 2909, 2869, 1737, 2000-1600, 1497, 1454, 1361, 1243, 1209, 1096, 910, 737, 698 cm$^{-1}$; $^1$H NMR($CDCl_3$): δ 7.31-7.26 (m, 35H, Ar-H), 7.16-7.14 (m, 1H, Ar-H), 6.49 (dd, 1H, *J* 10.73, 5.1 Hz, OCH=), 5.67 (d, 1H, *J* 2.6 Hz, H-1), 4.93 (d, 1H, *J* 8.16 Hz, $CH_2$-Ph), 4.83 (d, 1H, *J* 8.19 Hz, $CH_2$-Ph), 4.77 (d, 1H, *J* 8.16 Hz, $CH_2$-Ph), 4.68-4.38 (m, 12H, $CH_2$-Ph and H-5′), 4.20 (dd, 1H, *J* 10.8 and 1.41 Hz, =$CH_{2trans}$), 4.18-4.15 (m, 1H, H-3′), 4.10-4.05 (m, 1H, H-4), 4.01 (dd, 1H, *J* 5.1 and 1.44 Hz, =$CH_{2cis}$), 3.97-3.92 (m, 1H, H-4′), 3.86-3.81 (m, 1H, H-6), 3.75 (m, 1H, H-6), 3.64 (t, 1H, *J* 7.17 Hz, H-2), 3.56-3.49 (m,

3H, H-5, H-1′ and H-6′), 3.42 (dd, 1H, *J* 7.1 and 1.2 Hz, H-6′). $^{13}$C NMR ($CDCl_3$): δ 151.6 (OCH=), 128.3 (Ph), 128.3 (Ph), 127.9 (Ph), 127.8 (Ph), 127.7 (Ph), 127.6 (Ph), 127,5 (Ph), 104.6 (C-2′), 90.2 (C-1), 86.9 (=$CH_2$), 83.9 (C-5′), 81.8 (C-3′), 81.2 (C-4′), 79.5 (C-3), 79.4 (C-5), 77.3 (C-2), 75.6 ($CH_2$-Ph), 74.8 ($CH_2$-Ph), 73.4 ($CH_2$-Ph), 72.9 ($CH_2$-Ph), 72.6 ($CH_2$-Ph), 72.3 ($CH_2$-Ph), 71.0 (C-6′), 70.6 (C-4), 69.1 (C-1′), 69.5 (C-6). Anal. Calcd for $C_{63}H_{66}O_{11}$: C, 75.73; H, 6.66. Found: C, 75.67; H, 6.78.

**General procedure for synthesis of sucrose monoesters 11-15**

**Conventional synthesis**

Sucrose (1.00 g, 2.92 mmol) was dissolved in anhydrous DMF (50 ml), and triphenylphosphine (1.03 g, 3.90 mmol) and the corresponding acid (1.1 eq) were added at room temperature. After complete dissolution, the mixture was cooled to 0 °C and DIAD (diisopropyl azodicarboxylate) (95% in benzene, 0.82 ml, 3.90 mmol) was slowly introduced. TLC (ethyl acetate/MeOH/water 5/2/1) showed the appearance of two new compounds after 24 h stirring at room temperature. After removal of DMF under reduced pressure, the residue was purified by flash column chromatography, eluent ethyl acetate/acetone/water 100/100/1, then 10/10/1 to afford the 6,6′-disubstituted derivative as the faster moving fraction, and monosubstituted derivative as the slower moving fraction. Yields: 45-50 % (see Table 1) in the form of pale yellow oils, which were in all aspects identical to the products obtained bellow.

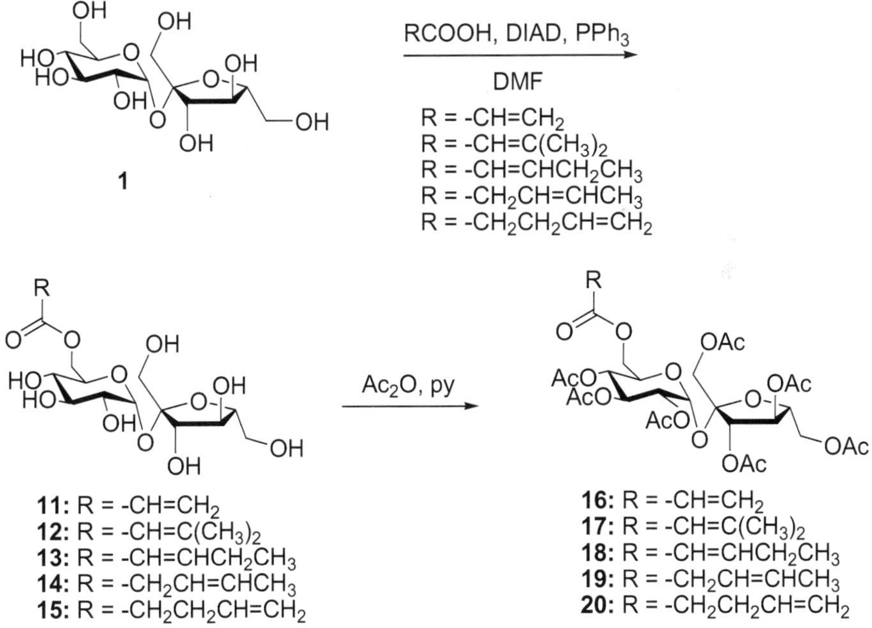


Fig. 9. Library of unprotected and acetylated double bond-containing sucrose monoesters.

**Microwave synthesis**

Sucrose (1.00 g, 2.92 mmol) was dissolved in anhydrous DMF (15 ml), and triphenylphosphine (1.03 g, 3.90 mmol) and the corresponding acid (1.1 eq) were added at room temperature. After complete dissolution, the mixture was cooled to 0 °C and DIAD (95% in benzene, 0.82 ml, 3.90 mmol) was slowly introduced. The reaction mixture was placed in the microwave cavity, and subjected to microwave irradiation (max 300W at constant temperature 145 °C) for 10 min. After removal of DMF under reduced pressure, the residue was purified by flash column chromatography, eluent ethyl acetate/acetone/water 100/100/1, then 10/10/1 to afford the corresponding disubstituted derivative as the faster moving fraction, and monosubstituted derivative as the slower moving fraction.

**6-*O*-acryloyl sucrose 11**

Yield: 51 %; $^1$H NMR: $\delta_H$(400 MHz, $D_2O$): 6.41-6.27 (1H, t, $J_{CH2b\text{-}CH2a=CH}$=16.7 Hz, =$CH_{2a}$), 6.19 (1H, m, =*CH*), 5.95-5.85 (1H, t, $J_{CH2a\text{-}CH2b=CH}$=11.4 Hz, =$CH_{2b}$), 5.32-5.25 (1H, d, $J_{1\text{-}2}$=3.5 Hz, $H_1$), 4.39-4.19 (2H, m, $H_{6a,b}$), 4.14-4.08 (1H, d, $J_{3'\text{-}4'}$=8.5 Hz, $H_{3'}$), 4.05-3.86 (2H, m+t, $J_{3'\text{-}4'\text{-}5'}$=8.7 Hz, $H_5$, $H_{4'}$), 3.84-3.75 (1H, m, $H_{5'}$), 3.75-3.61 (3H, m, $H_{6'a,b}$, $H_3$), 3.54 (2H, s, $H_{1'a,b}$), 3.49-3.41 (1H, dd, $J_{1\text{-}2}$=3.6 Hz, $J_{2\text{-}3}$=9.9 Hz, $H_2$), 3.38-3.30 (1H, t, $J_{3\text{-}4\text{-}5}$=9.9 Hz, $H_4$). $^{13}$CNMR: $\delta_C$(100 MHz, $D_2O$): 168.6 (-COO-), 133.0 ($CH_2$=), 127.7 (CH=), 104.1 ($C_{2'}$), 92.4 ($C_1$), 81.8 ($C_{5'}$), 76.7 ($C_{3'}$), 74.5 ($C_{4'}$), 72.7 ($C_3$), 71.3 ($C_2$), 70.6 ($C_5$), 69.8 ($C_4$), 63.9 ($C_6$), 62.9 ($C_{6'}$), 61.8 ($C_{1'}$). Anal. Calcd for $C_{15}H_{24}O_{12}$: C, 45.46; H, 6.10. Found: C, 45.27; H, 6.28.

**6-*O*-(3,3-dimethylacryloyl) sucrose 12**

Yield: 42 %; $[\alpha]_D$= +53.27 (c=0.81, $CHCl_3$). $^1$HNMR: $\delta_H$(400 MHz, $D_2O$): 5.69 (1H, s, $(CH_3)_2C$=*CH*-), 5.32-5.27 (1H, d, $J_{1\text{-}2}$=3.5 Hz, $H_1$), 4.39-4.16 (2H, dq, $J_{6a\text{-}5}$=5.1 Hz, $J_{6a\text{-}6b}$=12.1 Hz, $J_{H6\text{-}COO}$=53.4 Hz, $H_6$), 4.14-4.07 (1H, d, $J_{3'\text{-}4'}$=7.3 Hz, $H_{3'}$), 4.02-3.94 (1H, m, $H_5$), 3.93-3.85 (1H, t, $J_{3'\text{-}4'\text{-}5'}$=8.6 Hz, $H_{4'}$), 3.82-3.75 (1H, m, $H_{5'}$), 3.71-3.62 (3H, d+t, $J_{6'\text{-}5'}$=5.1 Hz, $J_{2\text{-}3\text{-}4}$=9.1 Hz, $H_{6'}$, $H_3$), 3.54 (2H, s, $H_{1'}$), 3.50-3.43 (1H, dd, $J_{1\text{-}2}$=3.6 Hz, $J_{2\text{-}3}$=9.4 Hz, $H_2$), 3.39-3.30 (1H, t, $J_{3\text{-}4\text{-}5}$=9.5 Hz, $H_4$), 2.03 (3H, s, -$CH_3$), 1.84 (3H, s, -$CH_3$). $^{13}$CNMR: $\delta_C$(100 MHz, $D_2O$): 168.9 (-COO-), 161.5 (($CH_3)_2C$=), 114.7 (-CH=), 104.1 ($C_{2'}$), 92.3 ($C_1$), 81.8 ($C_{5'}$), 76.8 ($C_{3'}$), 74.6 ($C_{4'}$), 72.8 ($C_3$), 71.3 ($C_2$), 70.8 ($C_5$), 70.2 ($C_4$), 63.2 ($C_6$), 63.1 ($C_{6'}$), 61.8 ($C_{1'}$), 27.1 ($CH_3$), 20.3 ($CH_3$). Anal. Calcd for $C_{17}H_{28}O_{12}$: C, 48.11; H, 6.65. Found: C, 47.97; H, 6.79.

**6-*O*-(2-pentenoyl) sucrose 13**

Yield: 47 %; $[\alpha]_D$= +52.70 (c=1.04, $CHCl_3$), $^1$HNMR: $\delta_H$(400 MHz, $D_2O$): 7.13-7.02 (1H, dt, $J_{CH\text{-}CH2}$=6.3 Hz, $J_{CH=CH}$=15.7 Hz, $CH_2$*CH*=CH), 5.90-5.80 (1H, d, $J_{CH=CH}$=15.8 Hz, CH=*CH*-COO), 5.33-5.29 (1H, d, $J_{1\text{-}2}$=3.7 Hz, $H_1$), 4.40-4.21 (2H, dq, $J_{6a\text{-}5}$=5.3 Hz, $J_{6a\text{-}6b}$=12.3 Hz, $J_{H6\text{-}COO}$=44.4 Hz, $H_6$), 4.14-4.09 (1H, d, $J_{3'\text{-}4'}$=8.7 Hz, $H_{3'}$), 4.04-3.97 (1H, m, $H_5$), 3.94-3.87 (1H, t, $J_{3'\text{-}4'\text{-}5'}$=8.6 Hz, $H_{4'}$), 3.84-3.76 (1H, m, $H_{5'}$), 3.72-3.64 (3H, d+t, $J_{6'\text{-}5'}$=4.1 Hz, $J_{2\text{-}3\text{-}4}$=10.0 Hz, $H_{6'}$, $H_3$), 3.56 (2H, s, $H_{1'}$), 3.51-3.45 (1H, dd, $J_{1\text{-}2}$=3.8 Hz, $J_{2\text{-}3}$=9.9 Hz, $H_2$), 3.40-3.32 (1H, t, $J_{3\text{-}4\text{-}5}$=9.6 Hz, $H_4$), 2.23-2.12 (2H, p, $J_{CH\text{-}CH2\text{-}CH3}$=6.9 Hz, $CH_2$), 1.00-0.92 (3H, t, $J_{CH2\text{-}CH3}$=7.4 Hz, $CH_3$). $^{13}$CNMR: $\delta_C$(100 MHz, $D_2O$): 169.3 (-COO-), 154.8 ($CH_2$CH=CH), 119.0 (-COO-CH=CH), 104.1 ($C_{2'}$), 92.3 ($C_1$), 81.8 ($C_{5'}$), 76.8 ($C_{3'}$), 74.6 ($C_{4'}$), 72.7 ($C_3$), 71.3 ($C_2$), 70.8 ($C_5$), 70.1 ($C_4$), 63.8 ($C_6$), 63.1 ($C_{6'}$), 61.8 ($C_{1'}$), 25.4 ($CH_2$), 11.6 ($CH_3$). Anal. Calcd for $C_{17}H_{28}O_{12}$: C, 48.11; H, 6.65. Found: C, 47.99; H, 6.74.

**6-*O*-(3-pentenoyl) sucrose 14**

Yield: 48 %, $[\alpha]_D^{25}$=+46.98 (c=0.94, $CHCl_3$); $^1$HNMR: $\delta_H$(400 MHz, $D_2O$): 5.64-5.53 (1H, m, $J_{CH\text{-}CH3}$=6.6 Hz, CH=*CH*-$CH_3$), 5.49-5.38 (1H, m, $J_{CH\text{-}CH2}$=7.0 Hz, CH=*CH*-$CH_2$), 5.32-5.27 (1H, d, $J_{1\text{-}2}$=3.2 Hz, $H_1$), 4.35-4.15 (2H, dq, $J_{6a\text{-}5}$=4.6 Hz, $J_{6a\text{-}6b}$=13.2 Hz, $J_{H6\text{-}COO}$=53.3 Hz, $H_{6a,b}$), 4.15-4.07 (1H, d, $J_{3'\text{-}4'}$=8.6 Hz, $H_{3'}$), 4.00-3.88 (2H, m+t, $J_{3'\text{-}4'\text{-}5'}$=8.8 Hz, $H_5$+$H_{4'}$), 3.84-3.76 (1H, m, $H_{5'}$), 3.74-3.69 (2H, d, $J_{6'\text{-}5'}$=6.7 Hz, $H_{6'}$), 3.69-3.61 (1H, t, $J_{2\text{-}3\text{-}4}$=9.7 Hz, $H_3$), 3.55 (2H, s, $H_{1'}$), 3.50-3.41 (1H, dd, $J_{1\text{-}2}$=3.6 Hz, $J_{2\text{-}3}$=9.7 Hz, $H_2$), 3.39-3.29 (1H, t, $J_{3\text{-}4\text{-}5}$=9.5 Hz, $H_4$), 3.03-3.00 (2H, m, $J_{CH\text{-}CH2}$=6.5 Hz, $CH_2$), 1.65-1.53 (3H, d, $J_{CH\text{-}CH3}$=6.0 Hz, $CH_3$). $^{13}$CNMR: $\delta_C$(100 MHz, $D_2O$): 175.5 (-COO-), 131.4 ($CH_3CH$=CH), 122.3 (-$CH_2$-CH=CH), 104.1 ($C_{2'}$), 92.4 ($C_1$), 81.8 ($C_{5'}$), 76.7 ($C_{3'}$), 74.5 ($C_{4'}$), 72.7 ($C_3$), 71.3 ($C_2$), 70.7 ($C_5$), 69.9 ($C_4$), 64.0 ($C_6$), 63.0 ($C_{6'}$), 61.7 ($C_{1'}$), 37.6 ($CH_2$), 17.6 ($CH_3$). Anal. Calcd for $C_{17}H_{28}O_{12}$: C, 48.11; H, 6.65. Found: C, 48.02; H, 6.88.

**6-*O*-(4-pentenoyl) sucrose 15**

Yield: 42 %,$[\alpha]_D^{25}$=+54.64 (c=0.91, $CHCl_3$) ; $^1$HNMR: $\delta_H$(400 MHz, $D_2O$): 5.86-5.73 (1H, ddt, $J_{CH\text{-}CH2}$=6.6 Hz, $J_{CH=CH2cis}$=10.6 Hz, $J_{CH=CH2trans}$=16.8 Hz, $CH_2$=*CH*-$CH_2$), 5.32-5.28 (1H, d, $J_{1\text{-}2}$=3.5 Hz, $H_1$),5.05-4.91 (2H, dd, $J_{CH\text{-}CH2cis}$=10.2 Hz, $J_{CH=CH2trans}$=17.7 Hz, *$CH_2$*=CH-$CH_2$), 4.35-4.15 (2H, dq, $J_{6a\text{-}5}$=4.6 Hz, $J_{6a\text{-}6b}$=11.8 Hz, $J_{H6\text{-}COO}$=47.4 Hz, $H_{6a,b}$), 4.15-4.08 (1H, d, $J_{3'\text{-}4'}$=8.7 Hz, $H_{3'}$), 4.00-3.89 (2H, m+t, $J_{3'\text{-}4'\text{-}5'}$=8.6 Hz, $H_5$+$H_{4'}$), 3.84-3.76 (1H, m, $H_{5'}$), 3.74-3.70 (2H, d, $J_{6'\text{-}5'}$=6.3 Hz, $H_{6'}$), 3.70-3.63 (1H, t, $J_{2\text{-}3\text{-}4}$=9.6 Hz, $H_3$), 3.56 (2H, s, $H_{1'}$), 3.50-3.43 (1H, dd, $J_{1\text{-}2}$=3.6 Hz, $J_{2\text{-}3}$=9.9 Hz, $H_2$), 3.40-3.31 (1H, t, $J_{3\text{-}4\text{-}5}$=9.6 Hz, $H_4$), 2.50-2.42 (2H, t, $J_{CH2\text{-}CH2}$=7.1 Hz, *$CH_2$*-COO), 2.33-2.23 (2H, q, $J_{CH2\text{-}CH2\text{-}CH}$=6.5 Hz, $CH_2$-*$CH_2$*-CH). $^{13}$CNMR: $\delta_C$(100 MHz, $D_2O$): 176.2 (-COO-), 137.4 ($CH_2$=CH), 115.8 ($CH_2$=CH), 104.1 ($C_{2'}$), 92.3 ($C_1$), 81.8 ($C_{5'}$), 76.7 ($C_{3'}$), 74.5 ($C_{4'}$), 72.7 ($C_3$), 71.3 ($C_2$), 70.7 ($C_5$), 69.9 ($C_4$), 63.7 ($C_6$), 63.0 ($C_{6'}$), 61.7 ($C_{1'}$), 33.3 (COO-$CH_2$), 28.6 ($CH_2$-CH=). Anal. Calcd for $C_{17}H_{28}O_{12}$: C, 48.11; H, 6.65. Found: C, 48.20; H, 6.69.

**General method for peracetylation**

Corresponding sucrose ester (**11-15**) (500mg) was dissolved in pyridine (5 mL) and 2.5 mL of acetic anhydride was added. The reaction mixture was placed in the microwave cavity, and subjected to MW irradiation (max 300W) at constant temperature (90 °C) for 5 min. The method afforded, after removal of the solvent and purification by flash column chromatography (eluent hexane/ethyl acetate 3/1, then 1/1), the corresponding acetylated sucrose monoesters (**16-20**) in 96-98 % yield.

**1′,2,3,3′,4,4′,6′-hepta-*O*-acetyl- 6-*O*-acryloyl sucrose 16**

$[\alpha]_D^{25}$= +45.96 (c=0.98, $CHCl_3$); $^1$HNMR: $\delta_H$(400 MHz, $CDCl_3$): 6.46-6.36 (1H, dd, $J_{CH2b\text{-}CH2a}$=5.3 Hz, $J_{CH2\ =CH}$=17.7 Hz, =*$CH_{2a}$*), 6.22-6.05 (1H, m, =*CH*), 5.88-5.80 (1H, t, $J_{CH2a\text{-}CH2b=CH}$=12.4 Hz, =*$CH_{2b}$*), 5.71-5.66 (1H, d, $J_{1\text{-}2}$=3.1 Hz, $H_1$), 5.49-5.40 (2H, m, $H_3$, $H_{3'}$), 5.38-5.33 (1H, t, $J_{3'\text{-}4'\text{-}5'}$=5.7 Hz, $H_{4'}$), 5.11-5.03 (1H, t, $J_{3\text{-}4\text{-}5}$=10.1 Hz, $H_4$), 4.90-4.83 (1H, dd, $J_{1\text{-}2}$=3.8 Hz, $J_{2\text{-}3}$=9.6 Hz, $H_2$), 4.35-4.24 (5H, m, $H_5$, $H_{6'a,b}$, $H_{6a,b}$), 4.24-4.08 (3H, m+s, $H_{5'}$, $H_{1'a,b}$), 2.18 (3H, s, $CH_3O$), 2.12 (9H, s, 3$CH_3O$), 2.10 (3H, s, $CH_3O$), 2.05 (3H, s, $CH_3O$), 2.02 (3H, s, $CH_3O$). $^{13}$CNMR: $\delta_C$(100 MHz, $D_2O$): 170.5-169.5 (8-COO-), 131.1 ($CH_2$=), 127.7 (CH=), 104.1 ($C_{2'}$), 89.9 ($C_1$), 79.1 ($C_{5'}$), 75.7 ($C_{3'}$), 75.0 ($C_{4'}$), 70.3 ($C_2$), 69.6 ($C_3$), 68.4 ($C_5$), 68.1 ($C_4$), 63.6 ($C_{6'}$), 62.8 ($C_{1'}$), 61.9 ($C_6$), 20.6 (7$CH_3CO$). Anal. Calcd for $C_{29}H_{38}O_{19}$: C, 50.44; H, 5.55. Found: C, 50.40; H, 5.51.

**1′,2,3,3′,4,4′,6′-hepta-*O*-acetyl- 6-*O*-(3,3-dimethylacryloyl) sucrose 17**

$[\alpha]_D^{25}$= +40.27 (c=1.04, $CHCl_3$), $^1$HNMR: $\delta_H$(400 MHz, $CDCl_3$): 5.74 (1H, s, $(CH_3)_2C$=*CH*-), 5.71-5.68 (1H, d, $J_{1\text{-}2}$=3.7 Hz, $H_1$), 5.49-5.40 (2H, t+d, $J_{3'\text{-}4'}$=5.3 Hz, $H_3$, $H_{3'}$), 5.39-5.33 (1H, t, $J_{3'\text{-}4'\text{-}5'}$=5.7 Hz, $H_{4'}$), 5.13-5.08 (1H, t, $J_{3\text{-}4\text{-}5}$=9.7 Hz, $H_4$), 4.90-4.84 (1H, dd, $J_{1\text{-}2}$=3.7 Hz, $J_{2\text{-}3}$=10.3 Hz, $H_2$), 4.39-4.33 (1H, dd, $J_{5\text{-}6}$=4.0 Hz, $J_{4\text{-}5}$=11.9 Hz, $H_5$), 4.33-4.09 (7H, m, $H_{6'}$, $H_6$, $H_{5'}$, $H_{1'}$), 2.18 (3H, s, $CH_3O$), 2.16 (3H, s, -$CH_3$), 2.12 (6H, s, 2$CH_3O$), 2.11 (3H, s, $CH_3O$), 2.10 (3H, s, $CH_3O$),

2.04 (3H, s, $CH_3O$), 2.01 (3H, s, $CH_3O$), 1.90 (3H, s, $-CH_3$). $^{13}$CNMR: $\delta_C$(100 MHz, $D_2O$): 170.5-169.4 (8-COO-), 157.9 (($(CH_3)_2C$=), 115.3 (-CH=), 104.1 ($C_{2'}$), 90.0 ($C_1$), 79.2 ($C_{5'}$), 75.8 ($C_{3'}$), 75.1 ($C_{4'}$), 70.3 ($C_2$), 69.8 ($C_3$), 68.6 ($C_5$), 68.4 ($C_4$), 63.6 ($C_{6'}$), 62.9 ($C_{1'}$), 60.7 ($C_6$), 27.4 ($CH_3$), 20.6 (7$CH_3$CO). Anal. Calcd for $C_{31}H_{42}O_{19}$: C, 51.81; H, 5.89. Found: C, 51.75; H, 5.91.

**1′,2,3,3′,4,4′,6′-hepta-*O*-acetyl- 6-*O*-(2-pentenoyl) sucrose 18**

$[\alpha]_D^{25}$= +76.47 (c=0.94, $CHCl_3$), $^1$HNMR: $\delta_H$(400 MHz, $CDCl_3$): 7.14-7.02 (1H, dt, $J_{CH\text{-}CH2}$=6.3 Hz, $J_{CH=CH}$=15.6 Hz, $CH_2$*CH*=CH), 5.91-5.82 (1H, d, $J_{CH=CH}$=15.7 Hz, CH=*CH*-COO), 5.73-5.68 (1H, d, $J_{1\text{-}2}$=3.7 Hz, $H_1$), 5.50-5.41 (2H, t+d, $H_3$, $H_{3'}$), 5.38-5.34 (1H, t, $J_{3'\text{-}4'\text{-}5'}$=5.5 Hz, $H_{4'}$), 5.15-5.08 (1H, t, $J_{3\text{-}4\text{-}5}$=9.8 Hz, $H_4$), 4.91-4.84 (1H, dd, $J_{1\text{-}2}$=3.5 Hz, $J_{2\text{-}3}$=10.3 Hz, $H_2$),4.38-4.32 (1H, dd, $J_{5\text{-}6}$=4.2 Hz, $J_{4\text{-}5}$=12.2 Hz, $H_5$), 4.32-4.14 (7H, m, $H_{6'}$, $H_6$, $H_{5'}$, $H_{1'}$), 2.30-2.20 (2H, p, $J_{CH\text{-}CH2\text{-}CH3}$=6.8 Hz, $CH_2$), 2.18 (3H, s, $CH_3O$), 2.12 (6H, s, 2$CH_3O$), 2.11 (3H, s, $CH_3O$), 2.10 (3H, s, $CH_3O$), 2.04 (3H, s, $CH_3O$), 2.02 (3H, s, $CH_3O$), 1.12-1.05 (3H, t, $J_{CH2\text{-}CH3}$=7.4 Hz, $CH_3$). $^{13}$CNMR: $\delta_C$(100 MHz, $D_2O$): 170.5 ($CH_3$CO), 170.1 ($CH_3$CO), 169.8 ($CH_3$CO), 169.6 ($CH_3$CO), 169.4 ($CH_3$CO), 166.3 (-COO-), 151.9 ($CH_2$CH=CH), 119.5 (-COO-CH=CH), 104.1 ($C_{2'}$), 90.0 ($C_1$), 79.2 ($C_{5'}$), 75.7 ($C_{3'}$), 75.1 ($C_{4'}$), 70.3 ($C_2$), 69.7 ($C_3$), 68.5 ($C_5$), 68.3 ($C_4$), 63.5 ($C_{6'}$), 62.8 ($C_{1'}$), 61.4 ($C_6$), 25.3 ($CH_2$), 20.6 (7$CH_3$CO), 12.0 ($CH_3$). Anal. Calcd for $C_{31}H_{42}O_{19}$: C, 51.81; H, 5.89. Found: C, 51.73; H, 5.92.

**1′,2,3,3′,4,4′,6′-hepta-*O*-acetyl- 6-*O*-(3-pentenoyl) sucrose 19**

$[\alpha]_D^{25}$= +40.71 (c=1.11, $CHCl_3$), $^1$HNMR: $\delta_H$(400 MHz, $CDCl_3$): 5.72-5.67 (1H, d, $J_{1\text{-}2}$=3.5 Hz, $H_1$), 5.65-5.52 (2H, m, $J_{CH\text{-}CH2}$, $J_{CH\text{-}CH3}$=5.2 Hz, CH=CH), 5.48-5.41 (2H, t+d, $J_{3'\text{-}4'}$=6.2 Hz, $H_3$, $H_{3'}$), 5.40-5.34 (1H, t, $J_{3'\text{-}4'\text{-}5'}$=6.0 Hz, $H_{4'}$), 5.12-5.04 (1H, t, $J_{3\text{-}4\text{-}5}$=9.8 Hz, $H_4$), 4.91-4.83 (1H, dd, $J_{1\text{-}2}$=3.6 Hz, $J_{2\text{-}3}$=10.4 Hz, $H_2$), 4.39-4.33 (1H, dd, $J_{5\text{-}6}$=4.1 Hz, $J_{4\text{-}5}$=11.8 Hz, $H_5$), 4.33-4.09 (7H, m, $H_{6'}$, $H_6$, $H_{5'}$, $H_{1'}$), 3.11-3.02 (2H, d, $J_{CH\text{-}CH2}$=5.0 Hz, $CH_2$), 2.17 (3H, s, $CH_3O$), 2.12 (3H, s, $CH_3O$), 2.11 (3H, s, $CH_3O$), 2.10 (3H, s, $CH_3O$), 2.10 (3H, s, $CH_3O$), 2.04 (3H, s, $CH_3O$), 2.02 (3H, s, $CH_3O$), 1.75-1.67 (3H, d, $J_{CH\text{-}CH3}$=5.2 Hz, $CH_3$). $^{13}$CNMR: $\delta_C$(100 MHz, $D_2O$): 171.9 (-COO-), 170.7 ($CH_3$CO), 170.5 ($CH_3$CO), 170.1 ($CH_3$CO), 169.9 ($CH_3$CO), 169.6 ($CH_3$CO), 129.7 (CH=CH), 122.3 (CH=CH), 104.1 ($C_{2'}$), 89.9 ($C_1$), 79.2 ($C_{5'}$), 75.7 ($C_{3'}$), 75.0 ($C_{4'}$), 70.3 ($C_2$), 69.6 ($C_3$), 68.5 ($C_5$), 68.2 ($C_4$), 63.5 ($C_{6'}$), 62.8 ($C_{1'}$), 61.8 ($C_6$), 37.5 ($CH_2$), 20.6 (7$CH_3$CO), 17.9 ($CH_3$). Anal. Calcd for $C_{31}H_{42}O_{19}$: C, 51.81; H, 5.89. Found: C, 51.83; H, 5.95.

**1′,2,3,3′,4,4′,6′-hepta-*O*-acetyl- 6-*O*-(4-pentenoyl) sucrose 20**

$[\alpha]_D^{25}$=+49.82(c=0.88, $CHCl_3$), $^1$HNMR: $\delta_H$(400 MHz, $CDCl_3$): 5.90-5.76 (1H, ddt, $J_{CH\text{-}CH2}$=6.4 Hz, $J_{CH=CH2cis}$=10.4 Hz, $J_{CH=CH2trans}$=16.5 Hz, $CH_2$=*CH*-$CH_2$), 5.71-5.67 (1H, d, $J_{1\text{-}2}$=3.6 Hz, $H_1$), 5.48-5.41 (2H, t+d, $J_{2\text{-}3\text{-}4}$=9.2 Hz, $J_{3'\text{-}4'}$=5.3 Hz, $H_3$, $H_{3'}$), 5.39-5.33 (1H, t, $J_{3'\text{-}4'\text{-}5'}$=5.7 Hz, $H_{4'}$), 5.12-5.06 (1H, t, $J_{3\text{-}4\text{-}5}$=9.5 Hz, $H_4$), 5.06-4.98 (2H, t, $J_{CH\text{-}CH2cis\text{-}CH2trans}$=9.6 Hz, *$CH_2$*=CH-$CH_2$), 4.89-4.84 (1H, dd, $J_{1\text{-}2}$=3.7 Hz, $J_{2\text{-}3}$=10.3 Hz, $H_2$), 4.38-4.32 (1H, dd, $J_{5\text{-}6}$=4.4 Hz, $J_{4\text{-}5}$=11.8 Hz, $H_5$), 4.32-4.28 (2H, t, $J_{6'a\text{-}6'b\text{-}5'}$=3.1 Hz, $H_{6'}$), 4.28-4.25 (2H, d, $J_{5\text{-}6}$=4.9 Hz, $H_6$), 4.23-4.19 (1H, t, $J_{4'\text{-}5'\text{-}6'}$=5.6 Hz, $H_{5'}$), 4.18 (2H, s, $H_{1'}$), 2.50-2.42 (2H, t, $J_{CH2\text{-}CH2}$=6.1 Hz, *$CH_2$*-COO), 2.42-2.34 (2H, q, $J_{CH2\text{-}CH2\text{-}CH}$=6.2 Hz, $CH_2$-*$CH_2$*-CH), 2.17 (3H, s, $CH_3O$), 2.12 (6H, s, 2$CH_3O$), 2.11 (3H, s, $CH_3O$), 2.10 (3H, s, $CH_3O$), 2.04 (3H, s, $CH_3O$), 2.02 (3H, s, $CH_3O$). $^{13}$CNMR: $\delta_C$(100 MHz, $D_2O$): 172.7 (-COO-), 170.5 ($CH_3$CO), 170.1 ($CH_3$CO), 169.9 ($CH_3$CO), 169.6 ($CH_3$CO), 169.5 ($CH_3$CO), 136.6 ($CH_2$=CH), 115.5 ($CH_2$=CH), 104.1 ($C_{2'}$), 90.0 ($C_1$), 79.2 ($C_{5'}$), 75.7 ($C_{3'}$), 75.0 ($C_{4'}$), 70.3 ($C_2$), 69.6 ($C_3$), 68.5 ($C_5$), 68.1 ($C_4$), 63.6 ($C_{6'}$), 62.8 ($C_{1'}$), 61.5 ($C_6$), 33.1 (COO-$CH_2$), 28.6 ($CH_2$-CH=), 20.6 (7$CH_3$CO). Anal. Calcd for $C_{31}H_{42}O_{19}$: C, 51.81; H, 5.89. Found: C, 51.92; H, 5.99.

## 3. Results and discussion

The synthesis of a library of new sucrose-containing monomers has been developed using microwave irradiation, focusing on the selectivity of the transformations. Synthesis of esters and ethers, substituted regioselectively at 6′-position of sucrose was achieved using protection-deprotection strategy. The regioselective formation of monounsaturated esters at 6-position, by applying the Mitsunobu conditions for esterification of sucrose has also been studied.

Sucrose is soluble in protic solvents such as water, methanol and ethanol, and reasonably soluble in dipolar aprotic solvents such as pyridine, DMF and DMSO, which are suitable as nucleophilic substitution media. DMF has comparatively high boiling point (153 °C) and a dipole suitable for the absorption of microwave radiation. Dichloromethane also proved suitable as a reaction media for the benzylated sucrose derivatives. The maximum temperature which the reaction mixtures were allowed to reach was chosen in every case according to the boiling points and stabilities of the reagents and solvent. The maximum irradiation power was set to 300 W for all the reactions in order to obtain comparable results, and the reaction time was optimized with a view to the best yield.

The reactions under microwave irradiation were performed using a monomodal microwave reactor MicroSynth Labstation (MileStone, USA)(www.milestonesrl.com) in open flasks equipped with temperature control sensor and magnetic stirring. This microwave synthesizer has a single mode cavity with temperature regulation and pressure control, which means that temperature runaway and explosion risk are avoided. It should be noticed that the reaction conditions are not expressed as a function of the magnetron power, as for most microwave-assisted reactions published, but by the reaction temperature.

### 3.1 Synthesis of benzylated sucrose-containing monomers

With the purpose of comparing the polymerization ability of monomers, in which the vinyl group is adjacent to the sucrose moiety, or is located further away, a small library was generated, using compound **4** as starting material.

Esterifications of the free 6′-hydroxyl of **4** were carried out by treating it in a mixture of dichloromethane and triethylamine, with the appropriate anhydride, to afford the expected compounds **5** and **6** in good yields (Table 1). The $^{1}$H NMR spectra of these compounds exhibit two signals which have been attributed to the double bond. For methacrylate **5** a singlet at 6.15 ppm and a doublet at 5.60 ppm and for crotonate **6** a mutiplet at 6.90 ppm and a doublet at 5.80 ppm. We assigned the allylic methyl groups of the molecules at 1.95 ppm for **5** and 1.70 ppm for **6**.

Conventional etherification was tested involving a primary halide, in our case allyl bromide or 4-vinylbenzyl chloride (Jhurry et al., 1992), optimizing the procedure for protected sucrose. In this case, the intermediate **4** was treated with NaH in DMF at 0°C, with the subsequent addition of the allyl bromide to form **7** in 59 % yield; or 4-vinylbenzyl chloride, to form **8** in 50 % yield under microwave irradiation (Fig. 8).

For obtaining vinyl sucrose ether **10**, we have adopted a two-step route via mixed acetals, based on the Gassman method, first reported in 1993 (Gassman et al., 1993), as it requires readily available reagents, mild conditions, and does not involve the use of heavy metal salt catalysts, such as mercury (Hughes et al., 2005). It consists in forming a vinyl group by the elimination of ethanol from mixed acetals with trimethylsilyl trifluoromethanesulfonate (TMS-triflate) in the presence of alkyl amines.

The mixed acetal **9** was prepared by treating **4** with ethyl vinyl ether in the presence of PPTS under microwave irradiation. No side products were observed and a mixture of diastereomers was produced, with no attempt at separation being made. After purification by flash column chromatography, the elimination reaction was performed by treating **9** with TMS-triflate and triethylamine under microwave irradiation, leading to **10** (Fig. 8). The yield of this step was decreased by the competing side reaction of formation of silyl ether, resulting of a complexation of the trimethylsilyl cation with the sucrose connected oxygen atom, as it was discussed by Gassman *et al.* (Gassman et al., 1993) and Hughes *et al.* (Hughes et al., 2005). Complexation of the trimethylsilyl cation with the ethoxy group oxygen produced the desired vinyl sucrose ether.

### 3.2 Functionalization of sucrose avoiding protecting group chemistry

The Mitsunobu reaction (Mitsunobu, 1981) is a convenient method for selective esterification. Such a process performed on free sucrose affords 6,1′,6′-triesters or 6,6′-diesters (Beraud et al., 1989; Bottle and Jenkins, 1984), establishing the reactivity of hydroxy groups as 6-OH > 6′-OH >1′-OH > secondary OH groups. Several interesting esters of sucrose, such as derivatives of fatty acids (Molinier et al., 2004) and 6-perfluoroalkanoates for biomedical uses (Abouhilale et al., 1991) have been prepared by this method. Treatment of the free sucrose with phthalimide under Mitsunobu conditions afforded derivative, in which the primary 6-OH and 6′-OH groups were replaced with phthalimide moieties, while the secondary ones at the C-3′ and C-4′ positions were converted into an epoxide (Amariutei et al., 1988).

In contrast to many related condensation reactions (Gu et al., 2008), Mitsunobu reactions proceed under mild, essentially neutral conditions and exhibit stereospecificity, functional selectivity and regioselectivity. Because of these features the Mitsunobu reaction has been employed in synthesis of macrolide antibiotics, nucleosides, nucleoside phosphates, amino acids, amino sugars, steroids and other natural products (Brain et al., 2001; Chong and Chu, 2000). The mechanism of the Mitsunobu reaction involves the formation of a phosphonium intermediate that reacts with the alcohol oxygen atom and the selectivity observed in these reactions is presumably a result of the preferential attack at the primary hydroxy groups (Grochowski et al., 1982; Itzstein and Jenkins, 1983). In the case of sucrose, the 1'-position being neopentyl-like, is considerably more hindered than the 6- and 6'-positions. The product has been shown to be the 6-substituted rather than the 6'-substituted by hydrolysis with invertase (Bottle and Jenkins, 1984; Guthrie et al., 1979).

Mitsunobu esterification conditions (Mitsunobu, 1981) are known to provide good selectivity, even in the case of a complex polyol like sucrose, thus allowing a rapid and efficient synthesis of sucrose esters. Although the atom economy of the procedure was poor, it gave excellent results on a small scale and allowed us to avoid the use of protecting group chemistry and multi-step procedures. Applying microwave irradiation to this procedure allowed us to reduce the reaction time to 10 min instead of 30 h at r.t., and provided better selectivity towards the 6-*O*-mono-ester over 6,6′-*O*-diester. Subsequent conventional acetylation afforded the corresponding acetylated products in quantitative yields. In the case of the monomer **11**, the products obtained were very reactive and tended to polymerize spontaneously upon concentration. This difficulty was overcome by adding hydroquinone to the mixture (Russo et al., 2007).

Our goal was to obtain monoesters of sucrose selectively (Fig. 9), which is why only one equivalent of the corresponding acid was used, and the irradiation was not more than 10 min at 145 °C. Thus it was possible to optimize the yields to about 50 % of regioisomerically pure unsaturated monoesters of sucrose. The corresponding sucrose 6,6′-*O*-diesters were isolated in smaller amounts than the monoesters, and it was not possible to avoid their formation.

| Comp. No. | Microwave conditions | Reagents | Time [min] | Yield [%] | Conventional conditions | Yield[a] [%], ref. |
|---|---|---|---|---|---|---|
| **5** | 35°C/300W | $(CH_2{=}C(CH_3)CO)_2O$, $NEt_3$, $CH_2Cl_2$ | 10 | 51 | r.t./ 4 h | 73 (Barros et al., 2004b) |
| **6** | 35°C/300W | $(CH_3CH{=}CHCO)_2O$, $NEt_3$, $CH_2Cl_2$ | 10 | 65 | r.t./ 4 h | 81 (Barros et al., 2004b) |
| **7** | 145°C/300W | $CH_2{=}CHCH_2Br$, NaH, DMF | 10 | 59 | r.t./ 3 h | 60 |
| **8** | 145°C/300W | $CH_2{=}CHC_6H_4CH_2Br$, NaH, DMF | 10 | 50 | 70°C/ 4 h | 46 (Crucho et al., 2008) |
| **9** | 35°C/300W | $CH_2{=}CHOEt$, PPTS, $CH_2Cl_2$ | 10 | 54 | r.t./ 2 h | 56 (Crucho et al., 2008) |
| **10** | 35°C/300W | TMSOTf, $NEt_3$, $CH_2Cl_2$ | 10 | 30 | r.t./ 2 h | 31 (Crucho et al., 2008) |
| **11** | 145°C/300W | $H_2C{=}CHCO_2H/PPh_3$, DIAD, DMF | 10 | 51 | r.t./ 24 h | 49 |
| **12** | 145°C/300W | $(CH_3)_2C{=}CHCO_2H/PPh_3$, DIAD, DMF | 10 | 42 | r.t./ 24 h | 46 |
| **13** | 145°C/300W | $CH_3CH_2CH{=}CHCO_2H/PPh_3$, DIAD, DMF | 10 | 47 | r.t./ 24 h | 50 |
| **14** | 145°C/300W | $CH_3CH{=}CHCH_2CO_2H/PPh_3$, DIAD, DMF | 10 | 48 | r.t./ 24 h | 47 |
| **15** | 145°C/300W | $H_2C{=}CHCH_2CH_2CO_2H/PPh_3$, DIAD, DMF | 10 | 42 | r.t./ 24 h | 45 |
| **16-20** | 90°C/300W | peracetylation $(CH_3CO)_2O$, pyridine | 5 | 96-98 | r.t./ 12 h | 96-98 |

Table 1. Reaction conditions and experimental results under microwave heating compared with conventional conditions.

The differentiation between the regioisomeric esters of sucrose can be verified by several methods (Guthrie et al., 1979), (Chauvin et al., 1993). However, the most common assignment is by high-resolution NMR, supported by HMBC (Zoete et al., 1999). For all the monoesters of unprotected sucrose in the $^1H$ NMR, differences in the chemical shifts of the protons corresponding to the glucose ring, compared to the spectra of unsubstituted sucrose, were observed. The signal for H-5 was moved downfield from 3.72 to 4.00 ppm, and the signal for H-6 (a and b) from 3.69 to 4.31 ppm and split from a singlet to a duplet and duplet of duplets (Fig. 10 - comparison of the sucrose region signals in the $^1H$ NMR spectra of sucrose and sucrose 6-*O*-monoesters). In the $^{13}C$ NMR spectrum a similar shift of the signals for the glucose ring was observed: C-4 from 69.6 in sucrose to 70.1 ppm in the ester, C-5 from 72.9 to 70.8 ppm, and most significantly, the C-6 from 60.4 to 63.8 ppm (Fig. 11). In the COSY spectrum, correlation between the proton at 4.02 ppm, which was assigned to H-5, and the methylene protons at 4.31 ppm, was observed assigning them to H-6. In the HMQC spectrum, the signal at 63.8 ppm gave a crosspeak with the H at 4.31 ppm, confirming it is C-6. This differentiation between the glucose and fructose rings and the three methylene signals of the protons and carbons have been confirmed by DEPT, COSY (Fig. 12), and HMQC spectra, and places the acyl-substituent at the C-6 position, but it is most certainly proved by HMBC technique (Fig. 13). Thus, a crosspeak between the C of the carbonyl group and H-6 was observed, confirming the place of the substituent at C-6, as well as the long range couplings between C-4 and H-6, between C-6 and H-4, and C-5 and H-6, which were consistent with structures **11-15**.

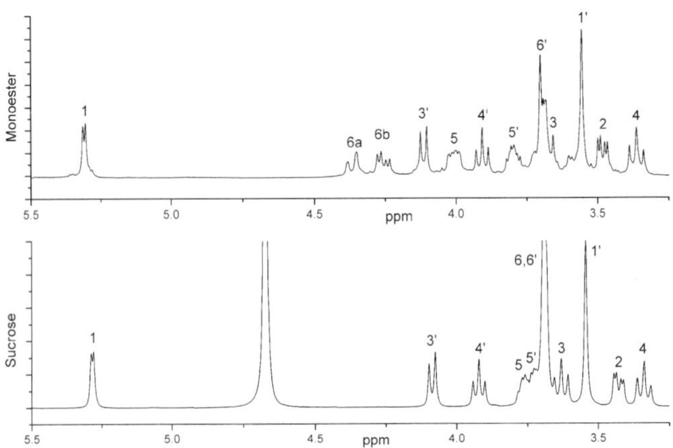


Fig. 10. Comparison of the sucrose region signals in the $^{1}$H NMR spectra of sucrose and sucrose 6-*O*-mono-esters.

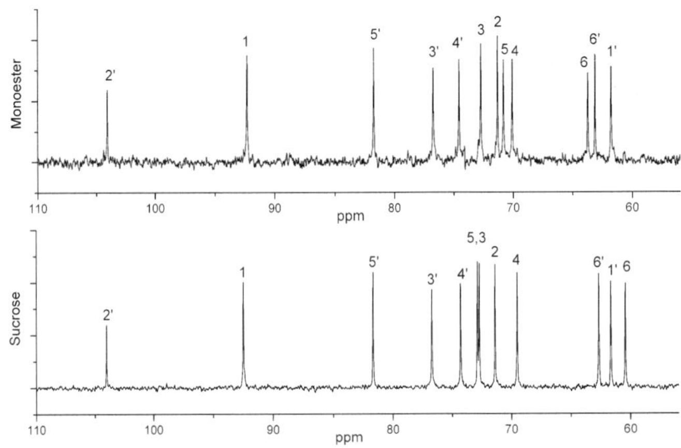


Fig. 11. Comparison of the sucrose region signals in the $^{13}$C NMR spectra of sucrose and sucrose 6-*O*-mono-esters.

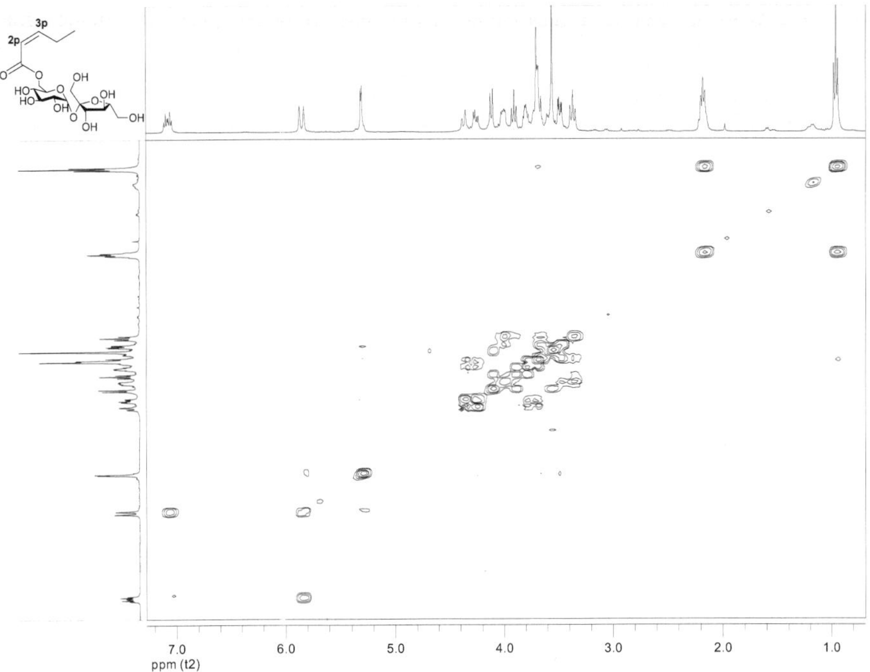


Fig. 12. COSY spectrum of 6-*O*-(2-pentenoyl) sucrose **13**.

To obtain the corresponding acetylated products **16-20**, conventional treatment with acetic anhydride in pyridine was performed. It is common opinion in the literature, that acetylation of the sugar derivatives facilitates and simplifies the NMR analysis (Jarosz and Mach, 2002; Zoete et al., 1999). However, in our case it proved impossible to distinguish the site of substitution in the acetylated products, as the attachment of an acyl group instead of an acetyl results in very small differences in chemical shifts for both the sugar proton and carbon signals, since both groups are connected through ester bonds. It was possible to obtain a complete assignment of the sugar signals (Sanders and Hunter, 1987), assuming the same position of the acyl substituents as in the unprotected derivatives – 6. In the HMBC spectrum all possible cross-peaks resulting from three-bond couplings between sugar protons and in the substituents groups are present. The possible cross-peaks between C-carbonyl groups and the sugar protons were observed, while the four-bond signals between the sugar carbons and acetyl $CH_3$ protons, reported in (Zoete et al., 1999), were not observed, and assignment for the site of the acyl substituent based on them was not possible.

### 3.2 Environmental assessment of the method

Two useful measures of the potential environmental acceptability of chemical processes are the *E* factor, defined as the mass ratio of waste to desired product, and the atom utilization, calculated by dividing the molecular weight of the desired product by the sum of the molecular weights of all substances produced in the stoichiometric equation (Table 2), (Sheldon, 2008). In this calculation the solvents were not taken into consideration as waste as they were easily recovered and reused. According to a study presented by Sheldon (Sheldon, 2008), these values are in the range reported for bulk chemicals, i.e. are lower than for the fine chemical industry. We could see from table 2 that the reactions involving formation of phosphorous intermediates (**11-15**) are less economic than the straightforward substitution, but one has to take into account the formation of the starting material **4** in three steps from sucrose. On the other hand, the advantage of the direct method stems from the improved regioselectivity and the avoidance of protecting group chemistry. The data shows that the one-step methods for the valorization of sucrose as a renewable natural feedstock were beneficial over multistep protection-deprotection strategies.

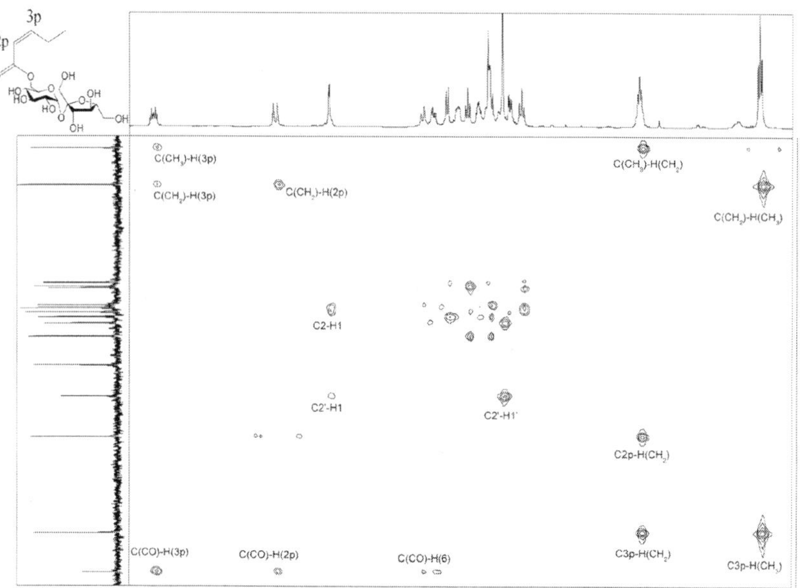


Fig. 13. HMBC spectrum of 6-*O*-(2-pentenoyl) sucrose **13**.

| Comp. No. | E factor (g waste/ g product) | Atom Utilization (M product/ M all compounds formed), [%] |
|---|---|---|
| **4 (from 1)** | 1.13 | 60 |
| **5** | 0.35 | 85 |
| **6** | 0.28 | 85 |
| **7** | 0.17 | 91 |
| **8** | 0.20 | 91 |
| **10** | 1.23 | 73 |
| **11** | 1.39 | 59 |
| **12** | 1.57 | 60 |
| **13** | 1.41 | 60 |
| **14** | 1.38 | 60 |
| **15** | 1.57 | 60 |

Table 2. Environmental acceptability and Percentage of atom utilization.

## 4. Conclusion

This chapter provides a number of mild and effective protocols for the synthesis of diverse monomeric carbohydrate building blocks. The utilization of microwave irradiation as a more efficient mode of heating leading to shorter reaction periods was evaluated. It has been shown that microwave radiation efficiently promoted the reactions and allowed a great reduction in reaction times and at the same time improving or maintaining the good yields achieved by conventional conditions. In the case of sucrose functionalization the elimination of solvent is not recommended as it led to significantly lower yields because of overheating. Microwave-assisted carbohydrate chemistry is, at the present time, experiencing considerable growth and has the potential to greatly improve the image of carbohydrate chemistry. Short reaction times and large rate enhancements as compared to conventional methods are observed, notably in reactions using mild reagents or involving steric hindrance. Yields are comparable or better than when using conventional methods, and sometimes much higher in reactions where the short reaction time prevents decomposition. The short reaction times in combination with easy reproducibility and workup, make these methods suitable for automation.

## 5. Acknowledgment

This work has been supported by Fundação para a Ciência e a Tecnologia (PTDC/ QUI/ 72120/ 2006).

## 6. References

Abouhilale, S., Greiner, J., and Riess, J. G., 1991, One-step preparation of 6-perfluoroalkylalkanoates of trehalose and sucrose for biomedical uses, Carbohydr. Res. 212:55-64.

Amariutei, L., Descotes, G., Kugel, C., Maitre, J. P., and Mentech, J., 1988, Sucrochimie IV.1 Synthèse régiosétective de dérivés aminés du saccharose via la réaction de Mitsunobu, J. Carbohydr. Chem. 7(1):21-31.

Andrade, M. M., and Barros, M. T., 2004, Facile conversion of O-sylil protected sugars into their corresponding formates using POCl3.DMF complex., Tetrahedron 60:9235-9243.

Andrade, M. M., Barros, M. T., and Rodrigues, P., 2007, Selective Synthesis Under Microwave Irradiation of Carbohydrate Derivatives Containing Unsaturated Systems, Eur. J. Org. Chem. :3655-3668.

Andrade, M. M., and Barros, M. T., 2009, Microwave-assisted 1,3-dipolar cycloaddition reactions of vinylic glycosides with aryl azides - unexpected synthesis of triazoles and acetyl group migration, Arkivoc:299-306.

Barros, M. T., Maycock, C. D., Rodrigues, P., and Thomassigny, C., 2004a, Improved anomeric selectivity for the aroylation of sugars., Carbohydr. Res. 339:1373-1376.

Barros, M. T., Maycock, C. D., Sineriz, F., and Thomassigny, C., 2000a, Fast galloylation of a sugar moiety: preparation of three monogalloylsucroses as references for antioxidant activity. A method for the selective deprotection of tert-butyldiphenylsilyl ethers., Tetrahedron 56:6511-6516.

Barros, M. T., Maycock, C. D., and Thomassigny, C., 2000b, Preparation of sucrose heptaesters unsubstituted at the C-1 hydroxy group of the fructose moiety via selective O-desilylation, Carbohydr. Res. 328:419-423.

Barros, M. T., Maycock, C. D., and Thomassigny, C., 2001, Bromine in Methanol: An Efficient Reagent for the Deprotection of the tert-Butyldiphenylsilyl Group, Synlett (7):1146-1148.

Barros, M. T., Petrova, K., and Ramos, A. M., 2004b, Regioselective Copolymerization of Acryl Sucrose monomers., J. Org. Chem. 69:7772-7775.

Barros, M. T., and Petrova, K. T., 2009, Ziegler-Natta catalysed polymerisation for the preparation of potentially biodegradable copolymers with pendant sucrose moieties, Eur. Polym. J. 45(1):295–301.

Barros, M. T., Petrova, K. T., and Ramos, A. M., 2007, Biodegradable Polymers based on α- or β-Pinene and Sugar Derivatives or Styrene, Obtained under Normal Conditions and Microwave Irradiation, Eur. J. Org. Chem.:1357-1363.

Barros, M. T., Petrova, K. T., and Singh, R. P., 2010a, Synthesis and biodegradation studies of new copolymers based on sucrose derivatives and styrene. , Eur Polym J 46:1151-1157.

Barros, M. T., Petrova, K. T., and Singh, R. P., 2010b, Synthesis of Hydrophilic and Amphiphilic Acryl Sucrose Monomers and Their Copolymerisation with Styrene, Methylmethacrylate and α- and β-Pinenes, Int. J. Mol. Sci. 11:1792-1807.

Barros, M. T., and Sineriz, F., 2002, Synthesis of optically active monomers and copolymers derived from protected 6'-O-acryloyl sucroses., Synthesis 10:1407-1411.

Beraud, P., Bourhim, A., Czernecki, S., and Krausz, P., 1989, Modification selective de mono- et de di-saccharides non proteges par l'intermediaire de liaisons ester et ether, Tetrahedron Lett. 30(3):325-326.

Bottle, S., and Jenkins, I. D., 1984, Improved synthesis of "Cord factor" analogues, J. Chem. Soc., Chem. Commun.:385-385.

Brain, C. T., Nelson, A., Tanikkul, N., and Thomas, E. J., 2001, An Approach to the Total Synthesis of Lankacidins: Synthesis of the Requisite Building Blocks., Tetrahedron Lett. 42:1247-1250.

Carneiro, M. J., Fernandes, A., Figueiredo, C. M., Fortes, A. G., and Freitas, A. M., 2001, Synthesis of carbohydrate based polymers, Carbohydr. Polym. 45:135-138.

Chauvin, C., Thibault, P., Plusquellec, D., and Banoub, J., 1993, Differentiation of Regioisomeric Esters of Sucrose by Ionspray Tandem Mass Spectrometry, J. Carbohydr. Chem. 12(4&5):459-475.

Chen, J., and Park, K., 2000, Synthesis of fast-swelling, superporous sucrose hydrogels, Carbohydr. Polym. 41:259-268.

Chong, Y. G., G., and Chu, C. K., 2000, A Divergent Synthesis of D- and L-Carbocyclic 4'-Fluoro-2',3'-dideoxynucleosides as Potential Antiviral Agents., Tetrahedron: Asymmetry 11:4853-4875.

Corsaro, A., Chiacchio, U., Pistarà, V., and Romeo, G., 2004, Microwave-assisted Chemistry of Carbohydrates, Curr. Org. Chem. 8:511-538.

Crucho, C. C., Petrova, K. T., Pinto, R. C., and Barros, M. T., 2008, Novel Unsaturated Sucrose Ethers and Their Application as Monomers, Molecules 13:762-770.

Davis, B. G., and Fairbanks, A. J., 2002, Carbohydrate Chemistry, University Press, Oxford.

Dordick, J. S., Linhardt, R. J., and Renthwish, D., 1994, Chemtech:33-39.

Fanton, E., Fayet, C., Gelas, J., Jhurry, D., Deffieux, A., and Fontanille, M., 1992, Ethylenic acetals of sucrose and their copolymerization with vinyl monomers, Carbohydr. Res. 226:337-343.

Ferreira, L., Vidal, M. M., Geraldes, C. F., and Gil, M. H., 2000, Preparation and characterisation of gels based on sucrose modified with glycidyl methacrylate, Carbohydr. Polym. 41:15-24.

Galgali, P., Varma, A. J., Puntambekar, U. S., and Gokhale, D. V., 2002, Towards biodegradable polyolefines: strategy of anchoring minute quantities of monosaccharides and disaccharides onto functionalized polystyrene, and their effect on facilitating polymer biodegradation., Chem. Commun. :2884-2885.

Gassman, P. G., Burns, S. J., and Pfister, K. B., 1993, Synthesis of cyclic and acyclic enol ethers (vinyl ethers). J.Org.Chem. 58:1449-1457.

Grochowski, E., Hilton, B. D., Kupper, R. J., and Michejda, C. J., 1982, Mechanism of the triphenylphosphine and diethyl azodicarboxylate induced dehydration reactions (Mitsunobu reaction). The central role of pentavalent phosphorus intermediates, J. Am. Chem. Soc. 104(24):6876-6877.

Gu, Y., Azzouzi, A., Pouilloux, Y., Jerome, F., and Barrault, J., 2008, Heterogeneously catalyzed etherification of glycerol: new pathways for transformation of glycerol to more valuable chemicals, Green Chem. 10:164-167.

Guthrie, R. D., Jenkins, I. D., Rogers, P. J., Sum, W. F., Watters, J. J., and Yamasaki, R., 1979, Studies of invertase (β-D-fructofuranosidase), Carbohydr. Res. 75:C1-C4.

Hughes, K. D., Nguyen, T.-L. N., Dyckman, D., Dulay, D., Boyko, W. J., and Giuliano, R., 2005, Synthesis of vinyl alpha-D-glucopyranosides from mixed acetal glycosides., Tetrahedron: Asymmetry 16:273-282.

Itzstein, M. v., and Jenkins, I. D., 1983, The mechanism of the Mitsunobu reaction. II. Dialkoxytriphenylphosphoranes, Aust. J. Chem. 36(3):557 - 563.

Jarosz, S., and Mach, M., 2002, Regio- and Stereoselective Transformations of Sucrose at the Terminal Positions, Eur. J. Org. Chem.:769-780.

Jhurry, D., Deffieux, A., and Fontanille, M., 1992, Sucrose based polymers. Linear polymers with sucrose side-chains., Makromol. Chem. 193:2997-3007.

Karl, H., Lee, C. K., and Khan, R., 1982, Synthesis and reactions of tert-butyldiphenylsilyl ethers of sucrose, Carbohydr. Res. 101:31-38.

Khan, R., 1984, Chemistry and New Uses of Sucrose:How Important?, Pure & Appl. Chem. 56(7):833-844.

Klein, J., M., K., and J., K., 1990, Poly(vinylsaccharide)s, 7 New surfactant polymers based on carbohydrates, Makromol.Chem. 191:517-528.

Kobayashi, K., 2001, Baiosaiensu to Indasutori 59(10):679-682.

Kobayashi, K., Sumitomo, H., and Ina, Y., 1985, Synthesis and Functions of Polystyrene Derivatives Having Pendant Oligosaccharides, Polymer Journal 17:567-575.

Lichtenthaler, F. W., and Mondel, S., 1997, Perspectives in the use of low molecular weight carbohydrates as organic raw materials, Pure & Appl. Chem. 69(9):1853-1866.

Lichtenthaler, F. W., and Peters, S., 2004, Carbohydrates as green raw materials for the chemical industry., C. R. Chim. 7:65-90.

Lidstrom, P., Tierney, J., Wathey, B., and Westman, J., 2001, Microwave-Assisted Organic Synthesis - a review., Tetrahedron 57:9225-9283.

Mach, M., Jarosz, S., and Listkowski, A., 2001, Crown Ether Analogs From Sucrose, J. Carbohydr. Chem. 20(6):485-493

Mitsunobu, O., 1981, The Use of Diethyl Azodicarboxylate and Triphenylphosphine in Synthesis and Transformation of Natural Products, Synthesis (1):1-28.

Molinier, V., Fitremann, J., Bouchu, A., and Queneau, Y., 2004, Sucrose esterification under Mitsunobu conditions:evidence for the formation of 6-O-acyl-3',6'-anhydrosucrose besides mono and diesters of fatty acids, Tetrahedron: Asymmetry 15:1753-1762.

Patil, D. R., Dordick, J. S., and Rentwisch, D., 1991, Chemoenzymatic synthesis of novel sucrose-containing polymers, Macromolecules 24:3462-3463.

Perrin, D. D., Armagedo, W. L. F., and Perrin, D. R., 1980, Purification of Laboratory Chemicals, Pergamon Press Ltd., New York, pp. 74-465.

Potier, P., Bouchou, A., Descotes, G., and Queneau, Y., 2000, Proteinase N-catalysed transesterifications in DMSO-water and DMF-water: preparation of sucrose monomethacrylate., Tetrahedron 41:3597-3600.

Queneau, Y., Fitremann, J., and Trombotto, S., 2004, The chemistry of unprotected sucrose: the selectivity issue, C. R. Chimie 7:177-188.

Queneau, Y., Jarosz, S., Lewandowski, B., and Fitremann, J., 2007, Sucrose chemistry and applications of sucrochemicals., Adv. Carbohydr. Chem. Biochem. 61:217-292

Russo, A., Maschio, G., and Ampelli, C., 2007, Reaction inhibition as a method for preventing thermal runaway in industrial processes, Macromol. Symp. 259 365-370.

Sanders, J. K. M., and Hunter, B. K., 1987, in: Modern NMR Spectroscopy: Chapter 10. Sucrose octa-acetate: a case history, Oxford University Press, New York, pp. 282-297.

Sheldon, R. A., 2008, E factors, green chemistry and catalysis: an odyssey, Chem. Commun. (29):3352-3365.

Soderberg, E., Westman, J., and Oscarson, S., 2001, Rapid Carbohydrate Protecting Group Manipulations Assisted by Microwave Dielectric Heating, J. Carbohydr. Chem. 20(5):397-410.

Varma, A. J., Kennedy, J. F., and Galgali, P., 2004, Synthetic polymers functionalized by carbohydrates: a review, Carbohydr. Polym. 56:429-445.

www.milestonesrl.com.

Zoete, M. S., Kneepkens, M. F., Waard, P., Osterom, M. W., Gotlieb, K. F., and Slagnek, T. M., 1999, Enzymatic synthesis and NMR studies of acylated sucrose acetates, Green Chem. 1(3):153-156.

# Part 6

# Applications

# 15

# Microwave Irradiation Effect in Water-Vapor Desorption from Zeolites

Hongyu Huang[1], Seiya Ito[2], Fujio Watanabe[2], Masanobu Hasatani[2] and Noriyuki Kobayashi[1]
[1]*Nagoya University*
[2]*Aichi Institute of Technology*
*Japan*

## 1. Introduction

In recent years, humidity control has been recognized as one of the important technologies in various fields; e.g., the system is required to maintain comfortable indoor air quality in household sector, and to improve the quality of products in industrial sector. In general, controlling the humidity through temperature, as in the case of conventional systems, appears to be an energy consuming process, and, depending to the operation conditions, does not assure the demand levels for humidity and temperature. Under these circumstances, desiccant humidity conditioner, which makes use of adsorption/desorption phenomena of porous adsorbent, has been gaining a great attention as an environmental friendly humidification/dehumidification system because of its advantages in the following points:

1. It consumes very little electrical energy, and for regeneration process it allows the use of solar energy and waste energy.
2. It is efficient when latent heat load is larger than the sensible load.
3. It is a clean technology, which can be used to condition the internal environment of buildings and operates without the use of harmful refrigerants.
4. The achieved control of humidity is better than that when using vapor compression systems.
5. In some cases the cost of energy to regenerate the desiccant is less than that when compared with the cost of energy to dehumidify the air by cooling it below its dew point.
6. Improvement in indoor air quality is more likely due to the normally high ventilation.
7. It has the capability of removing airborne pollutants.

The technology of adsorptive desiccant cooling presents interesting prospects as regards market penetration (Ando & Kodama, 2005; Davanagere et al., 1999; Elsayed et al., 2006, 2008; Ge et al., 2008; Halliday et al., 2002; Hamed, 2003; Kabeel, 2007; Kodama et al., 2001; Mavroudaki et al., 2002; Oshima et al., 2006).

The desiccants are natural or synthetic substances capable of absorbing or adsorbing water-vapor due to the difference of water-vapor pressure between the surrounding air and the desiccant surface. Typical adsorptive desiccant cooling process mainly consisting of a rotary dehumidifier (D-hum) and heat exchanger can be driven with low-temperature heat energy like solar energy or waste heat, and it has been expected to be alternative air conditioning considering various energy/environmental problems such as global warming. In the solid desiccant system, a desiccant, which is coated with silica gel or zeolite, is generally used as an adsorber/desorber. During the desorption process of the system, hot air is generally used as a heating medium, and the heating of air is carried out using a solar collector, electric heater, and exhaust heat in some industrial factories. However, indirect heating with hot air, especially with the one generated by an electric heater, results in consumption of large amount of energy for regeneration, because of the fact that, the heating of air is required to heat entire desiccant rotor consisting of an adsorbent and rotor matrix. As a result, the system is characterized by low heating efficiency. In addition, excess temperature rise in the rotor during desorption process causes low water adsorptivity of an adsorbent in the following adsorption process. Moreover, when regeneration is performed with lower thermal energy below 80°C, humidification/dehumidification performance greatly decreases due to insufficient water desorption. This problem has been discussed and the optimization of adsorption-desorption process has been examined in the field of the desiccant air-conditioning, but it has not got an essential solution yet (Hamamoto et al., 2004; Harshe et al, 2005; Kodama et al., 2005).

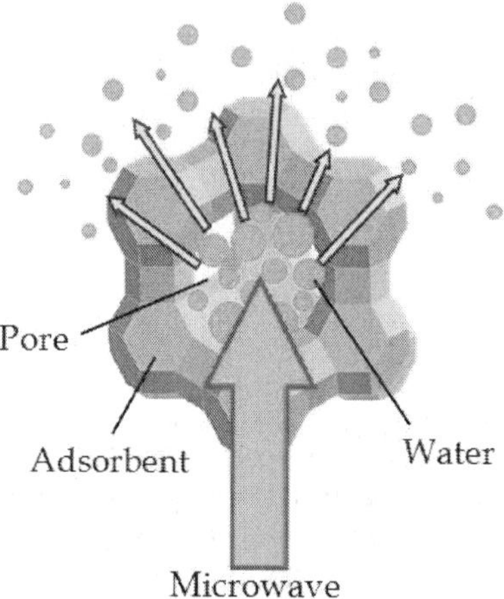


Fig. 1. Imaginative diagram of desorption by microwave selective heating

The method for application of the microwave irradiation as a regeneration heat source of the adsorption material is considered as a strategy to the above mentioned problems.Microwave irradiation has a great advantage of direct and rapid heating of material due to self-heating of a material under microwave irradiation, which consequently results in heating of only adsorbent containing adsorbed water without heating the surrounding air (as shown in Figure 1) (Bradshaw et al., 1997, 1998; Cherbanski & Molga, 2009; Kuo, 2008; Polaert et al., 2007, 2010; Yan et al., 2004, 2007).

Hence, in the process of water desorption, there is a possibility that adsorbed water is selectively heated by microwave irradiation rather than the adsorbent, resulting in an enhancement of desorption with lower energy consumption. Based on the above mentioned advantages of microwave heating, we have proposed the novel hybrid regeneration process, which combines microwave heating and conventional hot air heating. By combining both heating methods, the hybrid system shown in Figure 2 is expected to achieve highly energy efficiency for regeneration due to direct and rapid heating by microwave irradiation as well as low thermal energy utilization provided by hot air heating. The system also has a feature of promotion of lower heat utilization by assisting water desorption with an additional energy of microwave.

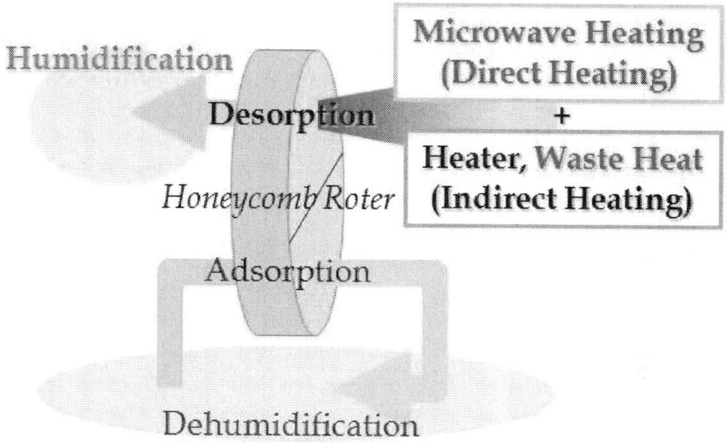


Fig. 2. Concept of the microwave heating hybrid system

Some researches have been carried out on the application of microwave heating to the regeneration of adsorbent for desiccant air conditioning. Ohgushi and Nagae have investigated heating and dehydration characteristics of various zeolites with microwave heating for the reusable desiccant in home, and reported that the mixture of Na-X and Ca-X was useful to prevent the thermal runaway (Ohgushi & Nagae, 2003). Concerning the durability of zeolite mixtures against microwave heating, they have also indicated 1.3% degradation of water adsorptivity after each MW irradiation (Ohgushi & Nagae, 2005).

In previous study, we paid attention to zeolite that showed strong water-vapor desorption capability as an adsorbent, and microwave irradiation effect was examined in water-vapor desorption of zeolite 13X (Saitake et al., 2007). As a result, the maximum desorption rate was found to be about 5 times higher for microwave heating at 800 W than that obtained for hot air heating. It was also observed that the amount of water desorbed from zeolite particles by microwave heating was 1.6-2.0 times larger than that by hot air heating, regardless of microwave power.

However, almost all experiments have been performed with powder or granular adsorbent, and there are very few researches on water desorption from desiccant rotor with microwave heating. Microwave desorption feature of these zeolite is almost unknown. Therefore, it is essential to grasp the influence of condition such as flow rate and temperature of air on desorption rate to establish the HM with microwave heating condition.

In this research, the examination of two items as follow was performed under the above-mentioned viewpoints that are; i) Experimental study was performed about the influence of adsorption equilibrium and pore architecture in desorption of microwave heating by three zeolites. ii) Experimental study was performed about the influence of gas flow rate and temperature in desorption of microwave heating.

## 2. Experimental

### 2.1 Adsorbents

3 kinds of zeolite samples (4A, OXYSIV-5 and DF-9, average particle size fraction of 500μm) (made by UNION SHOWA K.K., Japan) were used. The pore size of samples is 0.4nm, 0.8nm and 1.0nm. The water-vapor adsorption and desorption isotherms were measured by water-vapor adsorption device (Belsorp aqua3, BEL JAPAN, INC.) at 30°C and shown in Figure 3.

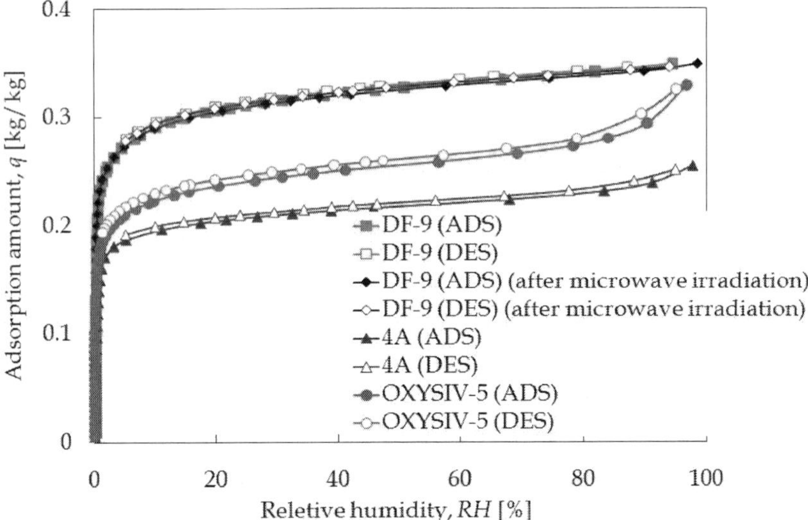


Fig. 3. Adsorption and desorption isotherms of water-vapor on zeolites at 30°C

The adsorption amount of water-vapor on 4A and OXYSIV-5 rise sharply in the range of the relative humidity, RH, below 5%, and then increased gradually. For 4A, the adsorption amount of water-vapor below 5% of RH accounts for 75% of the total adsorption amount. In addition, the adsorption amount of 4A is smaller than that of OXYSIV-5. For DF-9, the adsorption amount of water-vapor rise sharply in the range of the relative humidity, RH, below 10%, and then increased gradually almost similar to that of OXYIVE-5. The adsorption amount of water-vapor below 10% of RH accounts for 84% of the total adsorption amount which is 1.3 times larger than that of OXYSIV-5. In addition, the water-vapor adsorption and desorption of 4A and DF-9 showed desorption hysteresis, which is smaller than that of OXYSIV-5.

on, visual check and measurement of water-vapor adsorption and desorption isotherms was carried out after the microwave irradiation experiment. As a result, the damage and the transformation of the zeolites by microwave irradiation were not observed. Moreover, change of adsorption and desorption isotherms by the existence or nonexistence of microwave irradiation was not observed as shown as Figure. 3.

### 2.2 Experimental apparatus and method

The experimental apparatus, as shown in Figure 4, consisted of a microwave irradiator, a circulated packed adsorption, an evaporator, a microwave absorber, a heater and thermometers. To keep the apparatus on a constant temperature (30°C), the insulant was used to enclose the apparatus.

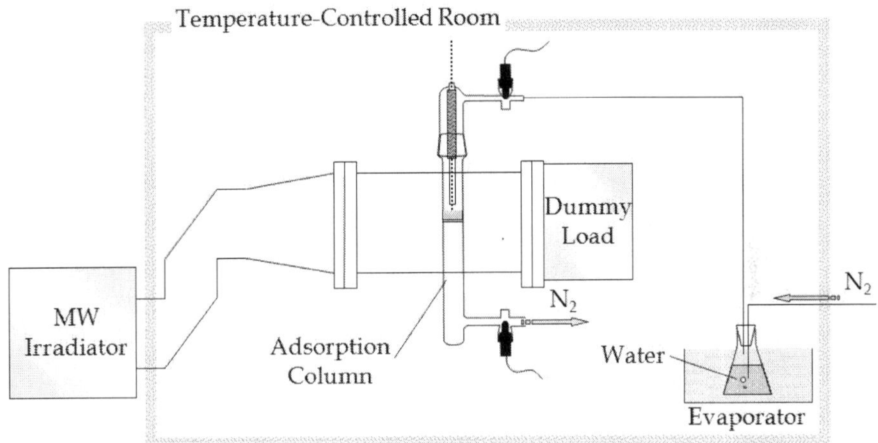


Fig. 4. Schematic diagram of experimental apparatus

Microwave was oscillated at 2450MHz, passed a cylindrical waveguide of 110mm in the diameter (TE11 mode) through a rectangular waveguide (TE01 mode), and was absorbed by the microwave absorber. The circulated type absorption column was vertically set at 140mm from the entrance and in the center of section of the circular waveguide. The position where the adsorption column was set up was the position where the electric field strength of microwave was the maximum. The humidity-temperature meters (PosiTector DPM, DeFelsko Corp.) were set up in each gateway of the circulation air was calculated from the measurement of temperature and humidity difference, and the amount of adsorption and desorption were calculated based on those. The fiber-optic thermometer was inserted in the center part of the adsorbent packed bed to measure the temperature, and this temperature was assumed to be a representative temperature of the adsorbent packed bed.

The samples bed (particle diameter: 0.3-1.0mm, adsorbent bed thickness: about 2.5mm (0.5g)) in adsorption column was set in the position where electric field intensity of microwave waveguide was the maximum. In this study, a micro heater (20W) is inserted in the sample bed upper which is a circulated packed adsorption column. The adsorbent bed temperature was adjusted with heating the circulation gas by electric power (shown in Figure 5). In microwave heating experiment, microwave irradiation was carried out with the supply of this measurement electric power.

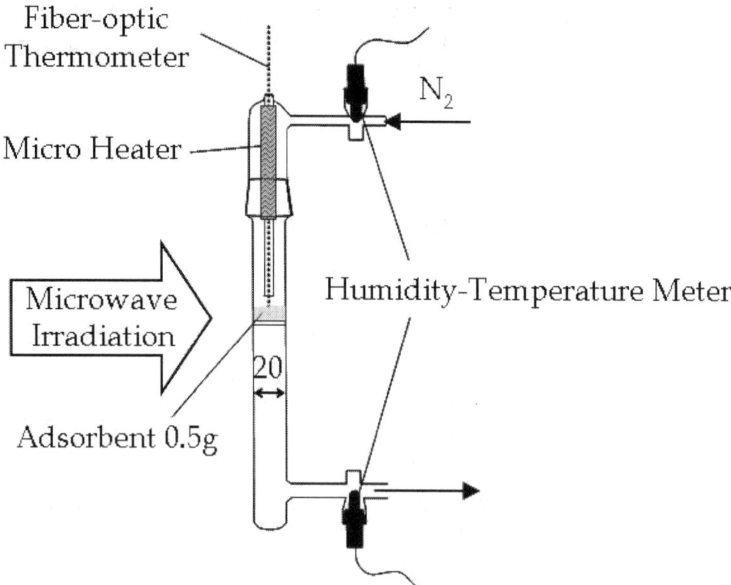


Fig. 5. Schematic diagram of adsorption column

The following two desorption experiments were carried out.

1. After zeolites were packed into the adsorption column, the adsorption column was heated and kept at 350°C. The $N_2$ gas of 99.99% was circulated enough to dry the adsorption column as a pretreatment. Adsorption process was carried out by circulating $N_2$ gas of relative humidity 40% to adsorption column at 3.18m/min. Desorption process was carried out by microwave heating under conditions of $N_2$ gas of 30°C with relative humidity 40%, gas flow rate of 1.62-6.36m/min and microwave power of 800W as comparison.
2. After the adsorption of DF-9 became equilibrium like 1) (flow rate: 3.18m/min), the desorption experiments was carried out by supplying power (electric power can heat 40 to 80°C of adsorbed bed achieving temperature.) and microwave heating (microwave output; 50W).

In experiments, temperature in the center of adsorbent bed and humidity of exit of adsorbent bed were measured, and the desorbed amount was calculated by using this result.

## 3. Results and discussions

### 3.1 Comparison of microwave heating desorption effect in different zeolites

As shown in Figure 6, for 4A and DF-9, heat and mass transfer behavior of microwave heating desorption process under the condition that adsorption column inlet temperature is 30°C and relative humidity is *RH*=40% showed similar behavior compared with previous study on OXYSIV-5. The good repeatability under the same experimental condition was confirmed.

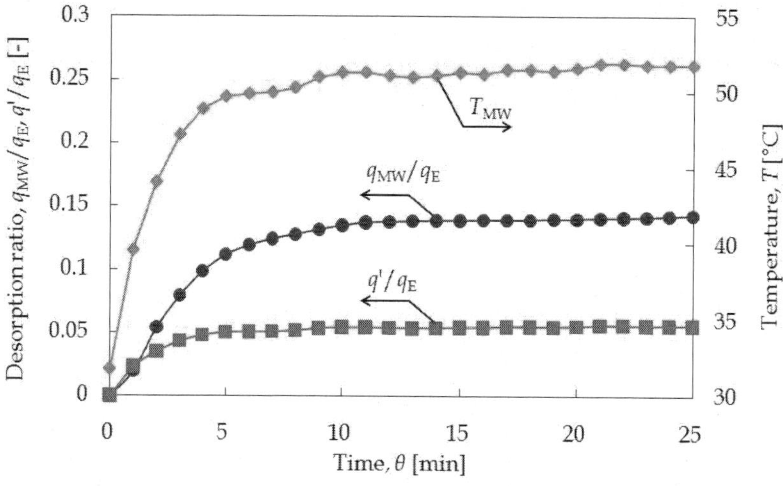


Fig. 6. Desorption ratio and temperature of zeolite DF-9 bed during microwave heating (microwave output: 800W, flow rate: 3.18m/min)

However, for the different type of the zeolites, temperature ($T_{MW}$) rise, maximum achieving temperature ($T_{MAX}$) of adsorbent bed and desorption amount ($q_{MW}$) by microwave irradiation heating were different. To compare the difference of desorption of the zeolites, equilibrium adsorption amount ($q_E$) at 30°C, $RH$=40%, $q_{MW}$, desorption ratio ($q_{MW}/q_E$) and $T_{MAX}$ after desorption begins to 15 minutes (microwave output 800W), hypothetical temperature desorption amount ($q'$), hypothetical temperature of heat source ($T'$), hypothetical temperature desorption amount ratio $R_q$(=$q_{MW}/q'$) and hypothetical temperature of heat source rise $T_D$(=$T'$−$T_{MW}$) calculated using adsorbed equilibrium relation of Figure 3 in each zeolite sample are shown in Table 1 together with the results of OXYSIV-5 in microwave irradiation time of 15min. Moreover, temperature rise rate ($\Delta T/\Delta\theta$) of adsorbent bed, relationship between desorption rate ($\Delta q_{MW}/\Delta\theta$) and adsorption ratio ($1-q_{MW}/q_E$) is shown in Figure 7 and Figure 8, respectively.

| Zeolites | 4A | DF-9 | OXSIV-5 |
|---|---|---|---|
| $q_E$ [kg/kg] | 0.218 | 0.322 | 0.242 |
| $q_{MW}$ [kg/kg] | 0.016 | 0.044 | 0.032 |
| $q_{MW}/q_E$ [-] | 0.074 | 0.136 | 0.132 |
| $T_{MAX}$ [°C] | 49.1 | 50.1 | 46.7 |
| $q'$ [kg/kg] | 0.007 | 0.017 | 0.019 |
| $T'$ [°C] | 62.4 | 93.7 | 61.4 |
| $R_q$ [-] | 2.22 | 2.59 | 1.57 |
| $T_D$ [°C] | 13.3 | 43.6 | 14.7 |

Table 1. Desorption amount and temperature rise

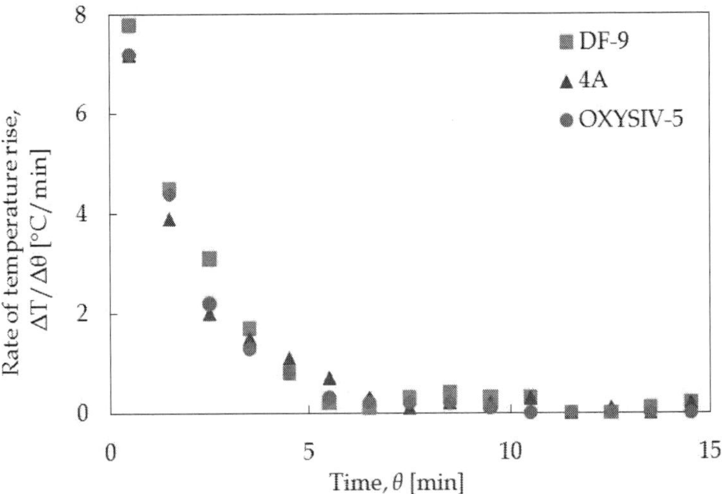


Fig. 7. Rate of temperature rise for zeolite bed

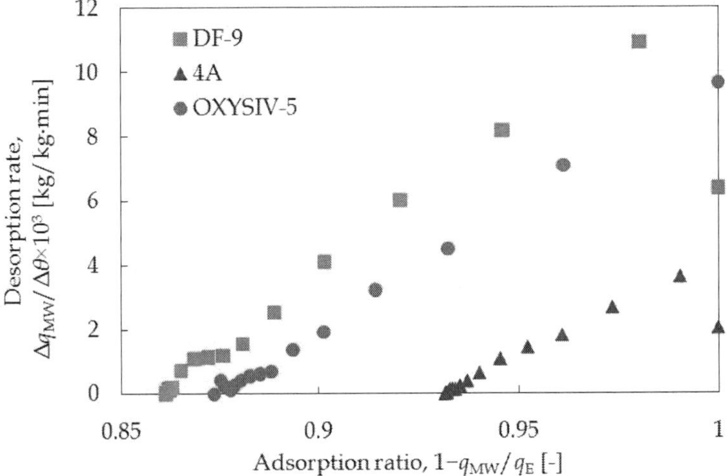


Fig. 8. Relationship of desorption rate and adsorption ratio

$q'$ and $T'$ are defined as follows. In this experiment, adsorbent temperature rises with microwave irradiation. On contrast, adsorbent bed inlet temperature are constant at 30°C and $RH$=40% (absolute humidity: H=0.0106kg/kg-air). Therefore, relative humidity in the bed decreases because adsorption bed temperature rises, $q'$ is equilibrium adsorption amount difference corresponding to this temperature rise on isotherms. In this calculation, Clausius-Clapeyron Equation (Arogba, 2001; Kolaczkiewicz & Bauer, 1985; Komdaurov, 2004) was used based on adsorption isotherm of two temperatures of various zeolites.

The following was observed from Table 1, Figure 7 and Figure 8.

1. Amount order of $q_{MW}$ is DF-9 > OXYSIV-5 > 4A. $q_{MW}$ of 4A and OXYSIV-5 are 0.36 times and 0.73 times of DF-9, respectively.
2. Amount order of $q_{MW}/q_E$ is also DF-9 > OXYSIV-5 > 4A. This value was slightly smaller than DF-9 in OXYSIV-5. In contrast, $q_{MW}/q_E$ of 4A was 0.54 times of DF-9.
3. Amount order of $T_{MAX}$ is DF-9 > 4A > OXYSIV-5.
4. $R_q$ of each zeolite samples shows more than 1. Rq of 4A and DF-9 was about 1.4 times and 1.6 times of OXYSIV-5. $T_D$ of 4A and DF-9 is about 0.9 times and 3.0 times of OXYSIV-5.
5. Temperature rise rate ($\Delta T/\Delta\theta$) is a little different with the type of the zeolites in desorption early time, but there is no big difference in change. The beginning desorption is the maximum, and then this value decreases afterwards.
6. On the other hand, adsorption rate shown the maximum in 0.98-1.0 of ($1-q_{MW}/q_E$) for DF-9 and 4A, and then decreased with the decrease of ($1-q_{MW}/q_E$) afterwards.

The above mentioned result of 1) shown that $q_{MW}$ of the adsorbent became small with small equilibrium adsorption amount under this experimental conditions. But, equilibrium adsorption amount of OXYSIV-5 is 0.76 times of DF-9, and this value is approximately same as 0.73 times in $q_{MW}$. However, although equilibrium adsorbed amount of 4A is 0.67 times of DF-9, $q_{MW}$ decreases greatly with 0.36 times.

In the range of this experiment, the water-vapor adsorption can be considered influenced by pore size with same shape of temperature rise rate of adsorbents (Figure 7). This is also shown in 6), the maximum desorption rate of 4A (pore size: 0.4nm) is 0.33 times of DF-9 (pore size: 0.8nm) and 0.38 times of OXYSIV-5 (pore size: 1.0nm). On the other hand, pore size of DF-9 is slightly small compared with that of OXYSIV-5, but desorption rate is considered depending on the adsorption amount, and the desorption rate is quick up with the large equilibrium adsorption amount.

For the result of $R_q$ in result 4), the same desorption effect of 4A and DF-9 are shown same as that of OXYSIV-5. And the effect is different with the type of zeolites. Concretely, $R_q$ is 1.4 times while beginning adsorbed amount of 4A when OXYIVE-5 is 0.88 times in the same experiment condition. This shows that desorption by microwave heating is more advantageous than by hot air heating in zeolite with small pore size. On the other hand, the beginning adsorption amount of DF-9 is 1.3 times and $R_q$ is 1.6 times compared with OXYIVE-5. This shows that microwave heating is effective in desorption of zeolite which is adsorbing water-vapor in large quantity.

To confirm desorption effect of the microwave more clearly, other experiments results of OXYSIV-5 are shown in Figure 9 and Figure 10. These figures shows the adsorbent bed temperature and the desorption amount change in microwave heating desorption (microwave output: 800W) and hot air heating desorption at the adsorption column inlet temperature of 30°C and $RH$=40%. $T_{MW}$ in figure is the adsorbent bed temperature by microwave heating. In addition, $T_{HE}$ is adsorbent bed temperature under hot air supply condition that performed hot air heating to show increased temperature as same as $T_{MW}$. And, $q_{HE}$ is desorption amount of water by hot air heating. $T_{HE}$ and $T_{MW}$ almost draw the same curve according to Figure 9. On the other hand, adsorption ratio of microwave heating ($1-q_{MW}/q_E$) is larger than adsorption ratio of hot air heating ($1-q_{HE}/q_E$) at all time. It is especially remarkable in beginning desorption. As for this, microwave heating causes desorption more than hot air heating desorption. It is shown that it is more effective under the condition with much adsorption water.

$T_D$ of 3) shows decreasing effect of heat source temperature by microwave heating. This value also changes with the type of the zeolites. This value is especially excellent in DF-9. Specifically, It corresponds to desorption amount when desorption of hot air temperature 93.7°C uses hot air temperature 50.1°C together with microwave heating, and the decrease of 43.6°C of temperature of heat source becomes possible.

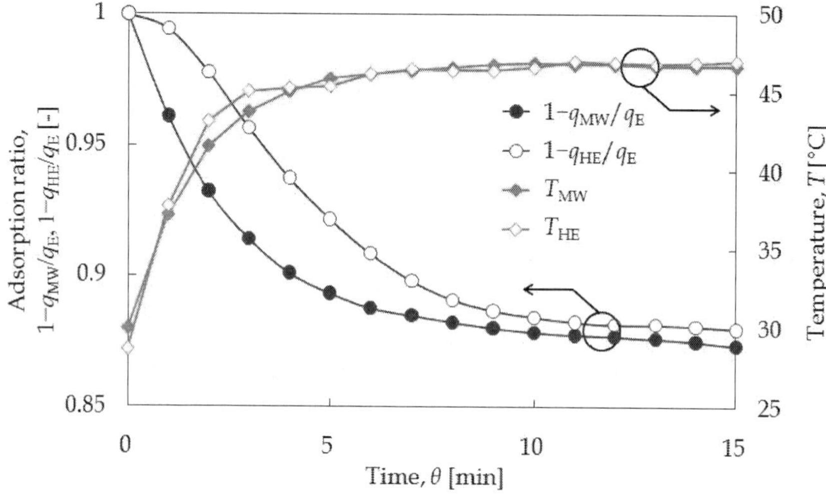


Fig. 9. Adsorption ratio and temperature during microwave irradiation and heating for zeolite OXYSIV-5 (microwave output: 800W, flow rate: 3.18 m/min)

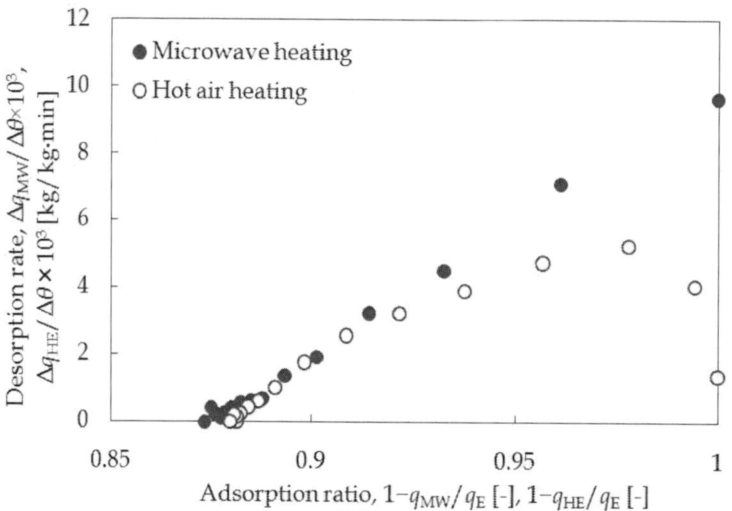


Fig. 10. Relationship of desorption rate and adsorption ratio

### 3.2 Influence of flow rate and hot air temperature exerted on microwave heating desorption

Desorption ratio ($q_{MW}/q_E$) and adsorbent bed highest achieving temperature ($T_{MAX}$) obtained from results of DF-9 under the conditions of three flow rates are shown in Table 2. It's observed that $q_{MW}$ shows the minimum by flow rate and $T_{MAX}$ decreases with increase of flow rate. It is considered as following, i) Desorption rate increases with increase of flow rate and adsorption temperature. ii) The increase of flow rate inhibits the temperature rise of adsorbent bed, and reduces the desorption rate.

| Flow rate [m/min] | $q_{MW}/q_E$ [-] | $T_{MAX}$ [°C] |
|---|---|---|
| 1.62 | 0.166 | 58.6 |
| 3.18 | 0.141 | 51.3 |
| 6.36 | 0.160 | 46.2 |

Table 2. Experimental results of desorption ratio and temperature for zeolite DF-9

In order to clarify the influence of hot air temperature exerted on microwave heating desorption (preset temperature: 45°C), adsorbent temperature change and desorption amount change of DF-9 desorption experiment by hot air and microwave hybrid system are shown in Figure 11. The following are observed from this figure.

1. Adsorbent bed centre temperature by hot air heating is almost adjusted to preset temperature. Moreover, adsorbent bed temperature $T_{MW}$ of microwave irradiation heating shows higher than $T_{HE}$.
2. The maximum amount change of hot air desorption shows up in about 4 minutes, and go back to the initial value in about 17 minutes. Desorption amount change of microwave heating begins to decrease after becoming the maximum in about 2 minutes, and the value is maintained for about 2 minutes longer, and appears larger than the initial value in 17 minutes.

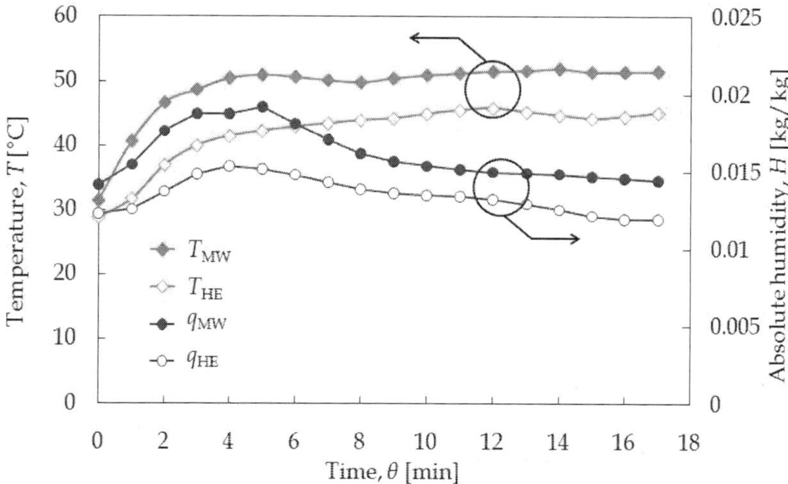


Fig. 11. Changes of absolute humidity and temperature for zeolite DF-9

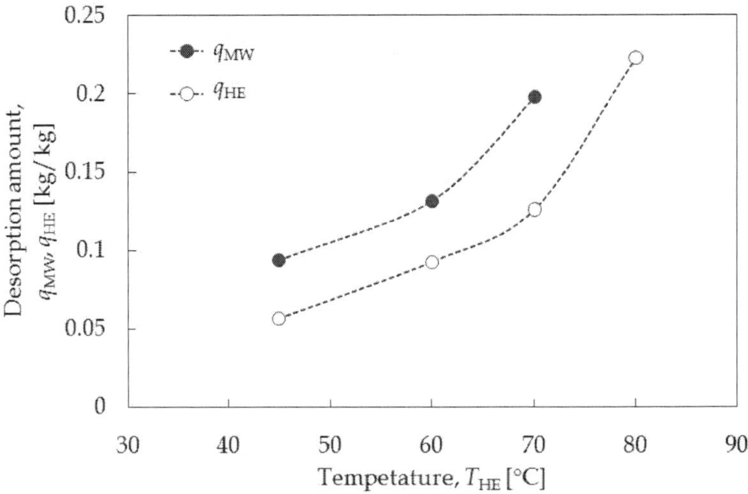


Fig. 12. Relationship between desorption amount and air temperature for zeolite DF-9

These tendencies were similarly observed through changing hot air temperature. Then the $q_{MW}$ and $q_{HE}$ corresponding to adsorbent bed temperature $T_{MW}$ and $T_{HE}$ were calculated and shown in Figure 12 after the desorption started 10 minutes. The temperature of the equal desorption amount is $T_{MW} < T_{HE}$. In addition, the difference of the temperature $\Delta T$ ($=T_{HE}-T_{MW}$) is equivalent to the temperature of heat source decrease effect. $T_{MW}$ shows the decrease effect of 16.5°C, 13.2°C and 8.6°C that obtained respectively at 45°C, 60°C and 70°C. Heat source temperature decrease effect decreased with the increase of $T_{MW}$ .The reason is considered as the loss coefficient of the microwave for the water decreased with increase of temperature (Koshijima et al., 2004). Moreover, $\Delta T$ is smaller than $T_D$. This depends on the microwave output being small.

## 4. Conclusions

The water-vapor desorption characteristics due to microwave heating adsorption equilibrium characteristics and pore size different types of zeolites were evaluated, and the following results were gained.

1. The effect of microwave irradiation was approved to be better than that of hot air heating in any zeolites. This effect is remarkable for zeolite that has small pore size (4A) or large adsorption amount of water-vapor (DF-9). The desorption amount of water-vapor from zeolites by microwave irradiation was 2.22 (4A) and 2.59 (DF-9) times larger than that by hot air heating.
2. The desorption rate increased along with large pore size of zeolites. Moreover, the desorption rate of similarly sized zeolites increased with the increase of the adsorption amount.
3. The desorbed effect of microwave irradiation according to the results of the temperature rise experiments during hot air heating was confirmed as same as the effect of microwave irradiation.

4. The minimum value of desorption ratio appeared with the air flow rate change.
5. On the same desorbed amount standard, the temperature of hot air and microwave irradiation hybrid type is lower than that of hot air heating. The microwave irradiation showed the effect of maximum 16°C decrease of the heat source.

As shown above, in this research, the speeding-up of desorption and the effect of heat source temperature decrease can be confirmed by microwave irradiation.

As challenges for the future research, it is necessity to study the uniform heating method of the adsorbent. So it's necessary to consider the shape of the adsorbent and microwave irradiation method.

When applying for microwave irradiation in the desiccant humidity conditioner, it's necessary to examine microwave irradiation method and the rotor shape etc. In addition, grasp of the microwave irradiation effect by low-humidity environment is necessary for upgrading the desiccant humidity conditioner.

## 5. Nomenclature

| | | |
|---|---|---|
| $H$ | absolute humidity | [kg-$H_2O$/kg-dry air] |
| $q$ | adsorption amount of water | [kg-$H_2O$/kg-zeolite] |
| $q'$ | desorption amount of water calculated from temperature rise of zeolite bed | [kg-$H_2O$/kg-zeolite] |
| $q_E$ | equilibrium adsorption amount of water | [kg-$H_2O$/kg-zeolite] |
| $q_{HE}$ | desorption amount of water by hot air heating | [kg-$H_2O$/kg-zeolite] |
| $q_{MW}$ | desorption amount of water by microwave heating | [kg-$H_2O$/kg-zeolite] |
| $RH$ | relative humidity | [%] |
| $R_q$ | ratio of $q_{MW}$ and $q'$ | [-] |
| $T$ | temperature | [°C] |
| $T'$ | temperature of zeolite bed calculated from desorption amount of water | [°C] |
| $T_D$ | $T' - T_{MW}$ | [°C] |
| $T_{MAX}$ | maximum temperature | [°C] |
| $T_{HE}$ | temperature of zeolite bed during hot air heating | [°C] |
| $T_{MW}$ | temperature of zeolite bed during microwave irradiation | [°C] |
| $\Delta T$ | $T_{HE} - T_{MW}$ | [°C] |
| $\theta$ | time | [min] |

## 6. References

Ando, K., & Kodama, A. (2005). Experimental Study on a Process Design for Adsorption Desiccant Cooling Driven with a Low-Temperature Heat. *Adsorption*, Vol. 11, (July 2005), pp. 631-636

Arogba, S.S. (2001). Effect of Temperature on the Moisture Sorption Isotherm of a Biscuit Containing Processed Mango (Mangifera Indica) Kernel Flour. *Journal of Food Engineering*, Vol. 48, No. 2, (May 2001), pp. 121-125

Bradshaw, S.M., Van Wyk, E.J., & De Swardt, J.B. (1997). Preliminary Economic Assessment of Microwave Regeneration of Activated Carbon for the Carbon in Pulp Process. *Journal of Microwave Power and Electromagnetic Energy*, Vol. 32, No. 3, pp. 131-144, ISSN 0832-7823

Bradshaw, S.M., Van Wyk, E.J., & De Swardt, J.B. (1998). Microwave Heating Principles and the Application to the Regeneration of Granular Activated Carbon. *Journal of The South African Institute of Mining and Metallurgy*, Vol. 98, No. 4, (July/August 1998), pp.201-210, ISSN 0038-223X

Cherbanski, R., & Molga, E. (2009). Intensification of Desorption Processes by Use of Microwaves—An Overview of Possible Applications and Industrial Perspectives. *Chemical Engineering and Processing: Process Intensification*, Vol. 48, No. 1, (January 2009), pp. 48-58

Davanagere, B.S., Sherif, S.A., & Goswami, D.Y. (1999). A Feasibility Study of a Solar Desiccant Air-Conditioning System—Part I: Psychrometrics and Snalysis of the Conditioned Zone. *International Journal of Energy Research*, Vol. 23, No. 1, (January 1999), pp. 7-21

Elsayed, S.S., Miyazaki, T., Hamamoto, Y., Akisawa, A., & Kashiwagi, T. (2006). Analysis of an Air Cycle Refrigerator Driving Air Conditioning System Integrated Desiccant System. *International Journal of Refrigeration*, Vol. 29, No. 2, (March 2006), pp. 219-228

Elsayed, S.S., Miyazaki, T., Hamamoto, Y., Akisawa, A., & Kashiwagi, T. (2008). Performance Analysis of Air Cycle Refrigerator Integrated Desiccant System for Cooling and Dehumidifying Warehouse. *International Journal of Refrigeration*, Vol. 31, No. 2, (March 2008), pp. 189-196

Ge, T.S., Dai, Y.J., Wang, R.Z., & Li, Y. (2008). Experimental Investigation on a One-Rotor Two-Stage Rotary Desiccant Cooling System. *Energy*, Vol. 33, No. 12, (December 2008), pp. 1807-1815

Halliday, S.P., Beggs, C.B., & Sleigh, P.A. (2002). The Use of Solar Desiccant Cooling in the UK: a Feasibility Study. *Applied Thermal Engineering*, Vol. 22, No. 12, (August 2002), pp. 1327-1338

Hamamoto, Y., Murase, S., Okajima, J., Matsuoka, F., Akisawa, A., & Kashiwagi, T. (2004). Analysis of Heat and Mass Transfer in a Desiccant Rotor. *Transactions of the Japan Society of Refrigerating and Air Conditioning Engineers*, Vol. 21, No. 1, (March 2004), pp. 63-75, ISSN 1344-4905

Hamed, A.M. (2003). Desorption Characteristics of Desiccant Bed for Solar Dehumidification/Humidification Air Conditioning Systems. *Renewable Energy*, Vol. 28, No. 13, (October 2003), pp. 2099-2111

Harshe, Y.M., Utikar, R.P., Ranade, V.V., & Pahwa, D. (2005). Modeling of Rotary Desiccant Wheels. *Chemical Engineering Technology*, Vol. 28, No. 12, (December 2005), pp. 1473-1479

Kabeel, A.E. (2007). Solar Powered Air Conditioning System Using Rotary Honeycomb Desiccant Wheel. *Renewable Energy*, Vol. 32, No, 11, (September 2007), pp. 1842-1857

Kodama, A., Watanabe, N., Hirose, T., Goto, M., & Okano, H. (2005). Performance of a Multipass Honeycomb Adsorber Regenerated by a Direct Hot Water Heating. *Adsorption*, Vol. 11, (July 2005), pp. 603-608

Kolaczkiewicz, J., & Bauer, E. (1985). Clausius-Clapeyron Equation Analysis of Two- Dimensional Vaporization. *Surface Science*, Vo. 155, No. 2-3, (June 1985), pp. 700-714

Kondaurov, V.I. (2003). The Clausius–Clapeyron Equations for Phase Transitions of the First Kind in a Thermoelastic Material. *Journal of Applied Mathematics and Mechanics*, Vol. 68, No. 1, pp. 65-79

Koshijima, T., Shibata, C., Toishi, T., Norimoto, K. & Yamada, S. (2004). *Microwave Heating Technology Collection*, NTS Inc., ISBN 978-4-86043-070-2, Tokyo, Japan

Kuo, C.Y. (2008). Desorption and Re-Adsorption of Carbon Nanotubes: Comparisons of Sodium Hydroxide and Microwave Irradiation Processes. *Journal of Hazardous Materials*. Vol. 152, No.3, (April 2008), pp.949-954

Mavroudaki, P., Beggs, C.B., Sleigh, P.A., & Halliday, S.P. (2002). The Potential for Solar Powered Single-Stage Desiccant Cooling in Southern Europe. *Applied Thermal Engineering*, Vol. 22, No. 10, (July 2002), pp. 1129-1140

Ohgushi, T., & Nagae, M. (2003). Quick Activation of Optimized Zeolites with Microwave Heating and Utilization of Zeolites for Reusable Desiccant. *Journal of Porous Materials*, Vol. 10, No. 2, (June 2003), pp. 139-143

Ohgushi, T., & Nagae, M. (2005). Durability of Zeolite Against Repeated Activation Treatments with Microwave Heating. *Journal of Porous Materials*, Vol. 12, No. 4, (October 2005), pp. 265-271

Oshima, K., Yamazaki, M., Takewaki, T., Kakiuchi, H., & Kodama, A. (2006). Application of Novel FAM Adsorbents in a Desiccant System. *Kagaku Kogaku Ronbunshu*, Vol. 32, No. 6, (December 2006), pp. 518-523, ISSN 0386-216X

Polaert, I., Ledoux, A., Estel, L., Huyghe, R., & Thomas, M. (2007). Microwave Assisted Regeneration of Zeolite. *International Journal of Chemical Reactor Engineering*, Vol. 5, A117

Polaert, I., Ledoux, A., Estel, L., Huyghe, R., & Thomas, M. (2010). Adsorbents Regeneration Under Microwave Irradiation Fordehydration and Volatile Organic Compounds Gas Treatment. *Chemical Engineering Journal*, Vol. 162, No. 3, (September 2010), pp.941-948

Yan, C.T., Shih, T.S., & Jen, J.F. (2004). Determination of Aniline in Silica Gel Sorbent by One-Step in Situ Microwave-Assisted Desorption Coupled to Headspace Solid-Phase Microextraction and GC–FID. *Talanta*, Vol. 64, No. 3, (October 2004), pp. 650-654

Yan, C.T., Jen, J.F., & Shih, T.S. (2007). Application of Microwave-Assisted Desorption/Headspace Solid-Phase Microextraction as Pretreatment Step in the Gas Chromatographic Determination of 1-Naphthylamine in Silica Gel Adsorbent. *Talanta*, Vol. 71, No. 5, (March 2007), pp. 1993-1997

Yoshida, H. (2005). *Handbook on Porous Adsorbents*, Fuji Technosystem, ISBN 978-4-93855-596-2, Tokyo, Japan

16

# Microwave Heating in Organic Synthesis and Drug Discovery

Hong Liu and Lei Zhang
*State Key Laboratory of Drug Research, Shanghai Institute of Materia Medica, Chinese Academy of Sciences China*

## 1. Introduction

Microwave dielectric heating was firstly reported on the use in organic chemistry by Gedye (Gedye, R. et al., 1986) and Giguere/Majetich (Giguere, R. J. et al., 1986) in the mid-1980s. However, it takes a long time before the wide application of this technology in the late 1980s and early 1990s. This has been principally attributed to the lack of controllability and reproducibility of the initial domestic microwave ovens implemented in organic chemistry, and partly due to a general lack of understanding of the basics of microwave dielectric heating. Since the year 2000, dedicated commercial microwave reactors for chemical synthesis have become available, promoting the wide spread of microwave energy in academic and industry to heat various chemical transformations. Nowadays, this enabling technology has been considerably exploited in organic synthesis and drug discovery.

Microwave technology triggers heating into the reacting system by either dipolar polarization or ionic conduction. When irradiated at microwave frequencies, electromagnetic waves pass through the dipoles or ions of the sample and cause the molecules to oscillate. In this process, energy is lost in the form of heat through molecular friction and dielectric loss. Because microwave radiation is introduced into the reaction system remotely without direct physical contact with reaction materials, this can lead to a rapid temperature increase throughout the sample causing less by-products or decomposition products. In contrast, conventional heating of organic reactions such as oil baths, sand baths or heating mantles are rather slow and create an inward temperature gradient, which may result in localized overheating and reagent decomposition when heated for prolonged periods.

The most prominent advantage of controlled microwave dielectric heating for chemical synthesis is the dramatic reduction in reaction time from days and hours to minutes. Moreover, microwave heating is able to reduce side reactions, increase yields, improve reproducibility, allow control of temperature and pressure, and even realize impossible reactions by conventional heating.

Since a large number of articles have been published in the area of microwave synthesis, including a number of general and specific review articles (Caddick, S. & Fitzmaurice, R., 2009; Mavandadi, F. & Lidström, P., 2004; Mavandadi, F. & Pilotti, Å., 2006; Kappe, C. O., 2004, 2008; Kappe, C. O. & Dallinger, D., 2006; Wathey, B. et al., 2002) and several books (Hayes, B. L., 2002; Kappe, C. O. et al., 2009; Kappe, C. O. & Stadler, A., 2005; Leadbeater, N.

E., 2010; Loupy, A., 2006; Polshettiwar, V., 2010; Tierney, J. P. & Lidström, P., 2005), in this chapter we only exemplify recent publications that demonstrate the impact of microwave heating on organic synthesis and drug discovery.

## 2. Microwave in general synthetic transformations

In early days, microwave heating was often used as an optional protocol when a particular reaction has failed to proceed under other conditions or requires exceedingly long reaction time or high temperature. Dedicated microwave chemistries have now made it obvious that many types of chemical transformations that require heating can be carried out successfully under microwave conditions. Microwave technology has also been applied in various formats ranging from the traditional solution-phase synthesis to solid-phase and solvent-free reactions.

### 2.1 Transition-metal-catalyzed reactions

Transition metal-catalyzed reactions are of enormous significance to form carbon–carbon (C-C) and carbon–heteroatom (C-X) bonds in organic chemistry. These reactions typically need hours or days to reach completion under traditional reflux conditions and often require an inert atmosphere. Microwave heating has been demonstrated over the past few years to significantly expedite these transformations, in most cases without an inert atmosphere. Furthermore, the inverted temperature gradients under microwave conditions may lead to an increased lifetime of the catalyst by elimination of wall effects (Kappe, C. O., 2004).

#### 2.1.1 Suzuki reactions

The Suzuki reaction, cross-coupling of aryl- or vinyl-boronic acid with an aryl- or vinyl-halide catalyzed by a palladium complex, is one of the most versatile reactions for the construction of carbon-carbon bonds, in particular for the formation of biaryls. Recent developments have expanded the possible applications of this reaction enormously. Microwave-assisted Suzuki reactions can now be performed in many different ways and have been incorporated into a variety of challenging synthesis.

Under microwave-heated condition, the Suzuki coupling of aryl chlorides with boronic acids was performed in an aqueous media using the air- and moisture-stable palladium catalyst (Scheme 1). The protocol developed in this publication allowed a drastic reduction of the reaction time to 15 min and the products were obtained in good yields (Miao, G. et al., 2005).

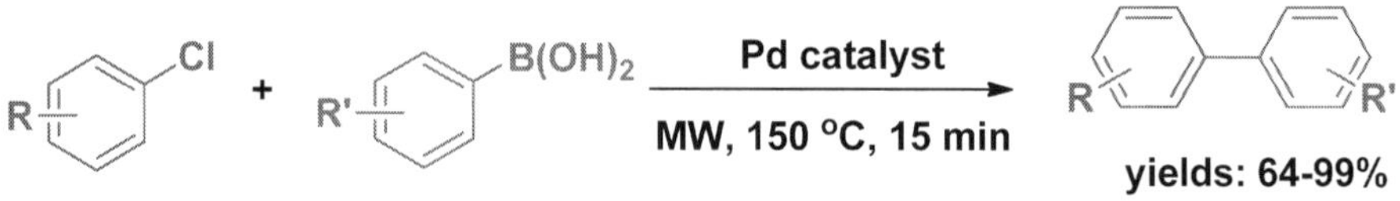

Scheme 1. Microwave-promoted Suzuki reactions of aryl chlorides.

Microwave activation enables a Suzuki coupling of boronic acids with aryl halides under solvent-free conditions using palladium catalyst system (Scheme 2). A large variety of boronic acids and bromo-, chloro- and iodoaryls could rapidly give biaryls in 10 minutes (Nun, P. et al., 2009).

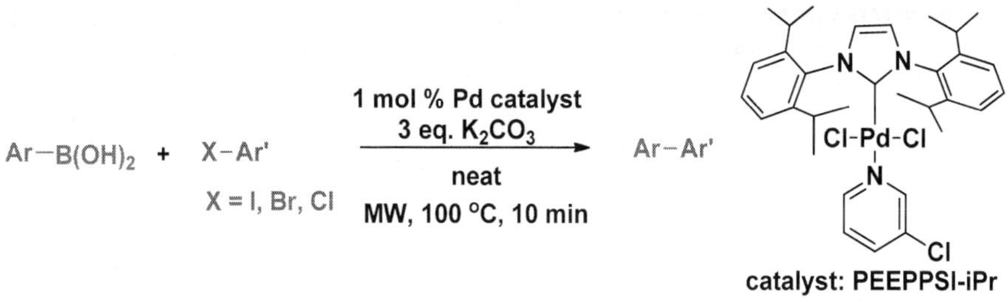

$Ar-B(OH)_2$ + X-Ar' (X = I, Br, Cl) → [1 mol % Pd catalyst, 3 eq. $K_2CO_3$, neat, MW, 100 °C, 10 min] → Ar-Ar'

catalyst: PEEPPSI-iPr

Scheme 2. Microwave-assisted solvent-free Suzuki reaction.

Replacement of the traditionally used Pd with less expensive Ni-based catalyst systems can significantly reduce costs for cross-couplings of this type especially when performed on scale. Kappe et al. reported a rapid and highly efficient microwave-assisted Ni-catalyzed Suzuki cross-coupling of aryl carbamates and sulfamates with boronic acid (Scheme 3). This protocol features coupling time of only 10 min. Also, the microwave conditions work exceedingly well with aryl chlorides as electrophilic cross-coupling partners and the scalability of the coupling process is demonstrated up to 700 mL scale with a multimode microwave reactor (Baghbanzadeh, M. et al., 2011).

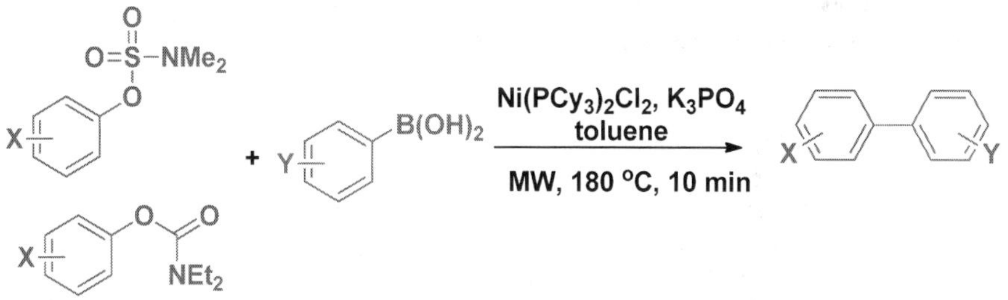

[$Ni(PCy_3)_2Cl_2$, $K_3PO_4$, toluene, MW, 180 °C, 10 min]

Scheme 3. Rapid nickel-catalyzed Suzuki reaction utilizing microwave heating.

We have successfully synthesized bromoenone of the Ni(II) complexes of $\beta$-alanine Schiff's base and developed a practical and highly efficient route to $\alpha$-aryl and $\alpha$-heteroaryl-substituted $\beta$-amino acids using the Suzuki coupling reaction (Scheme 4). The heterogeneous solution was stirred at 70 °C for 40 min under microwave irradiation. A broad range of aryl and heteroaryl substituents can be employed under the operationally simple and effective conditions (Ding, X. et al., 2009).

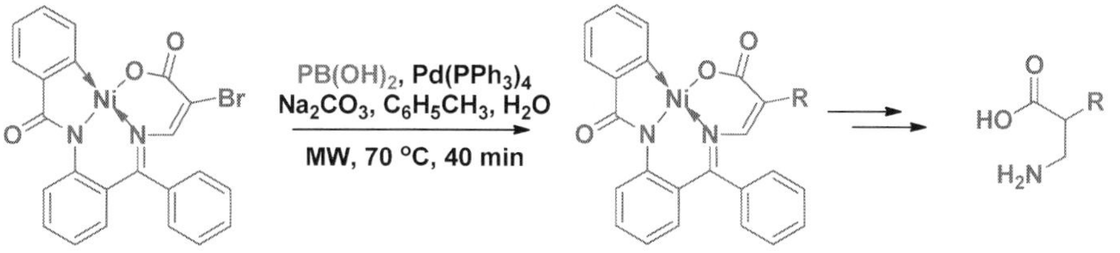

[$PB(OH)_2$, $Pd(PPh_3)_4$, $Na_2CO_3$, $C_6H_5CH_3$, $H_2O$, MW, 70 °C, 40 min]

Scheme 4. Microwave-assisted Suzuki couplings of Ni(II) complex with boric acids.

### 2.1.2 Heck reactions

Heck reactions between aryl halides and alkenes are also well established under microwave conditions. Regioselectivity is a key issue in the Heck reaction. Larhed described a regioselective Heck coupling employing different arylboronic acids with both electron-rich and electron-poor olefins (Scheme 5). Controlled microwave processing was used to reduce reaction time from hours to minutes both in small scale and in 50 mmol scale batch processes (Lindh, J. et al., 2007).

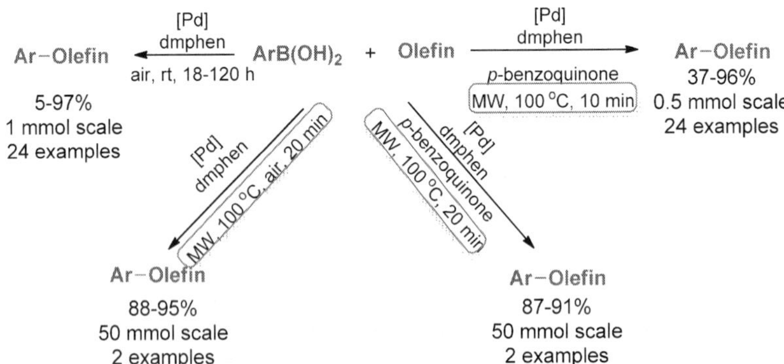


Scheme 5. Oxidative Heck reactions at room temperature and with microwave heating.

Another key issue in the microwave enhanced Heck reaction is asymmetric induction. Recently, a highly modular library of readily available phosphite-oxazoline ligands has been applied in the Pd-catalyzed asymmetric Heck reactions of several substrates and triflates under thermal and microwave conditions (Scheme 6). Both enantiomers of the Heck coupling products were obtained in excellent activities (conversion: >100% in 10 min), regioselectivities (>99%) and enantioselectivities (>99% ee) (Mazuela, J. et al., 2010).

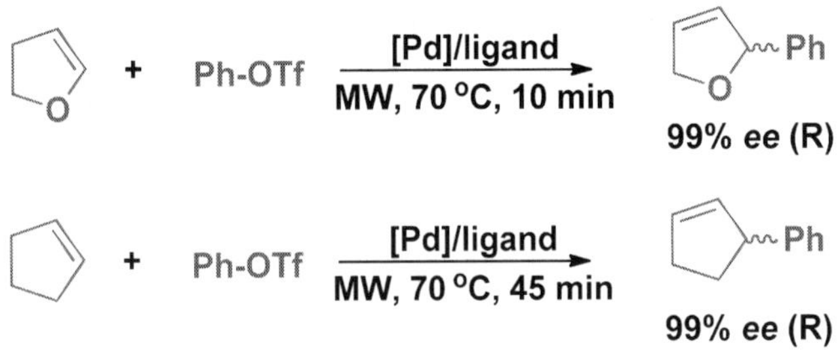

+ Ph-OTf [Pd]/ligand, MW, 70 °C, 10 min → Ph, 99% ee (R)

+ Ph-OTf [Pd]/ligand, MW, 70 °C, 45 min → Ph, 99% ee (R)

Scheme 6. Enantioselective Heck reactions under microwave-irradiation.

### 2.1.3 Buchwald-Hartwig cross coupling reactions

Buchwald-Hartwig chemistry, cross coupling of aryl halides or pseudohalides with primary or secondary amines, has become a powerful method for synthesizing arylamines. Conventionally, this reaction requires high temperature and long reaction time. Many fast and highly efficient applications have been developed in conjunction with microwave irradiation.

We have developed a practical, convenient strategy for the preparation of arylamines (Scheme 7). The process was efficiently promoted by readily available $Cu(OAc)_2$ in the presence of DBU under microwave irradiation. Both aromatic and aliphatic amines can be quickly coupled with various substituted aryl halides in good to excellent yields (Huang, H. et al., 2008a).

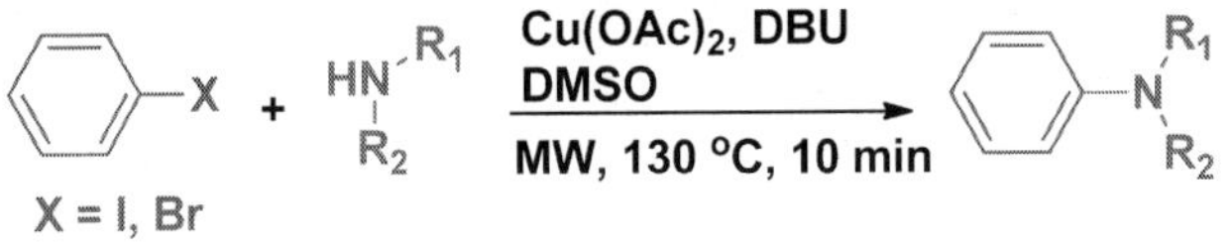

Cu(OAc)$_2$, DBU
DMSO
MW, 130 °C, 10 min
X = I, Br

Scheme 7. Microwave-assisted cross-coupling of aryl halides with various nitrogen derivatives.

### 2.1.4 Sonogashira coupling reactions

The Sonogashira reaction, cross-coupling of aryl or vinyl halides with terminal alkynes, is an important example of metal-catalyzed carbon-carbon bond-forming reactions. This reaction has also been subjected to microwave enhancement. Some weaknesses of the original procedure are long reaction time (hours to days), limited choice of aryl halides, and complicated workup procedures. We developed an efficient and effective microwave-assisted cross-coupling of terminal alkynes with various aryl chlorides including sterically hindered, electron-rich, electron-neutral, and electron-deficient aryl chloride (Scheme 8). It proceeded faster and generally gave good-to-excellent yields. This simple catalytic system was also effective for Suzuki coupling, Buchwald-Hartwig amination, and Heck coupling reactions with unactivated aryl chlorides. A broad spectrum of substrates coupled effectively under microwave irradiation to provide the desired products in minutes (Huang, H. et al., 2008b).

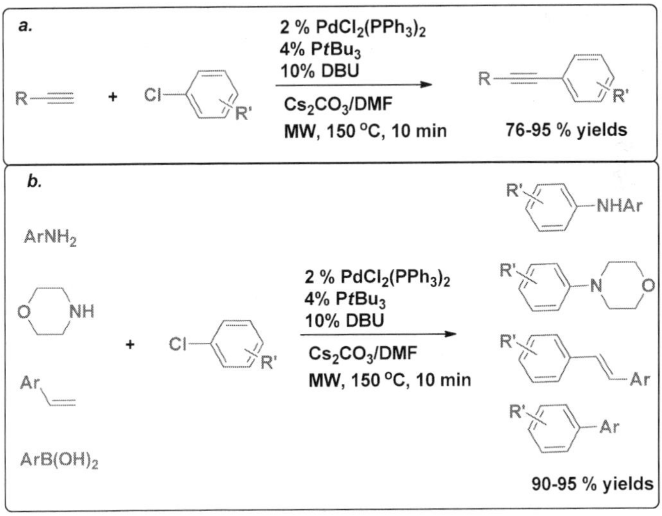

a.
2 % PdCl$_2$(PPh$_3$)$_2$
4% P*t*Bu$_3$
10% DBU
Cs$_2$CO$_3$/DMF
MW, 150 °C, 10 min
76-95 % yields

b.
ArNH$_2$
ArB(OH)$_2$
2 % PdCl$_2$(PPh$_3$)$_2$
4% P*t*Bu$_3$
10% DBU
Cs$_2$CO$_3$/DMF
MW, 150 °C, 10 min
90-95 % yields

Scheme 8. a) Microwave-assisted Sonogashira coupling of aryl chlorides with alkynes. b) Microwave-assisted C-C and C-N bond-forming reactions with aryl chlorides.

### 2.1.5 Other C-X bond forming reactions

Enhanced by microwave, an efficient coupling protocol of arylboronic acids with amines proceeded rapidly and conveniently to afford *N*-arylated amines in the presence of inexpensive $Cu(OAc)_2$ and DBU (Scheme 9). Compared with the former Ullmann reaction conditions, this methodology developed a much faster way to construct C-N bonds under microwave irradiation without using expensive ligand and palladium. A variety of substrates could participate in the process with higher purities and yields in short time (Chen, S. et al., 2008).

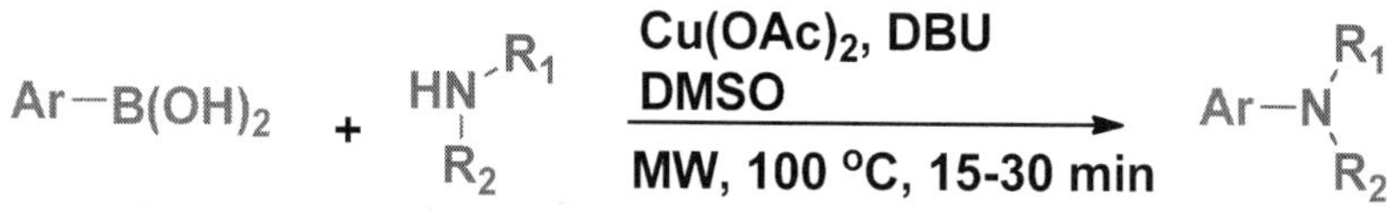

$$Ar{-}B(OH)_2 + HNR_1R_2 \xrightarrow[\text{MW, 100 °C, 15-30 min}]{Cu(OAc)_2,\ \text{DBU},\ \text{DMSO}} Ar{-}NR_1R_2$$

Scheme 9. Microwave-assisted *N*-arylation of amines with arylboronic acids.

An expeditious and efficient method was developed to prepare 2,6,9-substituted purines in higher purity and yields under microwave heating (Scheme 10). First, 2-chloro-6, 9-substituted purines were prepared via a one-pot two-step reaction, which involves a sequential $S_NAr$ displacement of the C6 chloro substituent with various anilines and amines, followed by *N*-alkylation and *N*-arylation at the N9 position with different organic halides and boronic acids. Second, $NaBF_4$ catalysis supports a $S_NAr$ substition of the C2 chloro displacement with high product conversion. All these reactions were carried out under microwave irradiation (Huang, H. et al., 2007). This protocol was applied in the preparation of novel purine derivatives with potent and selective inhibitory activity against c-Src tyrosine kinase (Huang, H. et al., 2010).

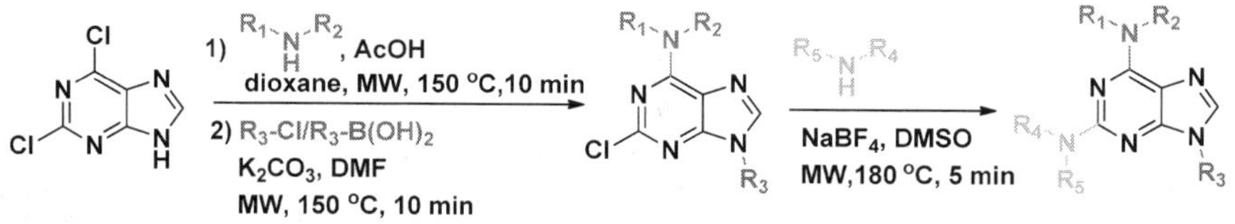

1) $R_1NHR_2$, AcOH, dioxane, MW, 150 °C, 10 min; 2) $R_3$-Cl/$R_3$-B(OH)$_2$, $K_2CO_3$, DMF, MW, 150 °C, 10 min; then $R_5NHR_4$, $NaBF_4$, DMSO, MW, 180 °C, 5 min

Scheme 10. Microwave-assisted rapid synthesis of 2,6,9-substituted purines.

## 2.2 Microwave-assisted multi-component reactions

Multi-component reactions (MCRs) hold considerable utility in diversity generating of functional molecule libraries. Microwave-assisted organic synthesis accelerates a variety of synthetic transformations with prominent advantages of short reaction time and high yields. Thus, combing rapid microwave enhancement with diversifying MCRs can minimize reaction time, the number of steps, energy consumption, and maximize synthetic efficiency. Many different microwave-assisted MCRs has received considerable exploration, including Mannich, Ugi, Biginelli and other heterocycle-forming reactions.

### 2.2.1 Microwave-assisted Mannich reactions

Mannich reactions are of great importance to synthesize $\beta$-amino ketones traditionally via three-component condensation of a substituted methyl ketone, an aldehyde, and an amine.

Bolm reported a proline-catalyzed direct asymmetric Mannich reaction under microwave irradiation (Scheme 11). After a short period of time, the Mannich products were obtained with only 0.5 mol % of catalyst and up to 98% ee. (Rodríguez B. & Bolm, C., 2006).

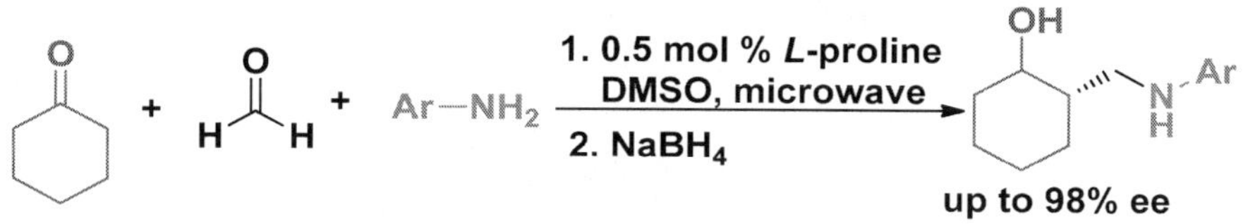

1. 0.5 mol % *L*-proline
DMSO, microwave
2. $NaBH_4$
up to 98% ee

Scheme 11. Microwave-assisted asymmetric Mannich reaction.

Hao et al. described a mild microwave-assisted stereoselective Mannich reaction of aromatic aldehydes with 1,2-diphenylethanone and hetero-arylamines in water (Scheme 12). As opposed to conventional heating, this method proceeded faster and in higher yields under controlled microwave heating (Hao, W. et al., 2009).

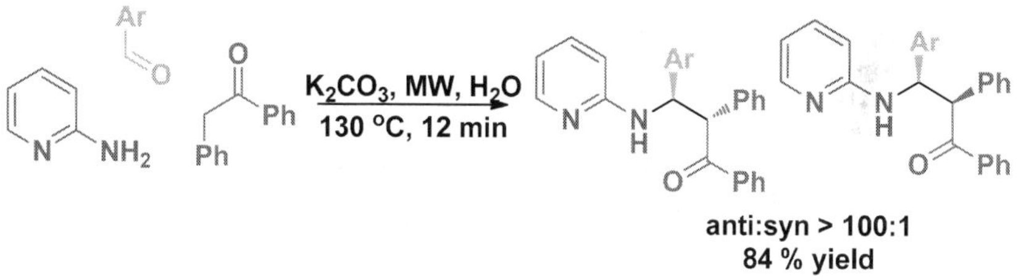

$K_2CO_3$, MW, $H_2O$
130 °C, 12 min
anti:syn > 100:1
84 % yield

Scheme 12. Microwave-assisted Mannich reaction in water.

### 2.2.2 Microwave-assisted Ugi reactions

The Ugi reaction is widely used in the pharmaceutical industry for preparing libraries of compounds. Classically, the reaction is performed at room temperature in methanol with reaction time up to 48 hours. Recently, a rapid solvent-free microwave synthesis of five- and six-membered lactams via a three-component Ugi reaction was developed (Scheme 13). The reaction was carried out in much shorter time (100 °C for 3 min) and the yields were improved in contrast to classical conditions (Jida, M. et al., 2010).

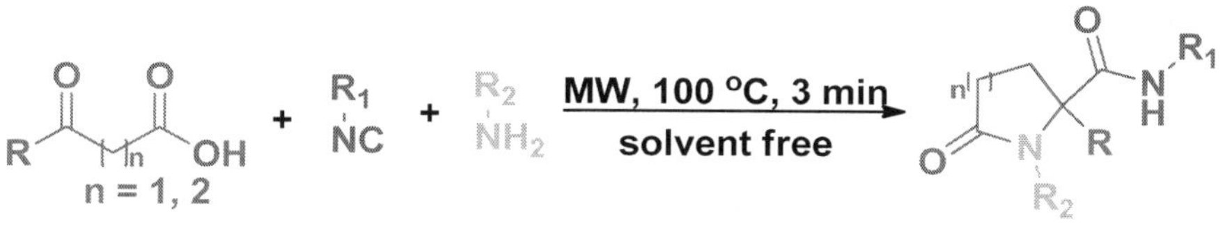

MW, 100 °C, 3 min
solvent free
n = 1, 2

Scheme 13. Microwave-assisted solvent-free Ugi reaction.

### 2.2.3 Microwave-assisted Biginelli reactions

The Biginelli reaction is of great importance to the synthesis of biologically active dihydropyrimidines. It involves the acid-catalyzed condensation of aldehydes, CH-acidic carbonyl components and urea-type building blocks. Kappe's group has made great contributions in the area of microwave-assisted Biginelli reactions. They have elaborately combined microwave irradiation with click chemistry (Khanetskyy, B. et al, 2004), parallel library-generation (Pisani, L. et al., 2007), solution- and solid-phase combinatorial chemistry based on this Biginelli reaction.

In an elegant example, Kappe et al. developed an efficient five-step linear protocol which involved an initial Biginelli multicomponent reaction for the generation of a variety of 2-substituted pyrimidines in high yields and comparatively short reaction time (Scheme 14). All five synthetic steps required for the synthesis of the target structures were carried out under microwave irradiation, both in solution phase and on solid phase (Matloobi, M. & Kappe, C. O., 2007).

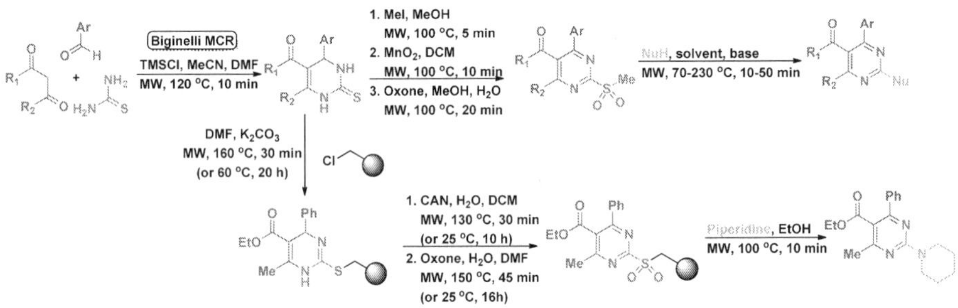

Scheme 14. Initial microwave-assisted Biginelli reaction in solution- and solid-phase synthesis.

### 2.2.4 Microwave-assisted synthesis of heterocycles

Heterocyclic compounds serve both as biomimetics and reactive pharmacophores of numerous drugs. The microwave enhancements have also been utilized in the synthesis of heterocyclic compounds.

Microwave-assisted synthesis allowed rapid preparation of small libraries of imidazole-based privileged heterocyclic structures in high yields and with clean and scalable reactions (Scheme 15) (Fantini, M. et al., 2010).

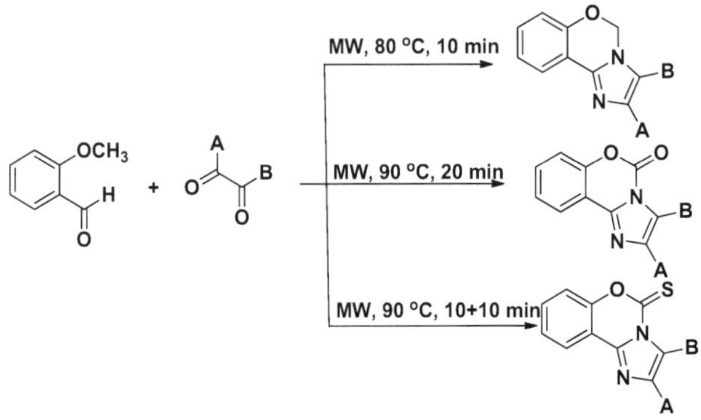

Scheme 15. Microwave synthesis of imidazole-based heterocycles.

Microwave heating was demonstrated to efficiently generate *N*-substituted 1,3-dihydrobenzimidazol-2-ones in good to excellent yields within minutes (Scheme 16). A variety of functional groups can be employed, rendering this method particularly attractive for the efficient preparation of biologically and medicinally interesting molecules (Li, Z. et al., 2008a).

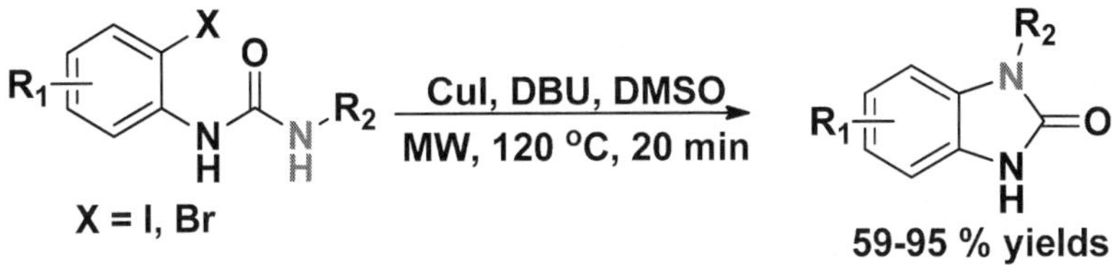

CuI, DBU, DMSO
MW, 120 °C, 20 min

X = I, Br

59-95 % yields

Scheme 16. Microwave in the synthesis of heterocycles.

We also developed a microwave-facilitated protocol for fast synthesis of 2*H*-1,4-benzoxazin-3-(4*H*)-ones via a cascade reaction with a nucleophilic substitution followed by a CuI/DBU-catalyzed coupling cyclization (Scheme 17). This method involved simple reaction conditions, short reaction time and a broad substrate scope in moderate to good yields (Feng, E. et al., 2009).

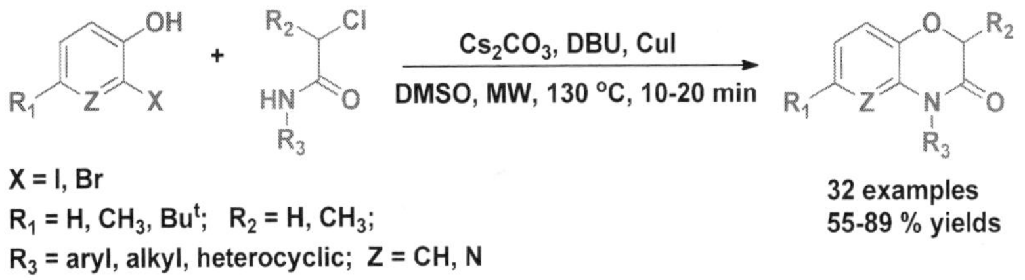

$Cs_2CO_3$, DBU, CuI
DMSO, MW, 130 °C, 10-20 min

X = I, Br
$R_1$ = H, $CH_3$, $Bu^t$; $R_2$ = H, $CH_3$;
$R_3$ = aryl, alkyl, heterocyclic; Z = CH, N

32 examples
55-89 % yields

Scheme 17. Microwave in the synthesis of heterocycles.

### 2.2.5 Aqueous microwave synthesis

The development of green synthetic protocols to reduce or eliminate the use and generation of hazardous substances constitutes a major challenge for chemists. In this context, the microwave-assisted organic reaction in water as a green alternative has become an important research area. The water interacts strongly with the microwave irradiation and is rapidly heated to high temperatures, enabling it to act as a less polar pseudo-organic solvent. Also, water is non-flammable, non-explosive and non-toxic, making it safe and ideally green. The combination of microwave chemistry and green chemistry offers great advantages such as dramatically reducing chemical waste and reaction time, rendering it an attractive and sustainable synthetic protocol.

There are many examples of successful application of microwave-assisted green chemistry. An efficient and convenient method was developed for the preparation of 2,4-(1*H*,3*H*)-quinazolinediones and 2-thioxoquinazolinones (Scheme 18). Substituted methyl anthranilate reacted with various iso(thio)cyanates in DMSO/$H_2O$ without any catalyst or base under microwave irradiation to generate diversity on the 2,4-(1*H*,3*H*)-quinazolinediones or 2-thioxoquinazolinones. A variety of substrates can participate in the process with good yields in 20 min. In particular, after the reaction mixtures were cooled to ambient temperature, a large amount of desired products precipitated and could easily be collected by filtration with higher purities, making this methodology suitable for library synthesis in drug discovery efforts (Li, Z. et al., 2008b).

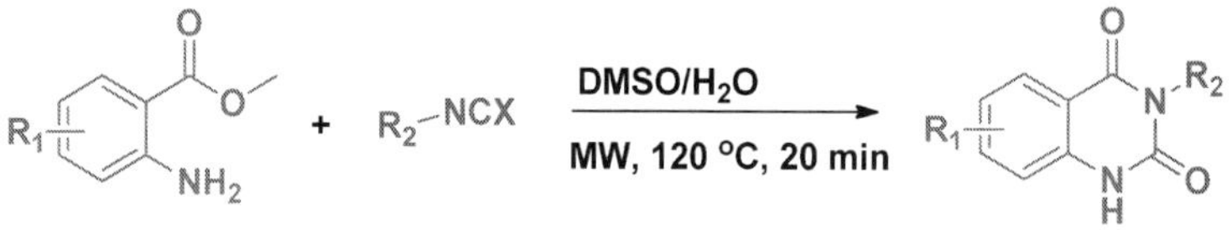


X =O or S
$R_1$ = H, 5-F or 4,5-dimethoxy
$R_2$ = substituted phenyl or alkyl

Scheme 18. Microwave-assisted rapid synthesis in DMSO/$H_2O$.

Fused tricyclic xanthines have been reported to be potent and selective antagonists of human $A_{2A}$ adenosine receptors, and have potential activities as anticonvulsants to treat chemically induced seizures. A simple, convenient and green synthetic approach to diverse fused tricyclic xanthines has been developed via gold(I) complex-catalyzed intramolecular hydroamination or silver(I)-catalyzed isomerization-hydroamination of terminal alkynes under microwave irradiation in water (Scheme 19). This reaction is atom economical and has high functional group tolerance (Ye, D. et al., 2009a).

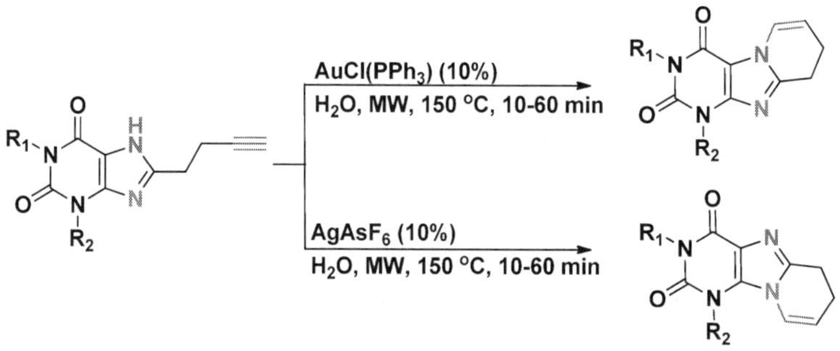


Scheme 19. Microwave-assisted chemo- and regioselective synthesis of fused tricyclic xanthines in water.

Using Au(I)-catalyzed 5-*endo-dig* cyclization in water under microwave irradiation, we developed a fast and green route to regioselectively prepare biologically interesting indole-1-carboxamides in moderate to high yields (Scheme 20). In comparison, the reaction required 3–48 hours when using traditional oil heating (Ye, D. et al., 2009b).

Under microwave heating, a mild and effective method for the construction of 3-hydroxyisoindolin-1-ones via a metal-free tandem transformation using a phase transfer catalyst proceeded in good yields with excellent regioselectivity (Scheme 21). The strategy presents an atom-economical and environmentally friendly transformation, in which one C–O bond and two new C–N bonds are formed in water from two simple starting materials (Zhou, Y. et al., 2010).

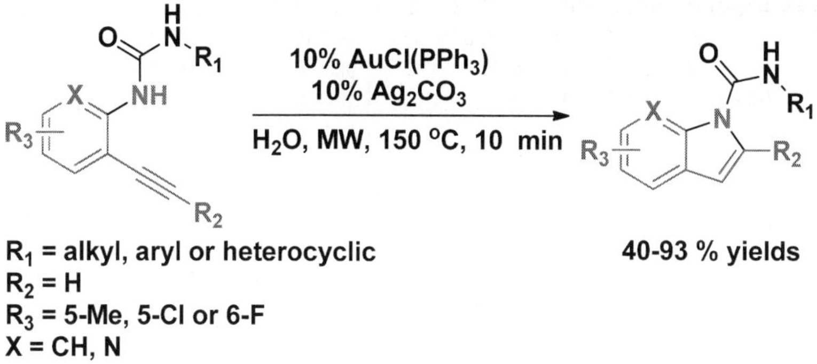

10% AuCl($PPh_3$)
10% $Ag_2CO_3$
$H_2O$, MW, 150 °C, 10 min

$R_1$ = alkyl, aryl or heterocyclic
$R_2$ = H
$R_3$ = 5-Me, 5-Cl or 6-F
X = CH, N

40-93 % yields

Scheme 20. Microwave-assisted regioselective synthesis of indole-1-carboxamides in water.

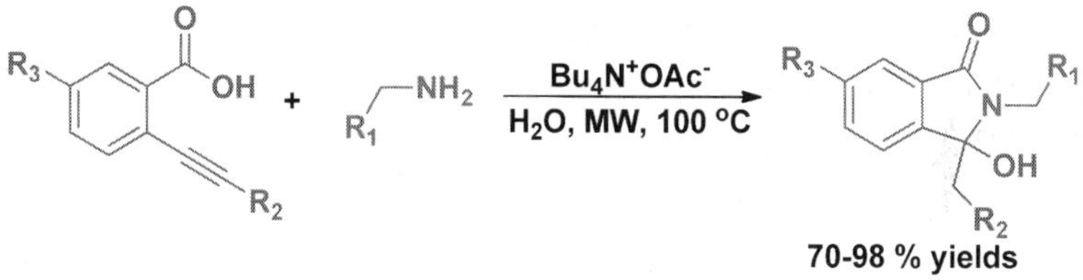

$Bu_4N^+OAc^-$
$H_2O$, MW, 100 °C

70-98 % yields

Scheme 21. Microwave-assisted regioselective synthesis in water.

## 3. Microwave synthesis in drug discovery

In drug discovery and development, rapid synthesis and quick construction of diverse libraries are the main objectives to speed up the process. For its specialty in high speed synthesis, microwave irradiation has now gained wide acceptance as an instrumental technology in the drug discovery programs, including target discovery, library generation for screening, lead generation and optimization endeavours, and process development.

### 3.1 Microwave synthesis used alone in drug discovery

Microwave heating used alone can considerably expedite the synthetic process in drug discovery compared with conventional heating.

#### 3.1.1 In molecular imaging

In the preclinical drug discovery, molecular imaging can help validate drug targets in assays of disease and select lead molecules in the chemistry optimization phase. A popular imaging method is positron emission tomography (PET) using radio-labelled pharmaceuticals to non-invasively elucidate biochemical mechanisms in animal and human subjects in vivo. The isotopes employed to label pharmaceuticals in imaging are mainly with short half-lives (carbon $^{11}C$ for 20 min and fluorine $^{18}F$ for 110 min). Therefore, short synthetic routes are crucial to obtain the final product in radio-labelling synthesis. Microwave technology can address the challenges of the rapid labelling of radiopharmaceuticals (Jones, J. R. & Lu, S., 2006).

Recently, several selective $mGlu_5$ ligands with subnanomolar affinity were labelled using [$^{18}$F]fluoride ion for imaging with PET (Siméon, F. G. et al., 2011). Under microwave irradiation, 2-chlorothiazoles were treated with [18F]fluoride ion in the presence of a potassium fluoride-Kryptofix 2.2.2 (K 2.2.2) complex in DMSO or MeCN, and the respective fluorine-18 labelled compounds were obtained in useful yields (Scheme 22).

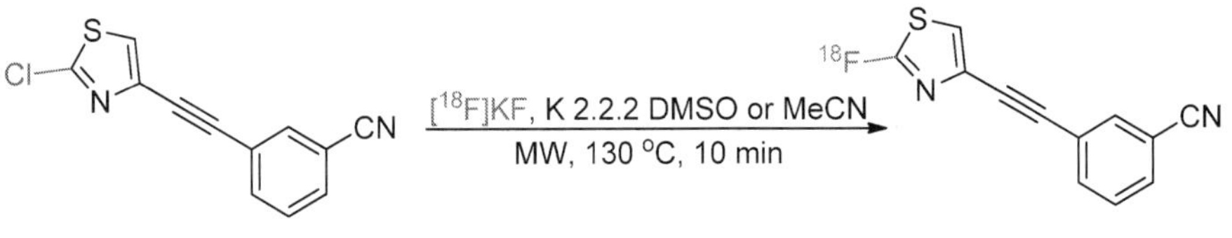

[$^{18}$F]KF, K 2.2.2 DMSO or MeCN
MW, 130 °C, 10 min

Scheme 22. Microwave-assisted rapid labelling for PET.

### 3.1.2 Lead generation and optimization

In the optimization of new HIV-1 protease inhibitors, microwave irradiation was introduced to accelerate a key step in the organic synthesis. With microwave enhancement, target compounds were synthesized using Stille or Suzuki cross-couplings at 120 °C for 30-50 min, all with retained (S)-configuration at the quaternary carbon. After biological evaluation, up to 56 times more potent compounds exhibiting excellent antiviral activities were obtained.

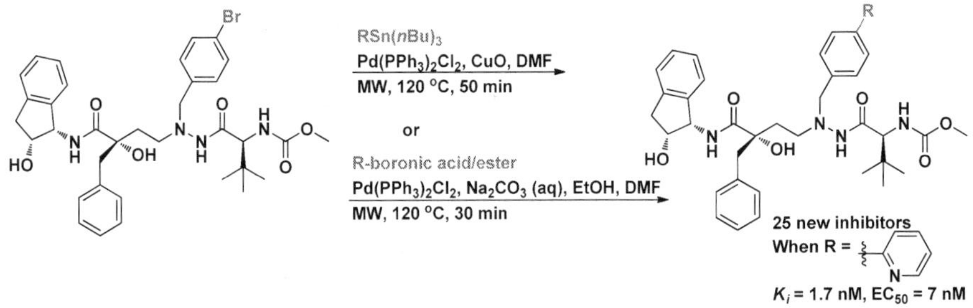

$RSn(nBu)_3$
$Pd(PPh_3)_2Cl_2$, CuO, DMF
MW, 120 °C, 50 min
or
R-boronic acid/ester
$Pd(PPh_3)_2Cl_2$, $Na_2CO_3$ (aq), EtOH, DMF
MW, 120 °C, 30 min

25 new inhibitors
When R =
$K_i$ = 1.7 nM, $EC_{50}$ = 7 nM

Scheme 23. Microwave-assisted rapid synthesis of target compounds.

### 3.2 Microwave synthesis in multidisciplinary approach

In the post genomic era, there is an ever-increasing demand to synthesize a myriad of substances in a short span of time for biological screening against the vast number of potential drug targets, thus quickly bringing new therapeutics to the market. A variety of strategies and techniques such as click chemistry, green chemistry, multi-component reactions, combinatorial synthesis, parallel synthesis, and automated library production to increase the output of pharmaceutically active chemical entities have been developed. All these techniques and strategies hold their distinguished advantages as well as shortcomings. On the other way, microwave-heating has proven its efficiency in dramatic reduction of reaction time, which is potentially important in drug discovery. Therefore, one of ideal options to accelerate the synthetic processes is to combine microwave chemistry with these techniques in drug discovery.

### 3.2.1 Combined with click chemistry

After rational docking calculations, a library of 15 compounds was subject to synthesis. Simone et al. took advantage of 1,3-dipolar cycloaddition reaction (click reaction) and subsequent Suzuki cross-coupling reaction to generate the library (Scheme 24). The microwave irradiation technique was introduced in the two steps to provide a faster way to obtain the desired final compounds in satisfactory yields. Biological evaluation of these compounds disclosed three new potential anti-inflammatory drugs (Simone, R. D. et al., 2011).

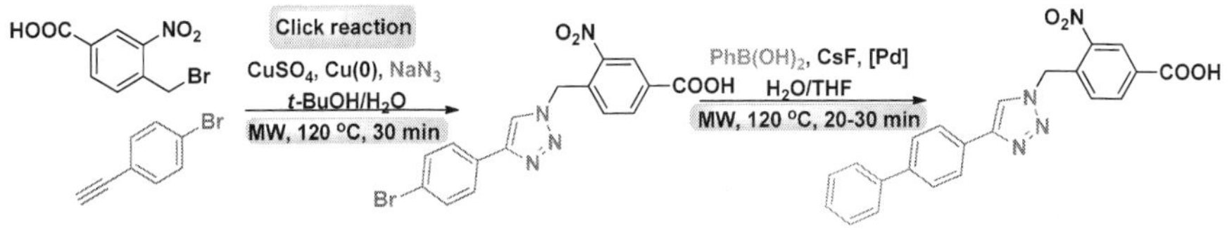

Click reaction: $CuSO_4$, Cu(0), $NaN_3$, *t*-BuOH/$H_2O$, MW, 120 °C, 30 min; $PhB(OH)_2$, CsF, [Pd], $H_2O$/THF, MW, 120 °C, 20-30 min

Scheme 24. Microwave synthesis combined with click chemistry.

### 3.2.2 Combined with solid-supported combinatorial chemistry

Combinatorial chemistry has made significantly positive contributions to generate libraries of structurally related compounds. Microwave-enhanced chemistry is ideally suited for combinatorial chemistry in drug discovery to rapid and automated preparation of a large number of compounds for optimization to novel therapeutic agents.

The popular solid-supported combinatorial chemistry simplifies workup and isolation of product which merely involves filtration of the resin and evaporation of the solvent. Its main drawback is the slow kinetics which makes reaction scouting and optimization tedious. Microwave heating has been shown to significantly accelerate the solution-phase combinatorial chemistry.

For example, Murray et al. adapted the microwave irradiation to solid-phase synthesis of $\beta$-peptides on polystyrene (PS) macrobeads. Microwave irradiation allowed rapid synthesis of a high-quality $\beta$-peptide combinatorial library via a split-and-mix approach (Scheme 25). Relative to conventional methods, the microwave approach significantly reduced synthesis time and amounts of reagents (Murray, J. K. et al., 2005).

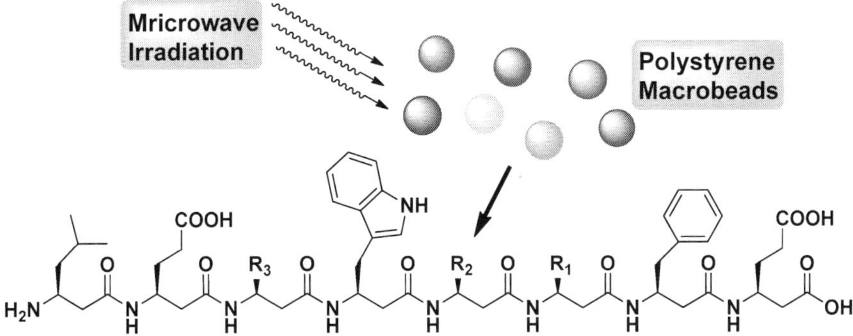


Scheme 25. Microwave synthesis combined with solid-supported combinatorial chemistry.

### 3.2.3 Combined with soluble polymer-supported combinatorial chemistry

Microwave heating has also been shown to significantly benefit the polymer-supported combinatorial chemistry. For example, the synthesis of indoline substituted nitrobenzene on a polyethylene glycol (PEG) support and its further elaboration to structurally diverse benzene-fused pyrazino/diazepino indoles is recently disclosed (Lin, P. et al., 2011). A reagent based diversification approach coupled with Pictet–Spengler type condensation reactions furnished these fused polycyclic scaffolds. Microwave irradiation was used as a means of rate acceleration for soluble polymer-supported reactions. The efficiency of these fused heterocyclic molecules to inhibit the vascular endothelial growth factor receptor 3 (VEGFR-3) was examined in vitro. A small set of potential drug candidates were identified as novel leads in this therapeutic area (Scheme 26).

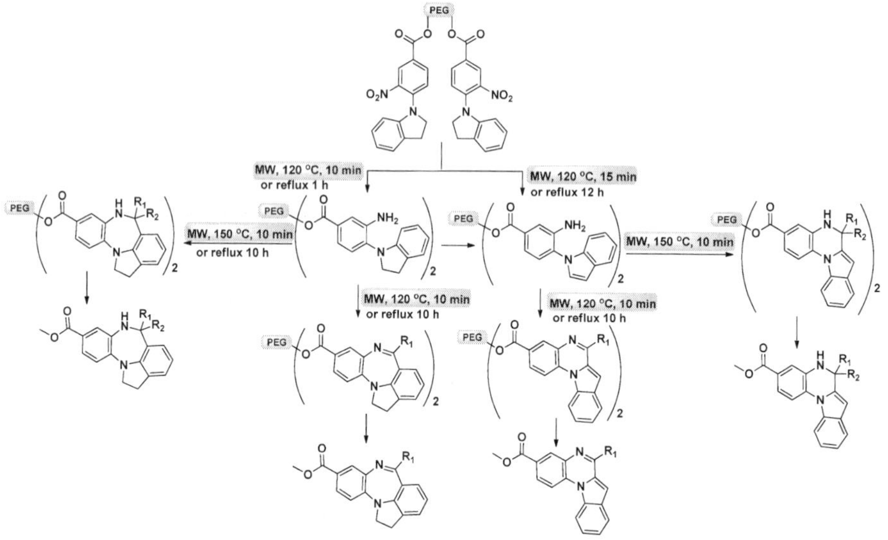

Scheme 26. Microwave irradiation in soluble polymer supported divergent synthesis of heterocyclic library for lead optimization.

### 3.2.4 Combined with MCR and parallel synthesis

In parallel synthesis, a highly reactive intermediate via a series of simple steps is usually synthesized and subsequently subject to a number of different reagents generating a library of compounds.

Zhou et al. demonstrated the efficiency of library synthesis involving a multi-component reaction, microwave heating, and fluorous separation in constructing four diversity points 1,4-benzodiazepine-2,5-diones (BZDs) library via a Ugi/de-Boc/cyclization/Suzuki strategy (Scheme 27). Under microwave heating, cyclohexylisocyanide and methyl 2-isocyanoacetate were used as convertible isocyanides for the cyclization reaction to form BZDs. Fluorous benzaldehyde was used as a tagged-component to simplify intermediates and final products purification by simple fluorous solid-phase extraction (Zhou, H. et al., 2010).

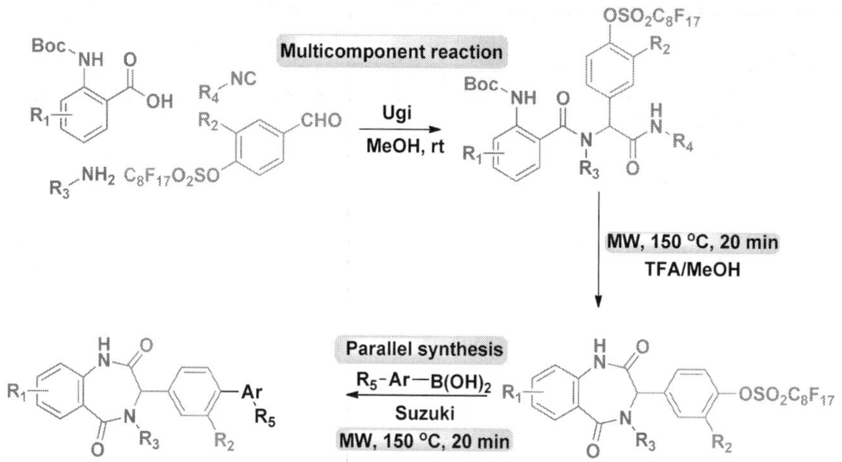

Scheme 27. Microwave-assisted solution-phase fluorous parallel synthesis.

### 3.2.5 Combined with MCR and polymer-support combinatorial chemistry

Hsiao et al. recently reported a new multidisciplinary synthetic approach comprising polymer-support synthesis, microwave-assisted synthesis, and multi-component condensation to facilitate synthesis of triaza-fluorenes library with a set of advantages such as rapid process, simple purification, and structural diversity in one shot (Scheme 28). Microwave irradiation greatly accelerates the rate of all reactions while polymer support facilitates purifications by simple precipitation technique. This strategy dramatically increases the efficiency of the overall multistep synthesis (Hsiao, Y. et al., 2010).

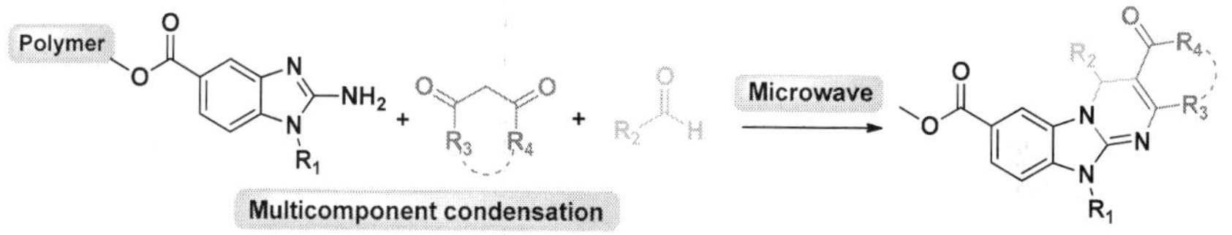

Scheme 28. Multidisciplinary synthetic approach for rapid access of molecules.

## 4. Conclusion and outlook

Microwave heating has gained a wide-spread acceptance both within academia and industry and blossomed into a useful technique for a variety of applications in organic synthesis and drug discovery. In organic chemistry, microwave irradiation allows to obtain rapid, reproducible, and scalable processes to synthesize new molecules in high yields. In drug discovery and development, microwave heating allows the medicinal chemists to get easy and rapid access to a larger number and libraries of novel biologically active compounds.

In spite of the exciting developments, there are still many challenges. The scalability, overall energy efficiency of microwave-heating (Moseley, J. D. & Kappe, C. O., 2011), and instrument cost remain unresolved.

Current microwave reactors are able to translate small scale microwave chemistry from milligram or gram scale to multi-kilogram scale using batch or continuous-flow processing. However, many of the benefits of small scale microwave chemistry are lost when the processes performed in larger batch reactors. Also, producing larger quantities of products in tons scale is still a discouraging bottleneck for present microwave reactors. All these challenges need more and dedicated input.

The second disadvantage is equipment cost. Although the cost of microwave reactors have dramatically reduced in recent years, the current price range is still many times higher than that of conventional heating equipment, considerably limiting the uptake of this valuable technology. Continued efforts to develop newer and more readily available microwave reactors as routine equipments of most laboratories are eagerly awaited.

Drug discovery is a challenging and demanding symphony, no one technique alone can whistle it. It is required to take a whole orchestra of strategies and techniques to play it well. In combination there is strength. Microwave chemistry has already been constantly combined with other enabling technologies and strategies such as multi-component reactions, solid-phase organic synthesis, or combinatorial chemistry. The combination of multidisciplinary approaches with microwave heating encourages scientists to initiate new and unexplored areas of complex pharmaceutical systems. To date, microwave irradiation still needs to find promising and tempting combinations with more techniques to satisfy the increased synthetic demands in industry and academia, for example microwave-assisted scale-up in water.

All in all, microwave heating is undoubtedly a bonanza for organic and medicinal chemistry researchers. This enabling technique has changed from the 'last sort' in early days to the 'first choice' nowadays for carrying out synthetic transformations requiring heat. In the future, with lower costs, microwave synthesizers will become an integral part and a standard technology in most synthetic laboratories, and will continually make valuable impact on both organic synthesis and drug discovery.

## 5. Acknowledgment

We gratefully acknowledge financial support from the National Natural Science Foundation of China (Grants 20721003, 81025017 and 20872153) and National S&T Major Projects (2009ZX09301-001, 2009ZX09501-001 and 2009ZX09501-010).

## 6. References

Baghbanzadeh, M.; Pilger, C. & Kappe, C. O. (2011). Rapid Nickel-Catalyzed Suzuki-Miyaura Cross-Couplings of Aryl Carbamates and Sulfamates Utilizing Microwave Heating. *The Journal of Organic Chemistry*, Vol. 76, No. 5, (January 2011), pp. 1507–1510.

Caddick, S. & Fitzmaurice, R. (2009). Microwave Enhanced Synthesis. *Tetrahedron*, Vol. 65, No. 17, (April 2009), pp. 3325–3355.

Chen, S.; Huang, H.; Liu,X.; Shen, J.; Jiang, H. & Liu, H. (2008). Microwave-Assisted Efficient Copper-Promoted *N*-Arylation of Amines with Arylboronic Acids. *Journal of Combinatorial Chemistry*, Vol. 10, No. 3, (April 2008), pp. 358–360.

Ding, X.; Ye, D.; Liu, F.; Deng, G.; Liu,G.; Luo, X.; Jiang, H. & Liu, H. (2009). Efficient Synthesis of $\alpha$-Aryl-/Heteroaryl-Substituted $\beta$-Amino Acids via Ni(II) Complex through the Suzuki Coupling Reaction. *The Journal of Organic Chemistry*, Vol. 74, No. 15, (June 2009), pp. 5656-5659.

Fantini, M.; Zuliani, V.; Spotti, M. A. & Rivara, M. (2010). Microwave Assisted Efficient Synthesis of Imidazole-Based Privileged Structures. *Journal of Combinatorial Chemistry*, Vol. 12, No. 1, (November 2010), pp. 181–185.

Feng, E.; Huang, H.; Zhou, Y.; Ye, D.; Jiang, H. & Liu, H. (2009). Copper(I)-Catalyzed One-Pot Synthesis of 2*H*-1,4-Benzoxazin-3-(4*H*)-ones from *o*-Halophenols and 2-Chloroacetamides. *The Journal of Organic Chemistry*, Vol. 74, No. 7, (March 2009), pp. 2846–2849.

Gedye, R.; Smith, F.; Westaway, K.; Ali, H.; Baldisera, L.; Laberge, L. & Rousell, J. (1986). The Use of Microwave Ovens for Rapid Organic Synthesis. *Tetrahedron Letters*, Vol. 27, No. 3, (January 1986), pp. 279-282.

Giguere, R. J.; Bray, T. L.; Duncan, S. M. & Majetich, G. (1986). Application of Commercial Microwave Ovens to Organic Synthesis. *Tetrahedron Letters*, Vol. 27, No. 41, (1986), pp. 4945-4948.

Hao, W.; Jiang, B.; Tu, S.; Cao, X.; Wu, S.; Yan, S.; Zhang, X.; Han, Z. & Shi, F. (2009). A New Mild Base-catalyzed Mannich Reaction of Hetero-arylamines in Water: Highly Efficient Stereoselective Synthesis of $\beta$-Aminoketones under Microwave Heating. *Organic & Biomolecular Chemistry*, Vol. 2009, No. 7, (February 2009), pp. 1410–1414.

Hayes, B. L. (2002). *Microwave Synthesis: Chemistry at the Speed of Light*, CEM, ISBN 0-9722229-0-1, USA.

Hsiao, Y.; Yellol, G. S.; Chen, L. & Sun, C. (2010). Multidisciplinary Synthetic Approach for Rapid Combinatorial Library Synthesis of Triaza-Fluorenes. *Journal of Combinatorial Chemistry*, Vol. 12, No. 5, (September 2010), pp. 723–732.

Huang, H.; Hong Liu, H.; Jiang, H. & Chen, K. (2008b). Rapid and Efficient Pd-Catalyzed Sonogashira Coupling of Aryl Chlorides. *The Journal of Organic Chemistry*, Vol. 73, No. 15, (July 2008), pp. 6037-6040.

Huang, H.; Liu, H.; Chen, K. & Jiang, H. (2007). Microwave-Assisted Rapid Synthesis of 2,6,9-Substituted Purines. *Journal of Combinatorial Chemistry*, Vol. 9, No. 2, (February 2007), 197-199.

Huang, H.; Ma, J.; Shi, J.; Meng, L.; Jiang, H., Ding, J. & Liu, H. (2010). Discovery of Novel Purine Derivatives with Potent and Selective Inhibitory Activity Against c-Src Tyrosine Kinase. *Bioorganic & Medicinal Chemistry*, Vol. 18, No. 13, (May 2010), 4615–4624.

Huang, H.; Yan, X.; Zhu, W.; Liu, H.; Jiang, H. & Chen, K. (2008a). Efficient Copper-Promoted *N*-Arylations of Aryl Halides with Amines. *Journal of Combinatorial Chemistry*, Vol. 10, No. 5, (July 2008), 617–619.

Jida, M.; Malaquina, S.; Deprez-Poulain,R.; Laconde, G. & Deprez, B. (2010). Synthesis of Five- and Six-membered Lactams via Solvent-free Microwave Ugi Reaction. *Tetrahedron Letters*, Vol. 51, No. 39, (September 2010), pp. 5109-5111.

Jones, J. R. & Lu, S. (2006). Chapter 18. Microwave-enhanced Radiochemistry, In: *Microwaves in Organic Synthesis*, 2nd Ed., Loupy, A. pp. 820-859, Wiley-VCH, ISBN 978-3-527-31452-2, Weinheim, Germany.

Kappe, C. O. & Dallinger, D. (2006). The Impact of Microwave Synthesis on Drug Discovery. *Nature Reviews Drug Discovery*, Vol. 5, (January 2006), pp. 51-63.

Kappe, C. O. & Stadler, A. (2005). *Microwaves in Organic and Medicinal Chemistry*, Wiley-VCH, ISBN 978-3-527-31210-8, Weinheim, Germany.

Kappe, C. O. (2004). Controlled Microwave Heating in Modern Organic Synthesis. *Angewandte Chemie International Edition*, Vol 43, No. 46, (November 2004), pp. 6250–6284.

Kappe, C. O. (2008). Microwave Dielectric Heating in Synthetic Organic Chemistry. *Chemical Society Reviews*, Vol. 37, No. 6, (April 2008), pp. 1127-1139.

Kappe, C. O.; Dallinger, D. & Murphree, S.S. (2009). *Practical Microwave Synthesis for Organic Chemists: Strategies, Instruments, and Protocols*, Wiley-VCH, ISBN 978-3-527-32097-4, Weinheim, Germany.

Khanetskyy, B.; Dallinger, D. & Kappe, C. O. (2004). Combining Biginelli Multicomponent and Click Chemistry: Generation of 6-(1,2,3-Triazol-1-yl)-Dihydropyrimidone Libraries. *Journal of Combinatorial Chemistry*, Vol. 6, No. 6, (November 2004), pp. 884-892.

Leadbeater, N. E. (2010). *Microwave Heating as a Tool for Sustainable Chemistry*, CRC Press, ISBN 978-1-4398-1270-9, Boca Raton, USA.

Li, Z.; Huang, H.; Sun, H.; Jiang, H. & Liu, H. (2008b). Microwave-Assisted Efficient and Convenient Synthesis of 2,4-(1*H*,3*H*)-Quinazolinediones and 2-Thioxoquinazolines. *Journal of Combinatorial Chemistry*, Vol. 10, No. 3, (April 2008), pp. 484–486.

Li, Z.; Sun, H.; Jiang, H. & Liu, H. (2008a). Copper-Catalyzed Intramolecular Cyclization to *N*-Substituted 1,3-Dihydrobenzimidazol-2-ones. *Organic Letters*, Vol. 10, No. 15, (July 2008), pp. 3263-3266.

Lin, P.; Salunke, D. B.; Chen L. & Sun, C. (2011). Soluble Polymer Supported Divergent Synthesis of Tetracyclic Benzene-fused Pyrazino/diazepino indoles: an Advanced Synthetic Approach to Bioactive Scaffolds. *Organic & Biomolecular Chemistry*, Vol. 9, No. 8, (2011), pp. 2925–2937.

Lindh, J.; Enquist,P.; Pilotti, Å.; Nilsson, P. & Larhed, M. (2007). Efficient Palladium(II) Catalysis under Air. Base-Free Oxidative Heck Reactions at Room Temperature or with Microwave Heating. *The Journal of Organic Chemistry*, Vol. 72, No. 21, (September 2007), pp. 7957-7962.

Loupy, A. (2006). *Microwaves in Organic Synthesis*, 2nd Ed., Wiley-VCH, ISBN 978-3-527-31452-2, Weinheim, Germany.

Matloobi, M. & Kappe, C. O. (2007). Microwave-Assisted Solution- and Solid-Phase Synthesis of 2-Amino-4-arylpyrimidine Derivatives. *Journal of Combinatorial Chemistry*, Vol. 9, No. 2, (March 2007), pp. 275-284.

Mavandadi, F. & Lidström, P. (2004) Microwave-Assisted Chemistry in Drug Discovery. *Current Topics in Medicinal Chemistry*, Vol. 4, No. 7, (2004), pp. 773-792.

Mavandadi, F. & Pilotti, Å. (2006). The Impact of Microwave-assisted Organic Synthesis in Drug Discovery. *Drug Discovery Today*, Vol. 11, No. 3-4, (February 2006), pp. 165-174.

Mazuela, J.; Pàmies, O. & Diéguez, M. (2010). Biaryl Phosphite–Oxazoline Ligands from the Chiral Pool: Highly Efficient Modular Ligands for the Asymmetric Pd-Catalyzed Heck Reaction. *Chemistry - A European Journal*, Vol. 16, No. 11, (March 2010), pp. 3434-3440.

Miao, G.; Ye, P.; Yu, L. & Baldino, C. M. (2005). Microwave-Promoted Suzuki Reactions of Aryl Chlorides in Aqueous Media. *The Journal of Organic Chemistry*, Vol. 70, No. 6, (February 2005), pp. 2332–2334.

Moseley, J. D. & Kappe, C. O. (2011). A Critical Assessment of the Greenness and Energy Efficiency of Microwave-assisted Organic Synthesis. *Green Chemistry*, Vol. 13, No. 4, (April 2011), pp. 794-806.

Murray, J. K.; Farooqi, B.; Sadowsky, J. D.; Scalf, M.; Freund, W. A.; Smith, L. M.; Chen, J. & Gellman, S. H. (2005). Efficient Synthesis of a $\beta$-Peptide Combinatorial Library with Microwave Irradiation. *Journal of the American Chemical Society*, Vol. 127, No. 38, (September 2005), pp. 13271-13280.

Nun, P.; Martinez, J. & Lamaty, F. (2009). Solvent-Free Microwave-Assisted Suzuki–Miyaura Coupling Catalyzed by PEPPSI-iPr. *Synlett*, No. 11, (December 2009), pp. 1761–1764.

Pisani, L.; Prokopcová, H.; Kremsner, J. M. & Kappe, C. O. (2007). 5-Aroyl-3,4-dihydropyrimidin-2-one Library Generation via Automated Sequential and Parallel Microwave-assisted Synthesis Techniques. *Journal of Combinatorial Chemistry*, Vol. 9, No. 3, (May 2007), pp. 415-421.

Polshettiwar, V. (2010). *Aqueous Microwave Assisted Chemistry: Synthesis and Catalysis*, The Royal Society of Chemistry, ISBN 978-1-84973-038-9, Cambridge, UK.

Rodríguez B. & Bolm, C. (2006). Thermal Effects in the Organocatalytic Asymmetric Mannich Reaction. *The Journal of Organic Chemistry*, Vol. 71, No. 7, (February 2006), pp. 2888-2891.

Siméon, F. G.; Wendahl, M. T. & Pike, V. W. (2011). Syntheses of 2-Amino and 2-Halothiazole Derivatives as High-Affinity Metabotropic Glutamate Receptor Subtype 5 Ligands and Potential Radioligands for in Vivo Imaging. *Journal of Medicinal Chemistry*, Vol. 54, No. 3, (January 2011), pp. 901–908.

Simone, R. D.; Chini, M. G.; Bruno, I.; Riccio, R.; Mueller, D.; Werz, O. & Bifulco, G. (2011). Structure-Based Discovery of Inhibitors of Microsomal Prostaglandin $E_2$ Synthase-1, 5-Lipoxygenase and 5-Lipoxygenase-Activating Protein: Promising Hits for the Development of New Anti-inflammatory Agents. *Journal of Medicinal Chemistry*, Vol. 54, No. 6, (February 2011), pp. 1565–1575.

Tierney, J. P. & Lidström, P. (2005). *Microwave Assisted Organic Synthesis*, Blackwell, ISBN 1-4051-1560-2, USA and Canada.

Wathey, B.; Tierney, J.; Lidström, P. & Westman, J. (2002). The Impact of Microwave-assisted Organic Chemistry on Drug Discovery. *Drug Discovery Today*, Vol. 7, No. 6, (March 2002), pp. 373-380.

Ye, D.; Wang, J.; Zhang, X.; Zhou, Y.; Ding, X.; Feng, E.; Sun, H.; Liu, G.; Jiang, H. & Liu, H. (2009b). Gold-catalyzed Intramolecular Hydroamination of Terminal Alkynes in Aqueous Media: Efficient and Regioselective Synthesis of Indole-1-carboxamides. *Green Chemistry*, Vol. 11, No. 8, (May 2009), pp. 1201–1208.

Ye, D.; Zhang, X.; Zhou, Y.; Zhang, D.; Zhang, L.; Wang, H.; Jiang, H. & Liu, H. (2009a). Gold- and Silver-Catalyzed Intramolecular Hydroamination of Terminal Alkynes: Water-Triggered Chemo- and Regioselective Synthesis of Fused Tricyclic Xanthines. *Advanced Synthesis & Catalysis*, Vol. 351, Iss. 17, (November 2009), pp. 2770–2778.

Zhou, H.; Zhang, W. & Yan, B. (2010). Use of Cyclohexylisocyanide and Methyl 2-Isocyanoacetate as Convertible Isocyanides for Microwave-Assisted Fluorous Synthesis of 1,4-Benzodiazepine-2,5-dione Library. *Journal of Combinatorial Chemistry*, Vol. 12, No. 1, (November 2009), pp. 206–214.

Zhou, Y.; Zhai, Y.; Li, J.; Ye, D.; Jiang, H. & Liu, H. (2010). Metal-free Tandem Reaction in Water: An Efficient and Regioselective Synthesis of 3-Hydroxyisoindolin-1-ones. *Green Chemistry*, Vol. 12, No. 8, (July 2010), pp. 1397–1404.